国际最IN建筑设计

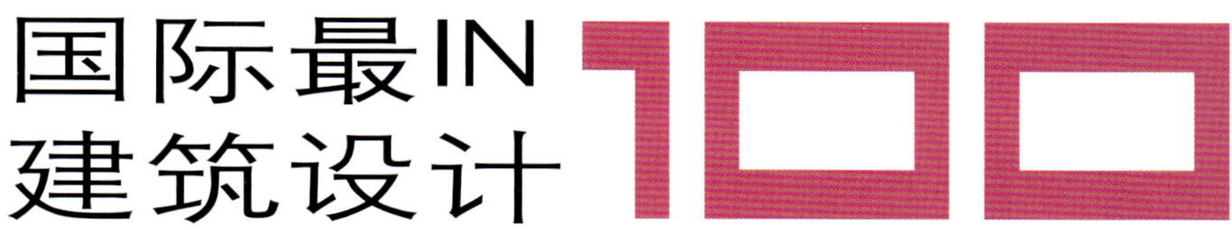

100 Global Architectural Schemes

范　悦/主编

于晓言 董洪兰 宋继红 王向红

王艺璇 计鑫 苗艳菲 徐丽 何雷 王思锐 张淼 王韶宁/译

大连理工大学出版社

图书在版编目(CIP)数据

国际最IN建筑设计100：汉英对照/范悦主编. —大连：大连理工大学出版社，2010.10
ISBN 978-7-5611-5863-0

Ⅰ.①国··· Ⅱ.①范··· Ⅲ.①建筑设计－作品集－世界－现代 Ⅳ.①TU206

中国版本图书馆CIP数据核字（2010）第201870号

出版发行：大连理工大学出版社
（地址：大连市软件园路 80 号 邮编：116023）
印 刷：利丰雅高印刷（深圳）有限公司
幅面尺寸：235mm×310mm
印 张：26.75
出版时间：2010 年 10 月第 1 版
印刷时间：2010 年 10 月第 1 次印刷
策 划：房 磊
责任编辑：张昕焱 杨 丹
封面设计：王志峰
责任校对：王 培

书 号：ISBN 978-7-5611-5863-0
定 价：348.00 元

发 行：0411-84708842
传 真：0411-84701466
E-mail: a_detail@dutp.cn
URL: http://www.dutp.cn

序

建筑设计的三态

一位建筑界的老者曾说过，建筑设计有“三态”：形态、生态和人态。形态指的是城市或建筑的空间形式问题，生态指的是建筑设计需要考虑的与自然共生的问题，而人态则意指建筑设计及实现过程中的决策管理及制度，或者可以引申为建筑各专业之间的协作。我认为这位老者非常准确地归纳了今天建筑设计的样态。今天，形态问题依旧是建筑设计的基本问题，但是围绕形态设计的“环境”却发生了不小的变化。首先，市场环境今非昔比。有谁见过比中国还大的建筑工地？记得12年前在德国柏林访问的时候，有感当时欧洲最大的建筑工地，并且有诸多世界知名的建筑师和事务所竞相献技，宛如建筑设计的欧洲武林大会。如今时过境迁，世界的比武大会东移来到中国，只是规模和热度不知要扩大了多少倍。标志性建筑在中国大地拔地而起，最具代表性的事件当属2008年的北京奥运和2010年的上海世博。当然，吸引民众眼球的还只是一部分的明星建筑，一方面，它让人们开始关注城市和建筑形态，关注建筑设计的创意和价值，另一方面，对于国内的设计院和大多数的建筑师来说，面对庞大而苛刻的设计任务，他们更关注设计水平与生产性的平衡。这方面，充分学习了解国外的优秀作品，总结国外先进而高效的建筑设计方法非常有必要。

今天，基于成熟的计算机解析技术的数字化设计及其环境有了极大的改善。这种设计技术的进步，正在改变建筑设计的过程和方式。即使从严格意义上讲，除了那些为数不多的明星建筑，参数化设计、数控制造等在一般建筑上的应用还只是个理想，但它却赋予了建筑师更多的可能性和更广阔的世界。可呼吸的表皮、可控的室内环境、新素材新结构的复合应用，开始渗透到各种规模和用途的建筑设计中，无疑改变了人们对于建筑和城市的观感。另外，将环境融入建筑中的设计理念和方法，以及对于旧的城市建筑遗产更好的延续和利用的设计，从另外一个角度反映了时代及环境的改变对于建筑设计方式的影响和作用。

其实，如果“人态”意指人的生活方式以及行为状态的话，从以人为本的角度考量，人态或可排到建筑设计三态的首位。从建筑设计的历史演变过程来看，20世纪现代建筑的主流确实是“形态”之争或者说为形态而设计的历史。但是，当形态的象征性，即建筑看起来像什么的问题变得不那么主要的时候，人们开始意识到空间形态说到底是为人的使用（包括物质和精神的）设计的，因此，形态的模糊，以及从人态或生态出发的设计也许可以作为21世纪新型建筑设计的重要特征吧。

本书汇聚了大量当今一流建筑师及其事务所的最新设计方案，具体阐述并表现了我以上的论述。无论作为建筑师的设计构思，或者专业研究的案例，以及作为建筑学的师生的欣赏与参考，其可读性和专业性都很强，希望能对大家的专业和设计有所帮助。

范悦

2010年10月

目录 CONTENTS

001

文化教育
CULTURE AND EDUCATION

商业办公
COMMERCE AND OFFICE

住宅建筑
RESIDENTIAL BUILDINGS

294—295

其他
OTHERS

350—351

国际最IN建筑设计

100

文化教育

CULTURE AND EDUCATION 1

未来城市图书馆与新媒体中心
URBAN LIBRARY OF THE FUTURE AND CENTRE FOR NEW MEDIA

建筑师：UNStudio

Ben van Berkel, Caroline Bos, Gerard Loozekoot with Jacques van Wijk, Wesley Lanckriet and Jordan Trachtenberg, Ren Yee, Wendy van der Knijff, Bartek Winnicki, Aurélie Kratoff, Patrik Noome, Marcin Koltunski, Joerg Lonkwitz, Miguel Noë, Imola Berczi, Elena Scripelliti

结构工程师：ABT, Antwerp and Netherlands

当地建筑师：Crepain Binst Architecture, Antwerp

甲方：CVBA Waalse Krook

项目地点：Gent, Belgium

建筑面积：19 498.6 m²

本案未来城市图书馆与新媒体中心位于比利时根特，设计有两个主要目的，一是创造动态、灵活、开放的学习环境，二是在引入一座独特的建筑的同时强化地域特色。

建筑外形呈流线型，与周围环境融合在一起。流线的外形在根据朝向的不同而不断变化的建筑外观上显而易见，从将建筑体量抬高的决定上也能看出与环境融合的想法。建筑抬高后，创造出了明亮、透明的空间，视野也更加广阔。但是分层的结构和较低的建筑体量保证了设计对城市环境造成的影响最小，同时还能看到根特富有特点的塔楼建筑。这样的结构还可以设置（绿色）屋顶露台，并保证只有较少的直射阳光穿透建筑进入室内。

The two main aims in the design for the Urban Library of the Future and Centre for New Media in Gent are to create a dynamic, flexible and open knowledge environment, whilst simultaneously strengthening the character of the location with the introduction of a building with a distinct architectural identity.

The building is both fluid in form and accommodating to its surroundings. This is evidenced by its appearance - which varies according to the orientation - as well as from the decision to lift the building volume above ground level, thereby creating light, transparency and expansive sightlines. However the layered structure and low construction volume ensure that the impact of the design on the urban profile is minimal and that views to the characteristic towers of Gent are preserved. The structure also makes it possible to introduce (green) roof terraces whilst also ensuring low levels of direct sunlight penetration.

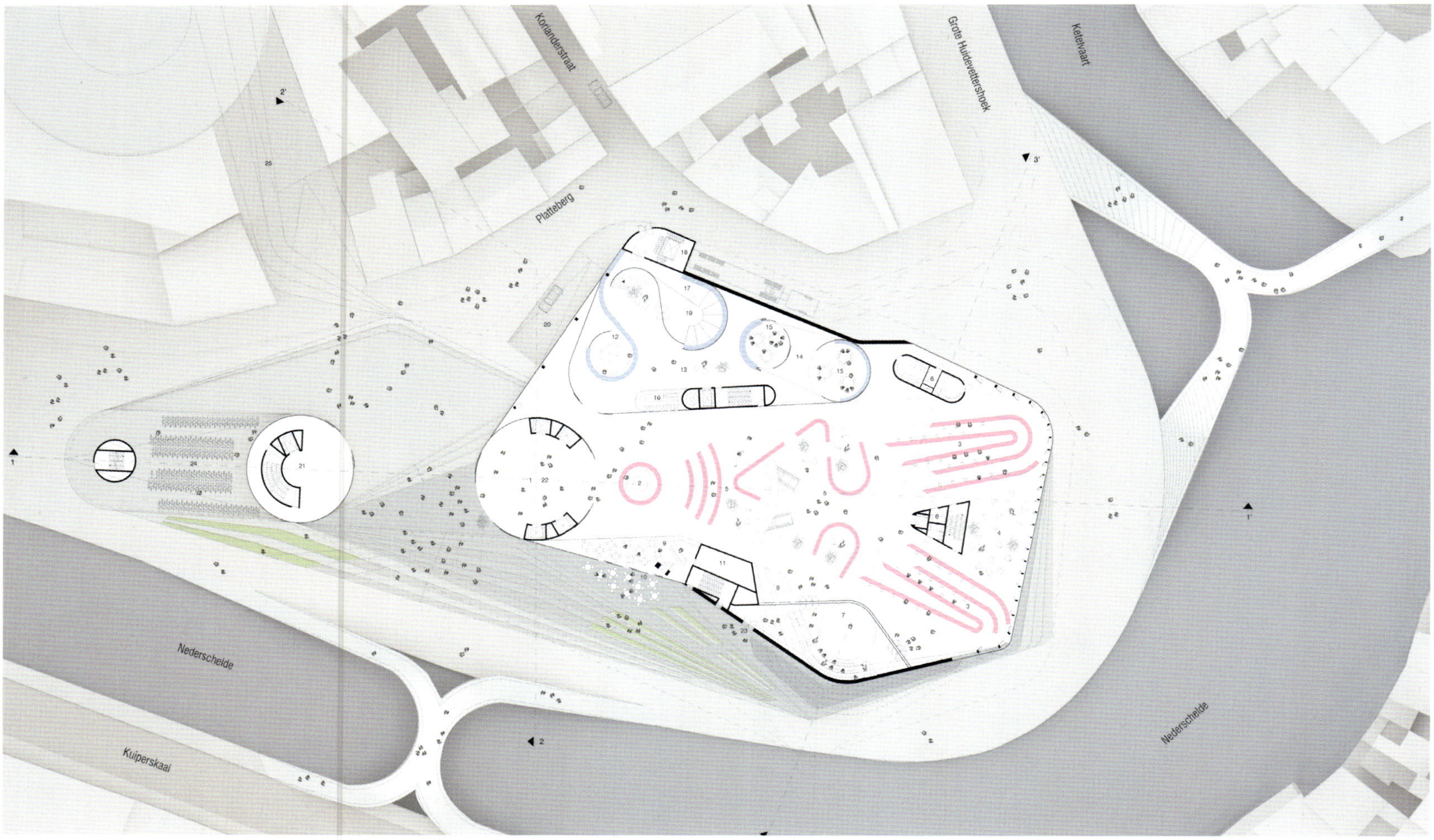

Korianderstraat
Grote Huidevettershoek
Ketelvaart
Platteberg
Nederschelde
Kuiperskaai
Nederschelde

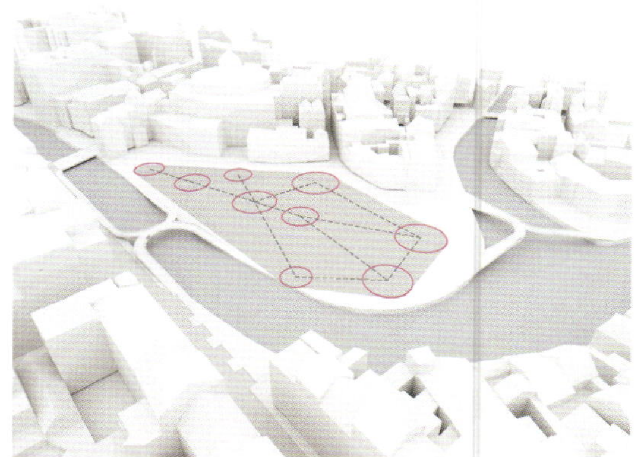

❶Interconnections and tangents
相互连接

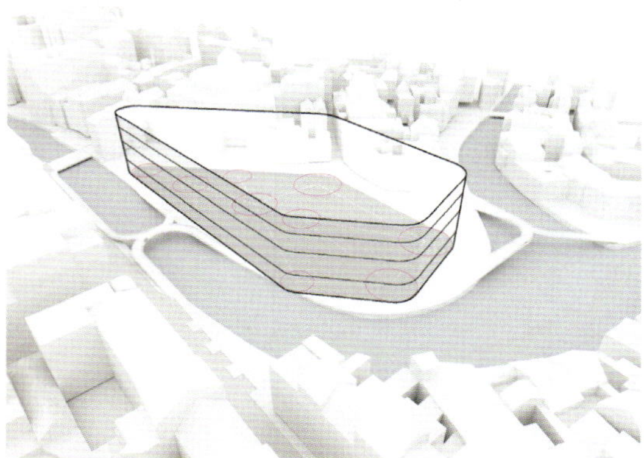

❷Maximum volume envelope
最大体量外围护结构

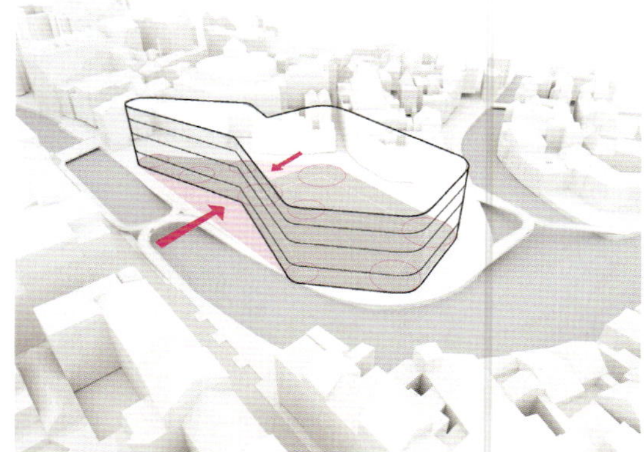

❸Introduction public squares and accentuation of views to the surroundings
引入公共广场，强调与周围的视觉联系

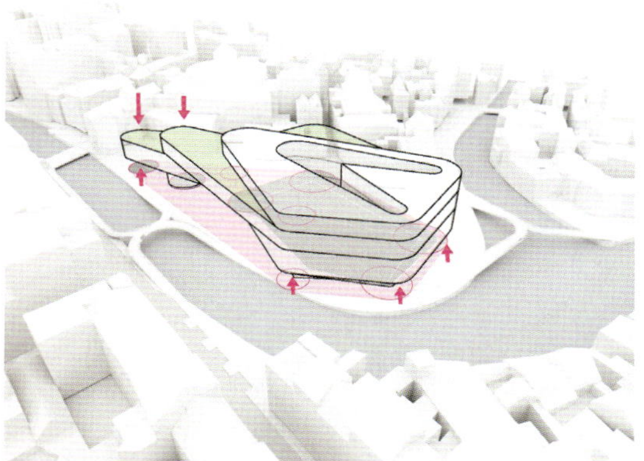

❹Footprint compensated with green roofs and addition of terraces in order to maintain views from and to the area. Raised corners to enable reactivation of the quay.
绿色屋顶和外加的露台弥补建筑足迹，以此来确保向内和向外的视线通畅。抬高的转角使码头充满活力

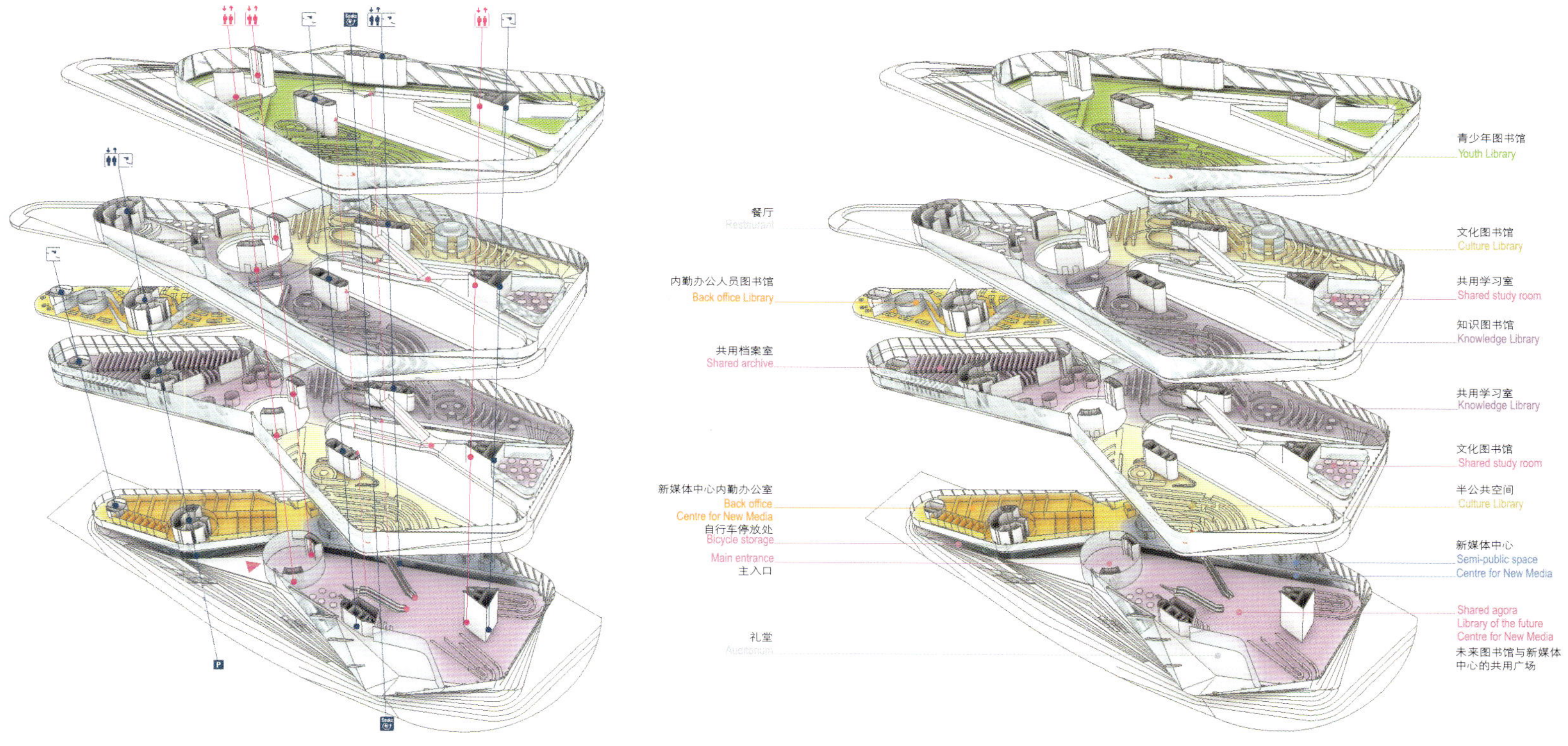

青少年图书馆
Youth Library
餐厅
Restaurant
文化图书馆
Culture Library
内勤办公人员图书馆
Back office Library
共用学习室
Shared study room
知识图书馆
Knowledge Library
共用档案室
Shared archive
共用学习室
Knowledge Library
文化图书馆
Shared study room
新媒体中心内勤办公室
Back office
Centre for New Media
半公共空间
Culture Library
自行车停放处
Bicycle storage
Main entrance
主入口
新媒体中心
Semi-public space
Centre for New Media
Shared agora
Library of the future
Centre for New Media
未来图书馆与新媒体中心的共用广场
礼堂
Auditorium

交通流线分解轴测图
Exploded & axonometric circulation

项目分界轴测图
Exploded & axonometric program

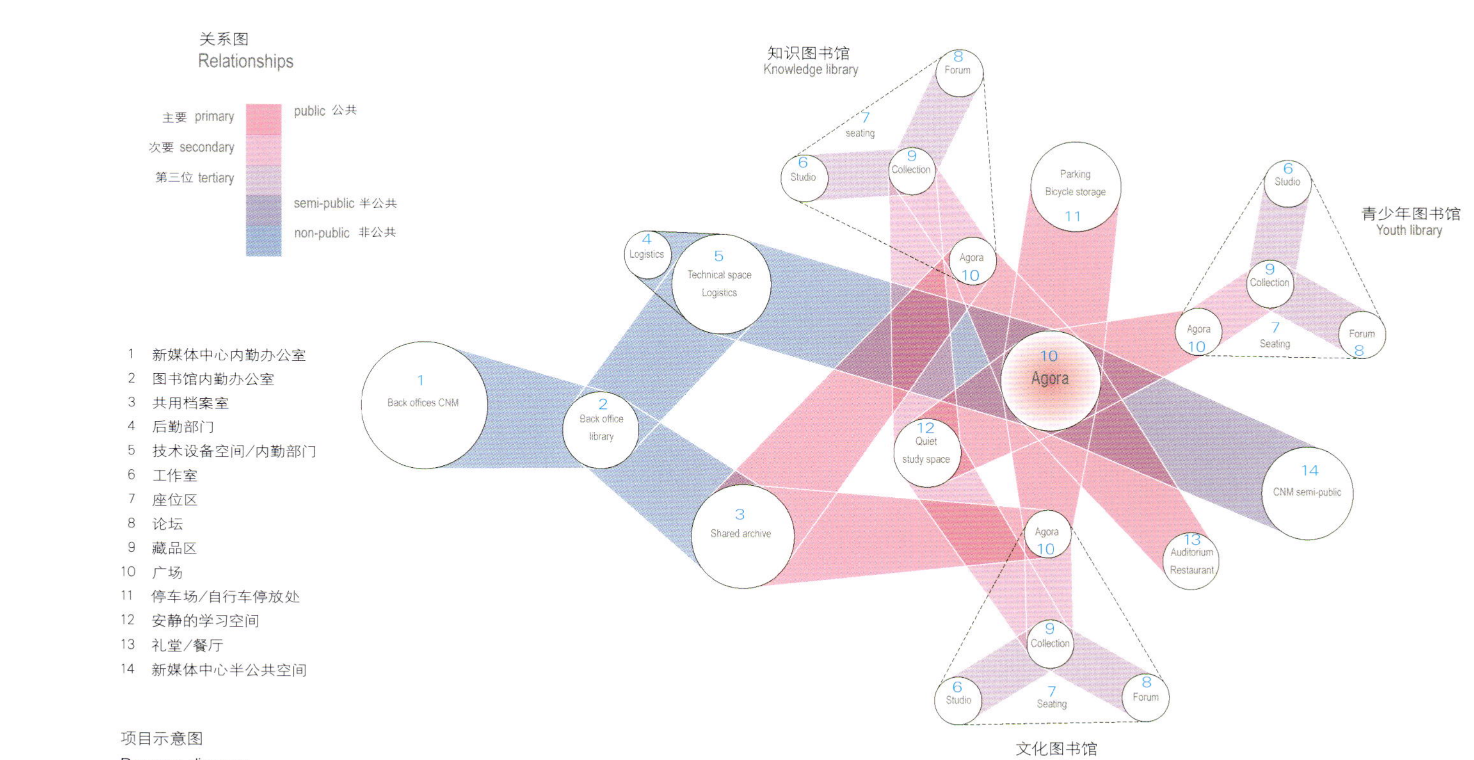

关系图
Relationships
主要 primary
次要 secondary
第三位 tertiary
public 公共
semi-public 半公共
non-public 非公共
1 新媒体中心内勤办公室
2 图书馆内勤办公室
3 共用档案室
4 后勤部门
5 技术设备空间/内勤部门
6 工作室
7 座位区
8 论坛
9 藏品区
10 广场
11 停车场/自行车停放处
12 安静的学习空间
13 礼堂/餐厅
14 新媒体中心半公共空间
知识图书馆
Knowledge library
8 Forum
7 seating
6 Studio
9 Collection
Parking
Bicycle storage
11
4 Logistics
5 Technical space Logistics
Agora 10
青少年图书馆
Youth library
6 Studio
9 Collection
Agora 10
7 Seating
Forum 8
1 Back offices CNM
2 Back office library
10 Agora
12 Quiet study space
14 CNM semi-public
3 Shared archive
Agora 10
13 Auditorium Restaurant
9 Collection
6 Studio
7 Seating
8 Forum
文化图书馆
Culture library

项目示意图
Program diagram

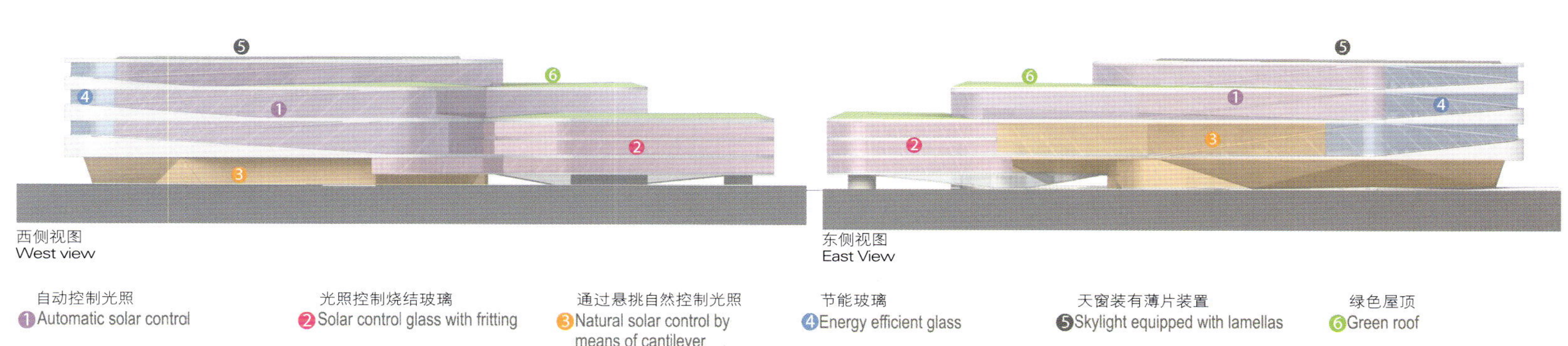

西侧视图
West view
东侧视图
East View
自动控制光照
1 Automatic solar control
光照控制烧结玻璃
2 Solar control glass with fritting
通过悬挑自然控制光照
3 Natural solar control by means of cantilever
节能玻璃
4 Energy efficient glass
天窗装有薄片装置
5 Skylight equipped with lamellas
绿色屋顶
6 Green roof

立面图
Facades

主要结构
Main construction

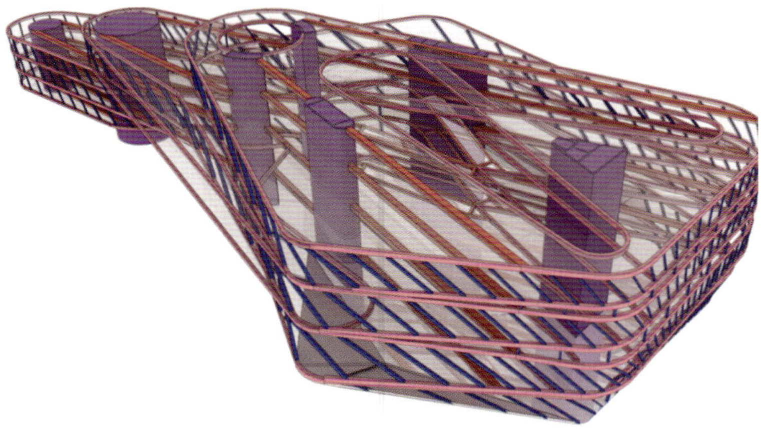

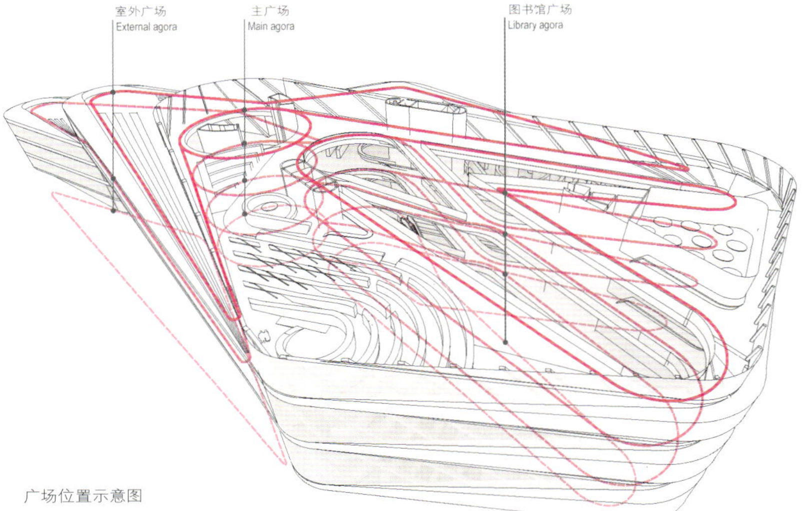

广场位置示意图
Agora diagram

室外广场
External agora
主广场
Main agora
图书馆广场
Library agora

主要楼层
Main floors

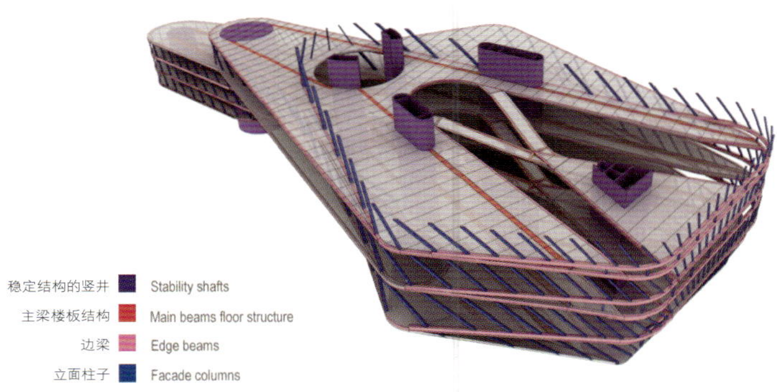

稳定结构的竖井 Stability shafts
主梁楼板结构 Main beams floor structure
边梁 Edge beams
立面柱子 Facade columns

结构示意图
Structural diagram

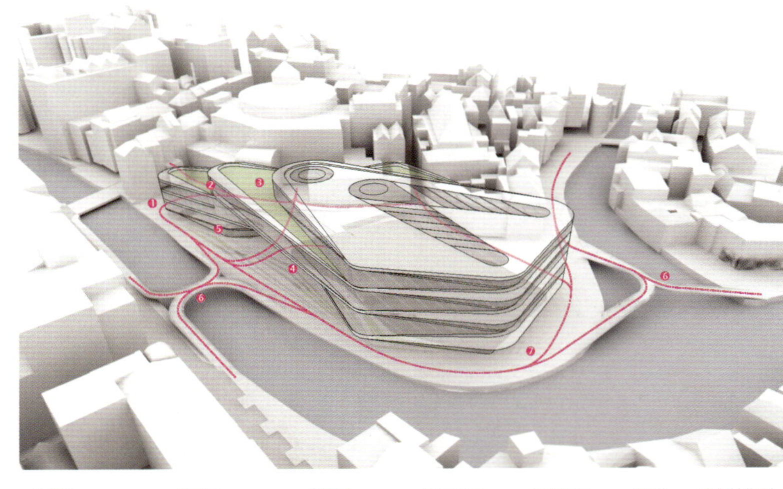

公共广场 ❶Public squares
高层餐厅 ❷High restaurant
绿色屋顶 ❸Green roofs
有顶入口区域 ❹Sheltered access area
自行车停放处 ❺Bicycle parking
连接桥 ❻Bridge connections
充满活力的码头 ❼Reactivating quays

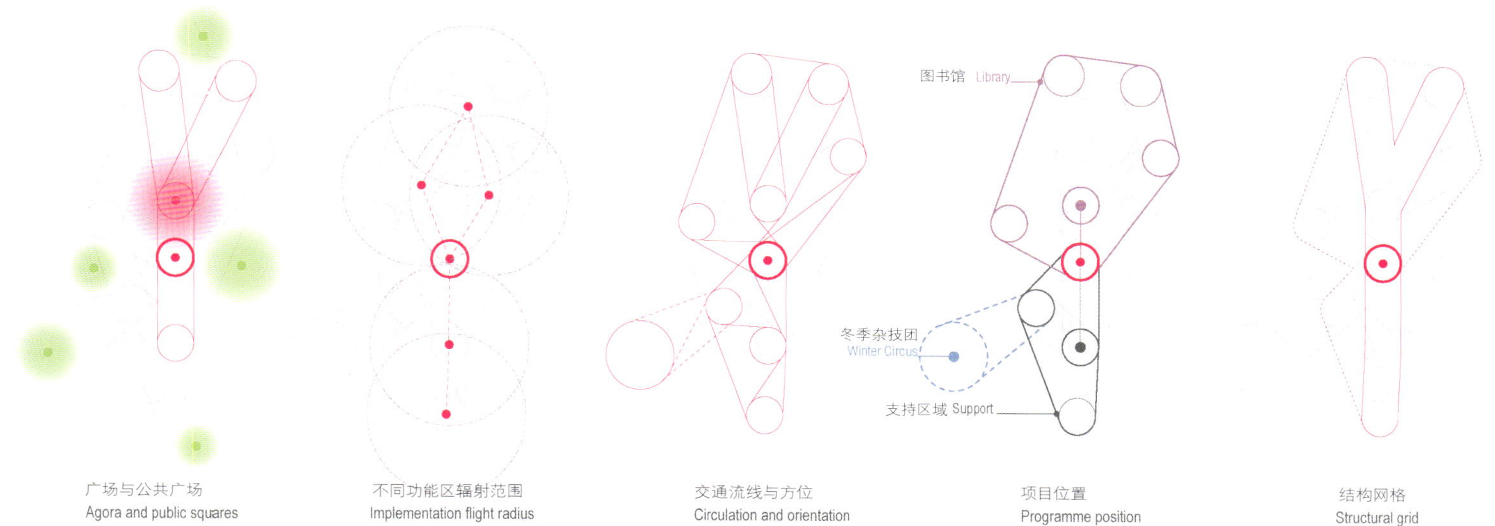

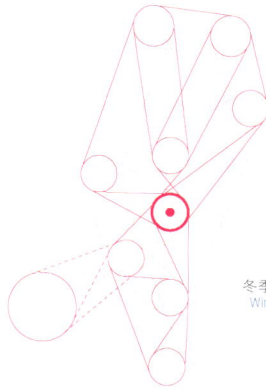

基地内的项目连接与系统纵览
Overview of Program Connections and systems within site parameters.

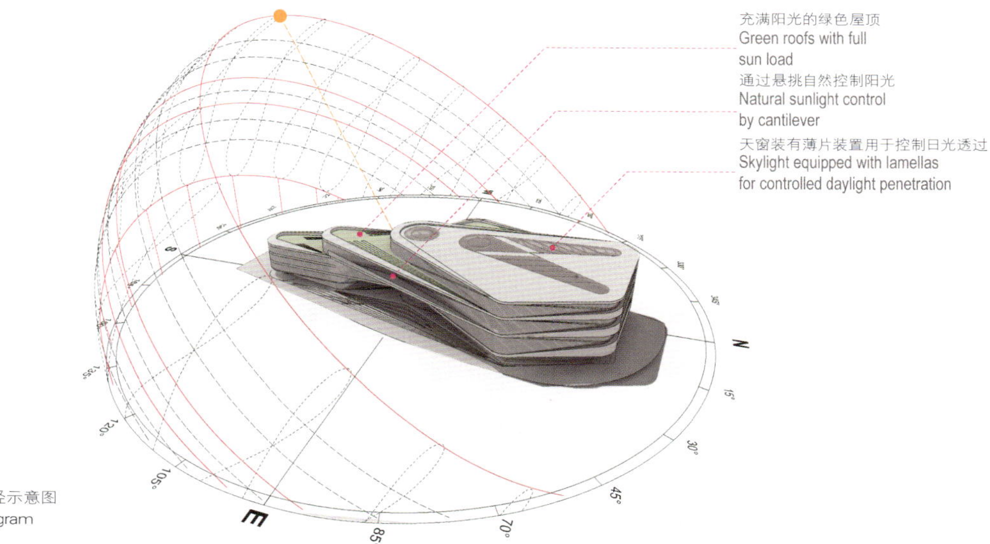

阳光照射路径示意图
Sun path diagram

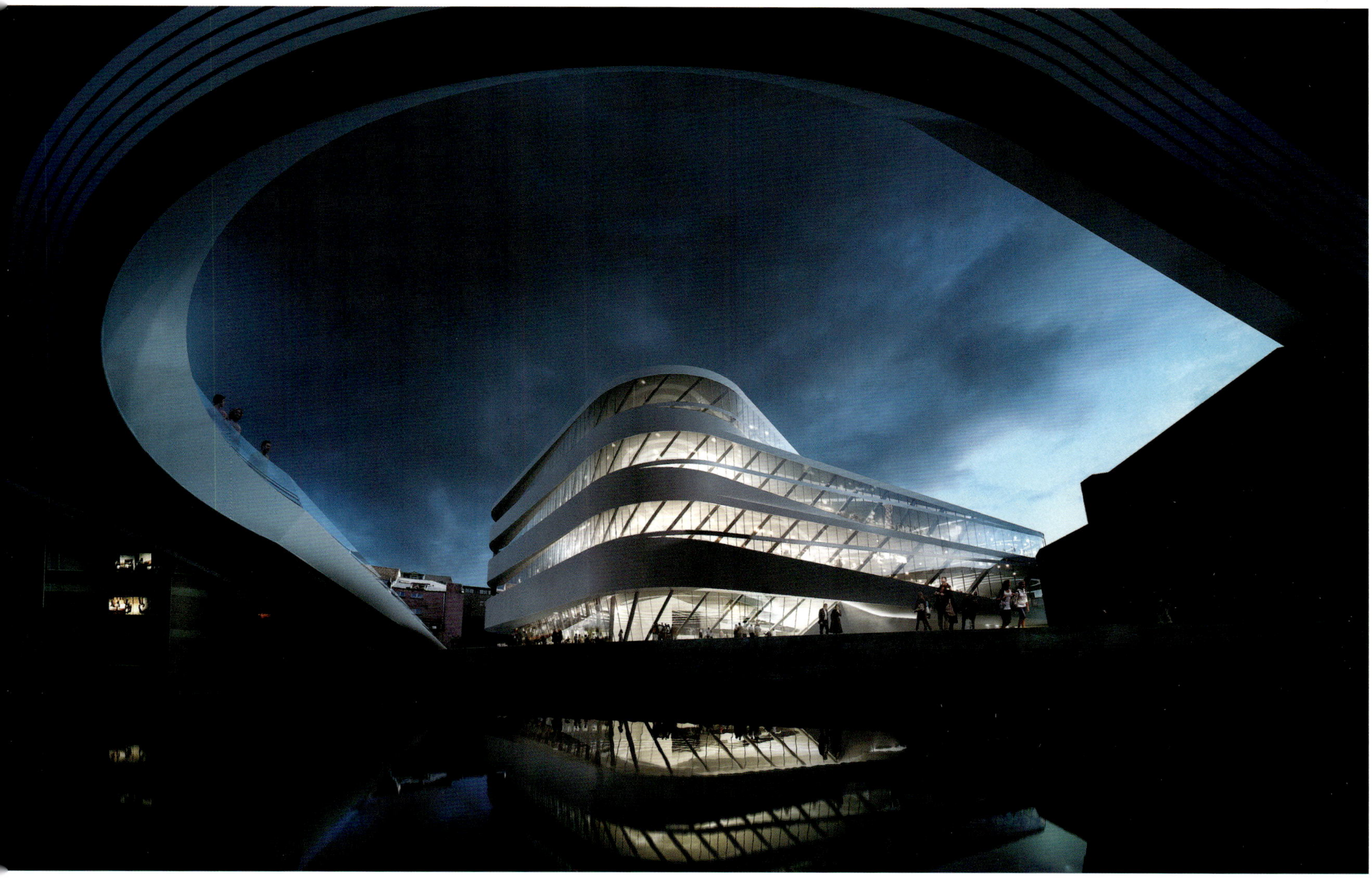

鄂尔多斯剧院设计
THEATRE IN ORDOS

建筑师：Yazdani Studio of Cannon Design
设计团队：Mehrdad Yazdani（设计主管）；Michael Tunkey（项目主管）；Philip Ra, AIA（高级设计师）；Johnson, FAIA（规划与设计）；Joe O'Neill, AIA, LEED AP（项目建筑师）；Andrew Wong（设计师）；Mimi Lam, LEED AP, Nadine Quirmbach, LEED AP, Ken Yip, Yanqian Lu, Weiwei Kuang, Xiaoshan Liu, Haizhen Song, Yiru Lu

为了能吸引像北京交响乐团这种大型乐团前来演出，需要设计一家世界级的剧院。设计的曲线婀娜多姿，酷似舞步，而且复制了"长袖舞"的动作。圆筒状的结构反映了长袖舞的复沓回转，其中分别容纳了1200个坐席的主音乐厅、335个坐席的表演剧场，以及100个坐席的黑盒子实验剧院。这种遍及建筑物及其周边的循环模式创建了一种共生关系，仿佛在观众和空间之间翩翩起舞。建筑中还包括一系列休息室、茶点区以及设备侧翼部分，它们如同长袖交织在剧场之间。

In anticipation of attracting the likes of the Beijing Symphony Orchestra to perform, a world-class Theatre needed to be designed. The design, with its undulating curves, mimics the footwork pattern of the dance and replicates the movement of the "long sleeves". In the dance, the looping of the long sleeves is reflected in circular drums that house the 1,200-seat main concert hall, 335-seat performance-theater, and 100-seat black box experimental theater respectively. The circulation throughout and around the building creates a symbiotic relationship - a dance of sorts - between the visitors and the space. Weaving like the sleeve, between the theatres, are a series of nodes with lounges, refreshment areas and service wings.

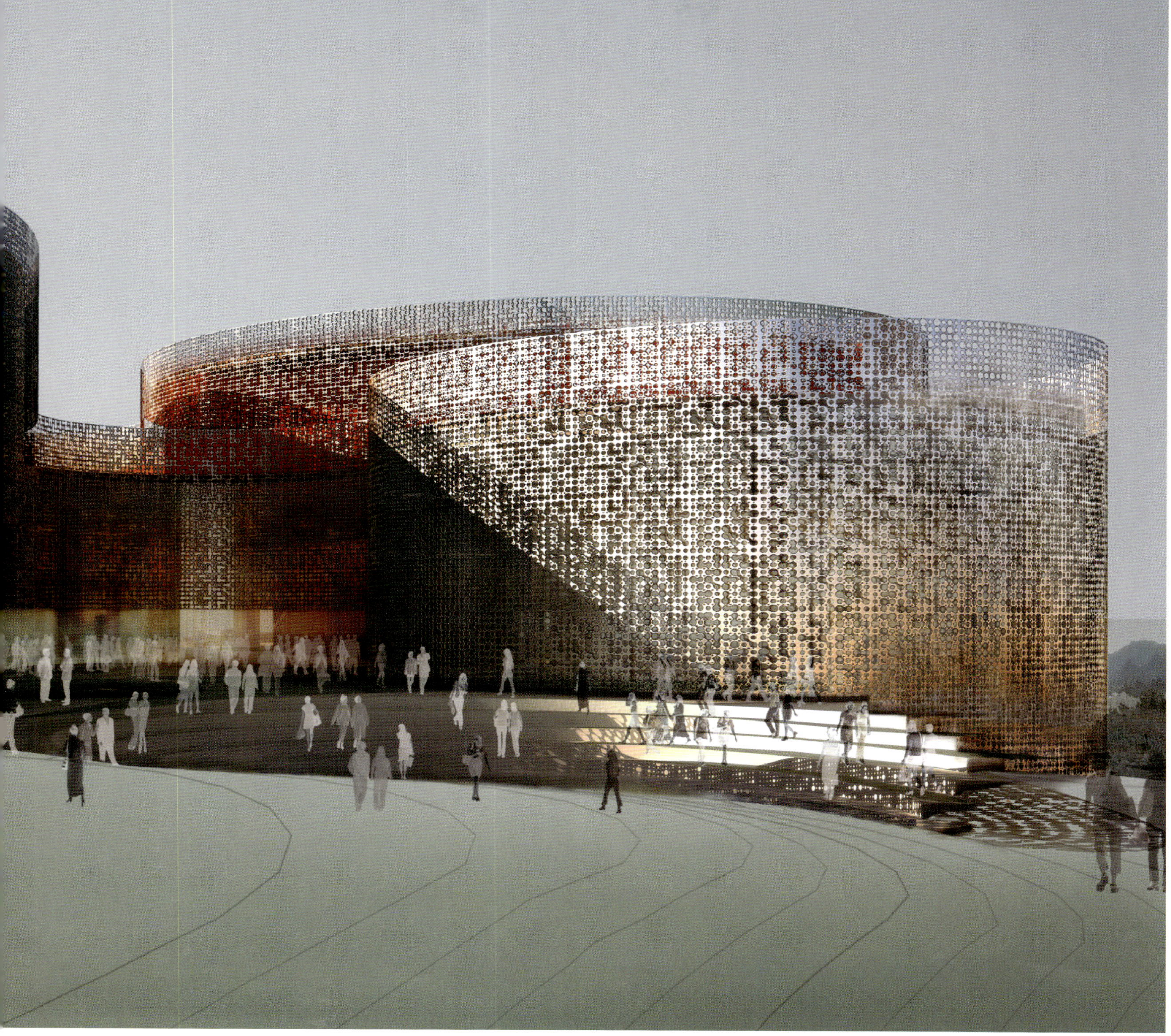

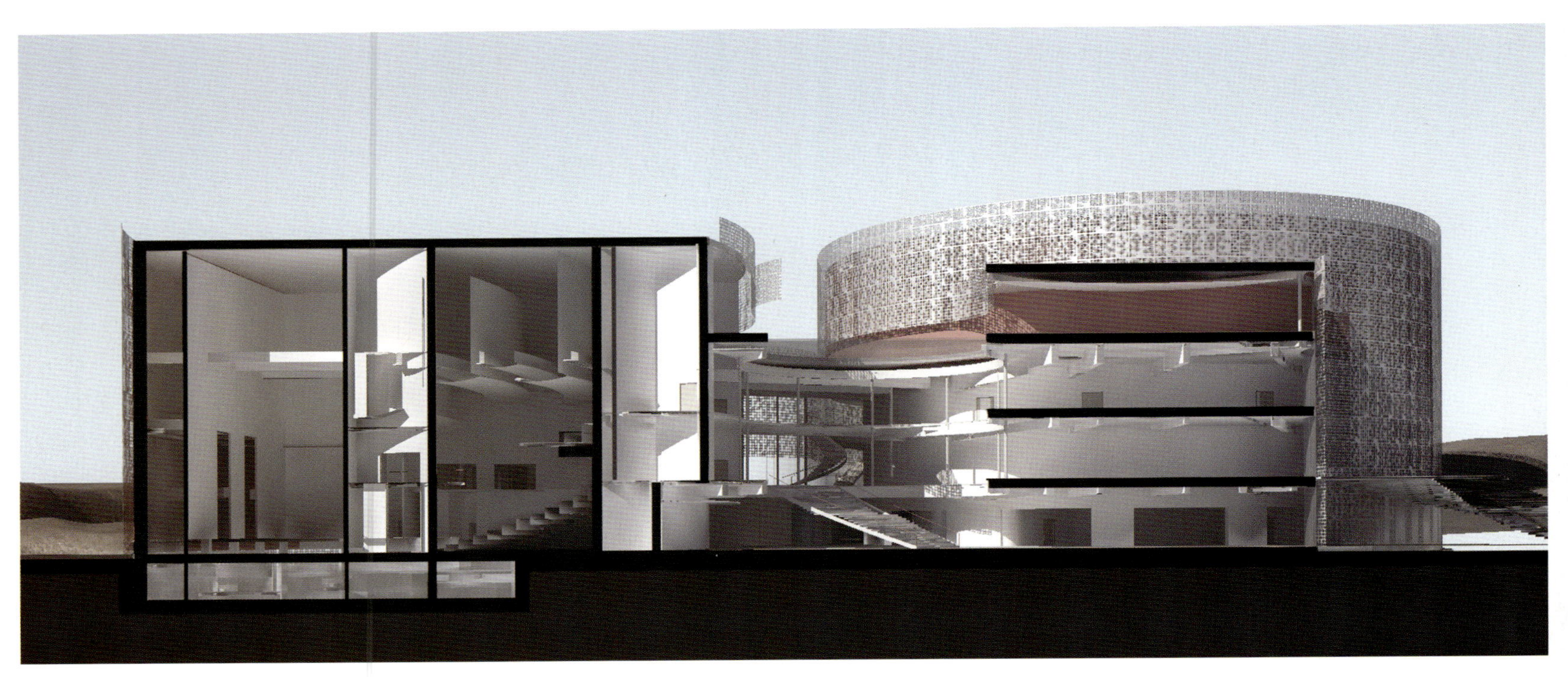

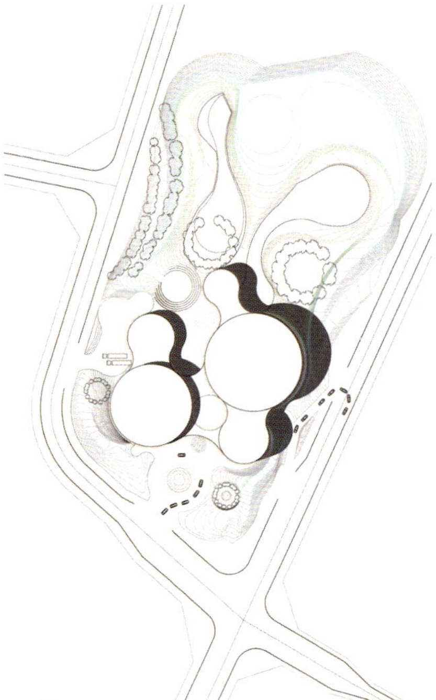

总平面
Site plan

1 下方入口广场
2 下方主入口
3 下方员工入口
4 花园入口
5 VIP入口
6 设备区
7 休息厅
8 门厅
9 500坐席的表演剧场
10 多功能剧场
11 1200坐席音乐厅
12 办公室入口
13 室外圆形露天剧场
1 Entry Plaza Below
2 Main Entry Below
3 Staff Entry Below
4 Garden Entry
5 VIP Entry
6 Service
7 Lounge
8 Lobby
9 500 Seat Performing Theatre
10 Multipurpose Theatre
11 1200 Seat Concert Hall
12 Office Entry
13 Outdoor Amphitheater
0 5 10 20m
二层平面
2nd floor

M.Y.
ORDOS 07

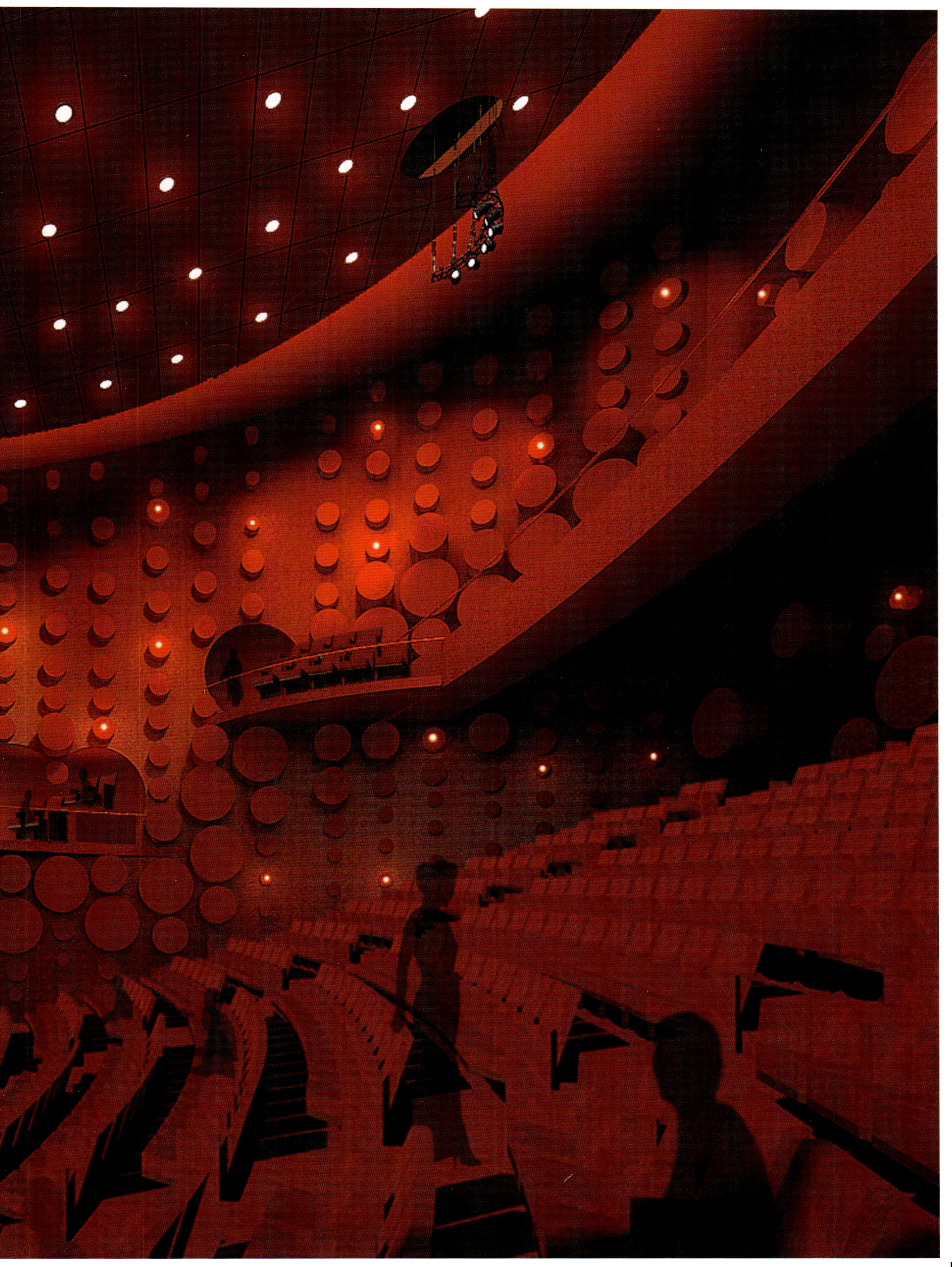

荷兰海牙的新舞蹈和音乐中心
NEW DANCE AND MUSIC CENTRE IN HAGUE, THE NETHERLANDS

甲方：City of The Hague
建筑师：Zaha Hadid Architects
设计：Zaha Hadid with Patrik Schumacher
项目合作人：Joris Pauwels
项目指导：Paulo Flores
当地建筑师：Bureau Bouwkunde（荷兰）
结构：AKT（英国）；设备：Max Fordham（英国）；剧院：Theatre Projects Consultants（英国）；
立面：Newtecnic（英国）；音响效果：Peutz（荷兰）

海牙新舞蹈和音乐中心的设计概念由场地独特的城市动态发展而来，形成的结构以微妙的体量姿态将人们从一层广场吸引到建筑的中心。建筑与公共领域的紧密结合进一步强调了建筑的亲和力。该建筑的同一屋檐下容纳着四个主要机构：皇家音乐学院、荷兰舞蹈戏剧院、Residential管弦乐队和中心的宾客功能空间。

建筑以优雅的弯曲屋顶轮廓线作为结束，将自己融入城市的天际线。

与基本的矩形体量相比，设计以水平百叶窗流动的力场为特色，在光影的作用下，百叶好像会移动一般。这种特征使立面具有了活泼的特质，突出了公共交通流线、门厅和具有雕刻特质的内部中庭，同时确保了建筑与外部广场的视线连接和怀着好奇心的路人瞥向室内的视线。

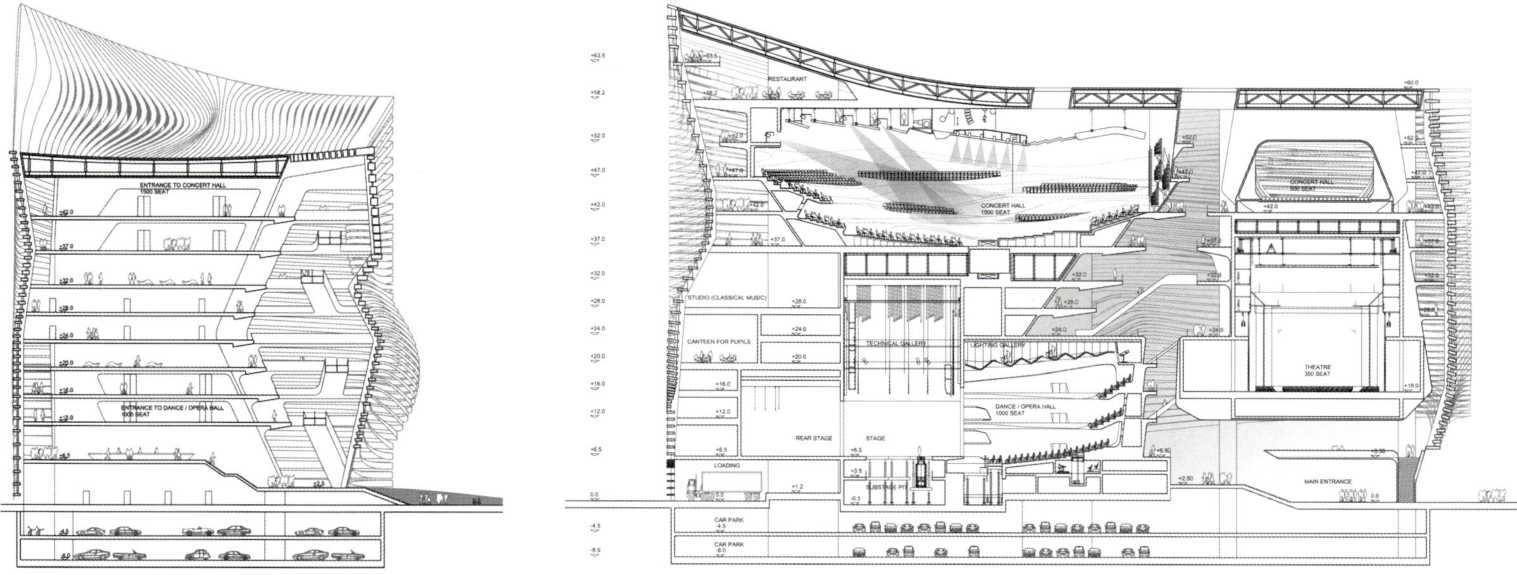

剖面图
Sections

The design concept for the New Dance and Music Centre in the Hague is developed from the unique urban dynamics of the site, resulting in a structure with subtle volumetric gestures that invite the public from the ground-level plaza into the heart of the building. The seamless continuation of the public domain into the structure further reinforces the cohesive character of the building, which combines four major institutions into one single envelope: the Royal Conservatory, the Netherlands Dance Theatre, the Residential Orchestra and the centre's Guest Programming. The building culminates in a gracious curving roofline that neatly nests itself within the city skyline.

In sharp contrast with basic rectangular geometry, the design features a fluid force field of horizontal louvers that seemingly moves when graced by light and shadow. This unique characteristic creates a playful language on the facade, articulating public circulation, the foyers and the sculpted inner atrium, while allowing visual connections outside to the square as well as exciting glimpses into the building.

台北表演艺术中心方案一
TAIPEI PERFROMING ARTS CENTRE

建筑师：DOS Architects Ltd.
3D效果图：Meshroom Ltd.

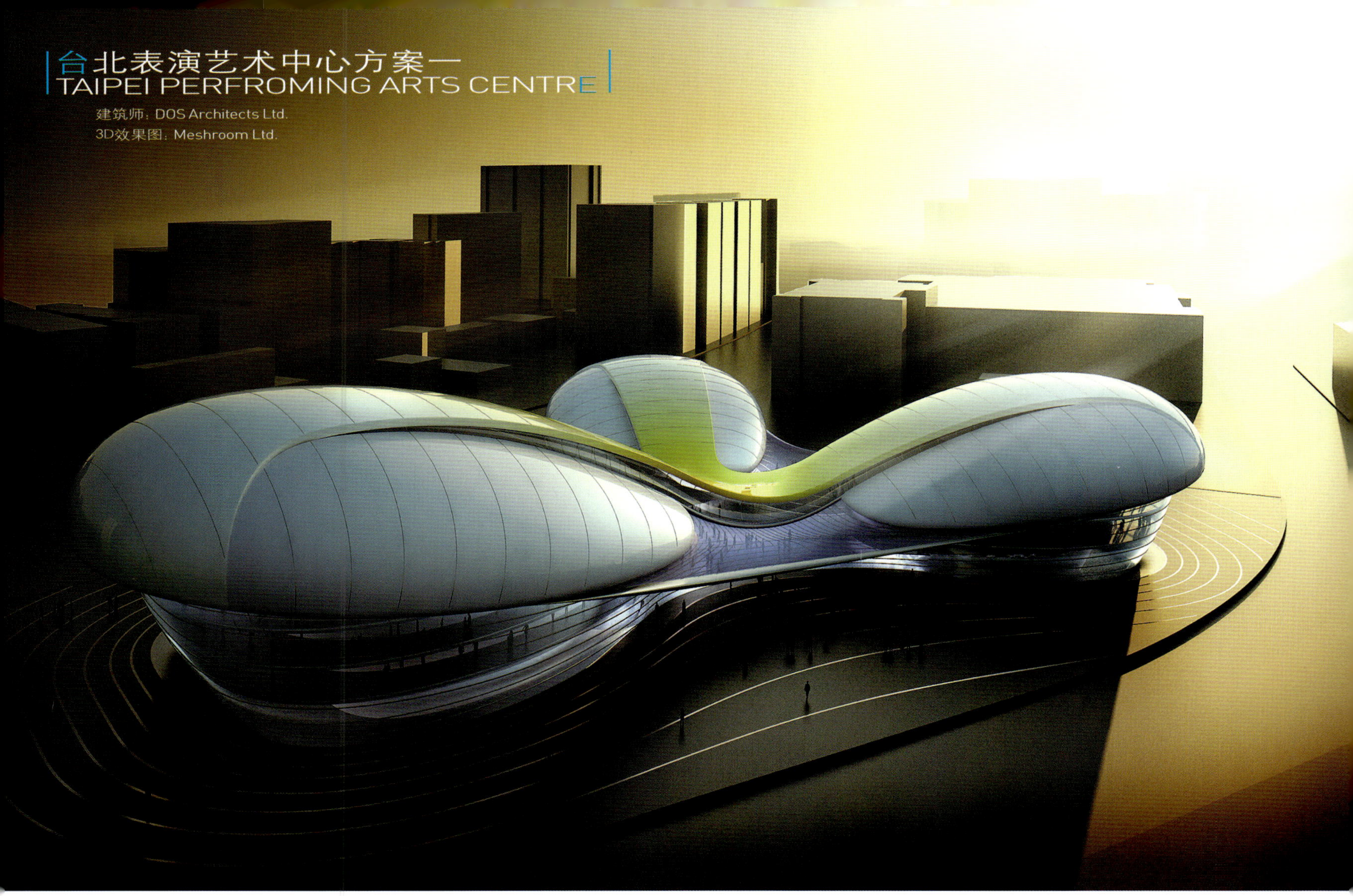

建筑师与Meshroom公司合作设计了台北表演艺术中心项目，它真正向人们证明了什么才是先进的建筑与设计。建筑师借用大自然中常见的语言，创造了这座优美的建筑，寓意为台北市一朵“含苞待放的花朵”，它也是一个新的城市图标。
建筑布局简单，三个礼堂的屋顶都好似花瓣，覆盖着整座建筑。面朝中心的大型公共空间吸引游人进入建筑，指引他们穿过建筑进入表演区域。

This project, done in collaboration with Meshroom, stands as a true testament to cutting edge Architecture and Design. By borrowing a language more common to the natural world, we have tried to create a building which represents a "budding flower" for the city of Taipei: A building which will stand as a new icon for the city.
The building has a simple arrangement on the site, with each of its three Auditoriums covered by a petal-like roof which sweeps over the whole building. A large communal space towards the centre draws visitors in and guides them through the building and towards the performance areas.

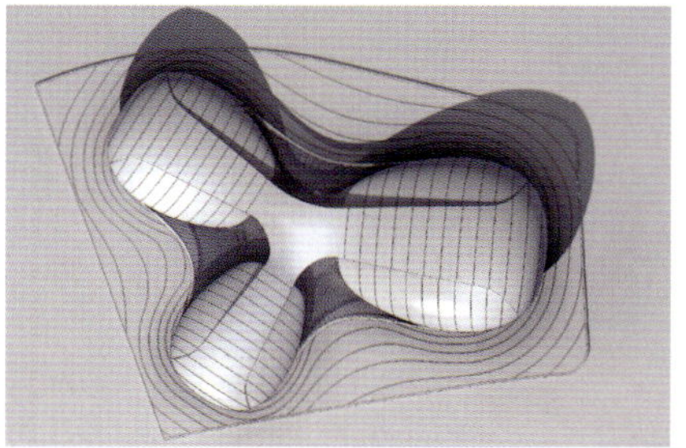

屋顶平面
Roof plan

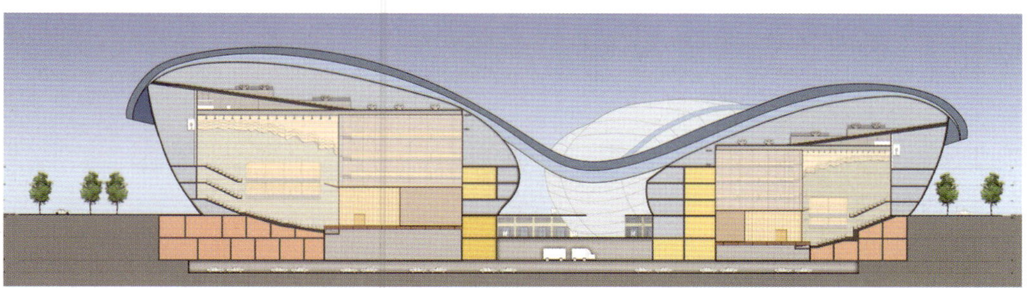

剖面图
Sections

购物区	shopping area
剧院门厅	theatre foyer
停车场和货物装卸区	car parking and truck loading
剧院礼堂	theatre auditorium
剧院舞台	theatre stages
服务区	service areas
服务区	service areas

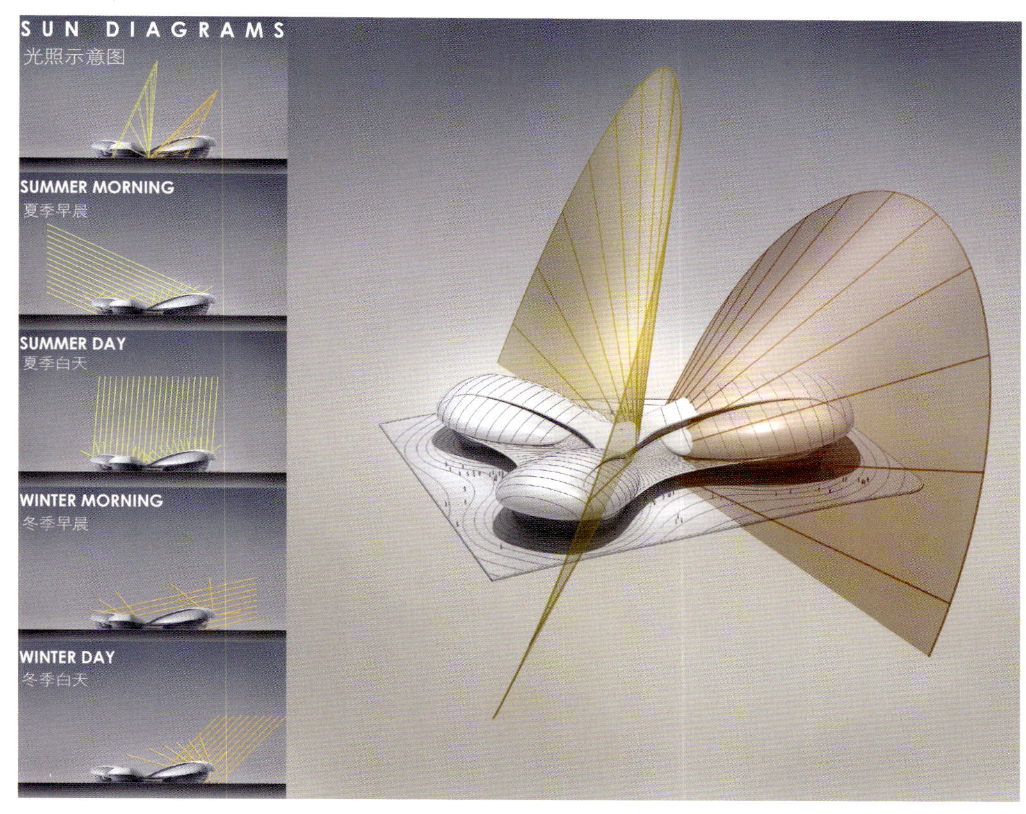

不同体量之间创造了很多公共空间

The volumes are placed so they create a number of spaces in-between that allow for public areas.

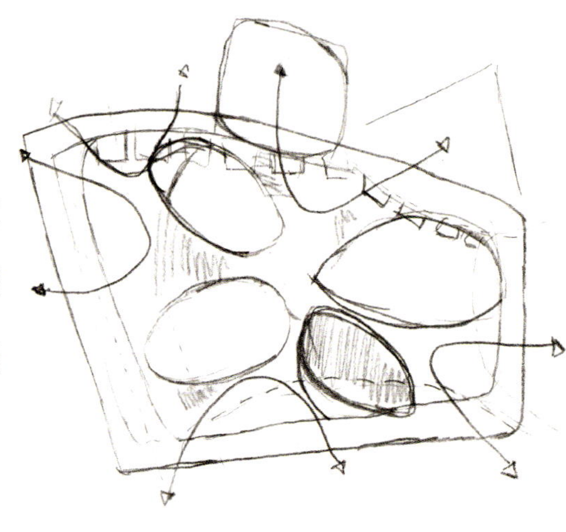

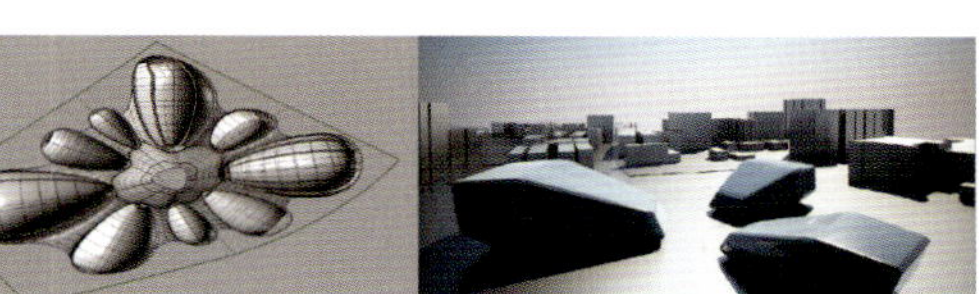

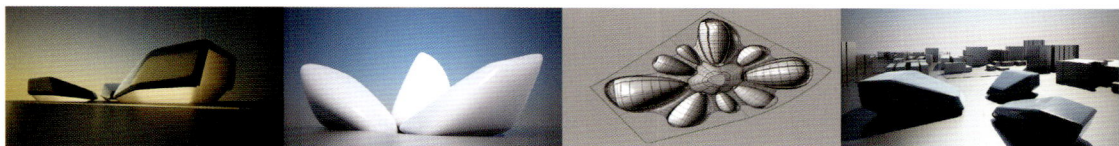

设计灵感
Inspiration

北立面
North elevation

南立面
South elevation

东立面
East elevation

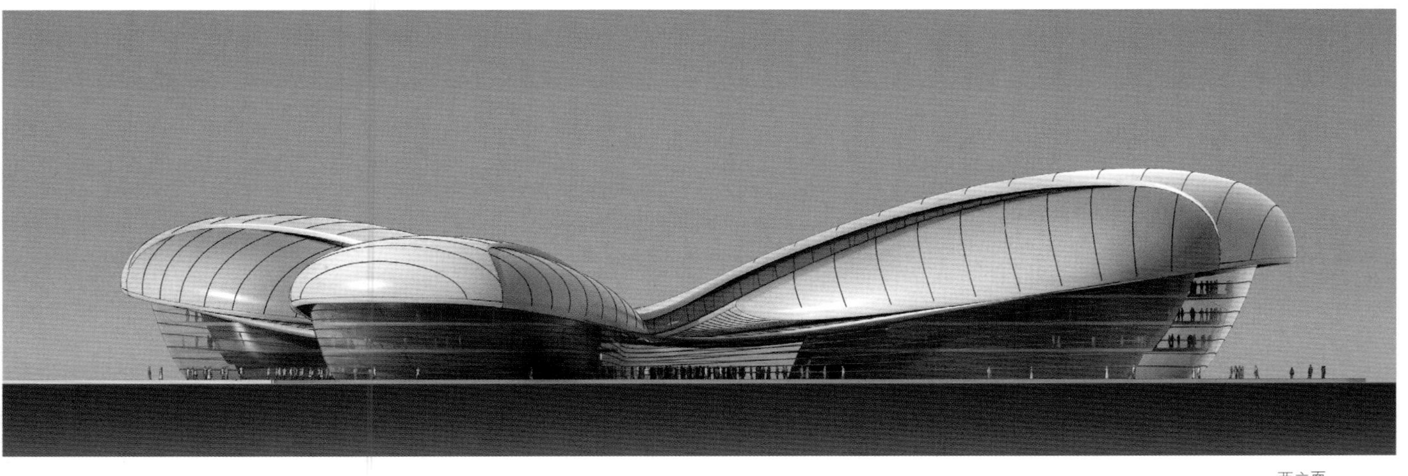

西立面
West elevation

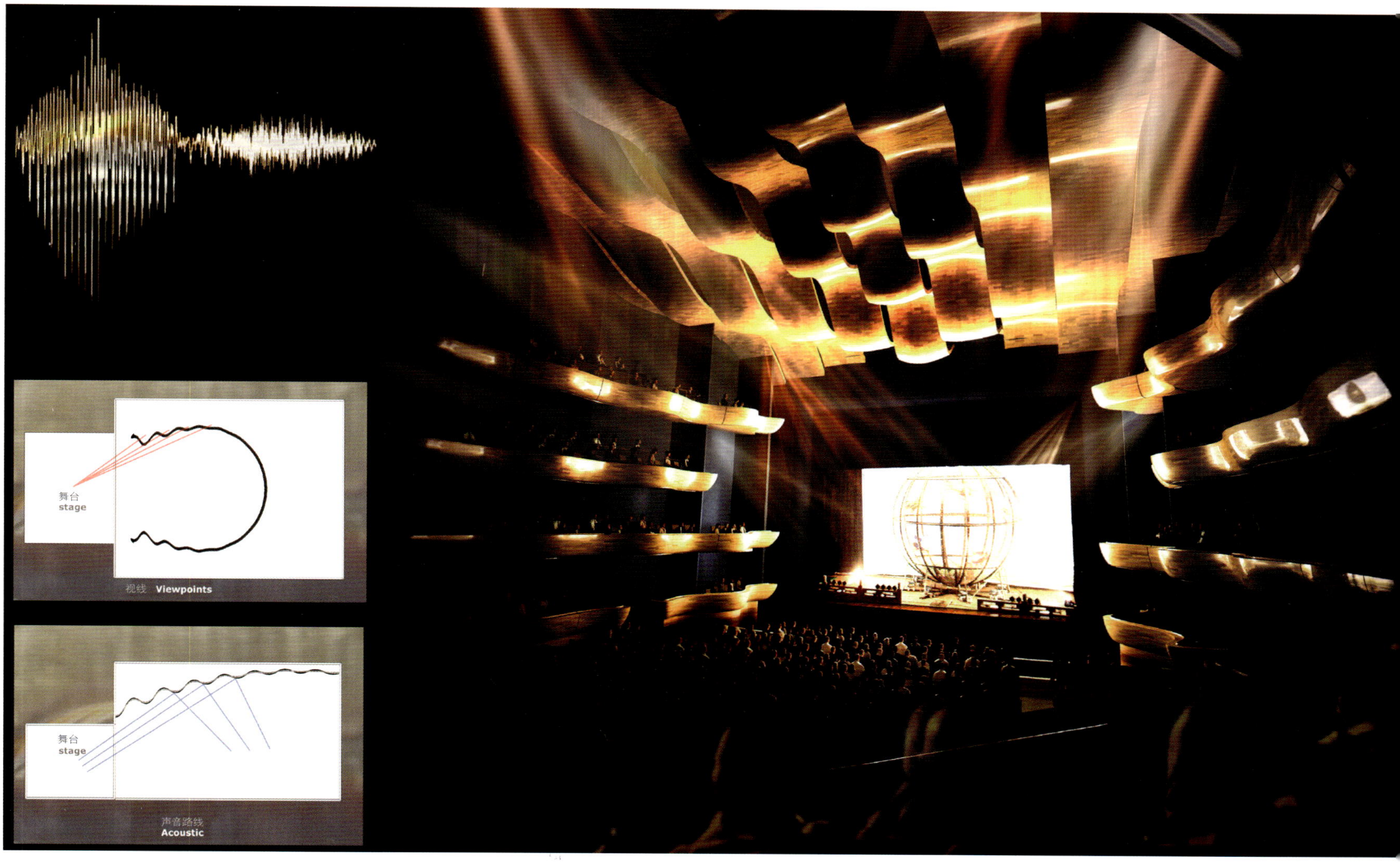

舞台
stage

视线 Viewpoints

舞台
stage

声音路线
Acoustic

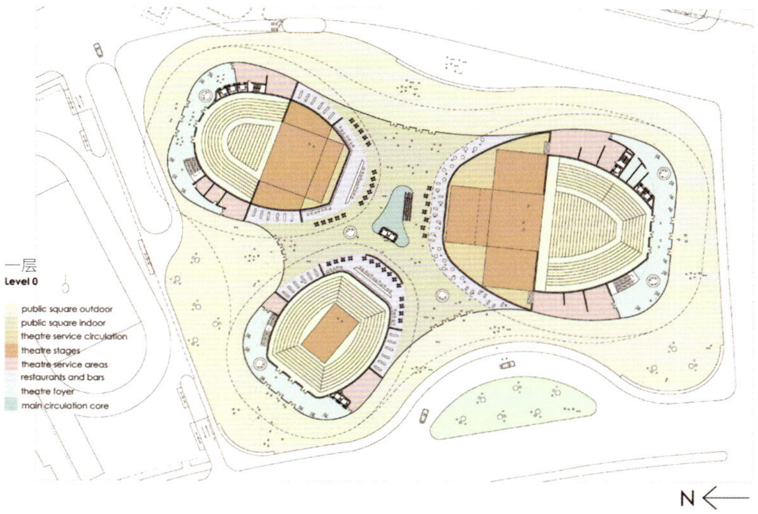

一层
Level 0

- public square outdoor
- public square indoor
- theatre service circulation
- theatre stages
- theatre service areas
- restaurants and bars
- theatre foyer
- main circulation core

N

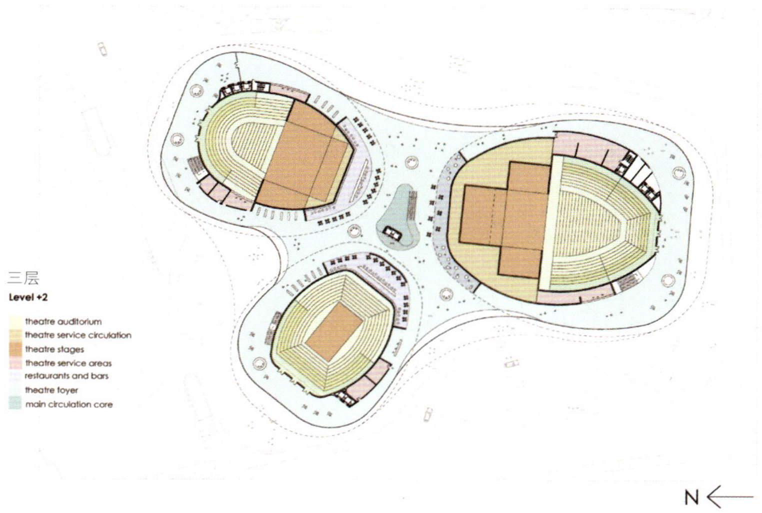

三层
Level +2

- theatre auditorium
- theatre service circulation
- theatre stages
- theatre service areas
- restaurants and bars
- theatre foyer
- main circulation core

N

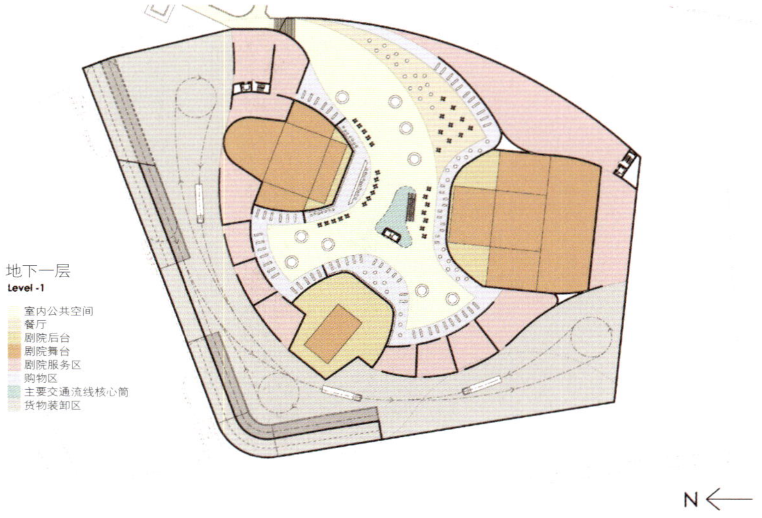

地下一层
Level -1

- 室内公共空间
- 餐厅
- 剧院后台
- 剧院舞台
- 剧院服务区
- 购物区
- 主要交通流线核心筒
- 货物装卸区

N

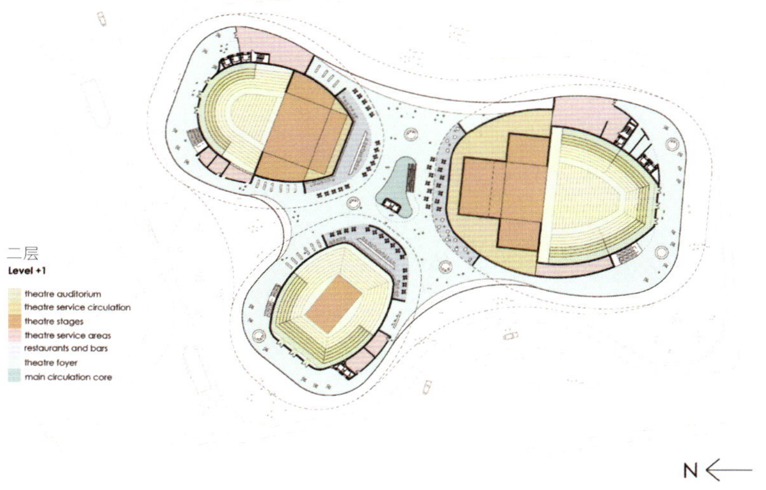

二层
Level +1

- theatre auditorium
- theatre service circulation
- theatre stages
- theatre service areas
- restaurants and bars
- theatre foyer
- main circulation core

N

巴统水族馆
BATUMI AQUARIUM

建筑师：Henning Larsen Architects
项目团队：Louis Becker (design director, partner) Anders Park (project manager), Viggo Haremst (design responsible), Michael Sørensen and Jaewoo Chun
甲方：Association A.T.U.
项目地点：Batumi in the Republic of Georgia
总楼面面积：2000 m^2
获奖情况：邀请赛一等奖

巴统水族馆的设计灵感来自巴统海滩上的鹅卵石。建筑将会坐落在巴统的乔治亚港口，就像一个岩石图标，无论从内陆还是海洋上都看得到。整个构造包括四个独立支撑的展览区，每个石头都代表一个海洋——地中海、黑海/红海、爱琴海和印度洋。四个独立设置的水族展览馆通过一个中央多功能空间连接，这个空间内有咖啡厅、礼堂、零售店，在这里可以看到作为背景的黑海和巴统海滩。参观者在开始参观之旅前可以在中央空间里聚集、娱乐、就餐、购物和放松。

Batumi Aquarium is inspired by the characteristic pebbles of the Batumi beach. The building will be situated in the Georgian port of Batumi and will stand out as an iconic rock formation – visible from both land and sea.The formation constitutes four self-supporting exhibition areas where each of the four stones represents a unique marine biotype – the Mediterranean, the Black Sea/Red Sea, the Aegean Sea and the Indian Ocean. The four dispersed aquarium exhibitions are connected by a central, multipurpose space including café, auditorium and retail functions with views of the Black sea and Batumi beach as scenic backdrop. Visitors gather in the central space to convene, play, eat, shop and relax before continuing their adventures through the exhibitions.

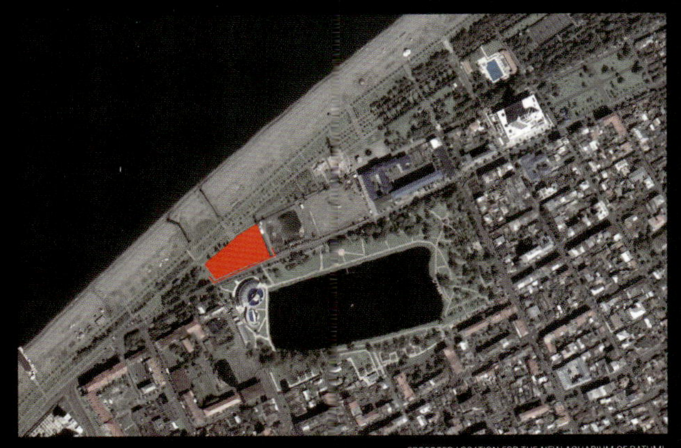

PROPOSED LOCATION FOR THE NEW AQUARIUM OF BATUMI
新巴统水族馆设计位置

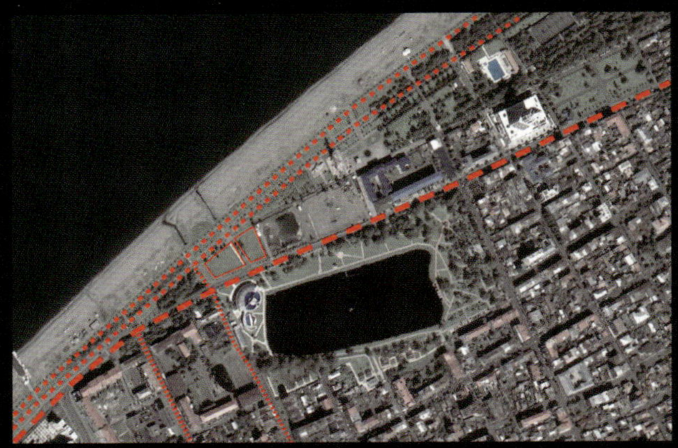

MAIN ROADS /PEDESTRIAN FLOWS
主要道路/人行流线

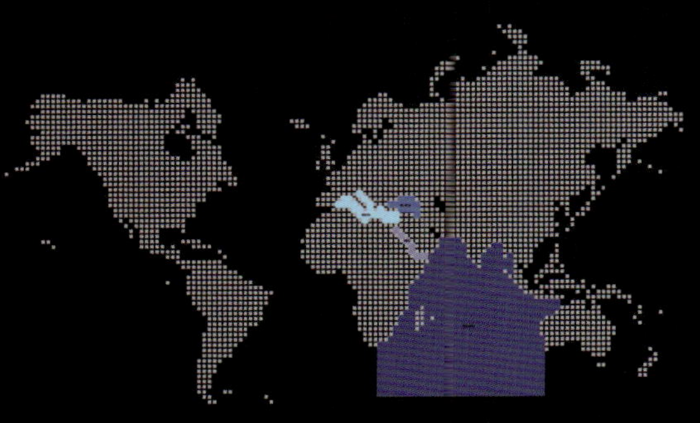

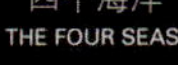

四个海洋
THE FOUR SEAS

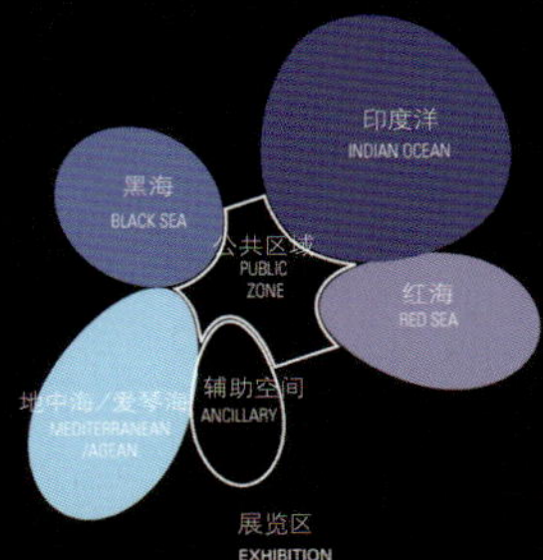

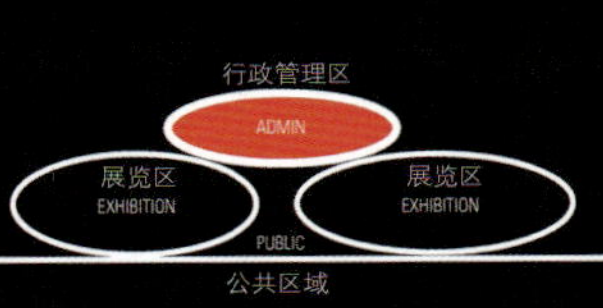

概念分区
CONCEPTUAL SECTION

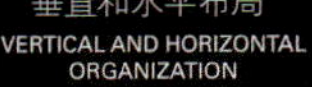

垂直和水平布局
VERTICAL AND HORIZONTAL ORGANIZATION

流线概念
FLOW CONCEPT

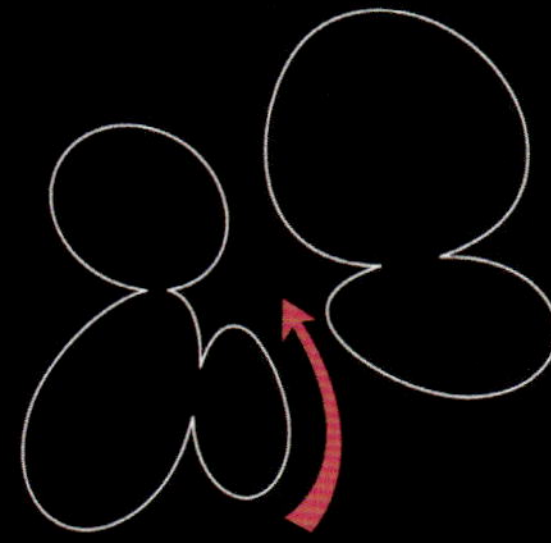

入口
ENTRANCE

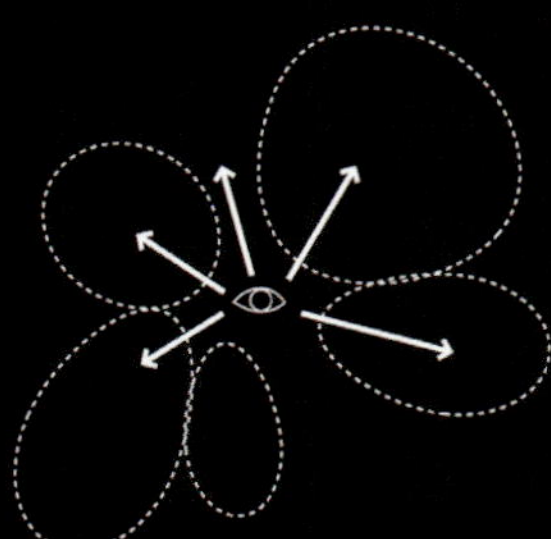

视觉联系
VISUAL CONNECTION

展览区流线
EXHIBITION FLOW

群岛和新的交点
ARCHIPELAGOS AND NEW MEETING POINTS

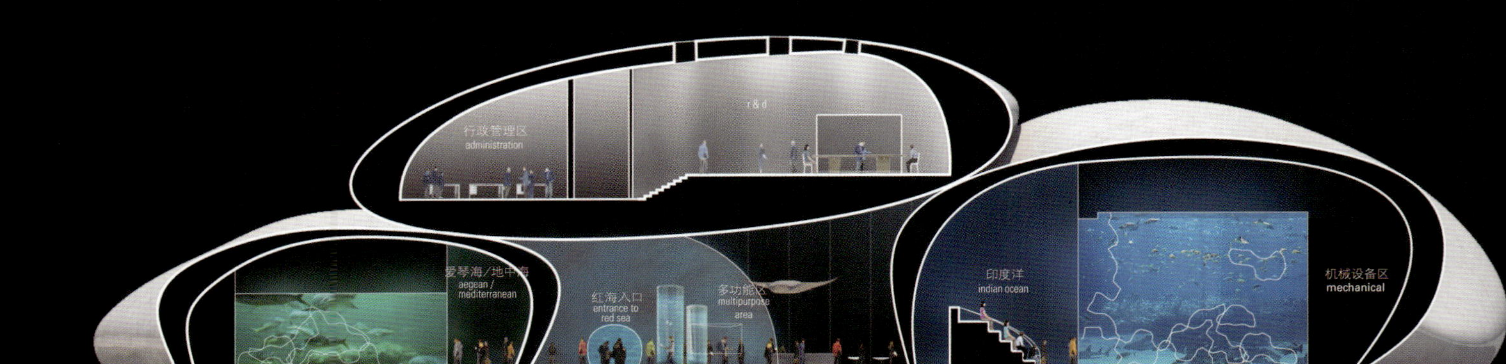

剖面图
CROSS SECTION

PARK VS CITY
公园vs城市

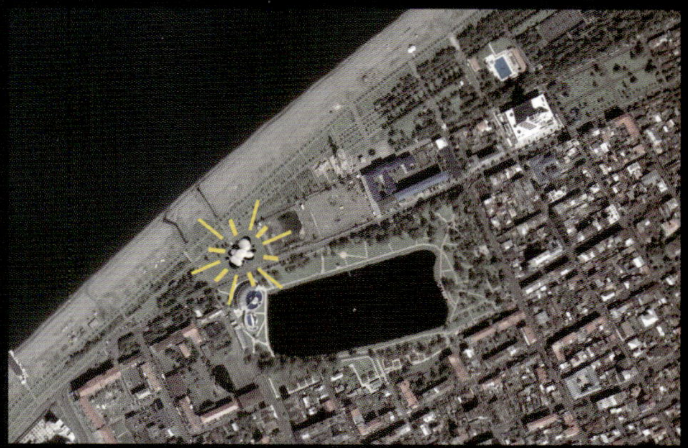

NOT ONLY A NEW ICON FOR THE BEACH, BUT FOR THE ENTIRE REGION AND CITY OF BATUMI
不仅是海滩的新图标，也是整个地区与巴统市的新图标

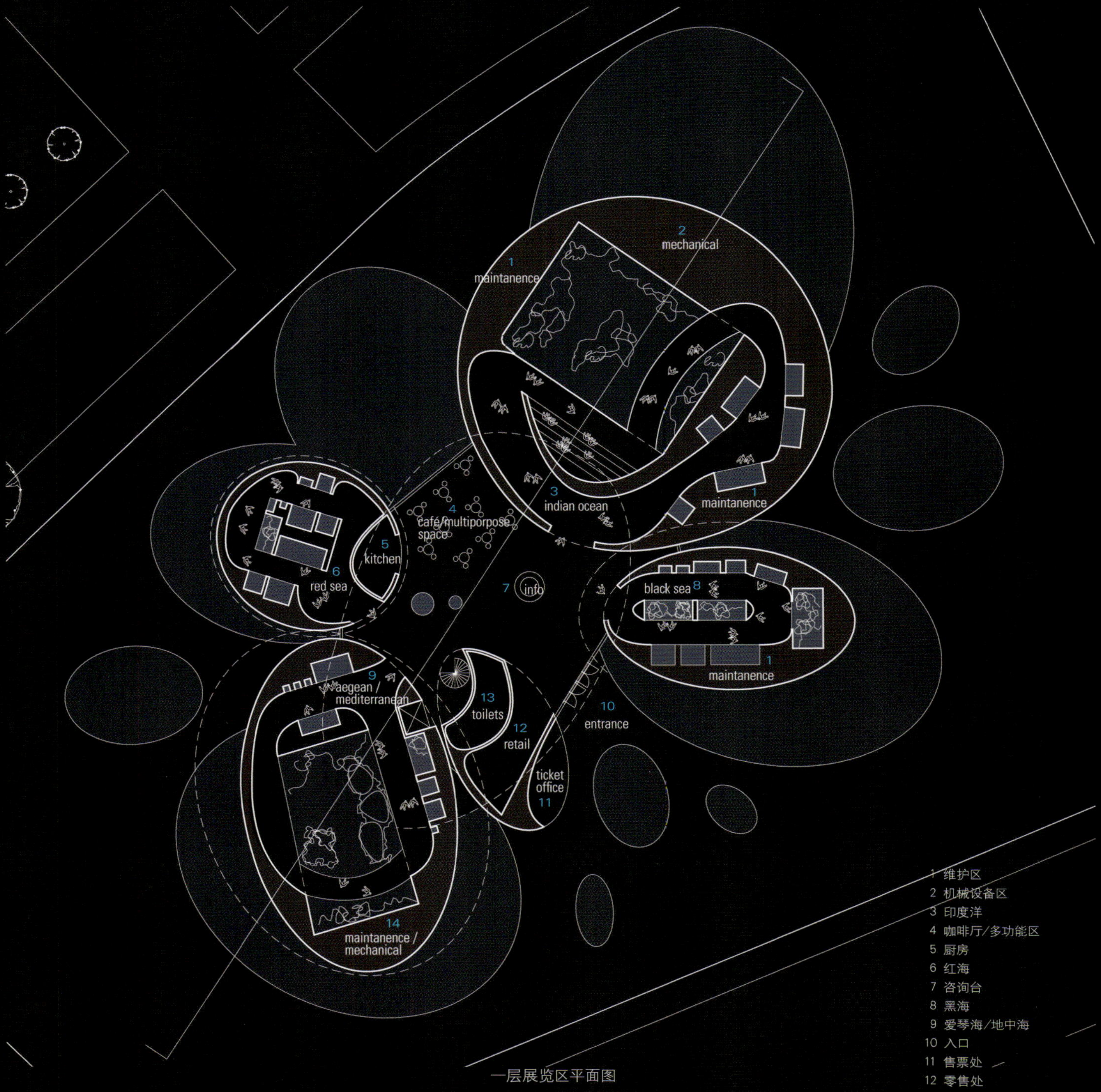

1 维护区
2 机械设备区
3 印度洋
4 咖啡厅/多功能区
5 厨房
6 红海
7 咨询台
8 黑海
9 爱琴海/地中海
10 入口
11 售票处
12 零售处
13 卫生间
14 维护区/机械设备区

一层展览区平面图
LEVEL 00_EXHIBITION

台湾高雄卫武营艺术中心
WEI-WU-YING CENTRE FOR THE ARTS, KAOHSIUNG, TAIWAN

建筑师：Mecanoo architecten, Delft, The Netherlands
甲方：Preparatory Office of the Wei-Wu-Ying Centre for the Arts of the Council for Cultural Affairs, Taiwan, China
当地建筑师：Archasia Design Group, Taipei, Taiwan, China
结构工程师：Supertech, Taipei, Taiwan, China
机械工程师：Yuan Tai, Taipei, Taiwan, China
3D指导：Lead Dao, Taipei, Taiwan, China

总平面图
Site plan

卫武营都会公园曾是一个军事设施旧址，现在成为台湾地区新卫武营艺术中心所在地。一个音乐厅、一座歌剧院、一间大剧院、一间演奏厅和一个大型户外座位区共同构成了艺术中心。这个新文化中心总共有6000个座位以及最先进科技的剧院设施，将会吸引来世界级的演奏艺术家和戏剧团体。高雄市拥有150万人口，是台湾地区第一大港口城市，也是世界最大的港口城市之一，而卫武营艺术中心将会成为这座城市的新图标。艺术中心的建造，标志着高雄市从一个港口城市提升为一个现代文化城市。高雄市与台北市之间的高速铁路已经竣工，而新城地铁系统也预计在两年内完工。周围65公顷的公园是整体设计的一部分。

The Wei-Wu-Ying Metropolitan Park, the site of a former military complex, is the location for the new Wei-Wu-Ying Centre for the Arts of Taiwan Area. The complex features a concert hall, an opera house, a theatre, a recital hall, and a large outdoor seating area. Hosting a total of 6,000 seats and the most technologically advanced theatre facilities, the new cultural complex will draw world class performing artists and theatre companies. The Wei-Wu-Ying Centre for the Arts will become the new icon of the city of Kaohsiung, the largest harbour city of Taiwan Area and one of the largest harbour cities in the world with 1.5 million inhabitants. By building the Centre for the Arts, the city will symbolize its evolution from a harbour city to a modern cultural city. The high-speed train between Taipei and Kaohsiung was completed recently and the new city metro system is slated for completion within two years. The surrounding 65 hectare park is an integral part of the design brief.

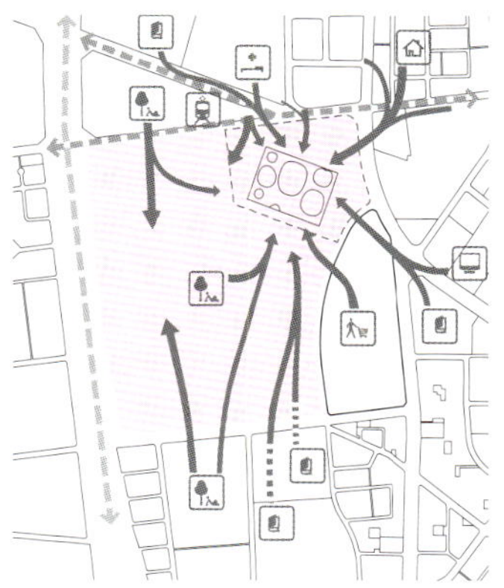

周围环境分析
Neibourhood analysis

城市绿网分析
City green network analysis

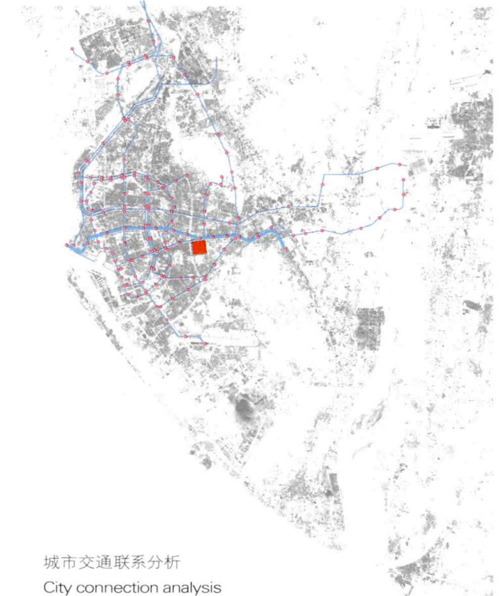

城市交通联系分析
City connection analysis

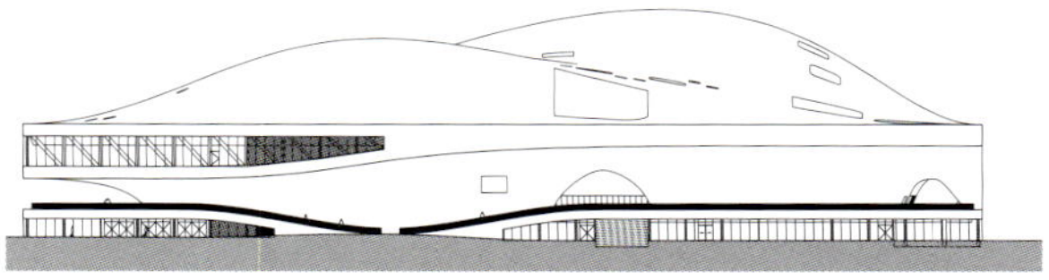

东立面
East elevation

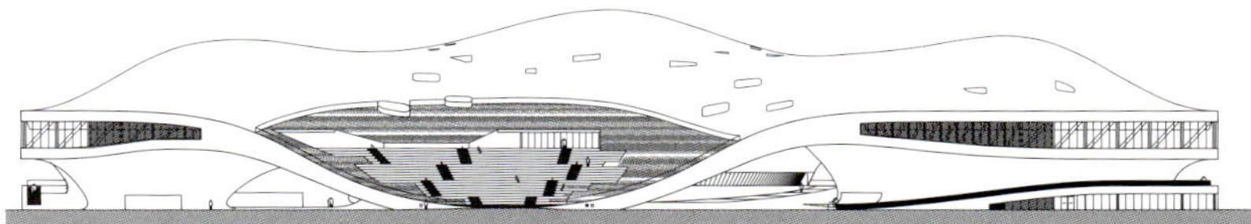

南立面
South elevation

西立面
West elevation

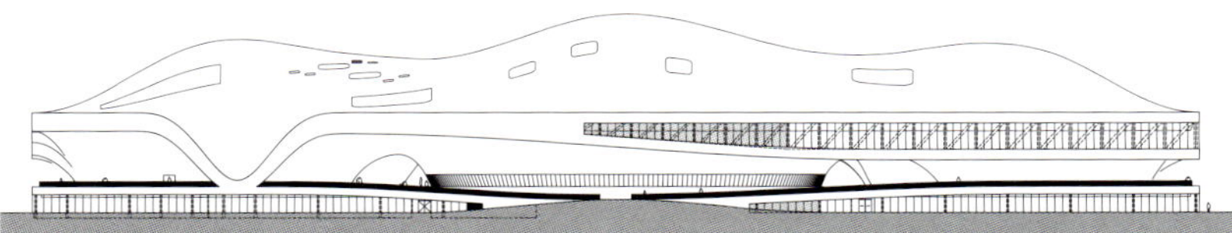

北立面
North elevation

舞蹈剧院
DANCE PALACE

建筑师：UNStudio

Ben van Berkel, Gerard Loozekoot with Christian Veddeler, Wouter de Jonge and Jan Schellhoff and Kyle Miller, Maud van Hees, Hans-Peter Nuenning, Arnd Willert, Nanang Santoso, Imola Berczi, Tade Godbersen, Patrik Noome

工程师：ARUP

剧院顾问：theateradvies bv, Amsterdam

Louis Janssen, Caroline Noteboom

效果图与线图：UNStudio and Rendertaxi, Aachen

甲方：PETERSBURG CITY LLC

项目地点：St. Petersburg, Russia

建筑面积：24 000 m^2

建筑高度：28 m

获奖情况：竞赛一等奖

荷兰著名建筑工作室UNStudio将舞蹈剧院设计成为一座开放的、极具魅力的剧院建筑，可以容纳1300名来宾（大礼堂可容纳1000人，小礼堂可容纳300人）。设计过程中重点考虑了公共门厅内宽敞的交通流线，以及与公共广场和整个城市的透明联系。建筑物通过其规模和变化的透明度与周围原有建筑融合在了一起。在规模上，它尊重并延续了圣彼得堡28m的典型屋顶高度；变化的透明度是通过设计了三角形的覆面板立面系统而获得的。不透明面板和多孔面板交替变化，根据功能区、视线和朝向的不同，具有不同的开放程度。

UNStudio's design for the Dance Palace presents an open and inviting theatre building with provision for 1300 guests (large auditorium 1000, small auditorium 300). Programmatic considerations focus on the spacious circulation of the public foyer and the transparent relationship to the surrounding public square and the city. Integration with the existing neighbouring buildings is achieved by both the scale of the building - which in elevation follows and respects St. Petersburg's typical 28m roofline – and the transformative transparency which is introduced by a facade system of triangular cladding panels. The variation between opaque and perforated panels creates a controlled openness, depending on programme, views and orientation.

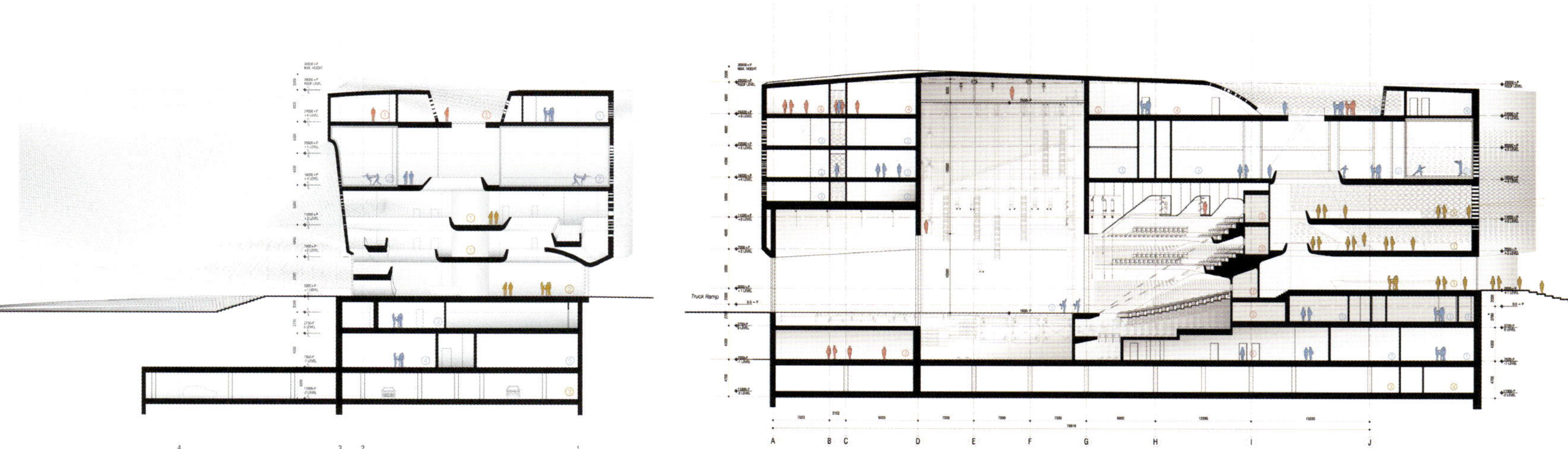

剖面图
Sections

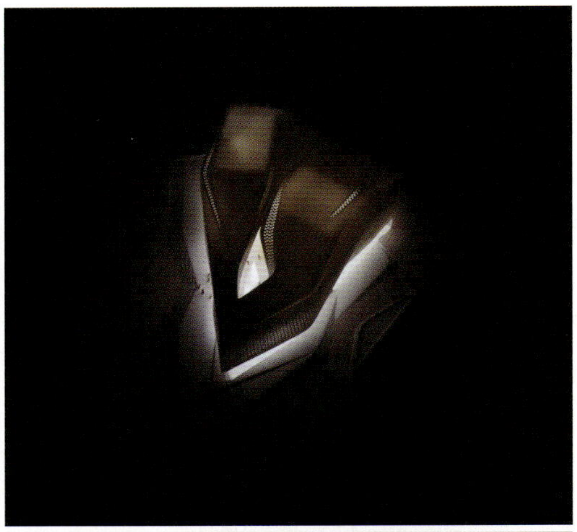

灵活的交叉类型

FLEXIBLE X-TYPE

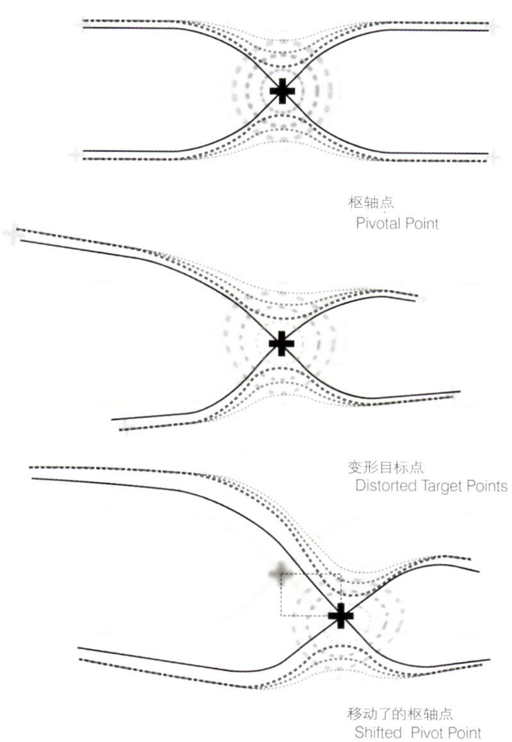

平面布局原理

PLAN PRINCIPLE

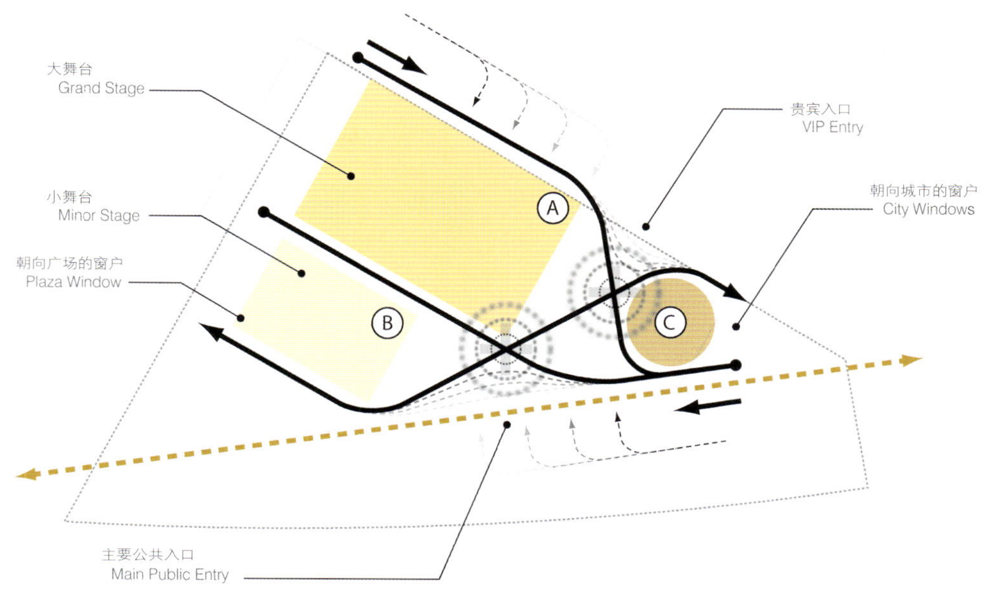

设计示意图

Design diagram

德国波恩的新贝多芬音乐大厅
THE NEW BEETHOVEN CONCERT HALL IN BONN, GERMANY

甲方：Deutsche Post AG, Deutsche Telekom, Deutsche Postbank
建筑师：Zaha Hadid Architects
结构工程师/立面顾问：Bollinger & Grohmann, Frankfurt, G

扎哈·哈迪德事务所的新贝多芬音乐大厅的城市设计理念为将波恩这座城市与莱茵河散步道联系在一起，发掘设计理念在充实河边公共生活方面的潜能。

建筑就像是莱茵河畔的一颗宝石。整个建筑体量布满了富有活力的孔洞，使整个建筑具有了渐变的光照效果。在夜晚，建筑变成了透明的，并从内部散发出光芒。

巨大的木制交响乐厅就像乐器的共鸣箱一样镶嵌在外围护结构中，隐约地向外界展示着自己。音乐厅内部散发出温暖热烈的氛围。与大音乐厅的外形相匹配的是小型的独唱和室内音乐会厅

音乐厅水晶般的造型与人工景观的改造相呼应。巨大的平台像岩石一样延伸到莱茵河畔，从大厅缓缓地下降到朝向莱茵河的公园。同样，有些景观也延伸到建筑的大厅中，使室内与周围的开拓地拥有相同的形式与空间逻辑。

A central priority in ZHA's urban design concept for a new Beethoven Festival Hall is linking the city of Bonn to the Rhine River promenade and leveraging that idea's potential to enrich public life on the river's edge.

A jewel is placed on the Rhine riverbank. Dynamic perforations wrap the volume, allowing for a gradient of light conditions throughout the building. At night, the object turns transparent, glowing from the inside.

The large wooden symphonic hall is inlayed like a musical instrument's resonating body within the exterior envelope, faintly expressing itself towards the outside. A warm glow resonates from within. The formal counterpoint to the large hall is constituted through the smaller recital and chamber music hall.

The concert hall's crystalline form is echoed in the artificial landscape's modulations. On the Rhine, large terraces stretch out like rocks, ramping down from the lobby into the park facing the Rhine. Similarly, certain landscape features extend into the building's lobby to provide the interior with the same formal and spatial logic as the surrounding land formations.

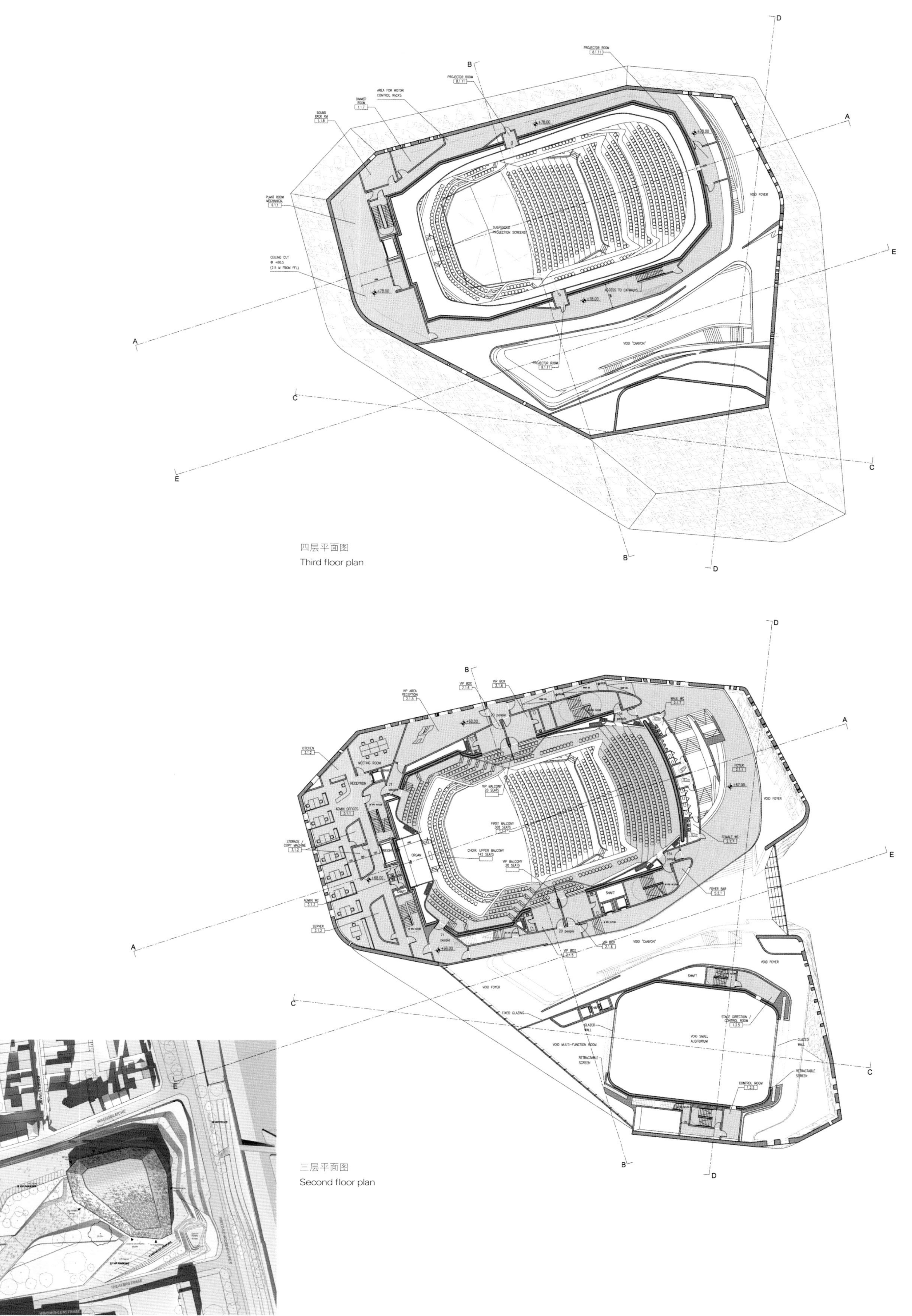

四层平面图

Third floor plan

三层平面图

Second floor plan

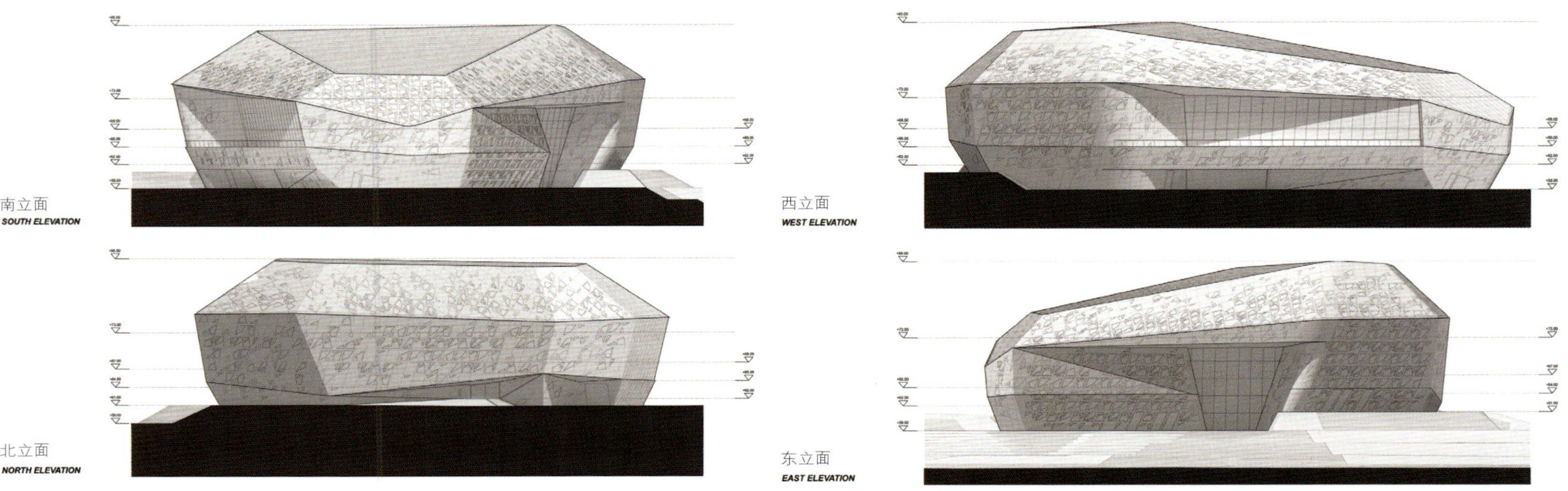
南立面
SOUTH ELEVATION
西立面
WEST ELEVATION
北立面
NORTH ELEVATION
东立面
EAST ELEVATION

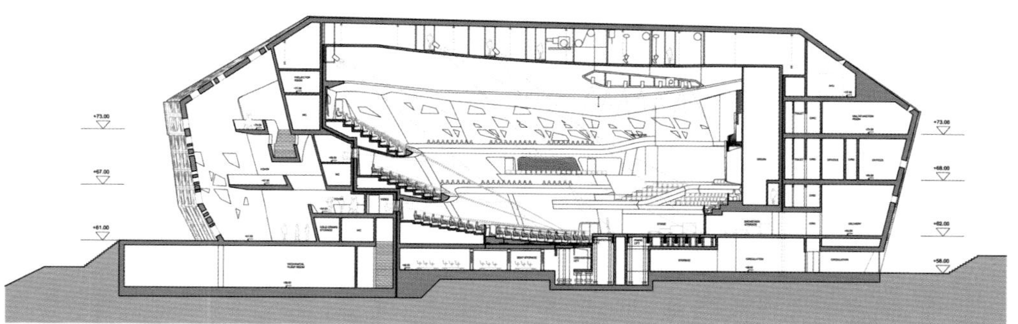

SECTION AA

AA剖面

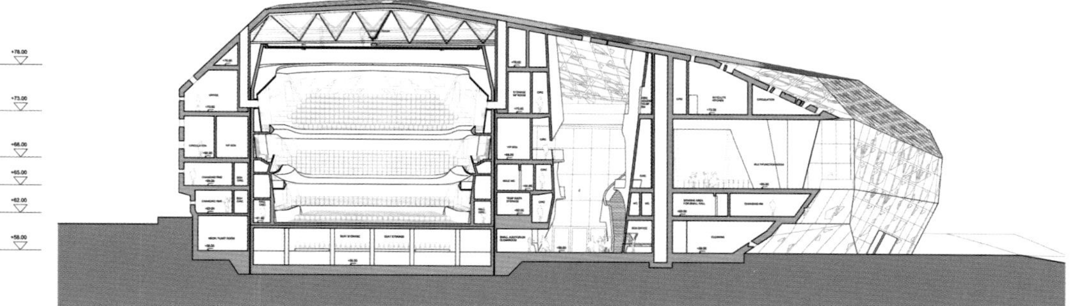

SECTION BB

BB剖面

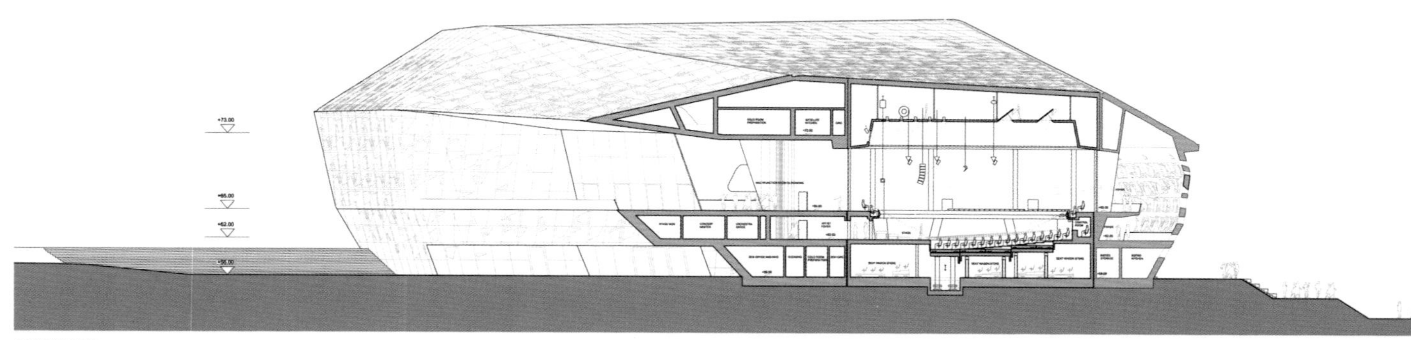

SECTION CC

CC剖面

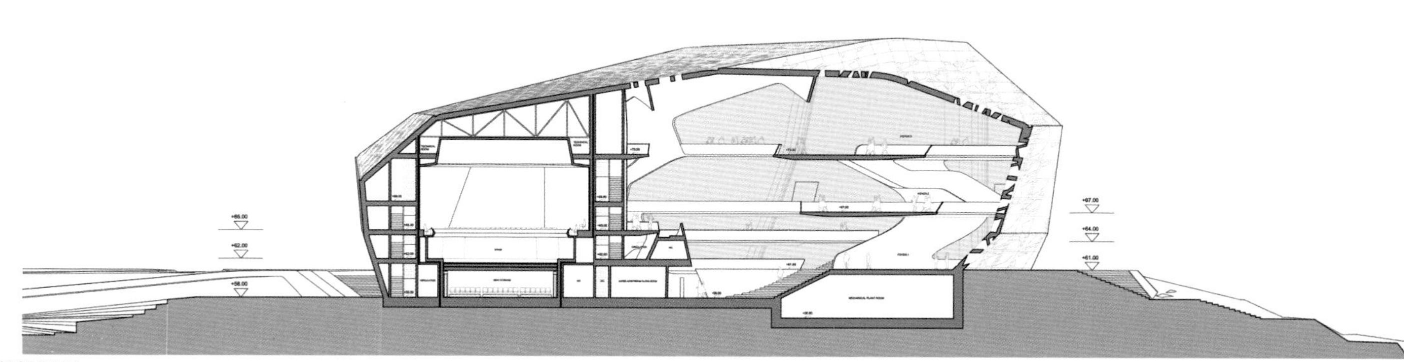

SECTION DD

DD剖面

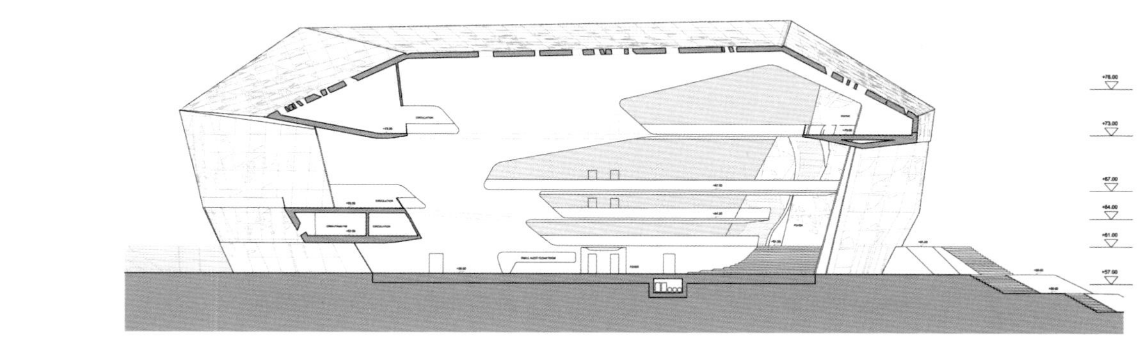

SECTION EE

EE剖面

阿斯坦纳国家图书馆
ASTANA NATIONAL LIBRARY

建筑师：BIG
合作者：ARUP, AGU
合伙负责人：Bjarke Ingels
项目负责人：Thomas Christoffersen
设计团队：Amy Campbell, Jakob Henke, Johan Cool, Jonas Barre, Daniel Sundlin
甲方：Kazakhstan Presidential Office
项目地点：Astana, Kazakhstan
面积：33 000 m^2
获奖情况：竞赛一等奖

新建的国家图书馆大约有33 000m^2。设计现代而充满理性，又融合了传统图书馆的经典建筑语汇。内部核心区档案室的圆形组织布局不但具有线性布局清晰的特点，还具有无限圆环方便的特性。阿斯坦纳市民、哈萨克斯坦人民以及国际游客在国家图书馆将了解哈萨克斯坦的历史、多元文化以及新首都和第一任总统。图书馆可以容纳各领域的人并使他们相互交流，包括公务员、政客、研究人员、学生、博物馆历史学家以及职员。

The new National Library encompasses an estimated 33 000m^2. The design was hailed as being both modern and rational and anchored in a classical vocabulary of traditional libraries. The circular organization of the archive at its inner core combines the clarity of a linear organization with the convenience of an infinite loop. The National Library will be the place where the citizens of Astana, the people of Kazakhstan as well as international visitors can come to explore the country's history, its diverse cultures, its new capital and its first president. The Library will accommodate and communicate with all segments of the population: civil servants, politicians, researchers, students, museum historians and staff.

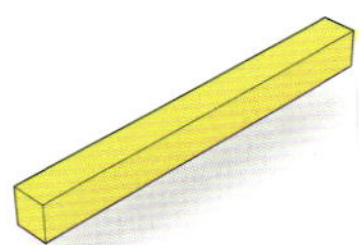

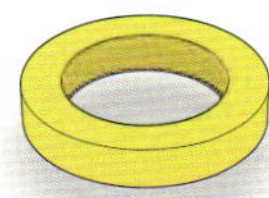

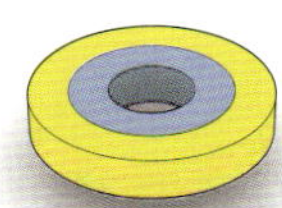

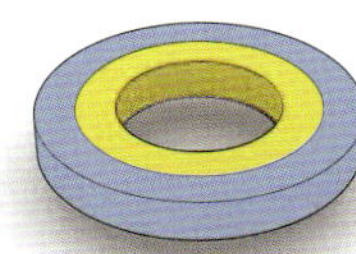

线性图书馆
Linear library

最简单的布局方式就是线性分组不同类的书籍，这样图书管理员就能根据合理的索引系统对书籍进行分类，例如从000到999的杜威十进分类法。线性布局的缺点就是有死角，空间质量较差。

The simplest organizational form would be a linear grouping of knowledge allowing librarians to archive all the collections according to a rational indexing system like the Dewy Decimal System from 000 to 999. The drawback of the linear organization is that it is dominated by dead ends and poor spatial quality.

完美的圆环
Perfect circle

档案室被布置成知识的圆环，两面都能接受到自然光和新鲜的空气。在外围能360°看到阿斯坦纳的景色，建筑的中央是冥想氛围的庭院，庭院上方就是蓝色的天空。无限圆环的简洁与完美使人们能更加清晰地在国家图书馆书架上数量惊人的书籍中找到自己的目标。

The archive is organized as a circular loop of knowledge, surrounded by light and air on both sides. On the periphery a 360 degree panorama of Astana - at the heart of the building a contemplative courtyard domed by the heavenly light blue of the celestial vault. The simplicity and perfection of the infinite circle allows for a crystal clear and intuitive orientation in the vast and growing collection that will populate the shelves of the National Library.

加衬
Lining

优点：沿着圆环的内侧加衬的一圈公共空间为会议室和礼堂提供了入口路线。
缺点：虽然这些空间对于集体活动和会议非常完美，但会使人感到封闭独立，与周围城市隔绝。

+Lining the circle along the interior with a ring of public programs provides accessible cluster of communal spaces for meeting rooms and auditoriums.

- While perfect for collective activities and conferences, the interiority could feel claustrophobic and isolated from the surrounding city.

围护结构
Envelopment

优点：利用公共空间包围圆形的图书馆，优化了内部的联系性。每一个公共空间都能直接通往图书馆，看到周围城市的景色。
缺点：虽然对于独立学习和观看城市全景非常有利，但缺点是缺少总体感和内部焦点。另外，环形的图书馆还减少了人们与外界的联系。

+By wrapping the circular library in public programs, internal connectivity is optimized. Each individual program gains direct access to the collection as well as a panoramic view of the surrounding city.

-Perfect for individual study and admiration of the urban panorama, the drawback is the lack of collectivity and internal focus. In addition the circular library could loose any contact with the outside world

墩座
Podium

将圆环抬高设置在由公共空间，大厅等组成的墩座上，保证了建筑内部与周围公共空间和花园的联系。但是这种堆叠效应降低了阅读图书与从上方观看美景的直接联系。

Lifting the circle up onto a podium of public programs, lobbies, etc ensures a solid integration with the surrounding public spaces and park, However, this stacking effect loses the benefits of direct relationships with the collections as well as the experience of panoramic overview from above.

圆形大厅
Rotunda

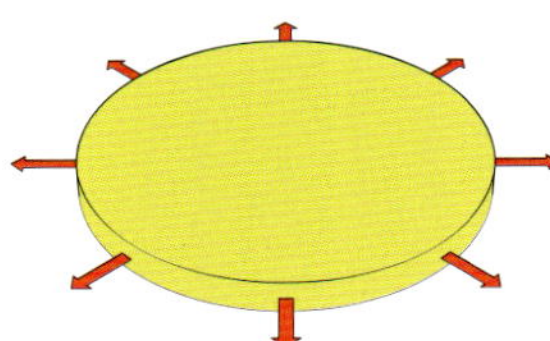

圆环（亚历山大图书馆）
Circle (Great Library of Alexandria)

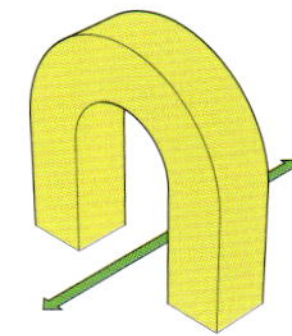

拱（巴黎凯旋门）
Arch (Arc de Triomphe)

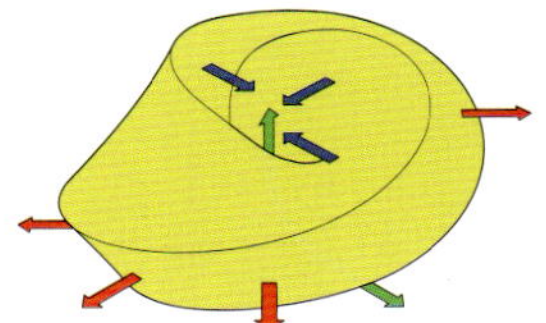

圆形大厅+圆盘+拱=莫比斯环
Rotunda + Disk + Archway = Möbius strip

档案室交通流线
Archive circulation

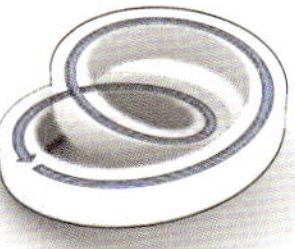

博物馆交通流线
Museum circulation

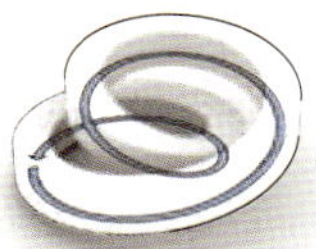

图书馆交通流线
Library circulation

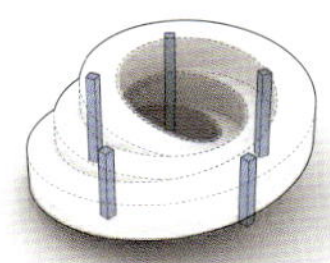

垂直核心筒
Vertical cores

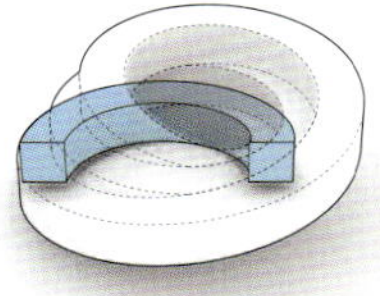

公共功能
Public Functions

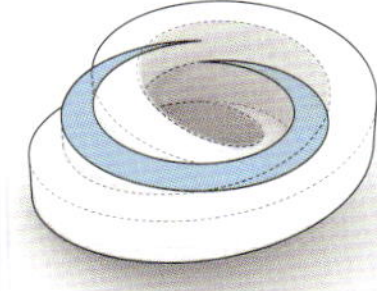

公共广场
Public Plaza

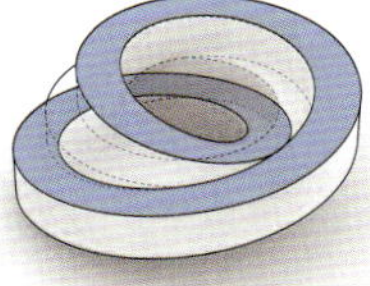

展览大厅
Exhibition Hall

阅览室
Reading Rooms

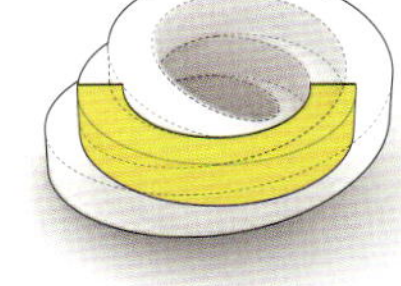

开放书架
Open Bookshelves

景观布局
Landscape

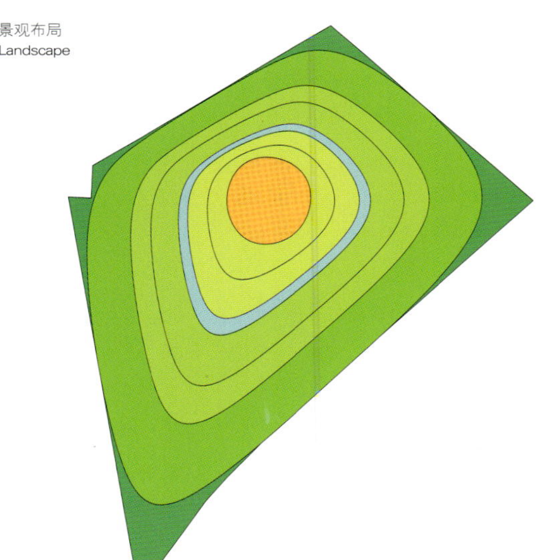

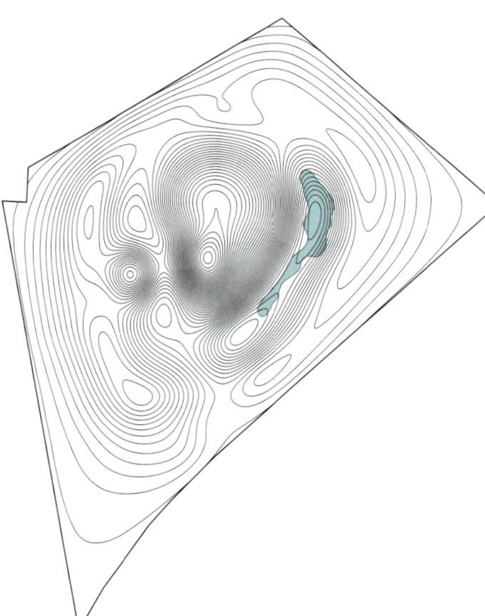

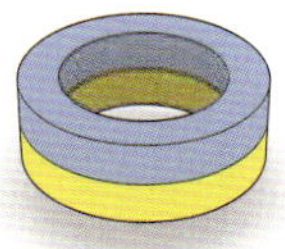

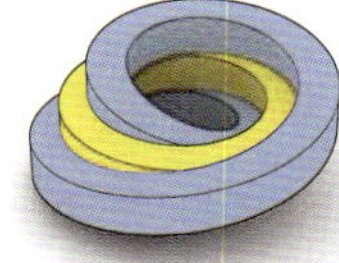

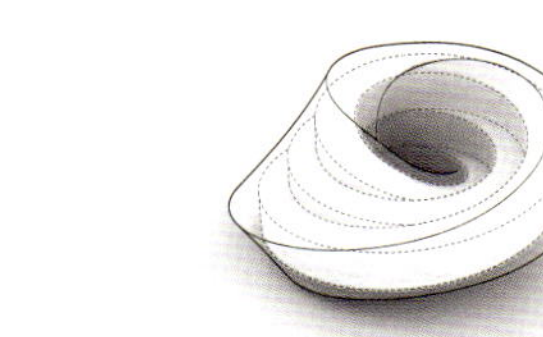

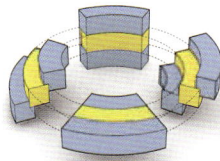

概况

Overview

通过在图书馆圆环墩座上方设置所有的公共功能空间，人们在最大程度上与周围环境结合的同时，还能看到更广泛的美景。

By reversely placing all the public functions on top of the circular podium of the library we will have the full panorama at the expense of integration with the surroundings.

公共螺旋（内＋外＋上＋下）

Public Spiral (Inside+outside+above+below)

在完美的圆环上增设结构最理想的方法就是设置一系列同时在内、外、上、下包围图书馆的公共空间。公共空间就像一个莫比斯环，从内到外，从下到上，无缝连接在一起，人们在这个螺旋的路线上可以看到周围城市和天际线的壮丽景色。

The ideal addition to the perfect circle will be a series of public programs that simultaneously wraps the library on the outside as well as the inside, above as well as below. Like a Möbius strip, the public programs move seamlessly from the inside to the outside and from ground to the sky providing spectacular views of the surrounding landscape and skyline.

莫比斯环

Möbius strip

完美的圆环和螺旋的公共路线这两个互锁的结构创造出了将水平布局转变为垂直布局的建筑。水平布局是将图书馆与起支持作用的空间比肩设置，垂直布局是通过斜向布局、垂直分层、水平连接和斜向的视线将它们垂直设置。

设计师用连续的表皮包围这些复杂的空间，创造出了莫比斯环，立面从内侧移动到外侧，然后又回到内侧。

The 2 interlocking structures: the perfect circle and the public spiral, create a building that transforms from a horizontal organization where library museum and support functions are placed next to each other, to a vertical organization where they are stacked on top of each other through a diagonal organization combining vertical hierarchy, horizontal connectivity and diagonal view lines.

By wrapping the transforming composition of spaces with a continuous skin we create a Möbius strip volume where the facades move from inside to outside and back again.

水平布局

Horizontal organization

无论是公共空间还是私人空间，无论是书籍还是人，都一个挨一个设置，具有最短的移动路径。

All programs, public and private, books and people are placed next to each other with direct shortcuts back and forth.

垂直布局

Vertical organization

各个空间一个设置在另一个上面，读者可以无阻碍地看到城市与庭院，保障了安全性，能够控制公共与私人空间之间的路线。

The programs are placed on top of each other, allowing the users to have uninterrupted views towards the city and the courtyard, and providing easy security and access control between public and private programs

斜向布局

Diagonal organization

博物馆、阅览室、餐厅、图书馆斜向设置在彼此的或上方或下方，可以直接看到彼此。

Museum, reading rooms, restaurants and book collections are placed diagonally above and below each other allowing direct sightlines, overview and glimpses across spaces, levels and programs.

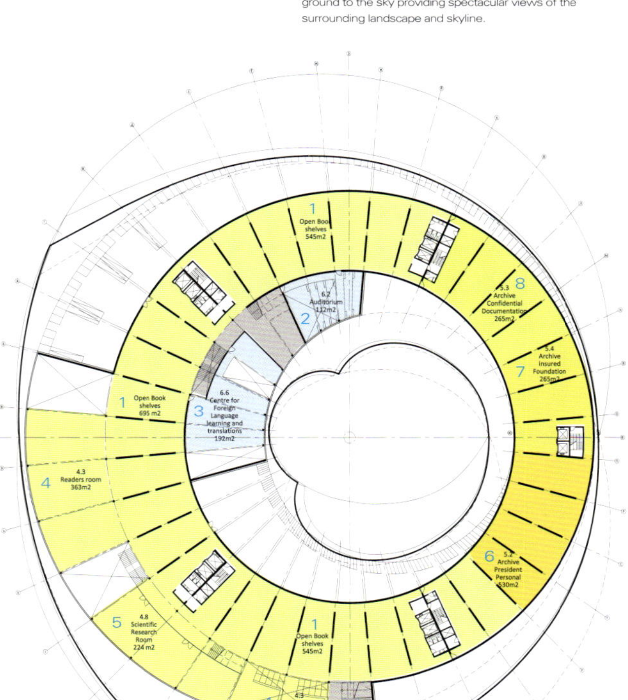

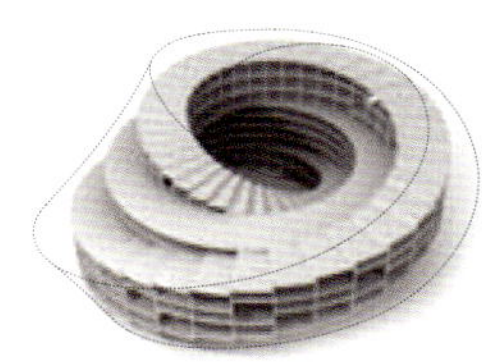

内部结构
Internal structure

外部结构
External structure

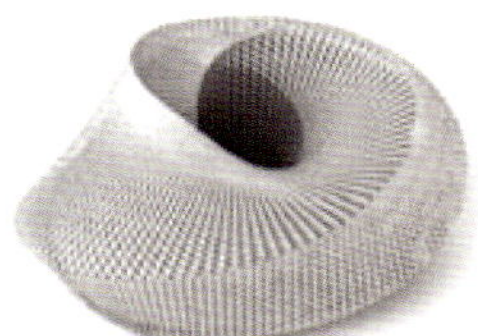

连续表面

国家图书馆的外围护结构与传统的建筑墙体和屋顶不同。像帐篷一样，墙体变成了屋顶，屋顶又变成了墙体。

Continuous surface

The envelope of The National Library transcends the traditional architectural categories such as wall and roof. Like a yurt the wall becomes the roof, which becomes the wall again.

七层平面图

Sixth floor plan

1　开放书架
2　礼堂
3　外语学习与翻译中心
4　阅览室
5　科研室
6　总统个人档案室
7　保险基金档案室
8　机密文件档案室

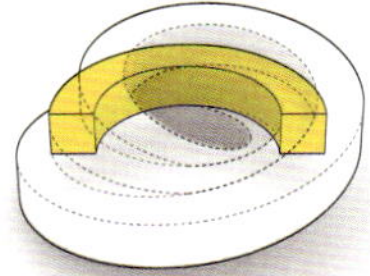

档案室
Archive

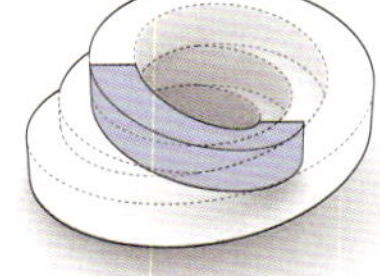

科研室
Scientific Research Room

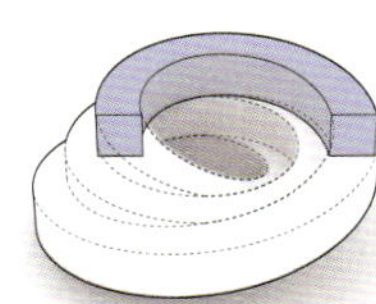

行政管理区域
Administration

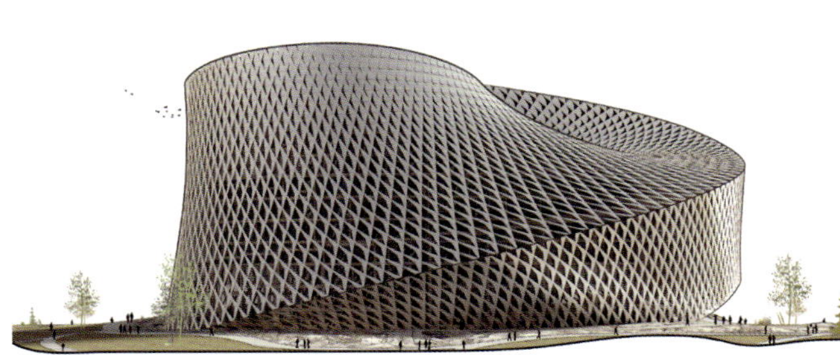

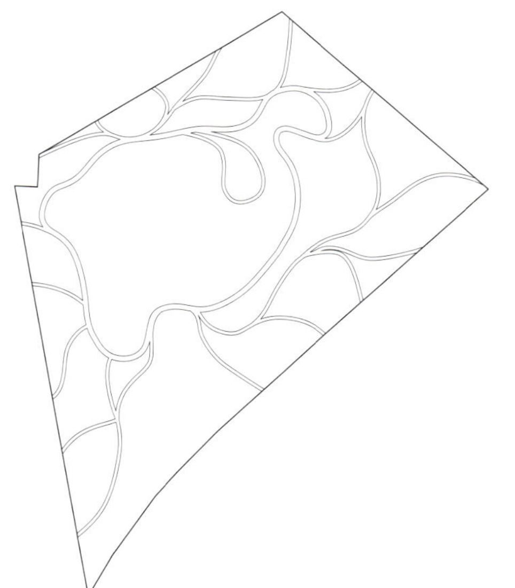

立面图
Elevations

1 展览厅
2 员工室
3 储存室
4 厨房
5 咖啡厅与餐厅
6 阅览室
7 图书档案室
8 会议室
9 媒体会议室
10 礼堂
11 停车场出口
12 室内冬季花园
13 档案信息室
14 主环形会议大厅
15 收藏室
16 入口大厅
17 期刊室
18 图书馆大厅
19 停车场
20 外语学习与翻译中心
21 计算机室
22 休息室
23 （档案室）走廊
24 走廊

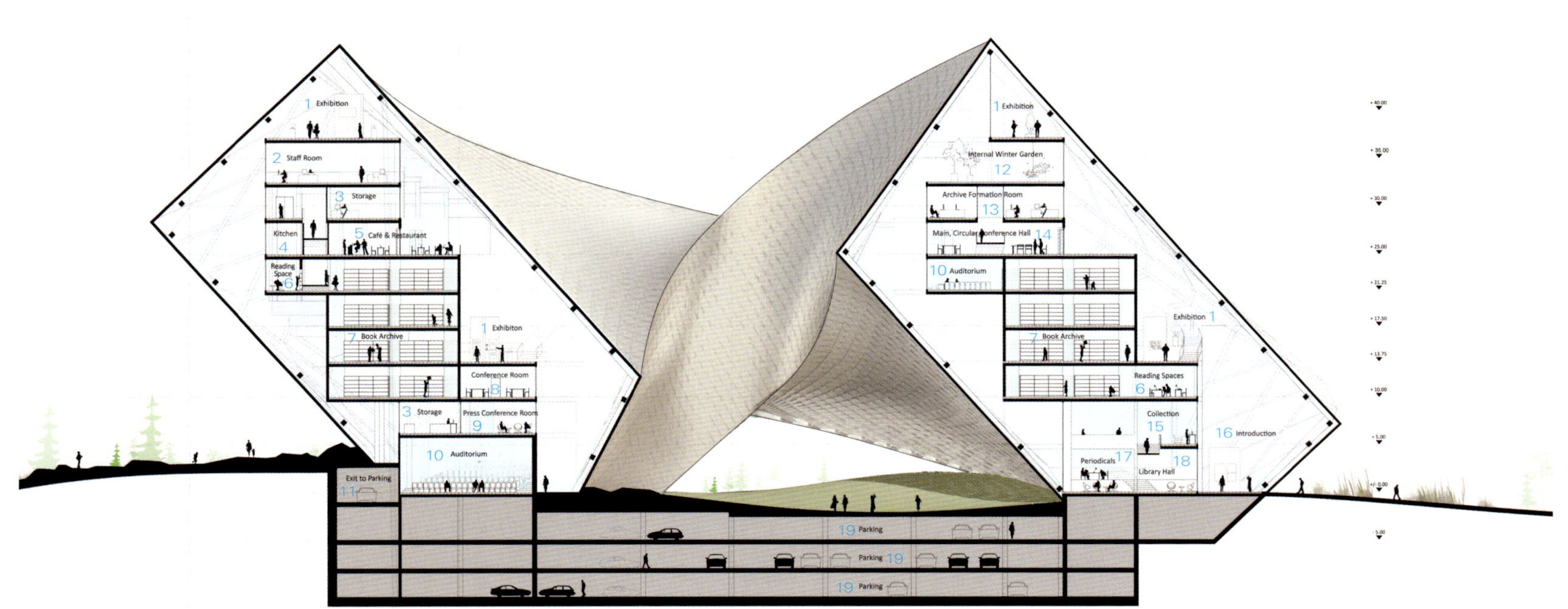

剖面图一
Section 1

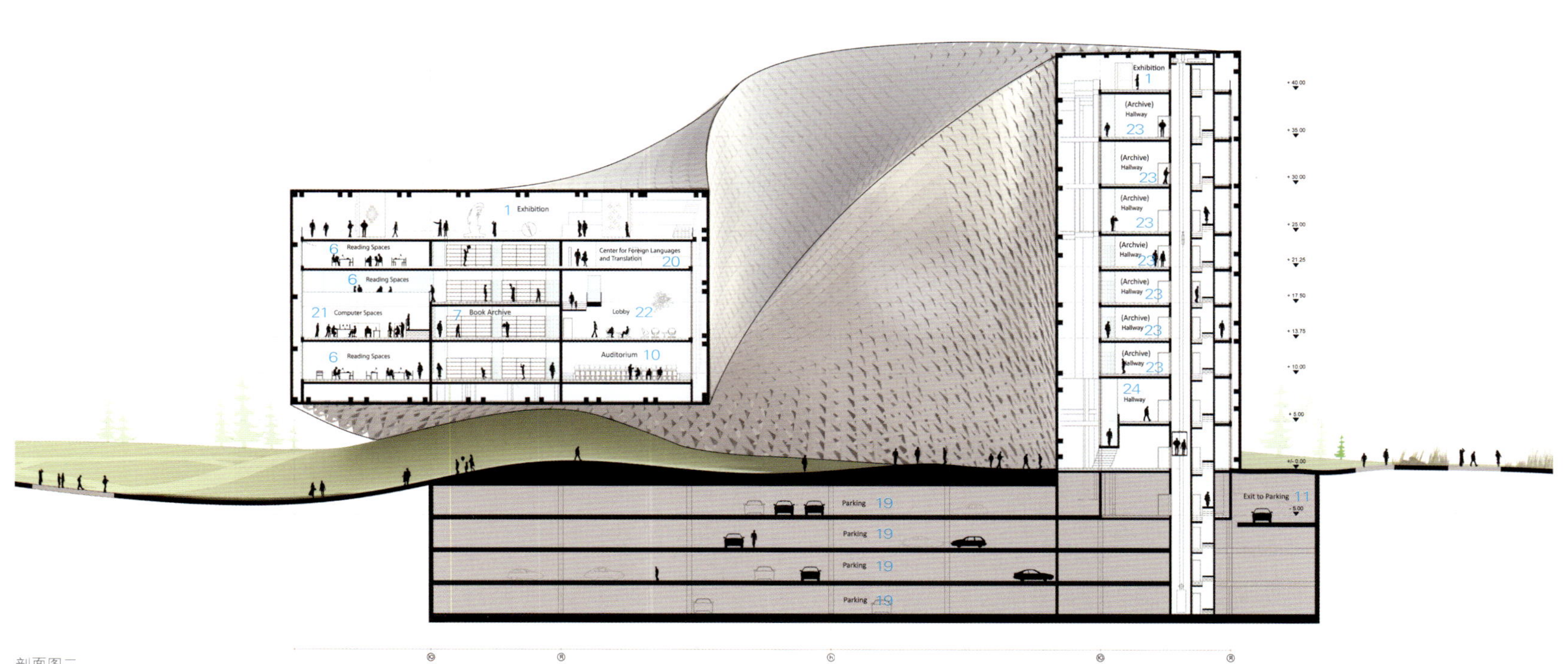

剖面图二

Section 2

奥尔胡斯大学植物园暖房
AARHUS UNIVERSITY HOTHOUSE IN THE BOTANICAL GARDEN

建筑师：C. F. Møller Architects
景观设计师：C. F. Møller Architects
工程师：Søren Jensen A/S
地点：Botanical Gardens, Aarhus, Denmark
甲方：Aarhus University / Danish University and Property Agency
规模：3300m²
获奖情况：2009年建筑竞赛一等奖
图片：C. F. Møller Architects

奥尔胡斯大学植物园中的蜗牛形暖房是暖房建筑的国际性标杆。它是由C. F. Møller建筑师事务所于1969年设计的。该建筑很好地融入周围的环境。因此，在设计新的暖房时，要将原有暖房的建筑价值铭记于心。

建筑体量采用有机的造型，人们可徜徉其中，并在树顶探索。它展现了植物的生物特征，并使人们可以游历不同气候带。这种游历方式将使暖房在未来成为整个欧洲暖房建筑的焦点。新暖房的设计以节能设计方案，材料，室内气候和技术知识为基础。

The snail-shaped hothouse in the Botanic Garden in Aarhus is a national icon in hothouse architecture. It was designed in 1969 by C. F. Møller, and is well adapted to its surroundings. Accordingly, it was important to bear the existing architectural values in mind when designing the new hothouse.

The organic form and the large volume, in which the public can go exploring among the tree-tops, present botany and a journey through the different climate zones in a way which will make the new hothouse in Aarhus a future attraction in a pan-European class in hothouse architecture. The design of the new hothouse is based on energy-conserving design solutions and on a knowledge of materials, indoor climate and technology.

Inclination
倾斜
Ellipse
椭圆形
Inflation
膨胀
设计/最优化的标准
Design- / optimization criteria
Reference model
(simple half-dome)
参考模型
(简单的半月丘形)
Variable parameters / Optimization of form
多变的参数/形式的优化
Optimal model
最优化模式
ETFE-cushion
四氟乙烯垫
East-West
tubular profile
东一西向管状构件
North-South
tubular profile
北一南向管状构件
Diagonal
cables
斜向交叉的锚索
Plinth
基座

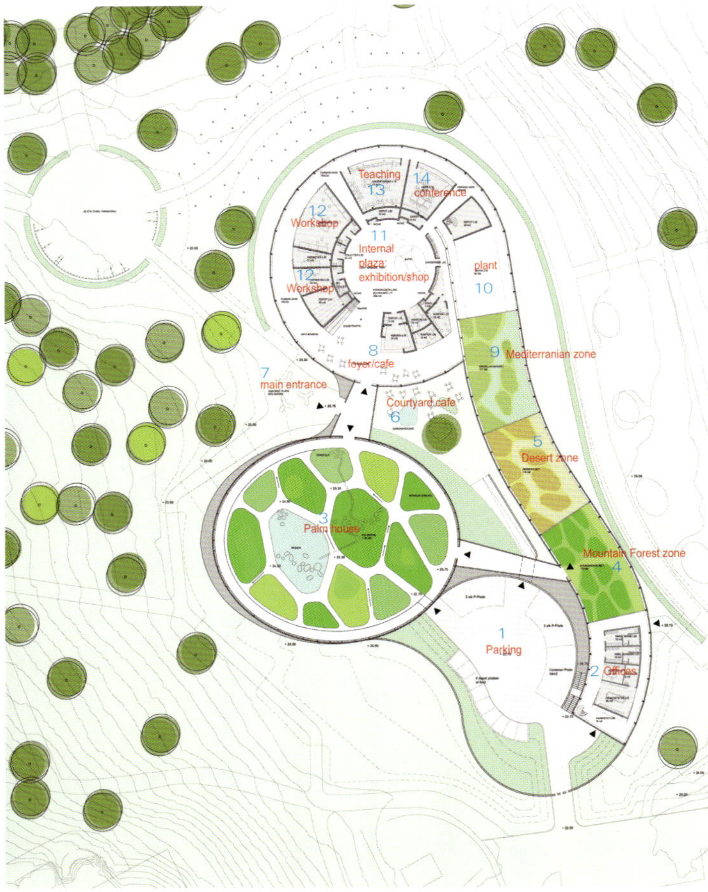

总平面图

Site plan

1 停车场
2 办公室
3 热带植物温室
4 山林区
5 沙漠区
6 庭院咖啡厅
7 主入口
8 休息室/咖啡室
9 地中海区域
10 设备区
11 室内广场展览区/商店
12 工作室
13 教学区
14 会议室

法罗群岛教育中心
FAROE ISLANDS EDUCATION CENTRE

建筑师：BIG, Fuglark

合作者：Lemming & Eriksson, Samal Johannesen, Martin E. Leo SP/F, KJ Elrad Radgevandi Verkfroedingar

合伙负责人：Bjarke Ingels

项目负责人：Jakob Lange

项目经理：David Zahle

项目建筑师：Johan Cool, Ole Elkjær Larsen

设计团队：Finn Nørkjær, Ken Aoki, Takumi Iwasawa, Oana Simionescu, Frederik Lyng, Christian Alvarez Gomez, Gaetan, Brunet, Todd Bennett

甲方：Mentamalaradid (Ministry of Culture) / Landsverk

项目地点：Torshavn, Faroe Islands

面积：19 200 m^2

获奖情况：竞赛一等奖

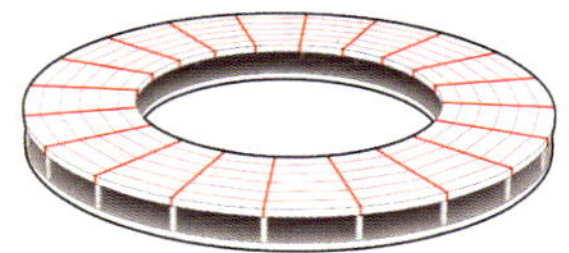

天花板构件

圆形的结构具有高度的重复性，由一系列完全相同的混凝土构件组成。

Ceiling Elements

The structure of the circular shape has a high degree of repetition, consisting of a series of identical concrete elements.

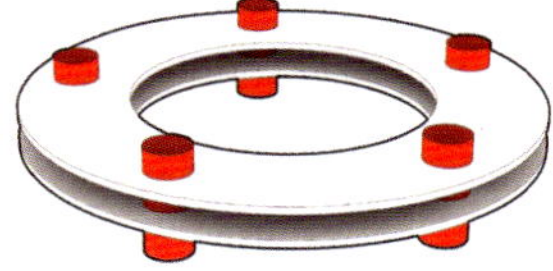

楼梯

五个楼梯稳定了建筑结构。

Staircases

The five staircases stabilise the building.

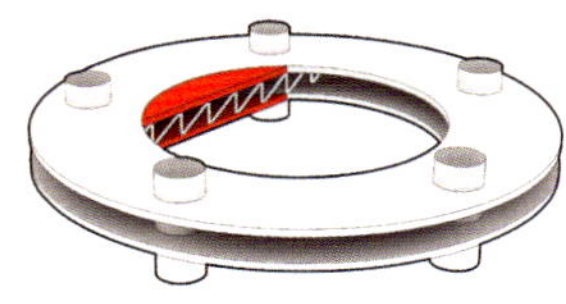

阳台

超宽的阳台由钢桁架支撑。

Balconies

The extra wide balconies are supported by a steel truss.

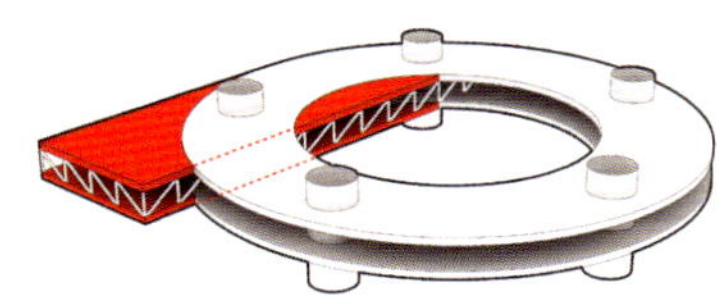

悬挑的教室

教室也由钢桁架支撑。

Cantilevering Classrooms

The classrooms are also supported by a steel truss.

结构示意图

Construction diagram

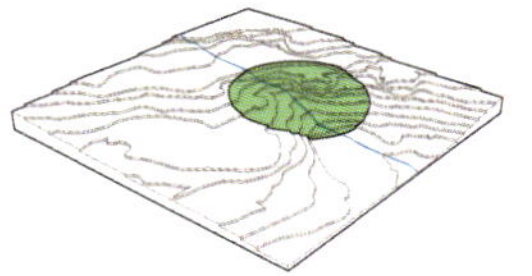

自助餐厅
多功能大厅
媒体中心
分组教室
图书馆

Cafeteria
Multi-purpose Hall
Media Centre
Group Room
Library

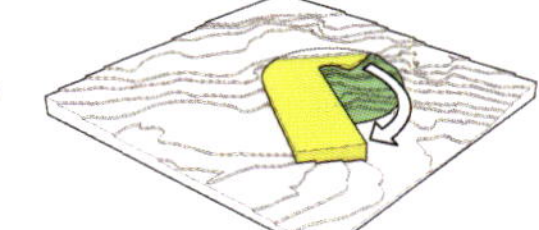

地下三层
运动

Floor Level -3
Athletics

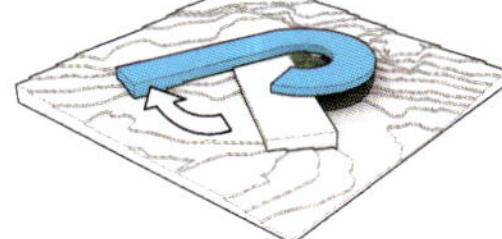

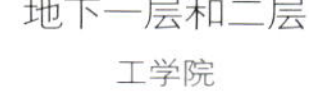

地下一层和二层
工学院

Floor Level -2 & -1
Technical School

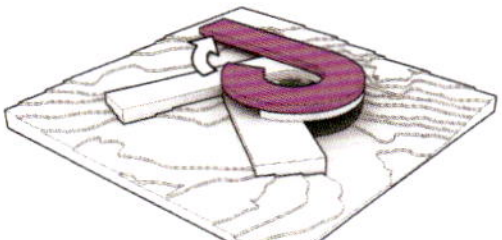

一层
行政管理
创造性核心

Floor Level 0
Administration
Creative Core

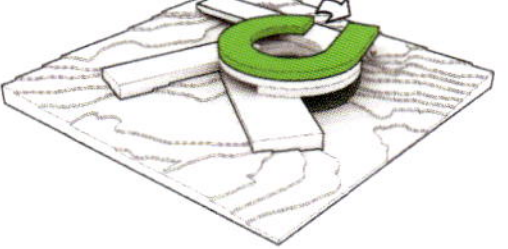

二层
商学院
风格与设计

Floor Level 1
Business College
Style & Design

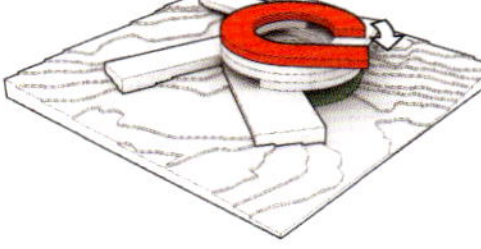

三层
体育馆
自然科学

Floor Level 2
Gymnasium
Natural Science

平面布局

Plan

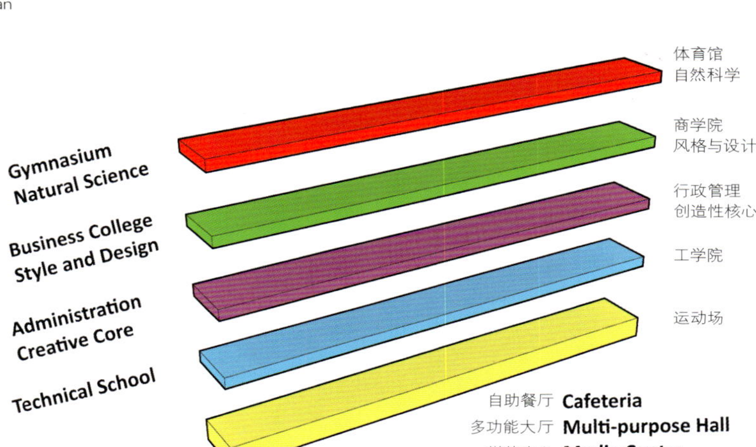

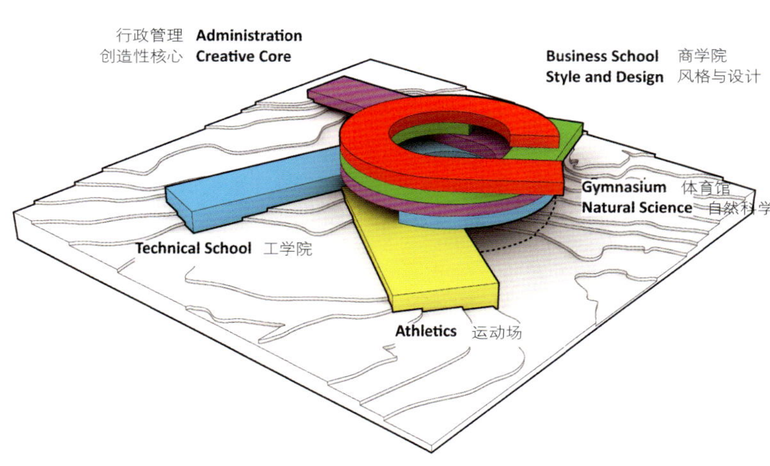

功能布局

Function organization

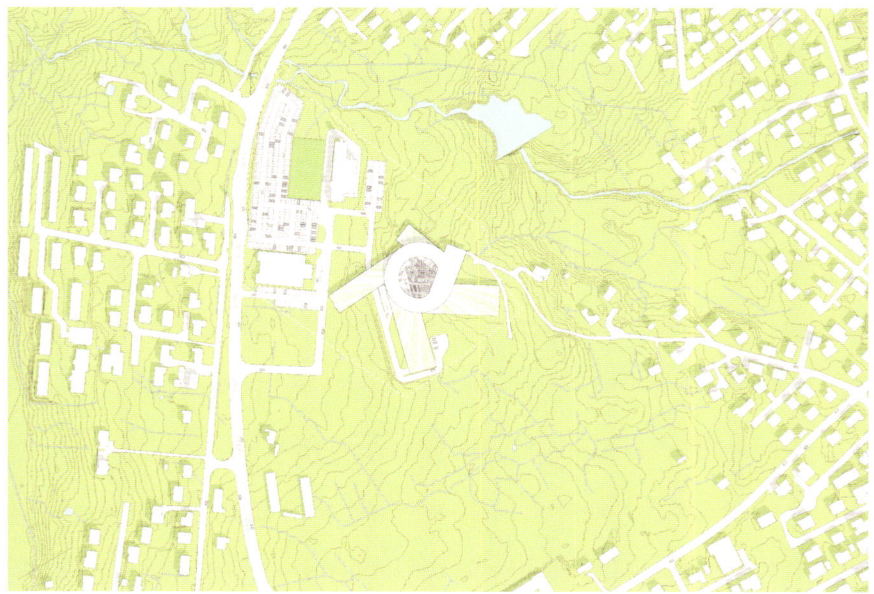

马克纳吉尔的新教育中心位于托沙芬郊区的半山腰，在所在区域成为所有教育项目协调和未来发展的基地。作为丹麦历史上最大的教育工程建筑，教育中心将法罗群岛体育馆、托沙芬工学院和法罗群岛商学院汇集一处，能够容纳1200名学生和300名老师。

The new Education Centre in Marknagil situated on a hillside on the outskirts of Torshavn, to serve as a base for coordination and future development of all educational programmes in the region. As the largest educational building project in the country's history, the institution combines Faroe Islands Gymnasium, Torshavns Technical College and Business College of Faroe Islands in one building, housing 1200 students and 300 teachers.

总平面图

Site plan

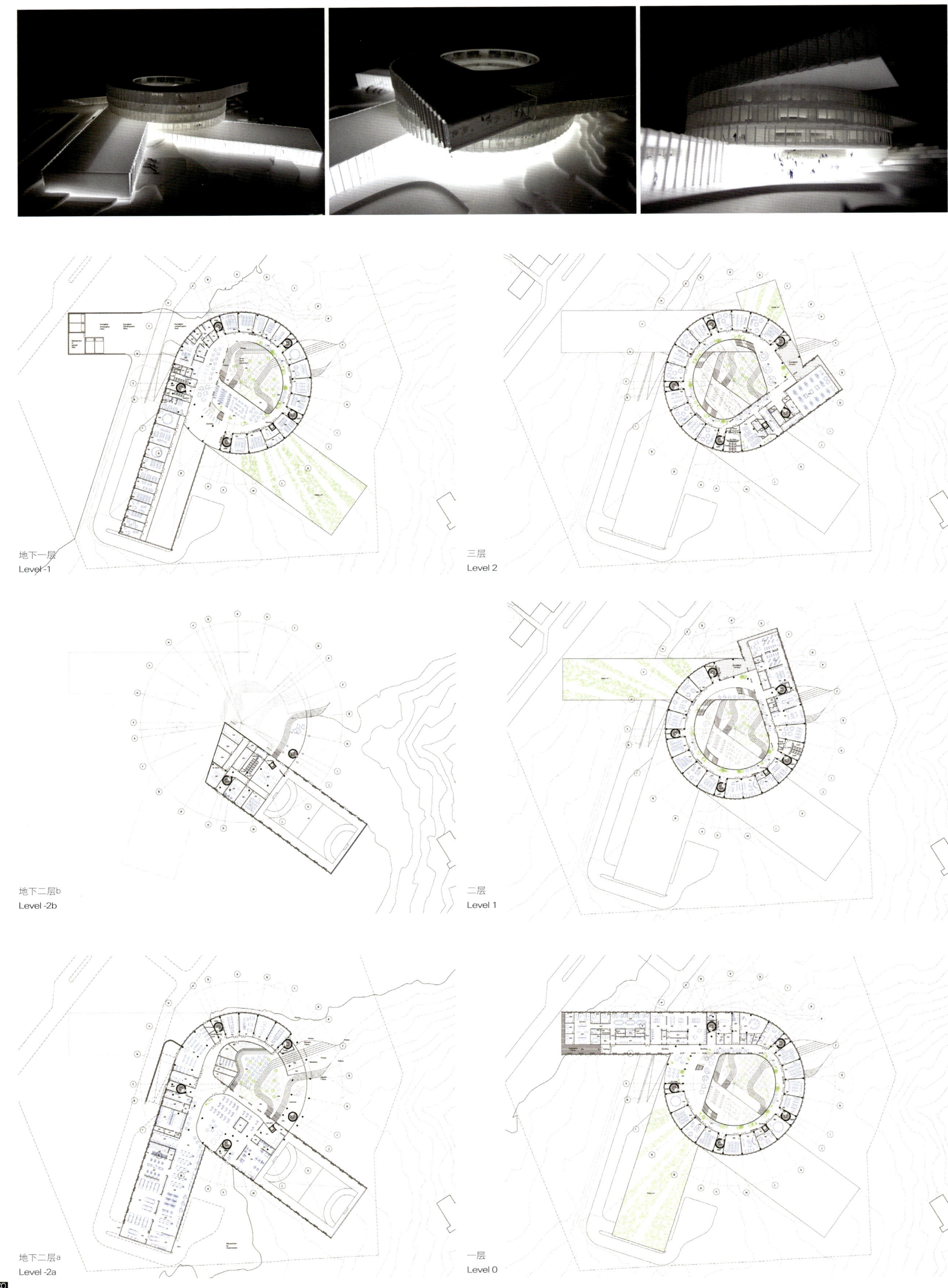
地下一层
Level -1
三层
Level 2
地下二层b
Level -2b
二层
Level 1
地下二层a
Level -2a
一层
Level 0

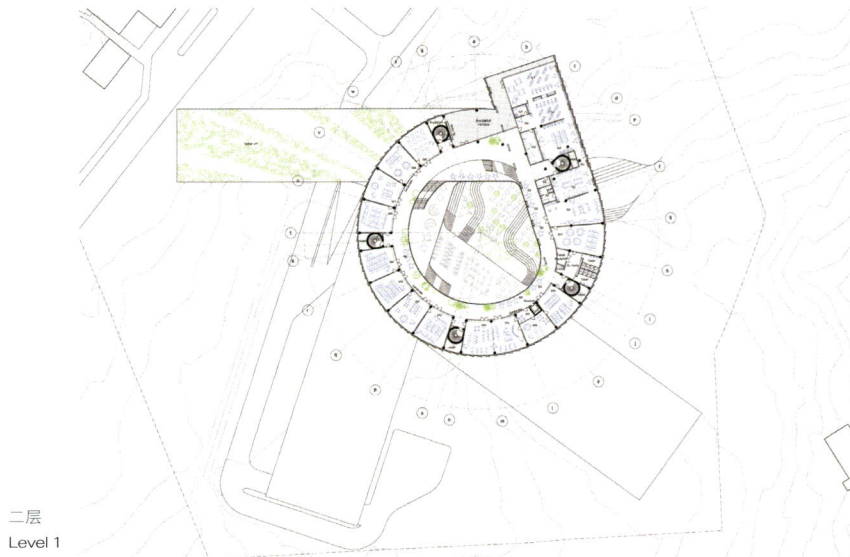

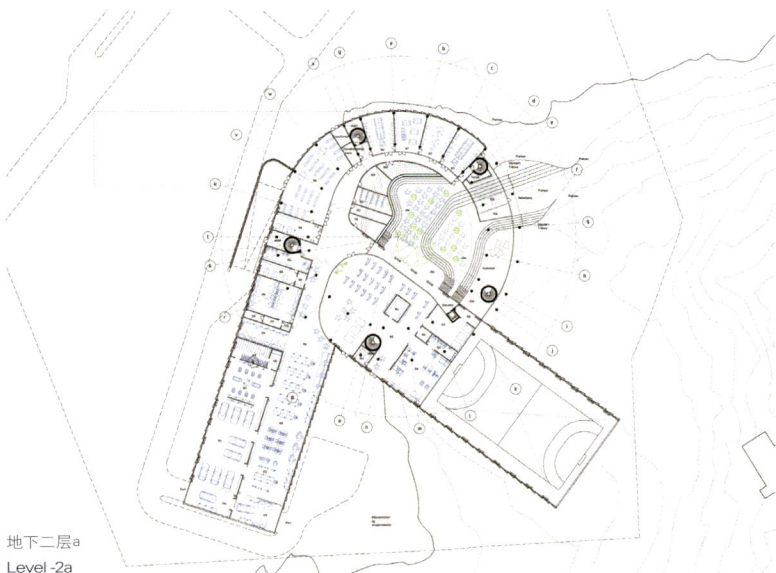

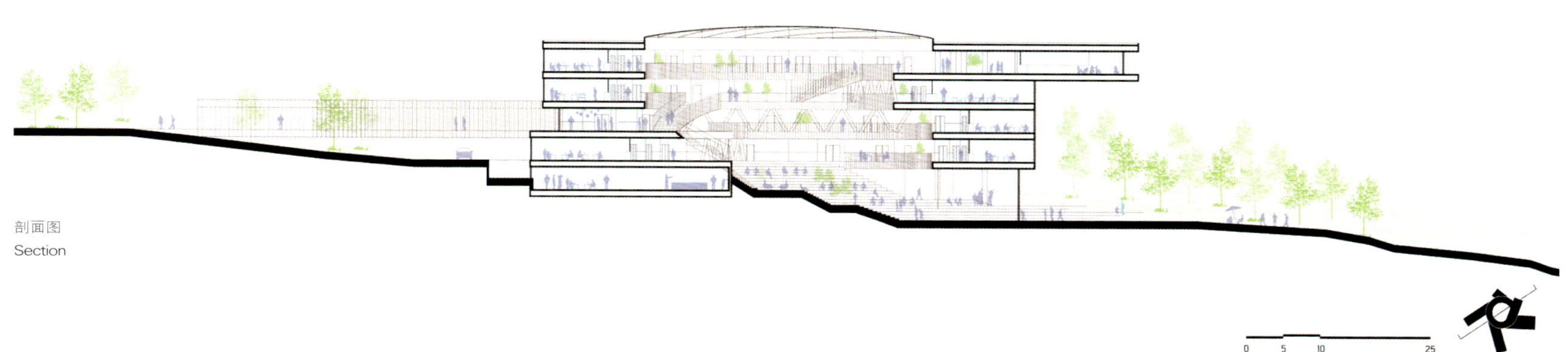

剖面图
Section

马里博尔美术馆
MARIBOR ART GALLERY

建筑师：Andrea Branzi - 2a+P/A (Gianfranco Bombaci, Matteo Costanzo)
参与设计建筑师：Tommaso Arcangioli, Valeria Bartolacci, Christian Galli, Flaminia Liberati, Giacomo Miola, Valentina Morelli, Consuelo Nuñez, Giulia Pastore
模型制作：Marco Galofaro (Modelab)
摄影师：Sebastiano Costanzo

该项目的主要目的在于试验将新建博物馆的各种活动与城市生活结合起来的可能性。林立的木柱填补了城市空间，凸显出由大而透明的天花板覆盖的公共区域。博物馆的支点是两个巨大的玻璃露台，在地下走廊交汇，并构成了通往博物馆地下结构的入口。第二个衔接元素是通往车库的地下通道，这里有一条入口坡度可进入广场西侧。这条道路可用于装卸艺术品，是去仓库的直接入口，并与东侧市场的停车场直接相连。室内人行道纵横交错，向公众展示了该博物馆的良好运作。

The project aims to experiment the possibilities of mixing the activities of the new museum with the city life. A forest of pillars invades the urban void, identifying a public space covered with a big transparent ceiling. The fulcrum of the museum is the two great glass patios, joined by an underground corridor, that constitute the access point to the underground museum structure. A second element of connection is the underground access street to the garage that with an entrance ramp leads to the west side of the square. This street ensures the admittance also to the loading and unloading area of the art pieces, a straight entrance to the deposits and the warehouses, and a direct link with the parking of the market on the east side. The internal pedestrian path crosses the vehicular one, showing to the public the functioning of the museum machine.

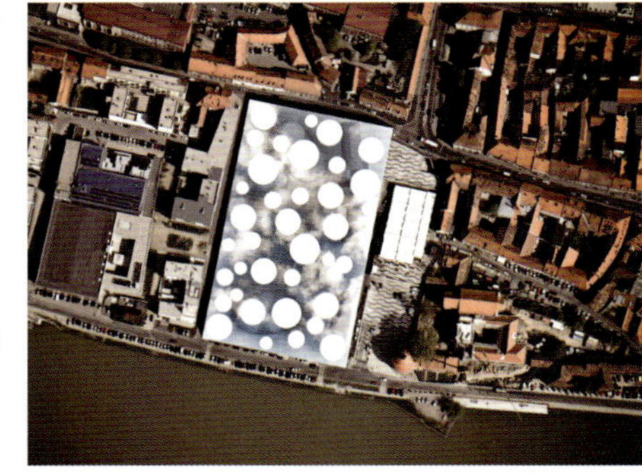

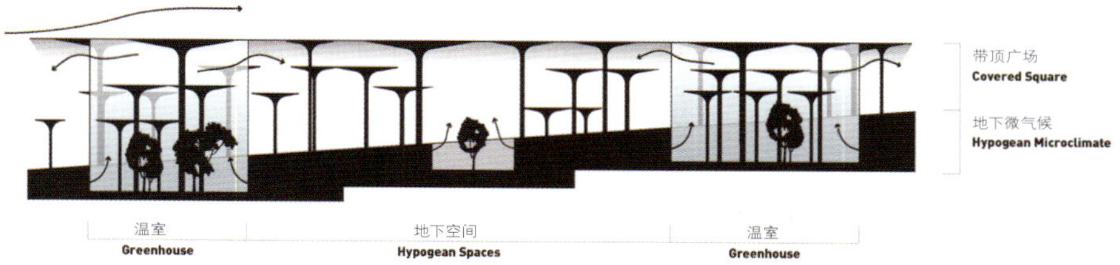

带顶广场
Covered Square

地下微气候
Hypogean Microclimate

温室
Greenhouse

地下空间
Hypogean Spaces

温室
Greenhouse

环境技术
ENVIRONMENTAL STRATEGY

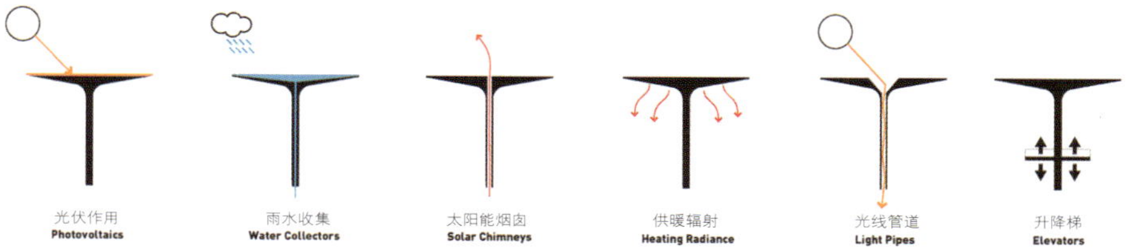

光伏作用
Photovoltaics

雨水收集
Water Collectors

太阳能烟囱
Solar Chimneys

供暖辐射
Heating Radiance

光线管道
Light Pipes

升降梯
Elevators

松树形装置
PINE DEVICES

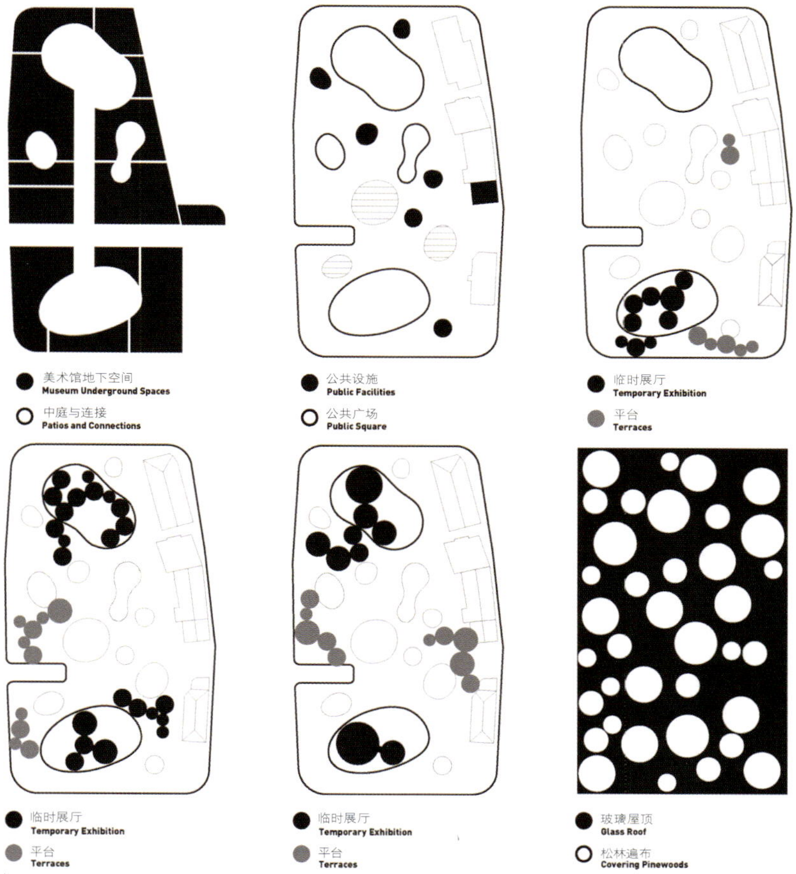

美术馆地下空间
Museum Underground Spaces

中庭与连接
Patios and Connections

公共设施
Public Facilities

公共广场
Public Square

临时展厅
Temporary Exhibition

平台
Terraces

临时展厅
Temporary Exhibition

平台
Terraces

临时展厅
Temporary Exhibition

平台
Terraces

玻璃屋顶
Glass Roof

松林遍布
Covering Pinewoods

项目开发
PROGRAM DEVELOPMENT

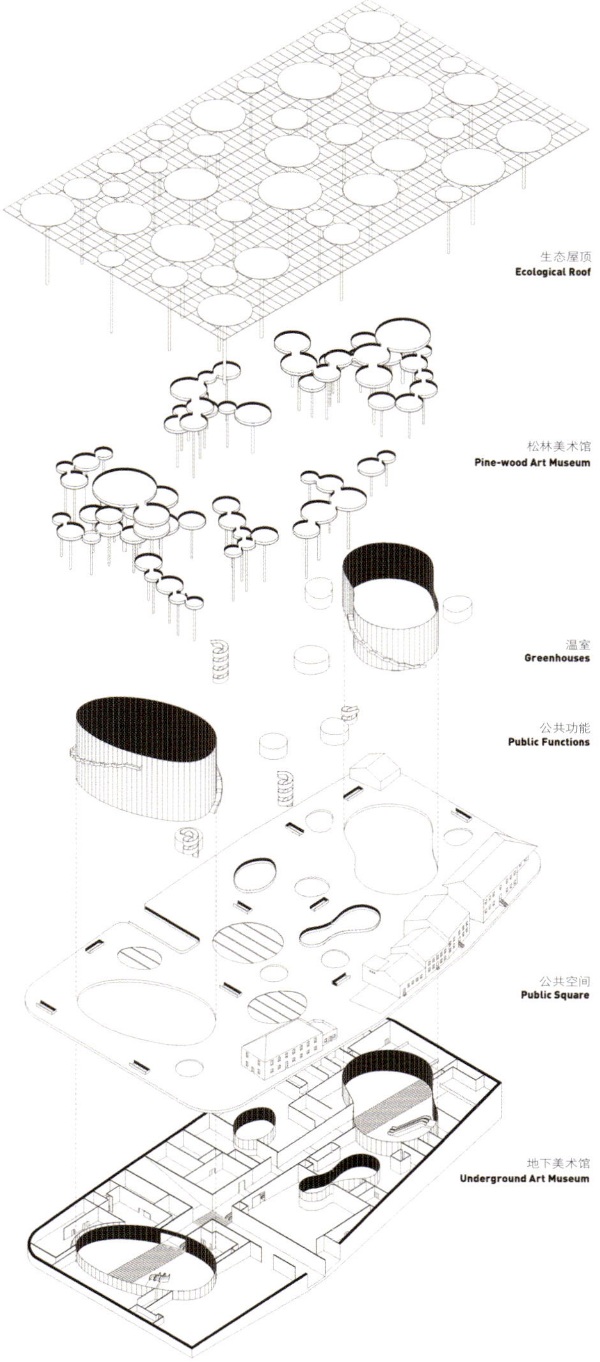

生态屋顶
Ecological Roof

松林美术馆
Pine-wood Art Museum

温室
Greenhouses

公共功能
Public Functions

公共空间
Public Square

地下美术馆
Underground Art Museum

轴测图
AXONOMETRIC VIEW

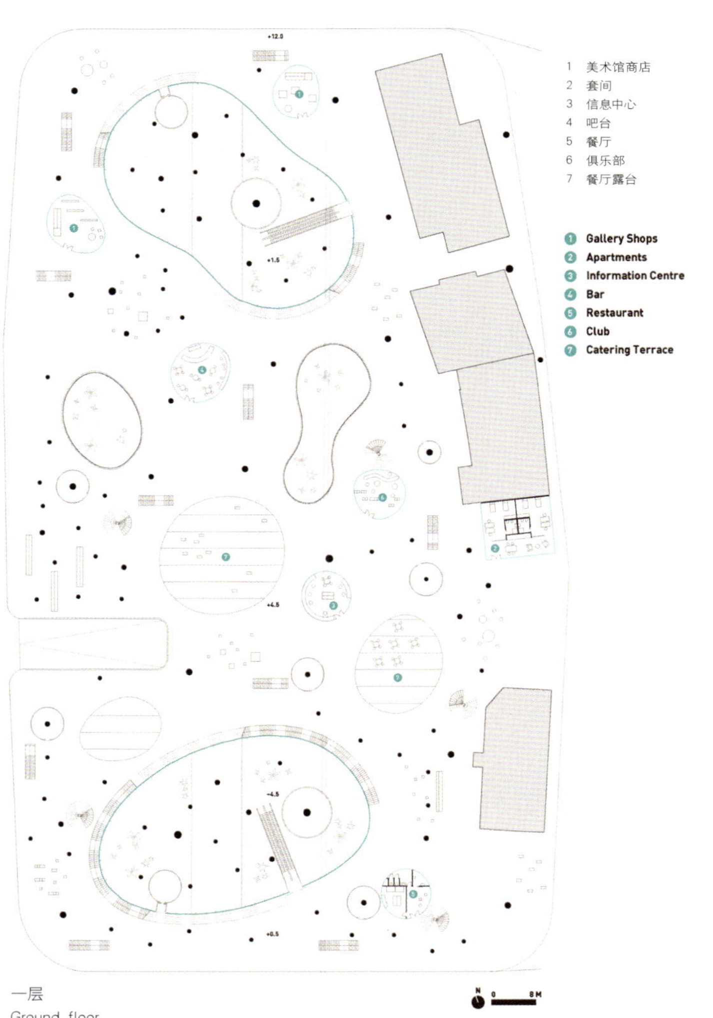

一层
Ground floor

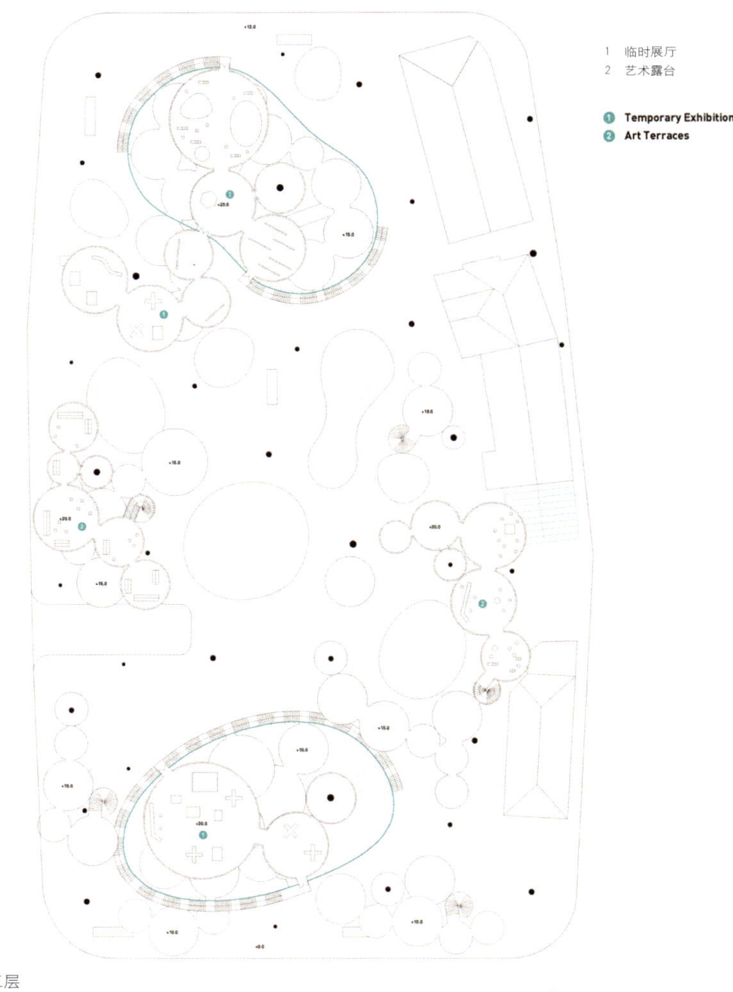

三层
Second floor

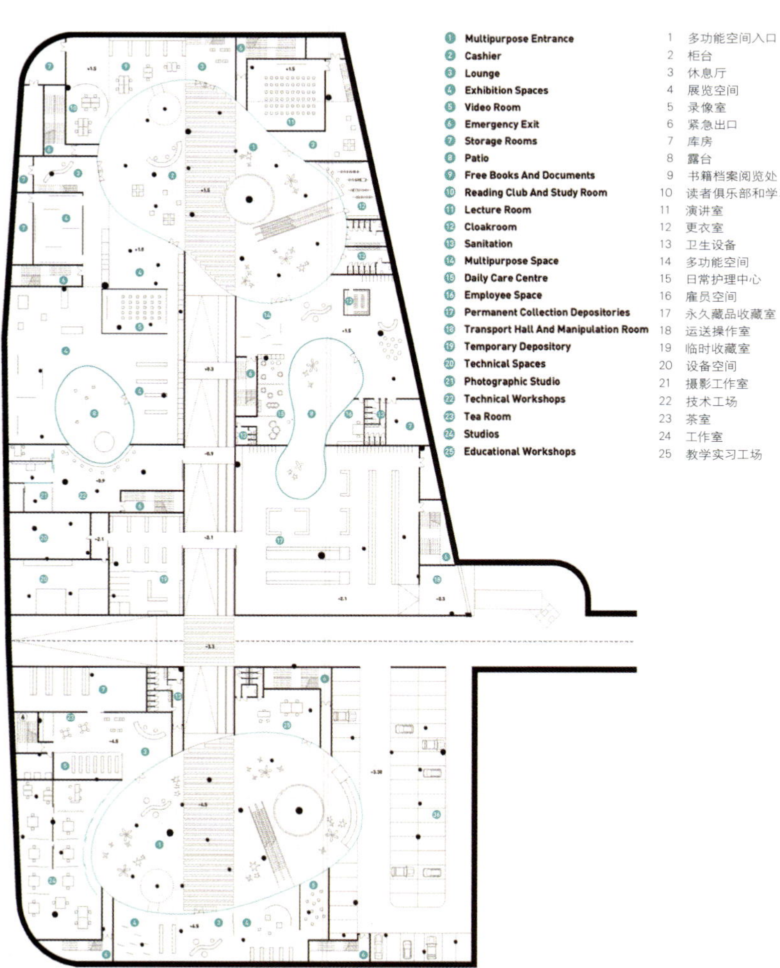

地下层
Basement plan

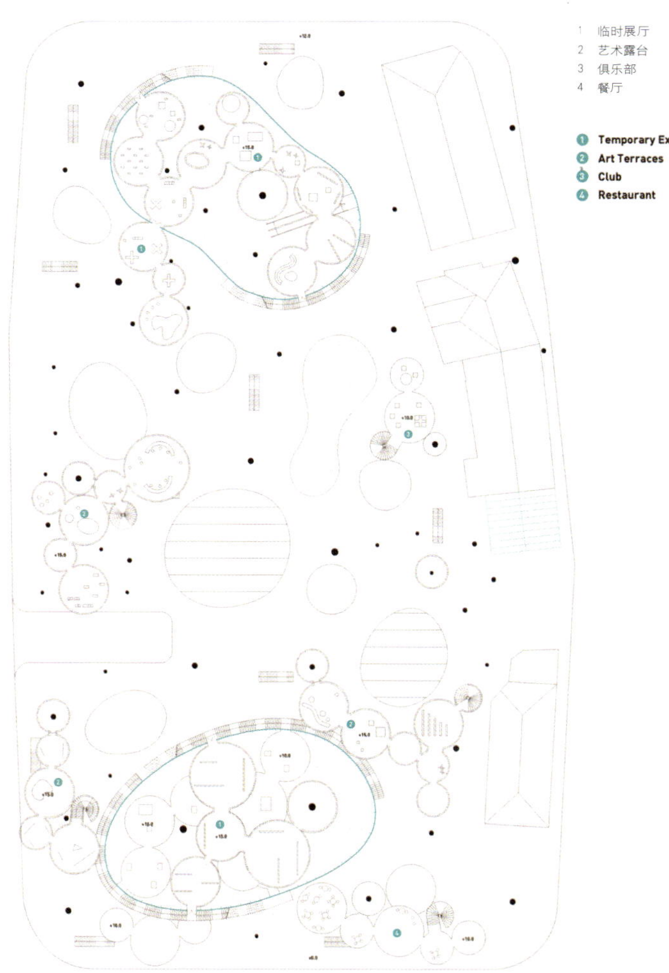

二层
First floor

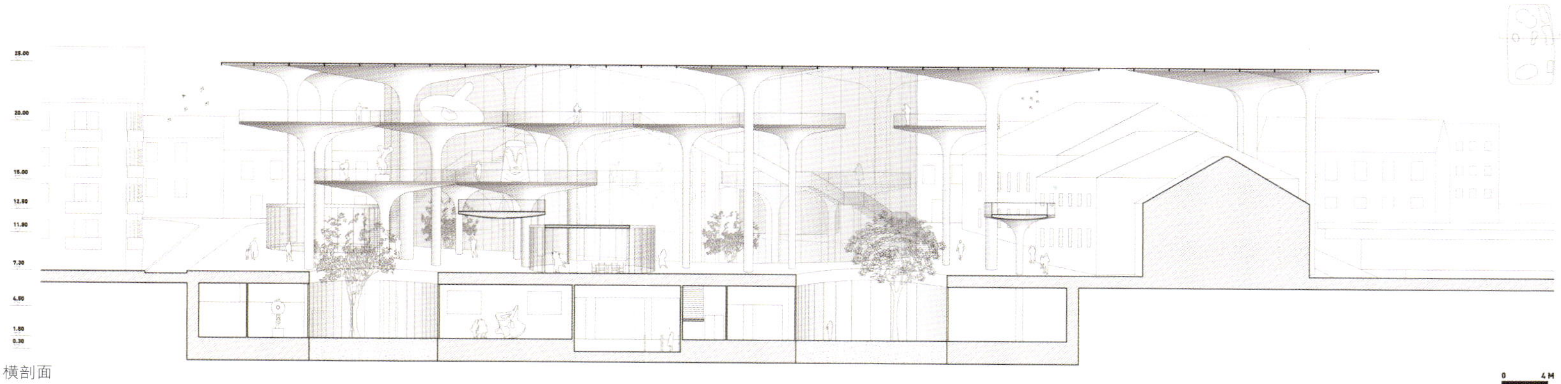

横剖面
Cross section

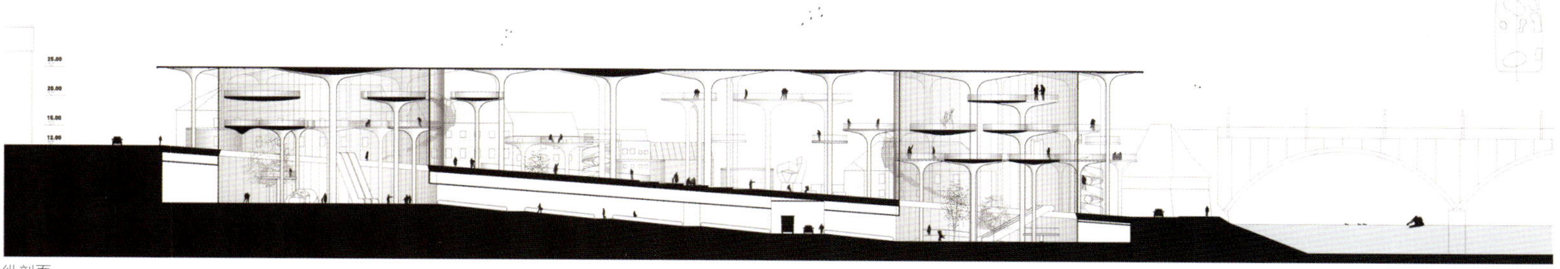

纵剖面
Longitudinal section

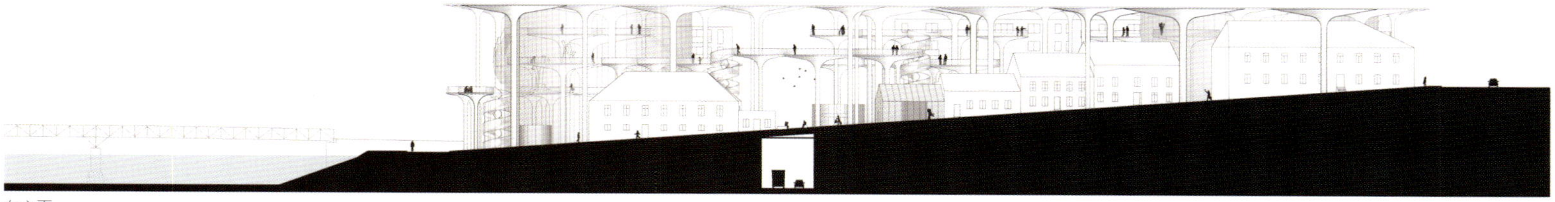

东立面
East elevation

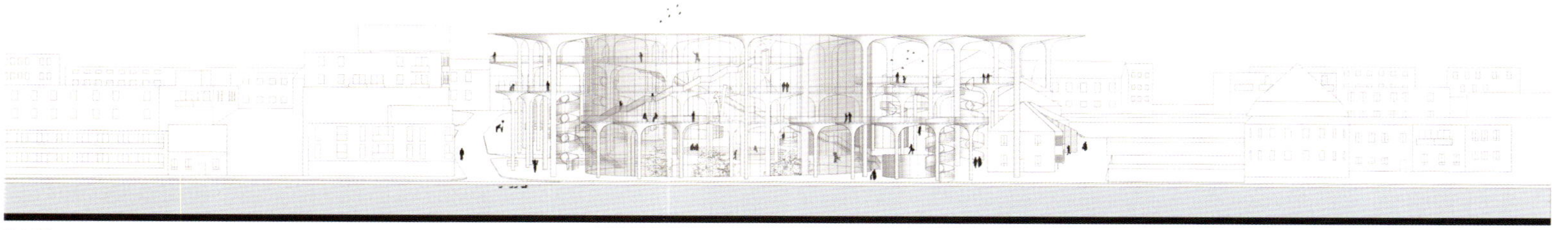

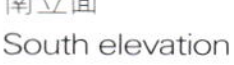

南立面
South elevation

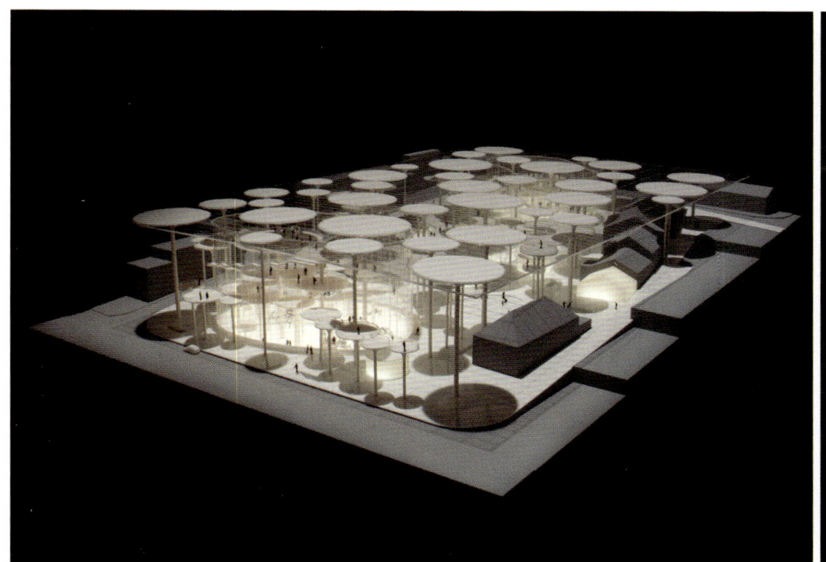

YOU HAVE TO DELIVER
HOPE
CUBISM
CAFETERIA
SHOP

埃及考古博物馆
ARCHAEOLOGICAL MUSEUM IN EGYPT

甲方：Österreichisches Archäologisches Institut
规划设计：COOP HIMMELB(L)AU
Wolf D. Prix / W. Dreibholz & Partner ZT GmbH
主设计师：Wolf D. Prix
参与项目建筑师：Karolin Schmidbaur
项目负责人：Andrea Graser
项目团队：Reinhard Hacker, Marcin Kurdziel, Steven Ma, Giulio Polita, Andrea Schöning
建模：Paul Hoszowski, Sebastian Buchta
效果图与线图：COOP HIMMELB(L)AU

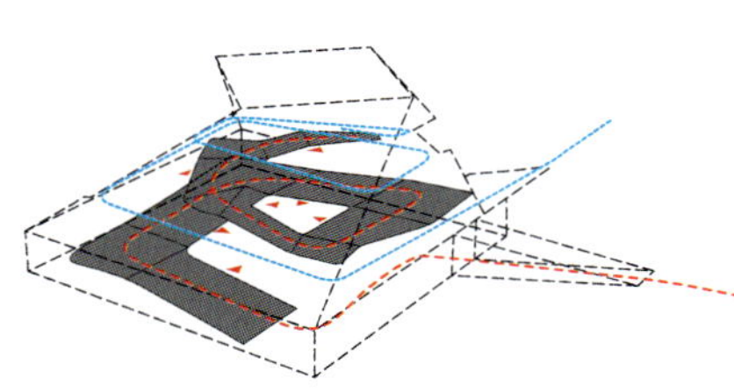

室内交通流线
INTERNAL CIRCULATION

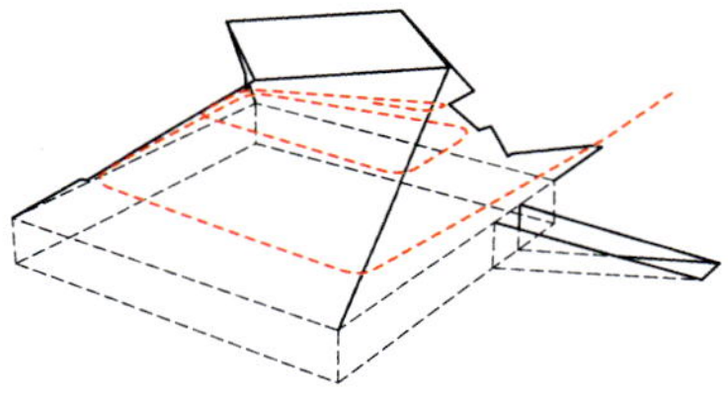

室外交通流线
EXTERNAL CIRCULATION

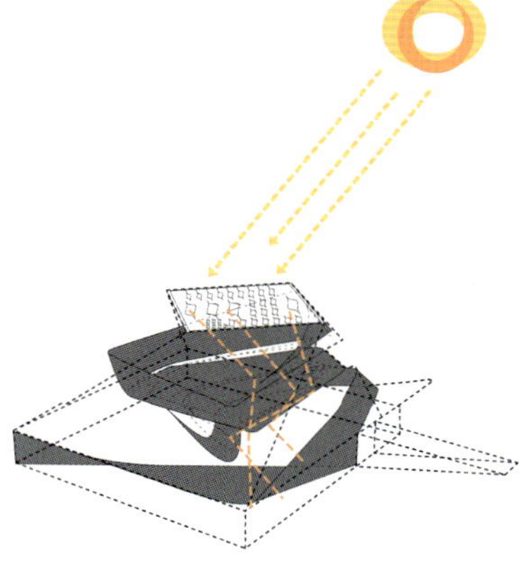

太阳照射示意图
SOLAR DIAGRAM

该项目坐落在尼罗河三角洲东部Tell el-Daba的发掘现场附近。新建筑的周围环绕着农业耕地，作为Elhosania Fakos河岸的标志性建筑，将会很引人注目。博物馆的设计风格以实用为主，并受场地的历史及其特殊的地形和气候条件的影响，形成了一种独一无二的建筑空间，与建筑用途及其所处的环境相得益彰。这种结构的发展来源于截顶金字塔的几何结构。该建筑就是一座适于步行的纪念碑。从入口广场的一条螺旋形坡路引领向上，如同镶嵌在倾斜外表面之中的圆圈环绕在博物馆周围，步移景异，视野逐渐打开，直到周围的风景悉数映入眼帘。这条通道一直通往顶层的咖啡厅和第二个入口，这让游客不仅可以在底层开始他们的博物馆之旅，也可以逆向而行。

This project is situated near the excavation site of Tell el-Daba in the eastern Nile delta. Surrounded by agricultural land, the new building will stand out as a landmark on the bank of the Elhosania Fakos River. The design of the museum, influenced by the history of the site and its specific conditions of topography and climate, and organized by the functional program, lead to a unique architectural space which in turn is responsive to its uses and the environment. The development of the form is derived from the geometry of a truncated pyramid. The building is a walkable monument. From the entrance plaza a spiral ramp leads upwards as a loop around the museum embedded in its sloped exterior surface, thereby offering a gradually changing and widening panoramic view to the surrounding landscape. The walkway ends at the top in front of a café and a secondary entrance, which allows the visitors to start their tour inside not only from ground floor, but also from high above.

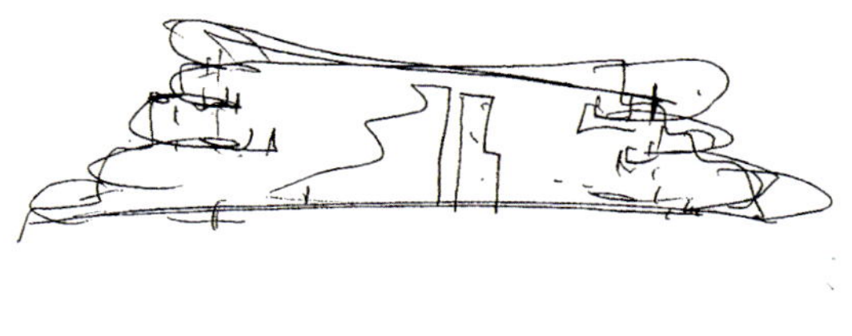

概念草图
CONCEPT SKETCH

概念模型
CONCEPT MODEL

意大利斯特朗格利美术馆
ART MUSEUM IN STRONGOLI, ITALY

美术馆馆长：Carla Piscitelli
规划设计：COOP HIMMELB(L)AU
Wolf D. Prix / W. Dreibholz & Partner ZT GmbH
主设计师：Wolf D. Prix
项目负责人：Andrea Graser
参与项目高级建筑师：Robin Heather
项目协调人：Giuseppe Zagaria
项目团队：Jenny Chow, Francesco Testa, Lam Le-Nguyen
结构工程：B + G Ingenieure, Bollinger & Grohmann GmbH, Frankfurt, Germany
能源与环境设计：ARUP Berlin, Brian Cody, Germany
三维效果图和线图：COOP HIMMELB(L)AU

从远处眺望意大利斯特朗格利城外的Motta Grande山顶，新斯特朗格利美术馆俨然成为卡拉布里亚区中心这座小城中的一个引人注目的地标。将三个功能体量彼此连接，就得出了该建筑的混合形式：公众入口处的标志性圆锥形结构、收藏展品的多功能大厅的主体，以及一个大胆悬挑的全景式餐厅。这种融合的形式犹如雕塑一般，外部被弧形的外皮所包围，建筑外皮是根据风和太阳的能量转换参数设计的。设计目的正是建造一座能产生更多能量的建筑物，而不仅仅满足于供建筑物自身使用。

Visible from far away on top of the Motta Grande, a hill just outside of Strongoli, the new Art Museum in Strongoli creates a remarkable landmark for the city in the heart of Calabria. The hybrid form of the building is developed from the connection of three functional bodies: an iconic, cone-shaped structure with the public entrance, the main volume of a multifunctional hall, which houses the exhibitions, and a daringly cantilevering panoramic restaurant. This sculptural merging of forms is enveloped by a curved outer skin that has been designed by energy transforming parameters driven by wind and sun. The aim of the proposal is to design a building that generates more energy than the building itself is using.

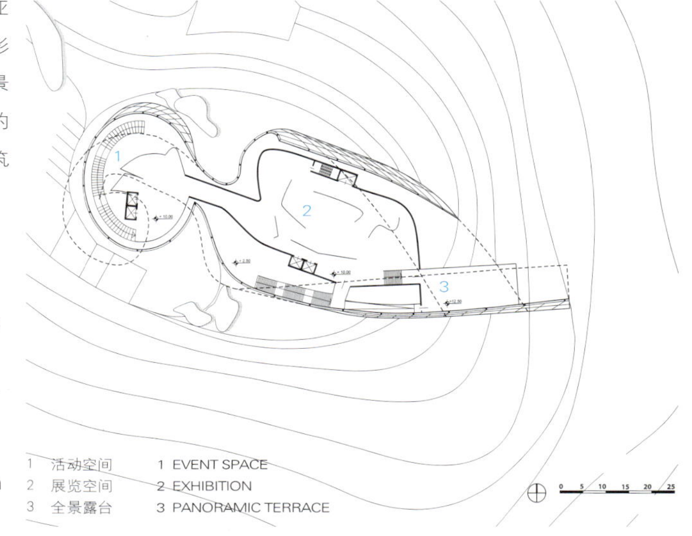

1 活动空间 1 EVENT SPACE
2 展览空间 2 EXHIBITION
3 全景露台 3 PANORAMIC TERRACE

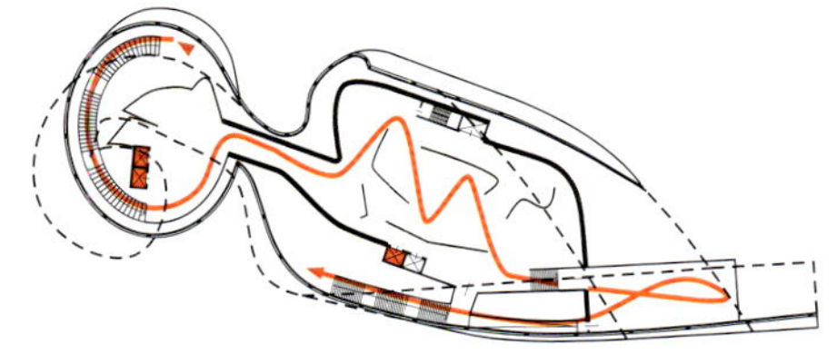
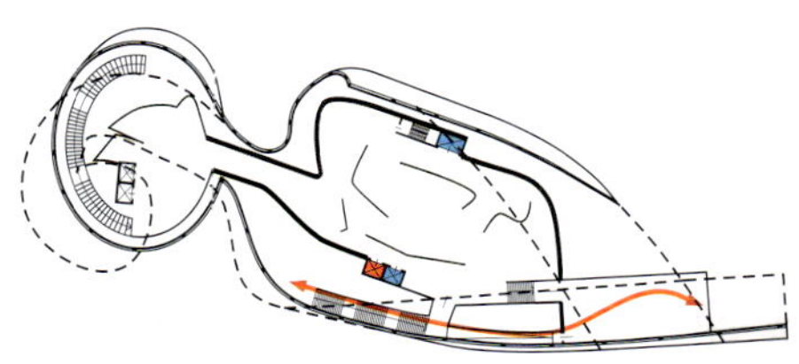
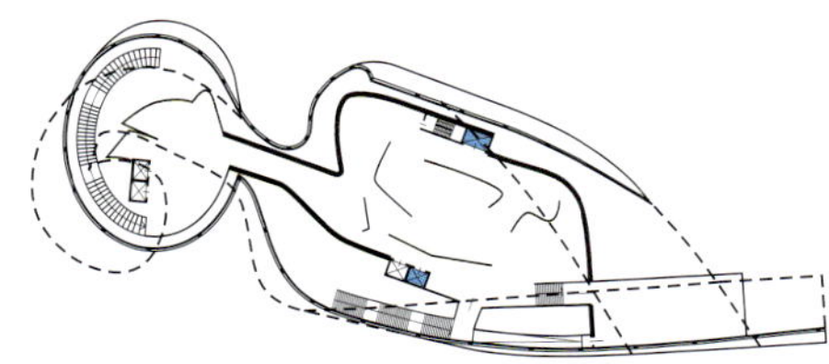

室内流线
INTERIOR CIRCULATION

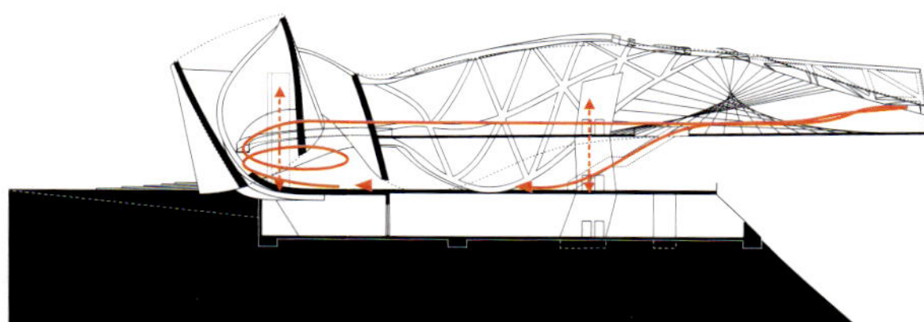

展览流线
EXHIBITION CIRCULATION

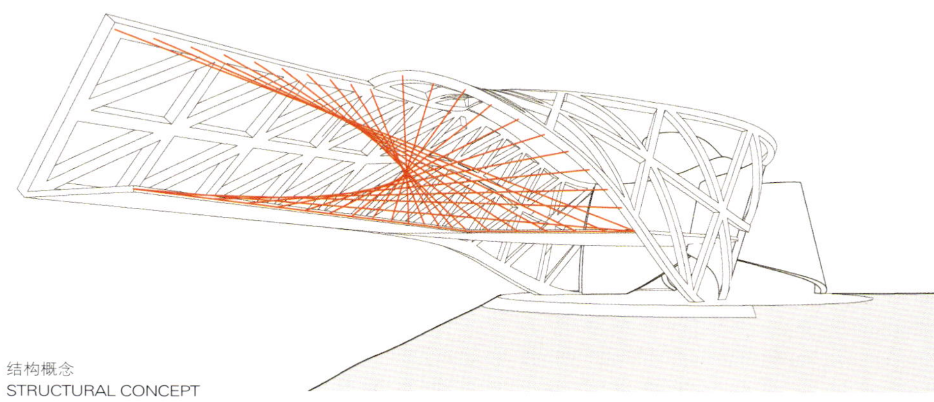

结构概念
STRUCTURAL CONCEPT

全景露台流线
PANORAMIC TERRACE CIRCULATION

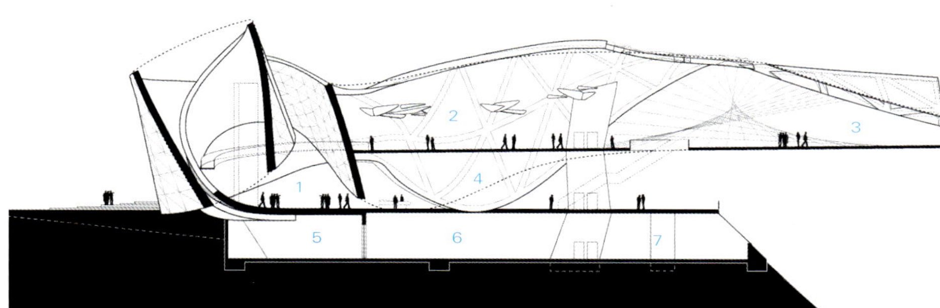

纵断面
LONGITUDINAL SECTION

1 活动空间 1 Event Space
2 展览空间 2 Exhibition
3 露台 3 Terrace
4 影像实验室 4 Video Lab
5 储藏室 5 Storage
6 工作坊 6 Workshop
7 行政办公室 7 Administration

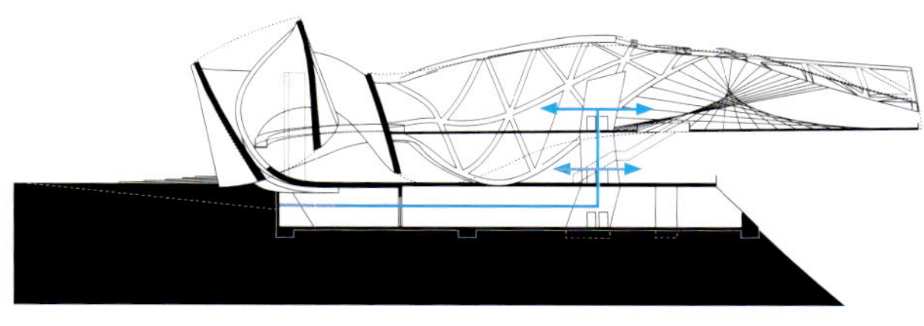

设备流线
SERVICE CIRCULATION

巴黎天空合流摩天大厦
VERTICAL CONFLUENCE, PARIS

建筑师：姜元+徐洋

项目地址：巴黎

获奖情况：2010 美国eVolo 摩天楼竞赛 特别奖

文化的交汇处

本方案处在巴黎核心区之外的塞纳河与马恩河的交汇处，巴黎地区的唯一河流交汇处，考虑利用此视觉景深通道，通过设置高度参照物，建立与巴黎中心区直接的视觉联系参照，消减巴黎环城高速内外发展不均衡造成的城市边界。

对应不同高度位置背景的三部分完全异化的平行空间：

顶部展览馆，俯瞰巴黎城市天际线，通过视觉处理将城市景观引入展馆空中庭院，成为此巴黎中国文化中心中展品的最佳背景——中法两种文化的交流。中部的多媒体图书馆，有着不同的视觉与光线环境，并面向巴黎最大绿地公园——凡仙公园。底部为表演中心，舞台前后设置，面向室内及室外沿河空间的露天看台。

外部公共空间，将塞纳河水位的变化纳入人流及景观设计元素考虑，营造随季节水位变化的动态水景环境，并联系北岸的体育场馆，西岸的公园绿化空间，与本方案的公共文化设施共同构成以此高层建筑为参照的区域中心。

我们认为此方案是作为历史城市发展更新的一种当代策略，它是对城市地区更新的重新思考，不仅可以恰当地融入现存的城市景观，同时也可作为强化历史城市机理节奏的当代节点。

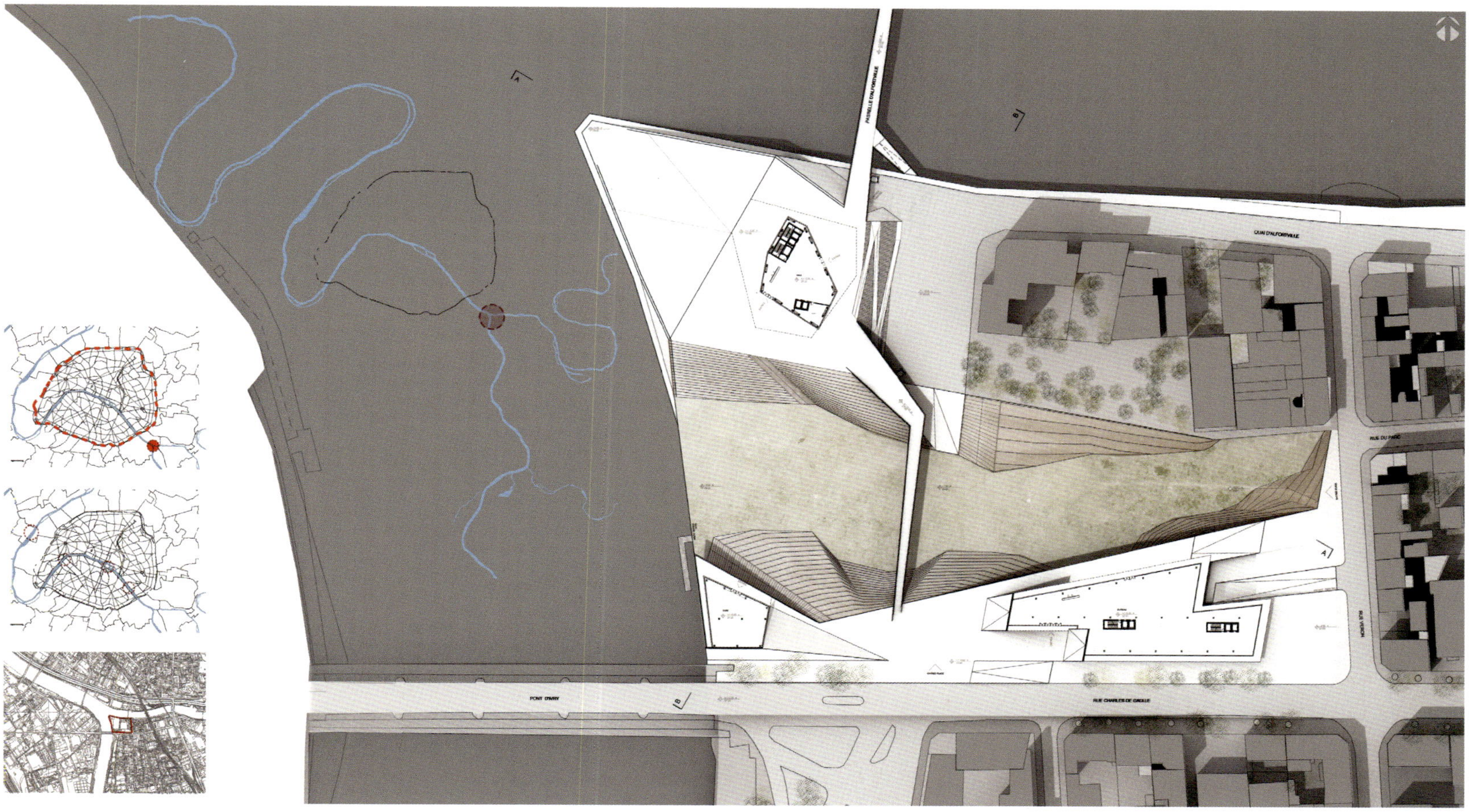

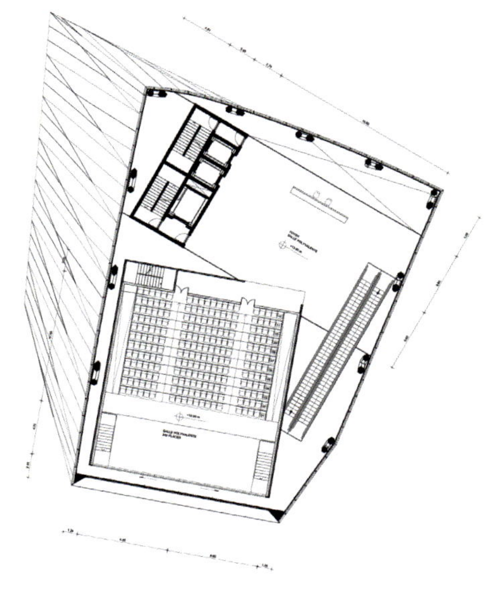

礼堂三层
3rd floor
auditorium

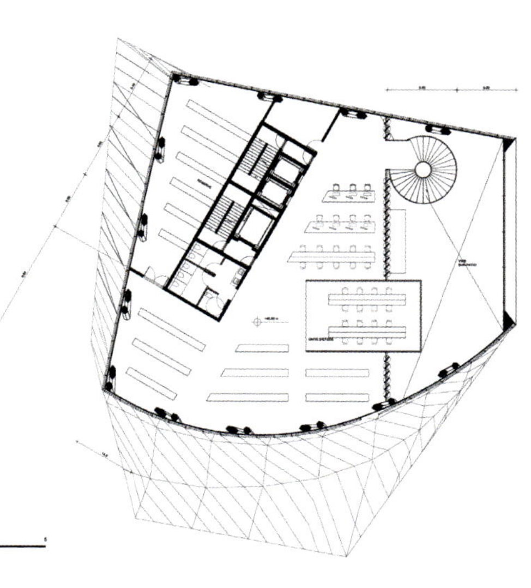

图书馆六层
6th floor
library

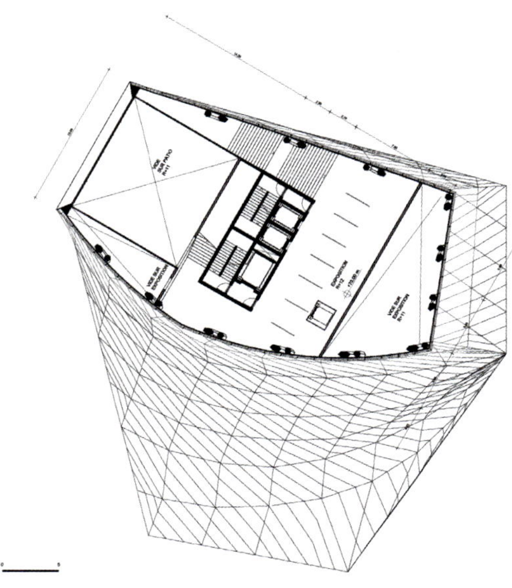

博物馆十一层
11th floor
museum

从两种河流走向看设计
View from two trends

塞纳河景
View from La Seine

东西剖面视图
East-west Section

A Confluence of Culture

The tour is located at the Confluence of la Seine and la Marne, it is one of the key points of the Parisian topography.
A skyscraper defined by three "frames" represents three rivers trends from the site: to the center of Paris, east of Paris (Vincennes) and south-east of Paris (Ivry-sur-Seine).
The building is designed to be the Chinese culture center in Paris. According to the different characteristic of the vertical context at different heights, three public programs are been proposed.
At the top of the tour, a museum, facing the full skyscraper of Paris through an open facade is created; while the background of the exhibition is meant to create a conversation between the different cultures.
At the middle of the tour, you can find a library facing the "bois de Vincennes", Paris's biggest green space.
Finally, in the bottom part of the tour, there is an auditorium facing the outside theatre, offering a showcase to both inside and outside.
With the arrangement of the local circulation, connecting the other facilities at the other sides of the confluence - a public green space and a sport center, at this moment, the skyscraper become a landmark building surrounded by the public facilities , gathering all the locale residents.
We believe that a skyscraper can be a contemporary answer to the historical city context, it can not only well intergrade in the existing urban tissue but also emphasize the rhythm of the historical urban structure.

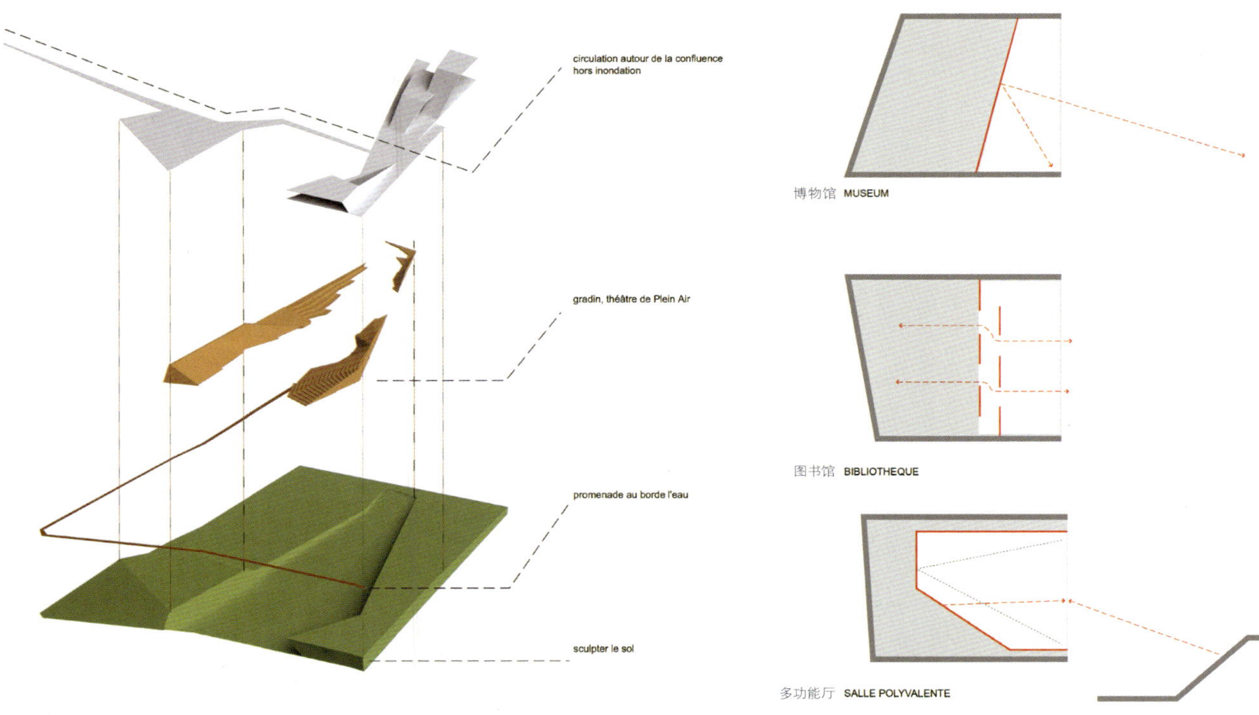

流线布置图
Circulation arrangement

出入口关系示意图
Relation ship of in-out

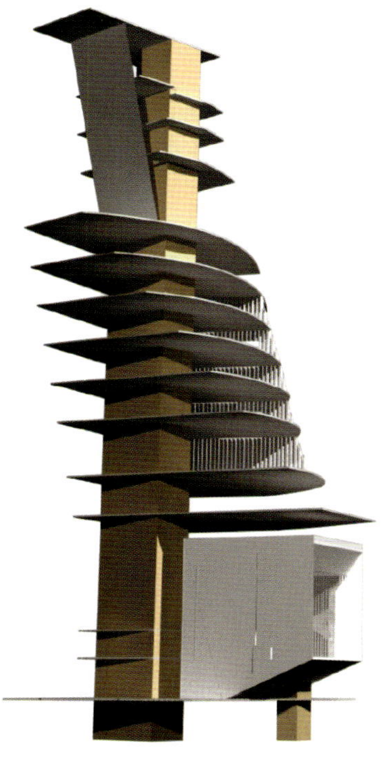

结构示意图
Structure

立面设计概念
Facade concept

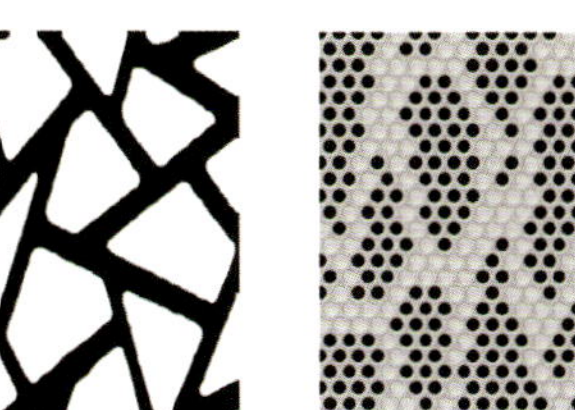

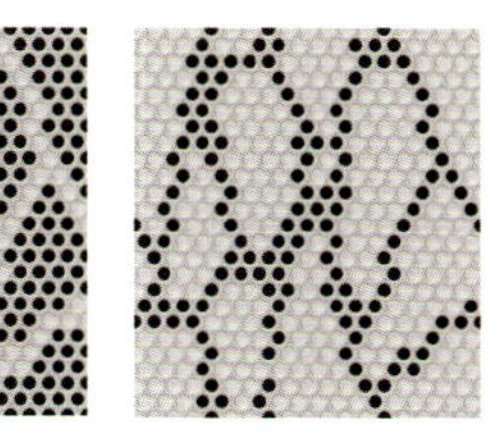

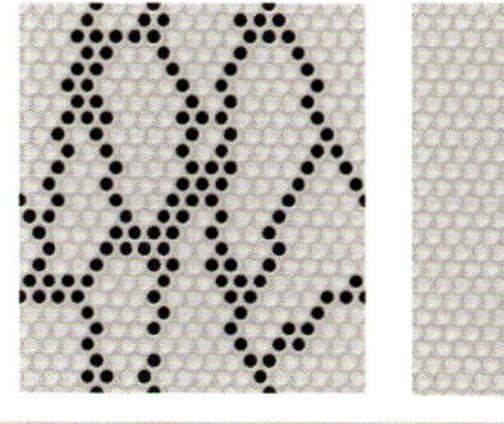

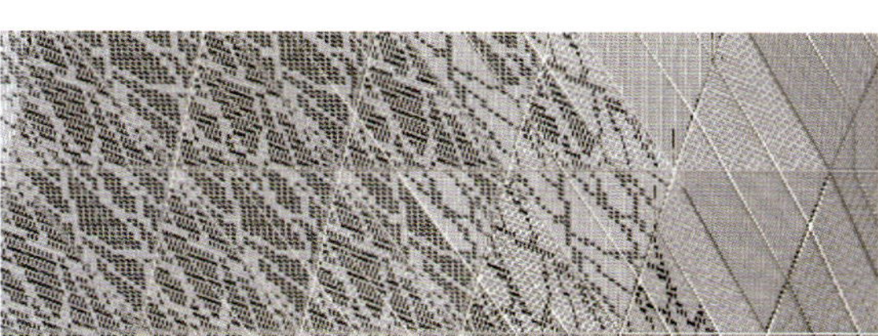

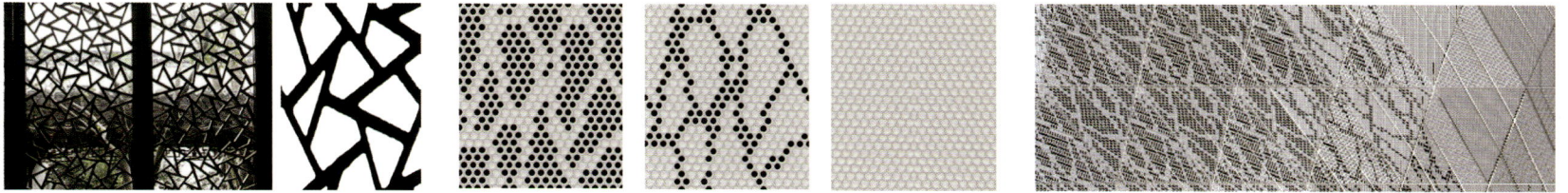

礼堂1
Auditorium 1

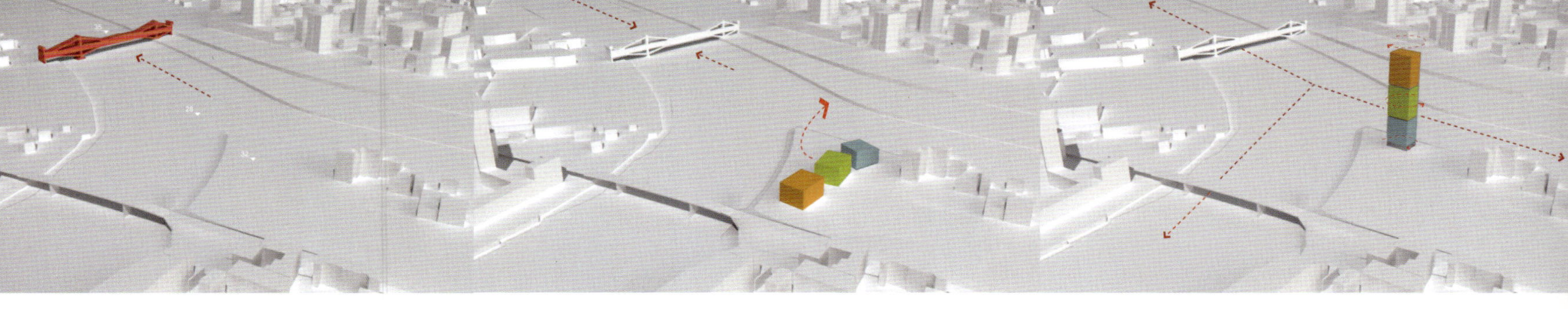

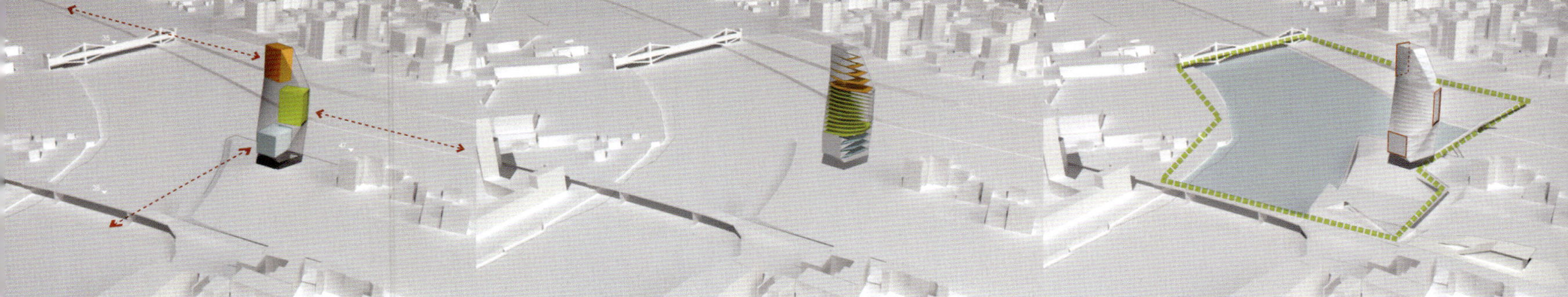

概念发展
Concept evolution

礼堂2
Auditorium 2

图书馆 1
Library 1

图书馆 2
Library 2

博物馆
Museum

Meudon-la-Forêt新文化中心
NEW CULTURAL CENTRE OF MEUDON-LA-FORÊT

设计团队：SERERO Architectes / David Serero, Bastien Casasoprana, Noriko Harada, Yoichi Ozawa, Fabrice Zaini
甲方：City of Meudon
项目地点：Meudon, France
景观设计师：OLM, BET
平面设计：Atelier 59
面积：1850m² + 6350 m² 室外
预算：5 970 000

本案的设计灵感来自大自然，更具体地说是来自树木，没有采用简单的象征性形式，而是将自然与建筑深深地联系在一起。建筑遵从了所在地区的正交网格，同时与地面空间联系紧密。该项目的组织布局是围绕中央核心区——一个300座的剧院而设。四周的外围护结构抬高，露出了大厅的内部空间和交通流线。建筑的四周设计了很多梯形的窗户，避免了位于城市中心的剧院传统的“黑盒子”效果。

The project is inspired by nature and more specifically by the tree, not in its simple symbolic shape, but in its deep link with architecture. The building follows the orthogonal grid of the district while developing a close connection with the ground space. The project is organized around a central nucleus created by a theater of 300 seats. Its peripheral envelope is lifted above the hall to reveal the inside spaces and circulation. Trapezoidal windows are placed on the whole surrounding of the building avoiding the traditional "black box" effect of a theater located in the middle of an urban environment.

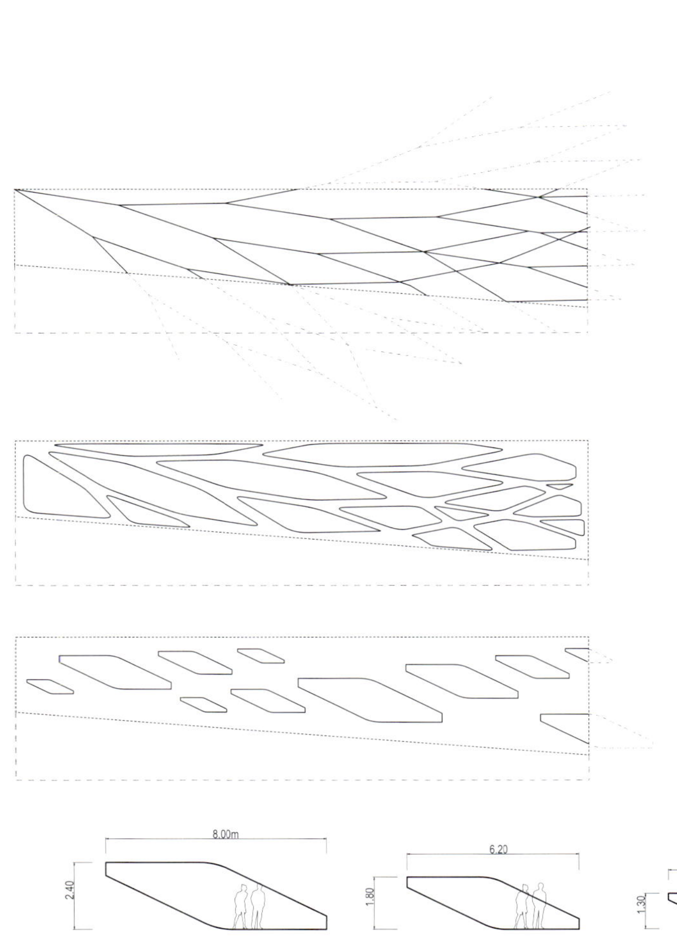

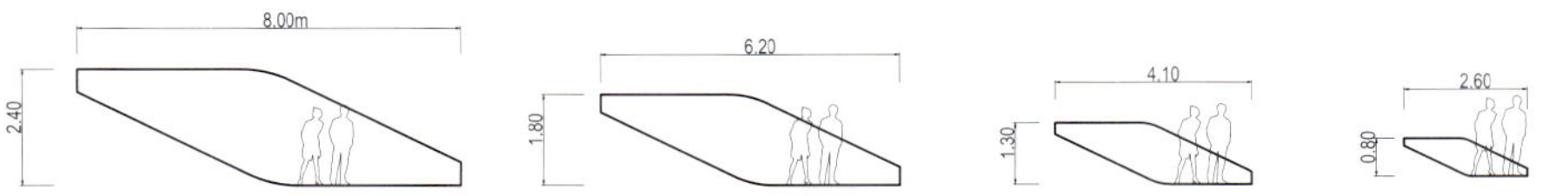

8.00m
2.40
6.20
1.80
4.10
1.30
2.60
0.80

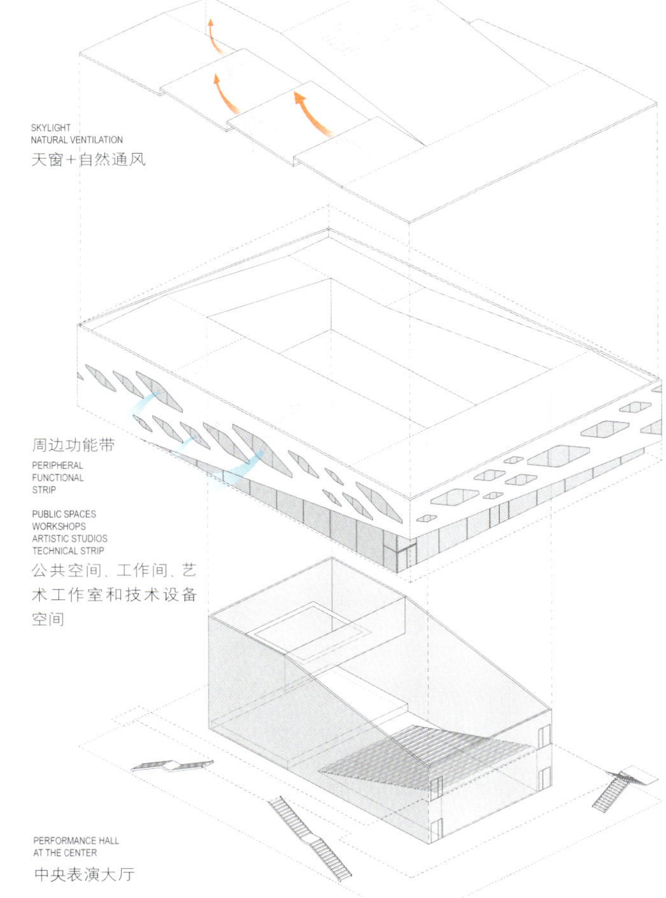

SKYLIGHT
NATURAL VENTILATION
天窗+自然通风
周边功能带
PERIPHERAL
FUNCTIONAL
STRIP
PUBLIC SPACES
WORKSHOPS
ARTISTIC STUDIOS
TECHNICAL STRIP
公共空间、工作间、艺术工作室和技术设备空间
PERFORMANCE HALL
AT THE CENTER
中央表演大厅

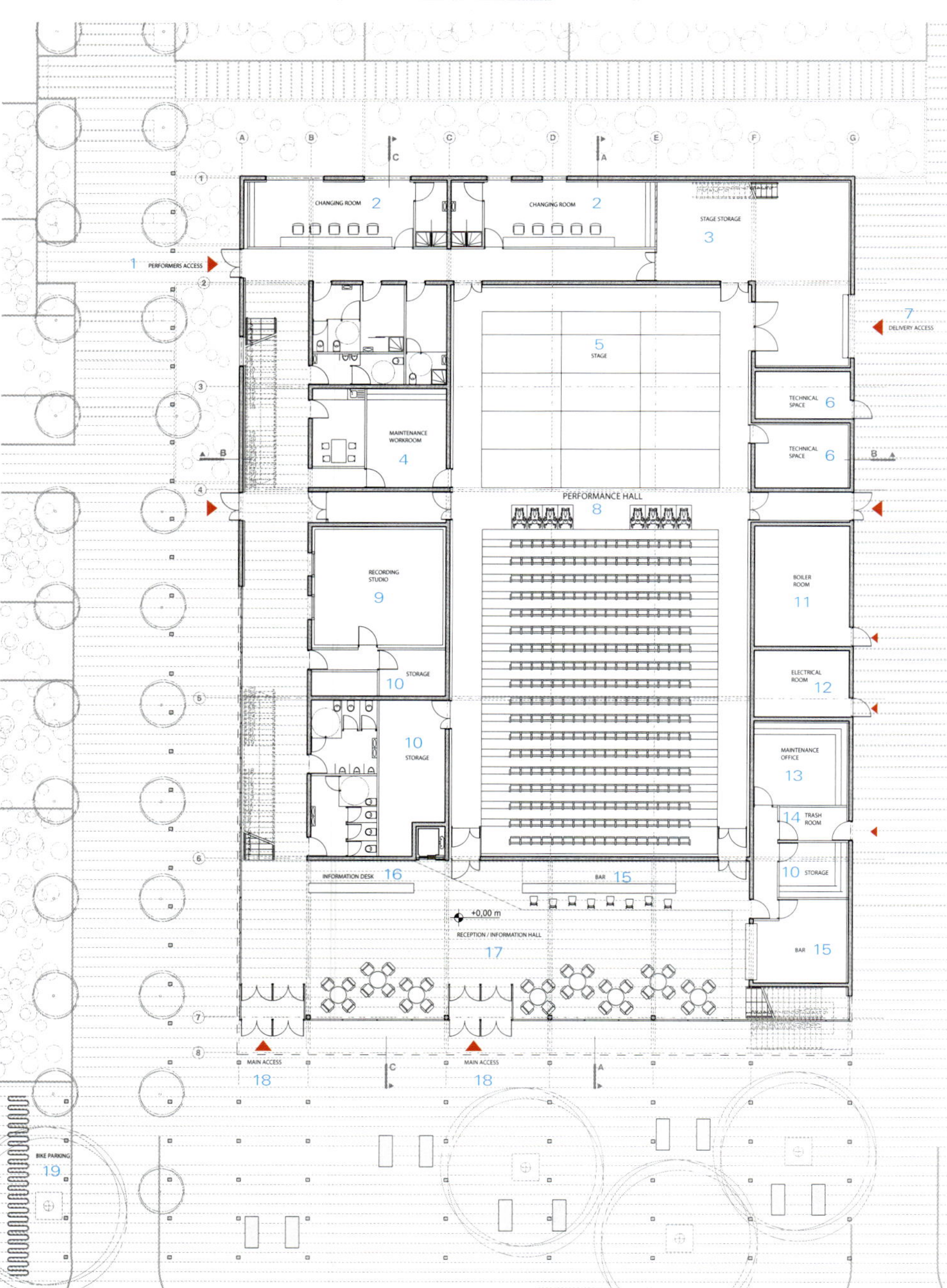

一层平面图
Ground floor plan

1 演员入口
2 化妆室
3 舞台用品存放室
4 维护工作室
5 舞台
6 技术设备室
7 送货入口
8 表演大厅
9 录音工作室
10 储存室
11 锅炉房
12 电力室
13 维护办公室
14 垃圾存放处
15 酒吧
16 咨询台
17 接待大厅/信息大厅
18 主入口
19 自行车停放处

水立方——海洋大厅及图书馆
THE WATER CUBE – THE HALL AND LIBRARY OF THE OCEANS

建筑师：MVRDV
甲方：Yeosu Expo 2012 Committee
主要建筑师：Winy Maas, Jacob van Rijs, Nathalie de Vries
项目团队：Wenchian Shi, Attilio Ranieri, Kyo Suk Lee, Ignacio Zabalo Martin
生态学顾问：ARUP Amsterdam

"生机勃勃的海洋及海岸线"，正是韩国丽水2012年世界博览会的主题，此次博览会旨在号召人们深度认识并重视海洋和海事资源及其对人类的重要性。建筑师对此次主题会馆的设计诠释了世博会的核心——将海洋浓缩在一个建筑空间内：中心的海洋大厅呈中空结构，由堆成一个大立方体的水箱环绕，而海水在玻璃立面和楼板间流动。每个水箱都展示了海洋独特的一面——深层海域、热带海洋、美洲红树、暗礁，这些元素共同将这个立方体量变为一座大百科博物馆、海洋图书馆。当世博会结束时，这种设计可以灵活转换功能。

"The Living Ocean and Coast", the theme of the 2012 International Exposition in Yeosu, South Korea, asks for a greater recognition and awareness of the oceans and marine resources, and their importance to mankind. The architects' design for the thematic pavilion interprets the Expo's focus by "extracting a block from the ocean": a central void, the Hall of the Oceans, is surrounded by water basins which are stacked in the shape of a cube, based on the structural capacity of the glass facades and floors. Each water basin displays a specific aspect of the ocean – the deep sea, the tropical waters, the mangroves, reefs - turning the cube into an encyclopaedic museum, the library of the oceans! The design allows for a flexible conversion once the Expo has ended.

展览概念

EXHIBITION CONCEPT

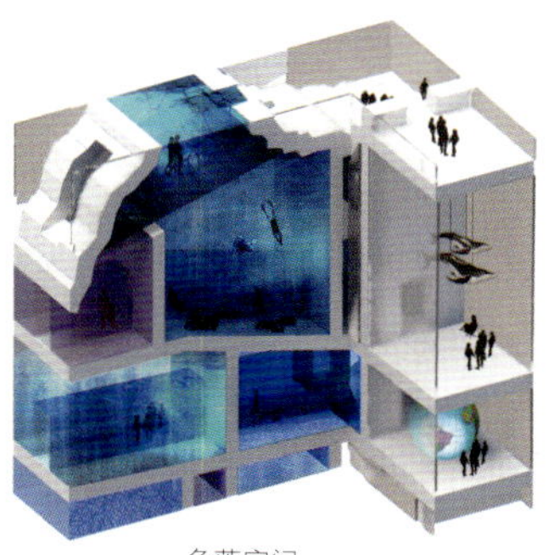

角落空间
Corn Space

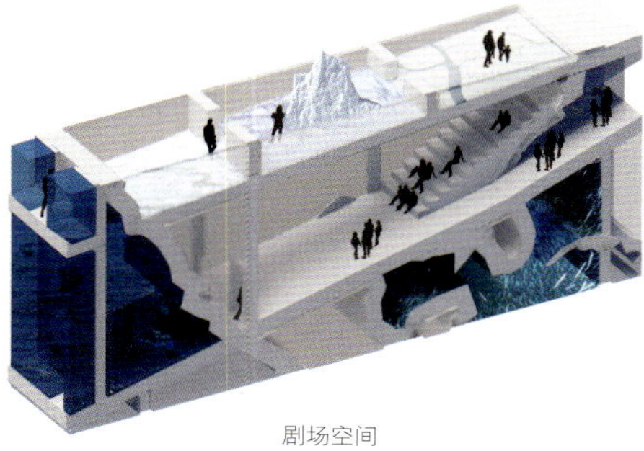

剧场空间
Theater Space

走廊空间
Corridor Space

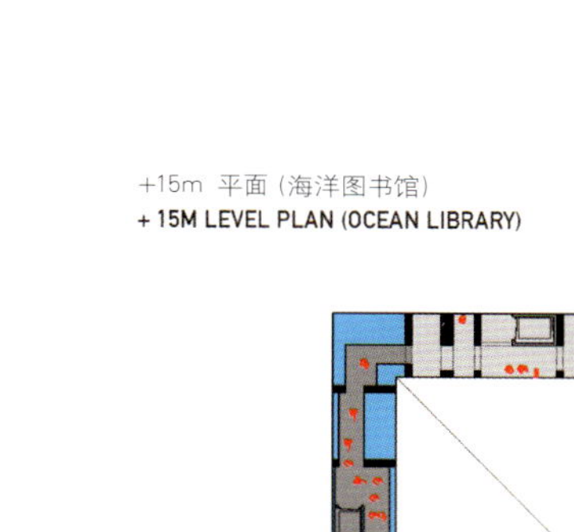

浮动的积水区
Floating Basin

概念

CONCEPT

海洋
Ocean

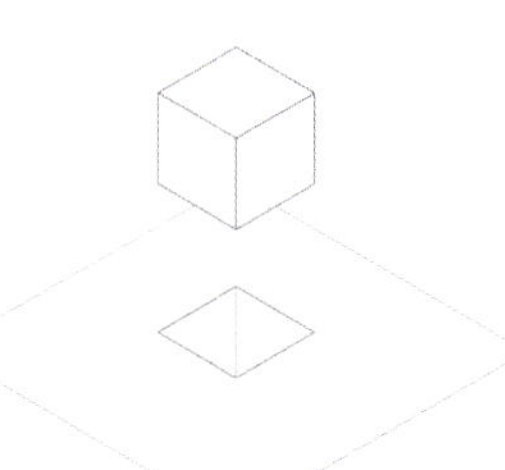

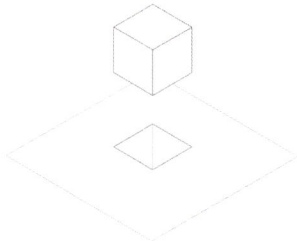

挖洞
Hollow out

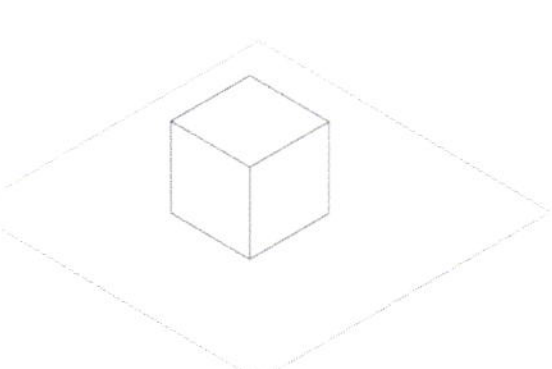

水立方
Water Cube

空心立方体
Hollow Cube

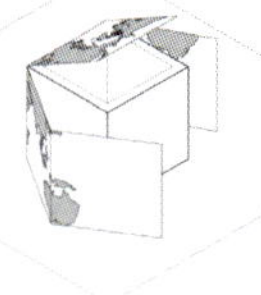

地球表皮
Globe Skin

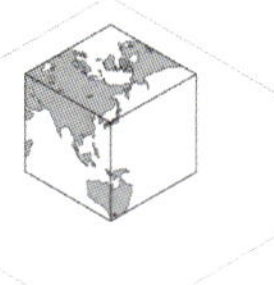

地球立方体
Globe Cube

场地布局

SITE LAYOUT

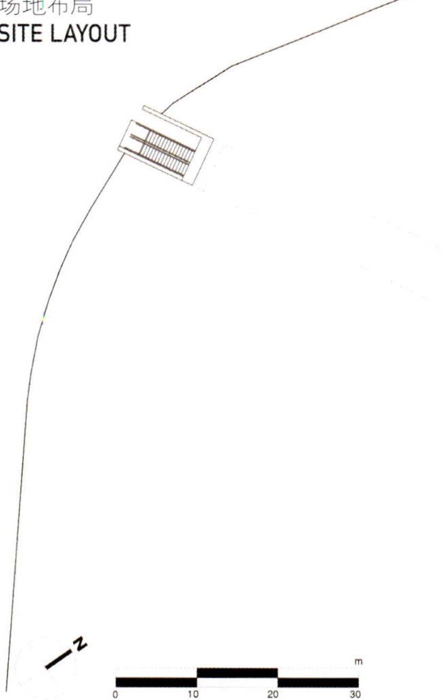

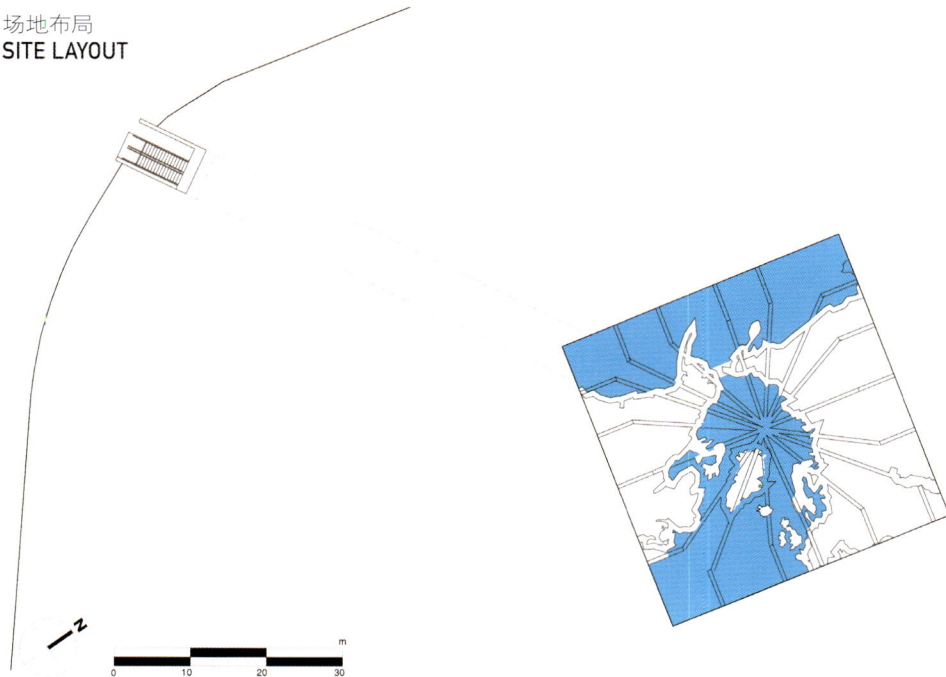

+35m平面（酒吧/屋顶）
+ 35M FLOOR PLAN (BAR/ROOF)

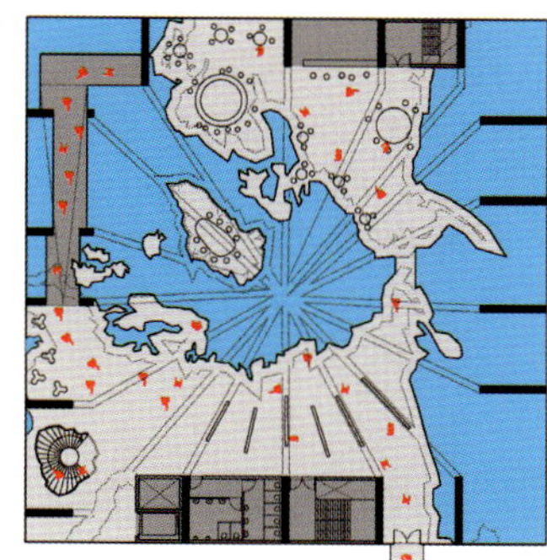

+15m 平面（海洋图书馆）
+ 15M LEVEL PLAN (OCEAN LIBRARY)

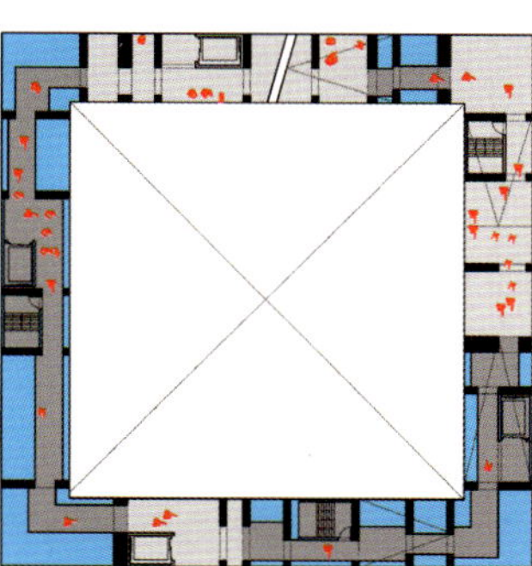

0.00m平面（展厅）
0.00M LEVEL PLAN (EXHIBITION)

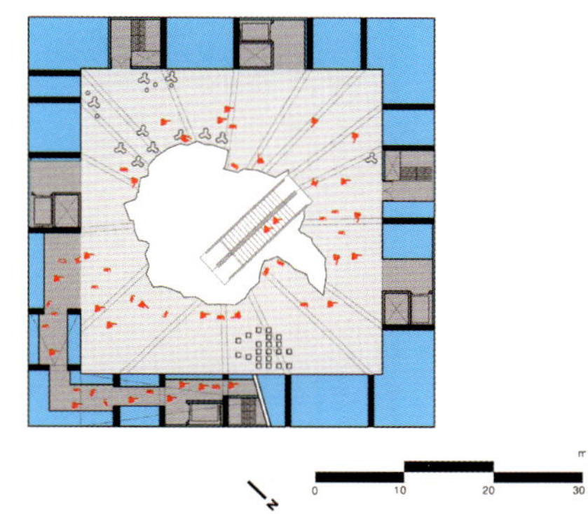

-5m平面（入口）
-5.00M LEVEL PLAN (ENTRANCE)

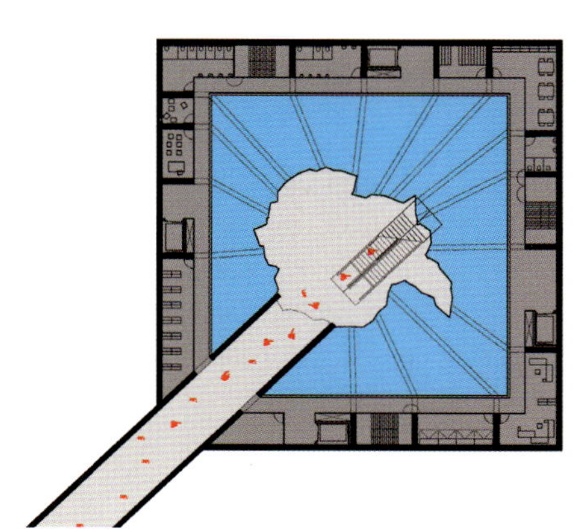

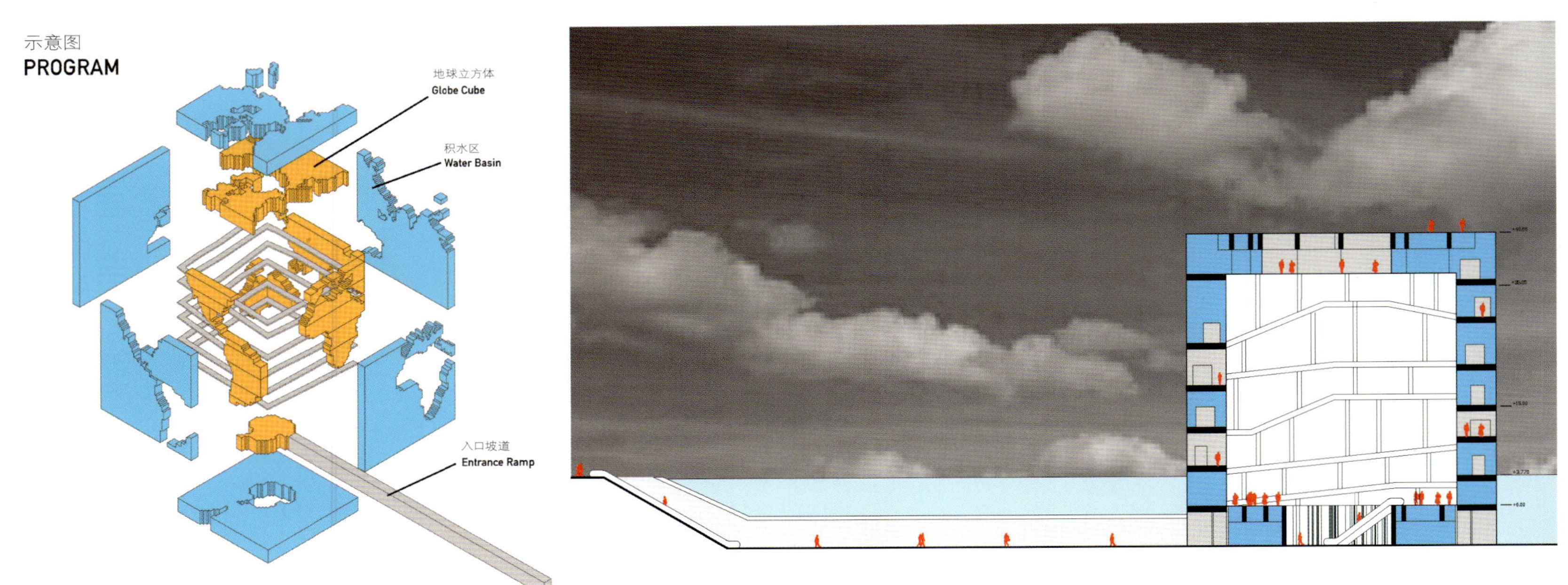
示意图
PROGRAM
地球立方体
Globe Cube
积水区
Water Basin
入口坡道
Entrance Ramp

环境概念

ENVIRONMENTAL CONCEPT

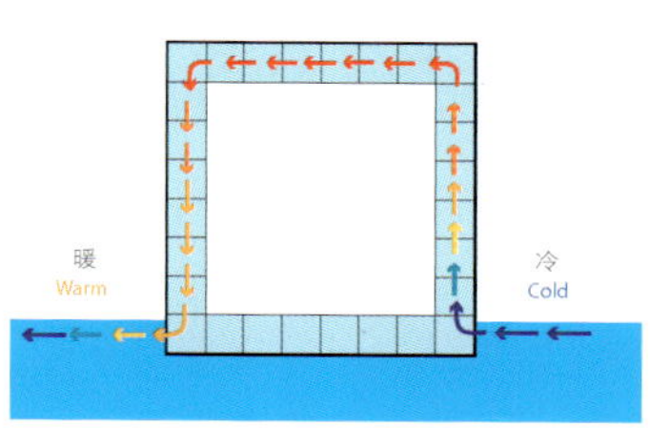

海水制冷系统
Sea Water Cooling System

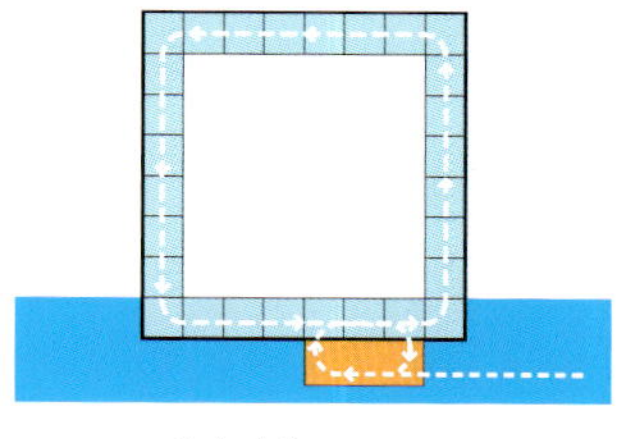

海水净化
Water Purification

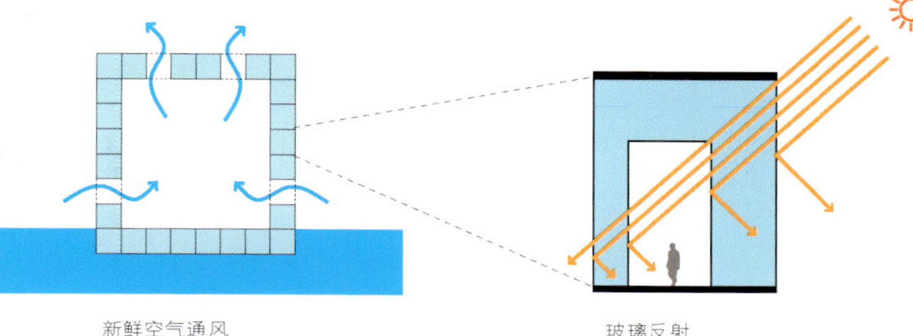

新鲜空气通风
Fresh Air Ventilation

玻璃反射
Glass Refelection

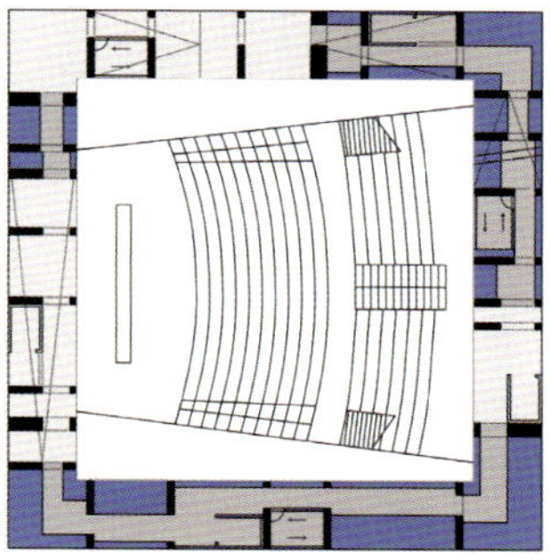

海洋剧场
Ocean Theater

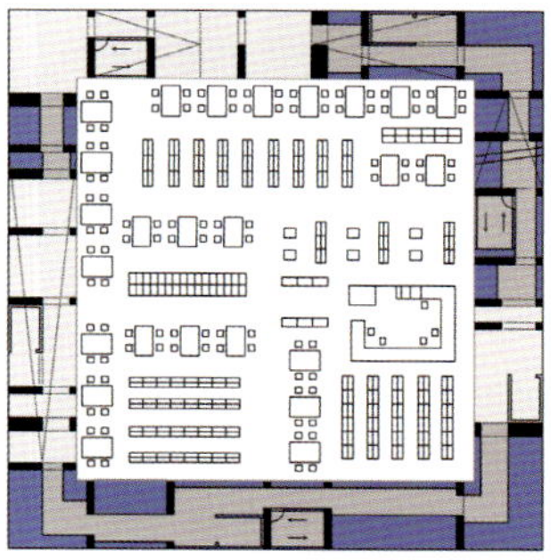

海洋图书馆
Ocean Library

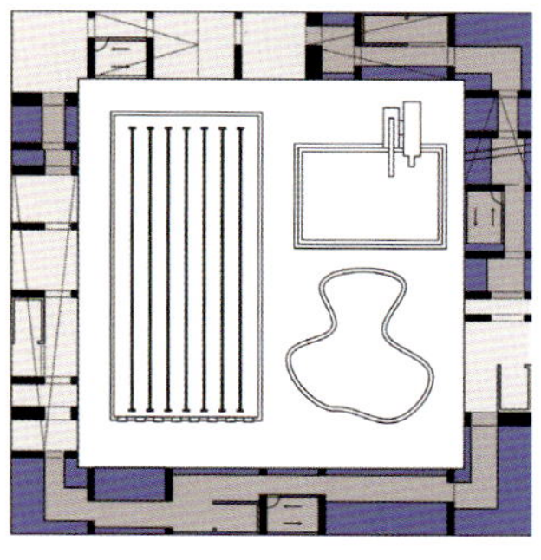

海洋游泳池
Ocean Swimming Pool

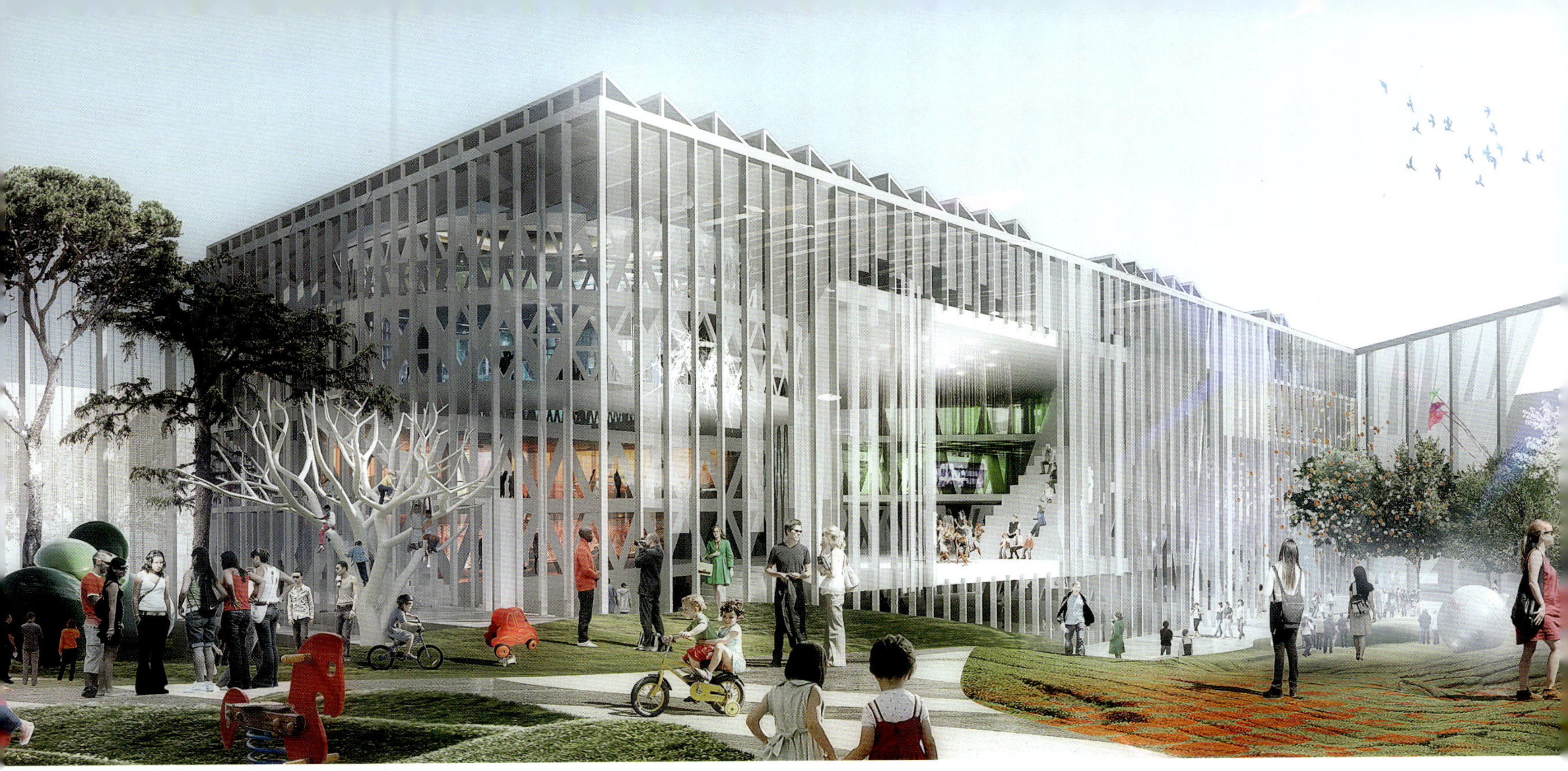

哥本哈根的运动文化馆
HOUSE OF CULTURE AND MOVEMENT, COPENHAGEN

建筑师：MVRDV, Rotterdam and ADEPT, Copenhagen
设计团队：Winy Maas, Jacob van Rijs and Nathalie de Vries with Fokke Moerel, Klaas Hofman, Attilio Ranieri, Chris Green, Kate van Heusen, Henryk Struski, Emanuela Gioffreda, Raymond van den Broek, Sanne van der Burgh
合作建筑师：ADEPT - Anders Lonka, Martin Krogh, Martin Laursen, Allan Nørregaard, Simon Poulsen, Joana Bastos, Tatyana Eneva, Umut Üsüdür, Jens Nielsen and Kostya Miroshnychenko
景观设计师：SLA, Copenhagen - Stig L. Andersson, Lene Dammand Lund, Helene Koch, Sofie K. Dybro, Martin Arfalk and Rasmus Astrup
结构设计与成本核算：Søren Jensen, Copenhagen (Anders Galsgaard)
气候与立面设计：Max Fordham, London (Guy Nevill, Stuart McKechnie)
程序与效果图制作：Imitio, Copenhagen (Christian Borch), Learning Spaces, Copenhagen (Winie Ricken), Ducks Scèno, Paris (Frans Swarte)
艺术印记：Luxigon, Paris

窗帘
Curtain

自行车停车场
Cycle Parking

艺术装置
Artistic Installation

灌溉
Irrigation

光影
Light

广告板
Advertisement

展示
Performance

垂直绿化
Vertical Green

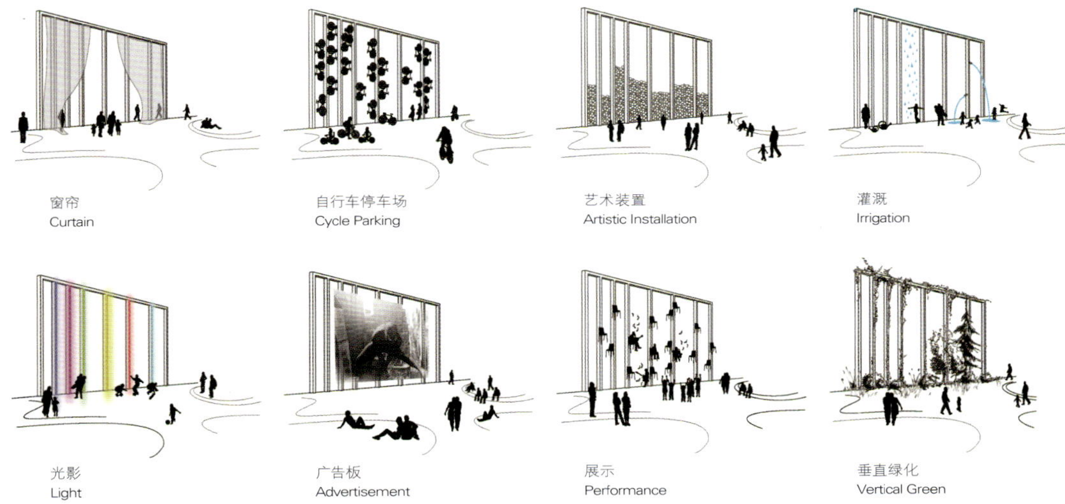

这座运动文化馆的核心目标就是为Flintholm地区提供一个充满活力的会所，让各个年龄段的人群都能参与到种类繁多的活动中。健康、文化、休闲及教育将很好地融合，最终给人们带来一种引人入胜的建筑体验。运动文化馆的主楼是一种矩形的玻璃空间，包含着六组完美组合的项目元素。其中的这些空间可以灵活设计为可举办多种活动和带有主要流通区的“游乐天地”。这些组合元素具有更独特的用途：剧场、健康中心、饮食中心、参禅区、学习区以及展览厅、健身及活动中心、全民健身中心以及行政区。

地下停车场平面
Parking Basement

The main ambition for the House of Culture and Movement is to offer the Flintholm neighborhood a dynamic meeting point for people of all ages taking part in a wide range of activities. Health, culture, leisure and education should smoothly blend together to create a spectacular architectural experience that will become a destination. The main building, the House of Culture and Movement is a rectangular glass volume containing six stacked ideal programmatic elements. The space in-between can be programmed flexibly as a "play zone" with various activities and main circulation. The stacked elements hold more specific uses: a theatre, a health zone, food zone, a zen area, a study centre and exhibition hall, fitness and activity centre, a wellness centre and an area for the administration.

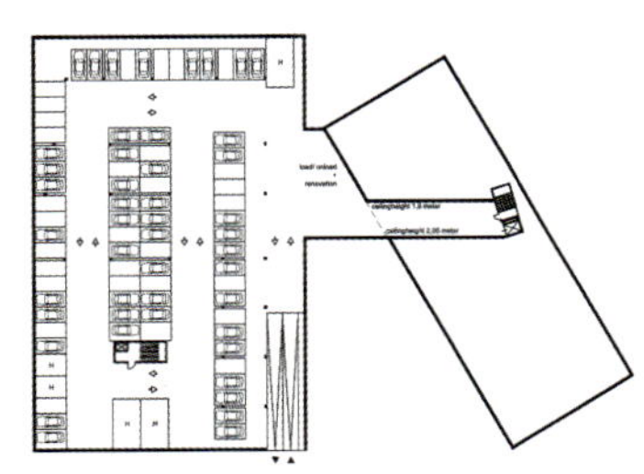

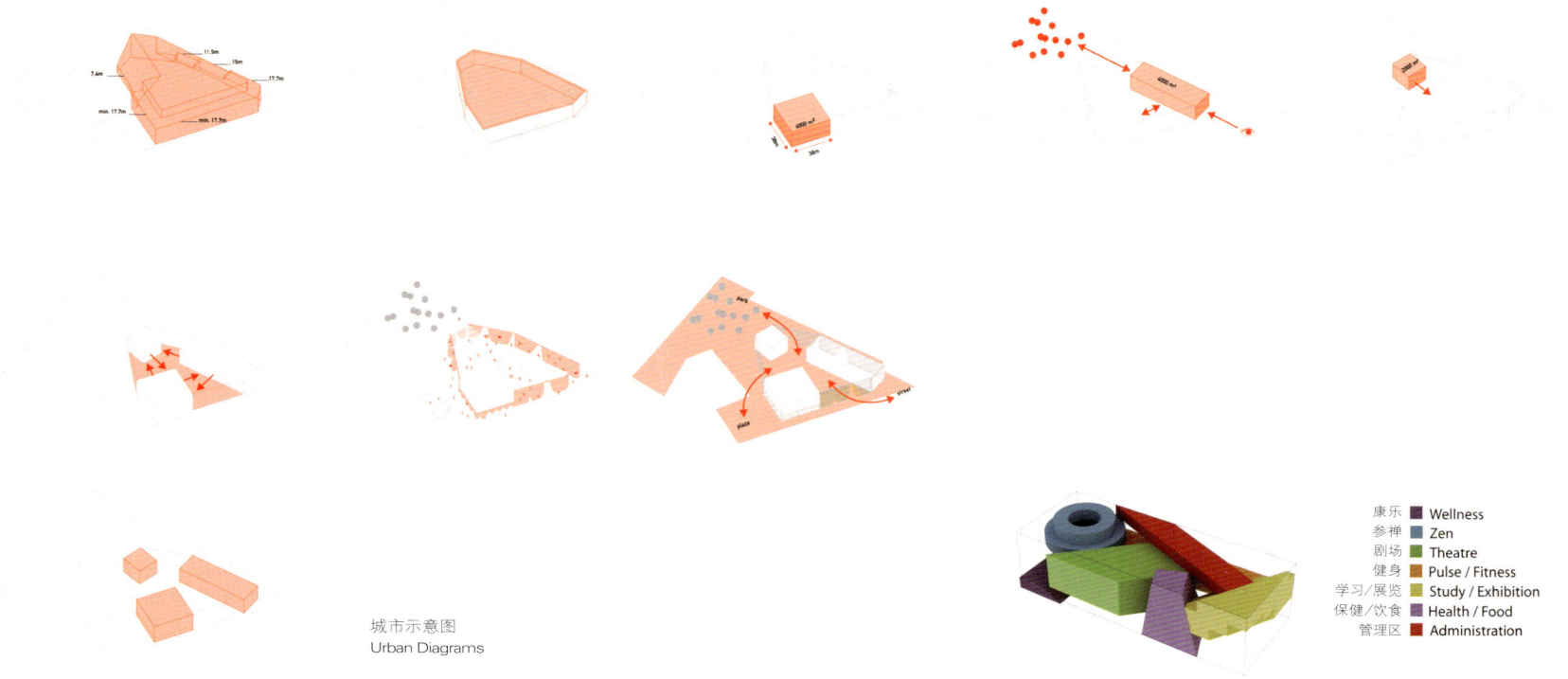

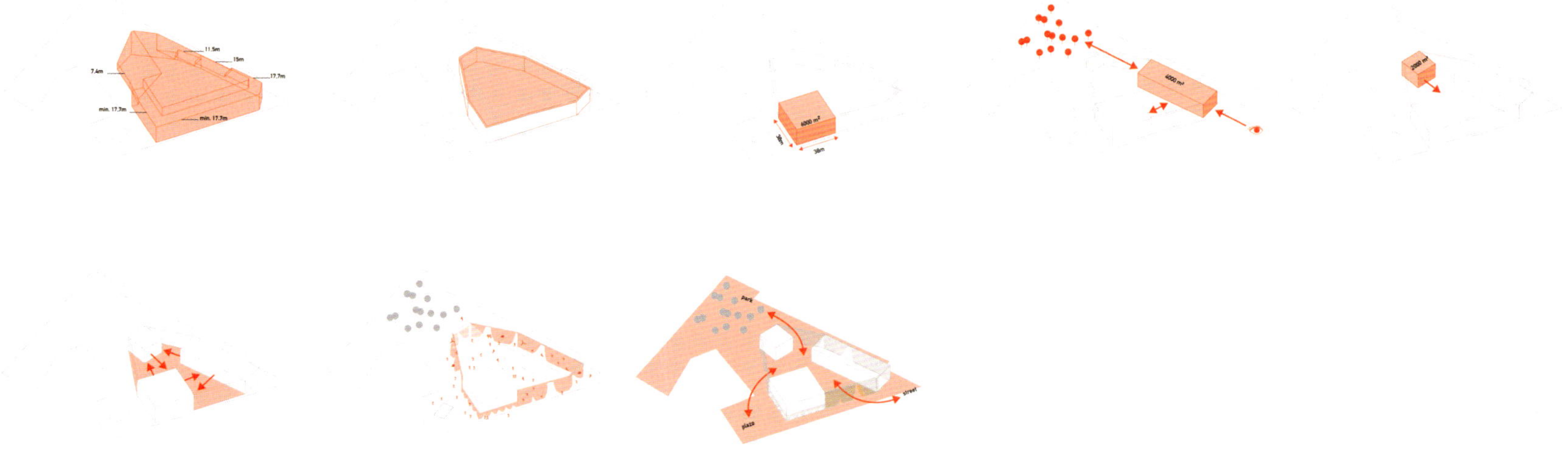

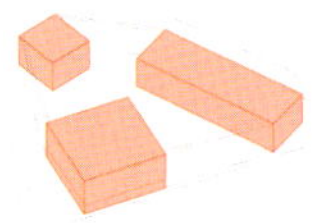

城市示意图
Urban Diagrams

康乐 Wellness
参禅 Zen
剧场 Theatre
健身 Pulse / Fitness
学习/展览 Study / Exhibition
保健/饮食 Health / Food
管理区 Administration

剖面图
Section

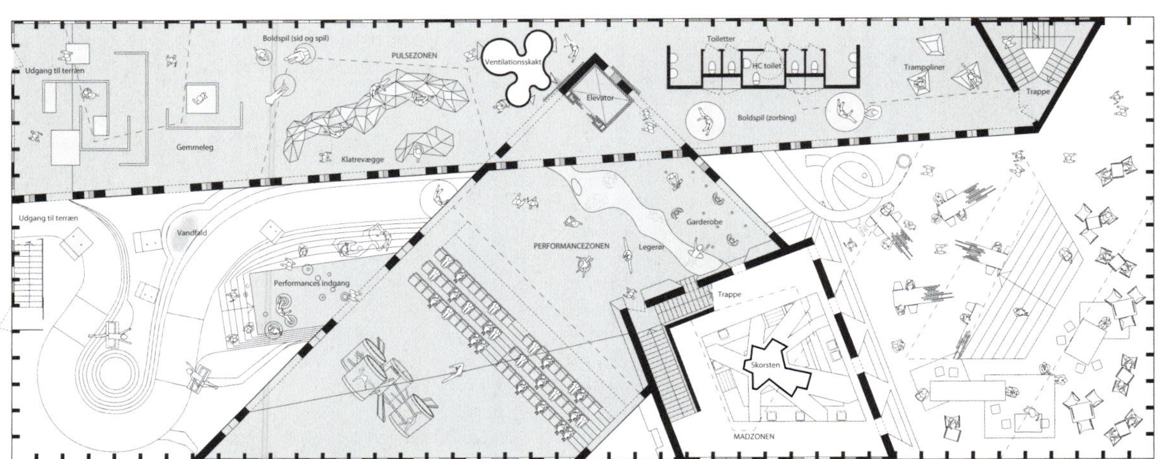

一层平面
Ground floor plan

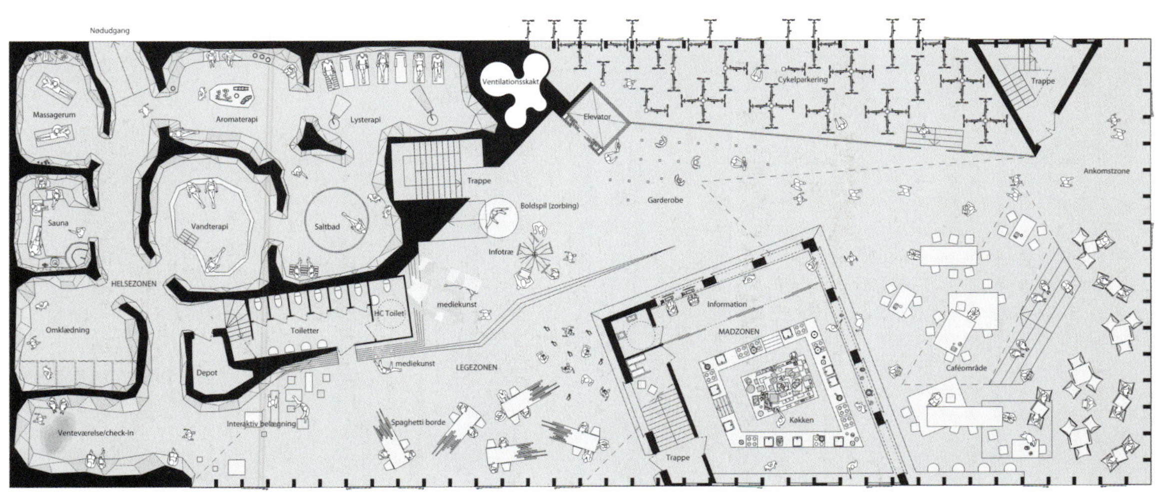

四层平面
3rd floor plan

场地平面图
Site plan

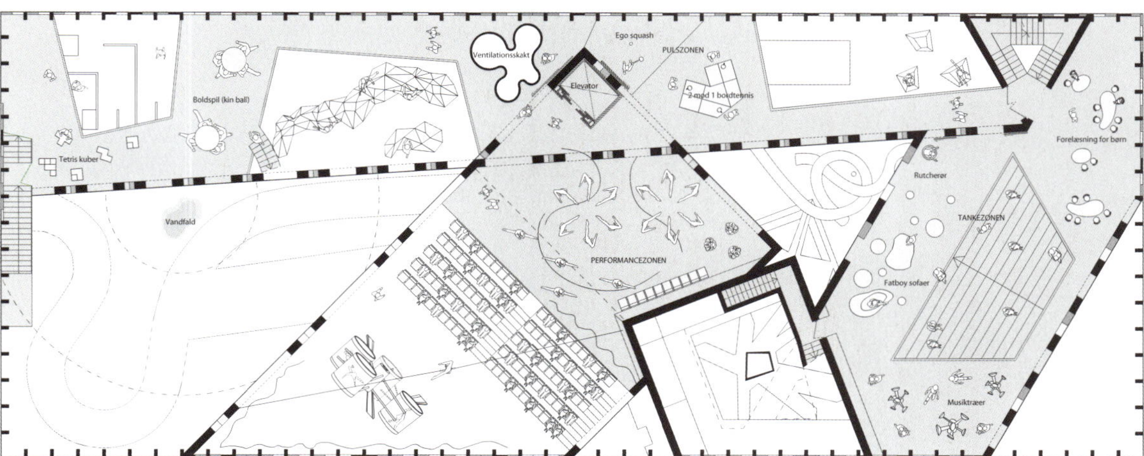

七层平面
6th floor plan

十层平面
9th floor plan

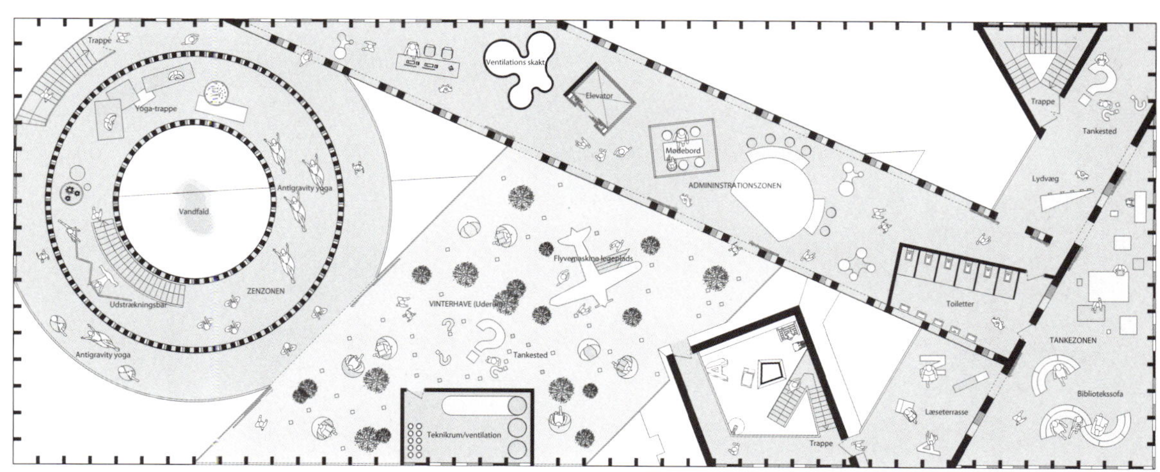

十三层平面
12th floor plan

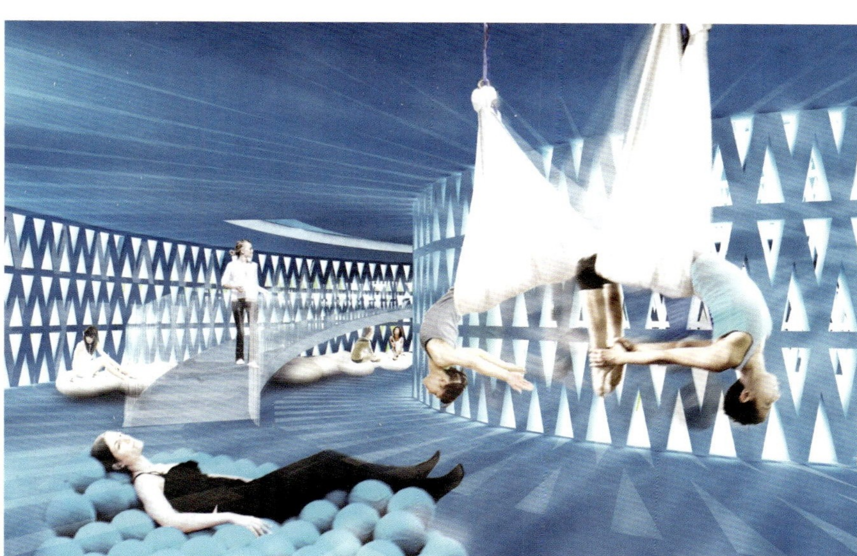

德国威斯巴登法律学校
WIESBADEN LAW SCHOOL, GERMANY

建筑师：3XN
甲方：Land Hessen and European Business School
工程师：Werner Sobek Engineering & Design（结构工程），HL-Technik Engineering Partner（环境气候工程），hhpberlin – Ingenieure für Brandschutz（防火设计）

该设计要求是以空间的多样性创造最适合学习的结构。这座新建筑以中心集会场所的形式，将自身与现存的历史结构连接起来，为周围的人群、建筑和城市之间的聚会提供便利。只有通过不同观点的碰撞，人们才能变得更加见多识广。这个集会区的功能是作为建筑物的“心脏”，在这里学生、教授及律师都能使用这所大学的会议设施。威斯巴登法律学校应成为这座城市的空间核心区——一个充满活力的区域，将自身与周围的环境融合，标志着稳固与开放。

The belief of the design is that spatial diversity creates the best framework for learning. The new building connects itself with the existing historic structure by way of a central marketplace, facilitating meetings between people, buildings and the city around it. It is through interaction that points of view and people become informed. The marketplace functions as the "heart" of the building where a synergy is established between students, professors and practising lawyers – all of whom are able to use the University's conference facilities. The vision is that Wiesbaden Law School should be a spatial attraction for the city – a dynamic gathering place that blends together with its surroundings signalling solidity and openness.

总平面图
Site plan

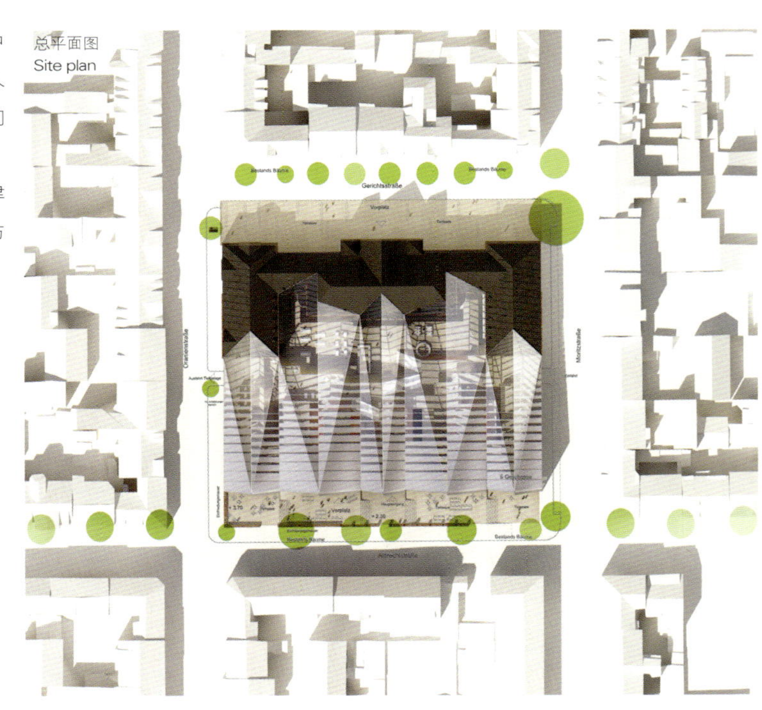

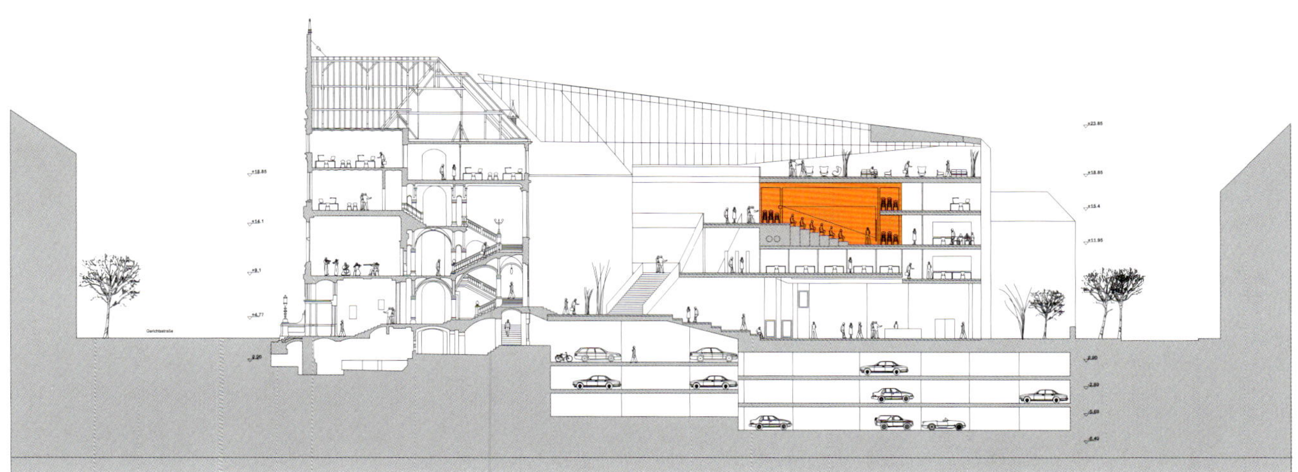

A剖面
Section A

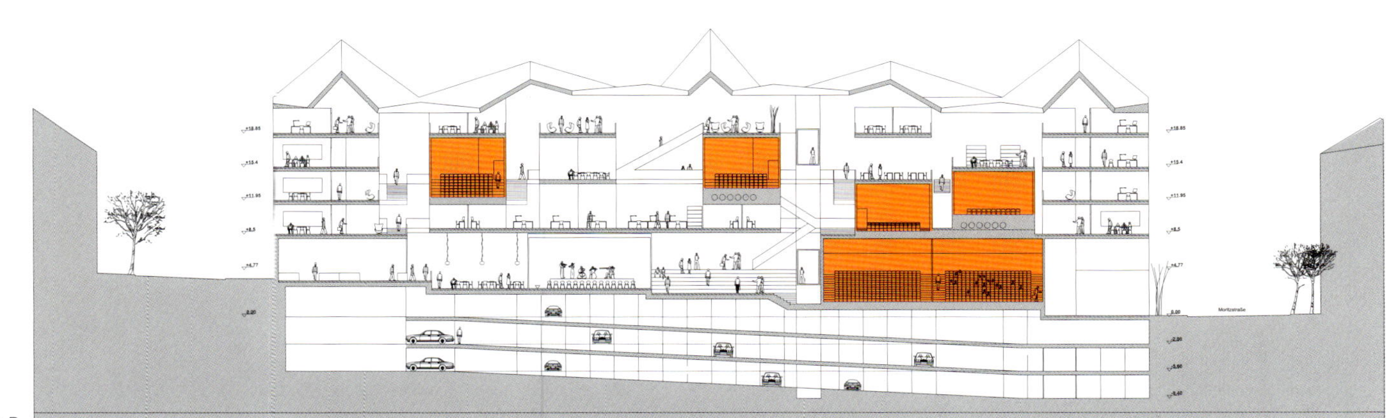

B剖面
Section B

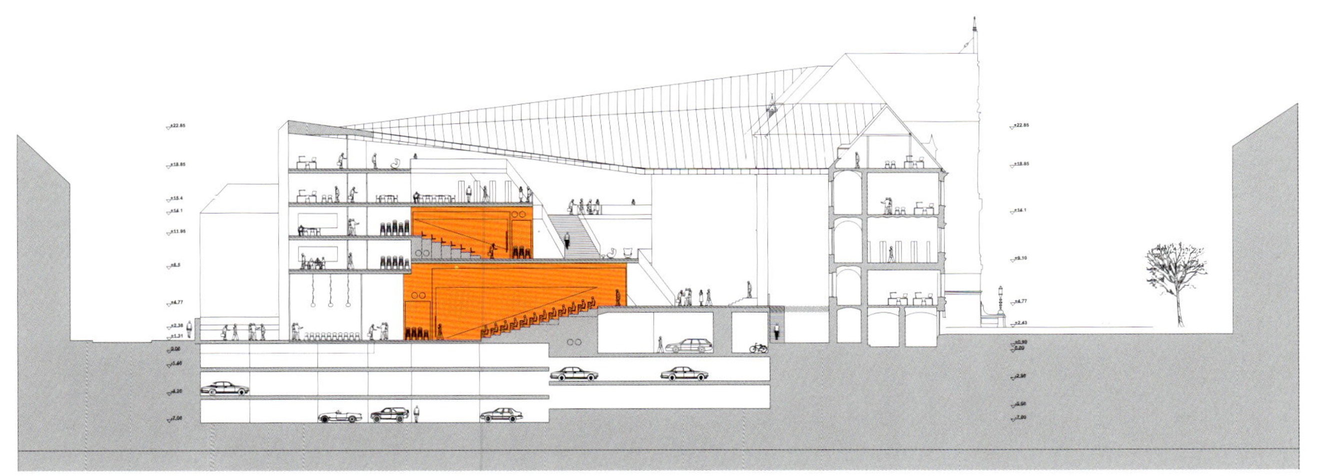

C剖面
Section C

东立面
East elevation

西立面
West elevation

南立面
South elevation

台北表演艺术中心方案二
TAIPEI PERFORMING ARTS CENTRE

建筑师：NL Architects: Pieter Bannenberg, Walter van Dijk, Kamiel Klaasse
项目负责人：Thijs van Bijsterveldt, Guus Peters
设计团队：Rebecca Eng, Joost Luub, Yuichi Tanaka, Yannick Vanhaelen, Murk Wymenga, Gen Yamamoto, Ivar van der Zwan
当地顾问：Wie-Nien Chen
构造：Arjan Habraken /ARUP Amsterdam

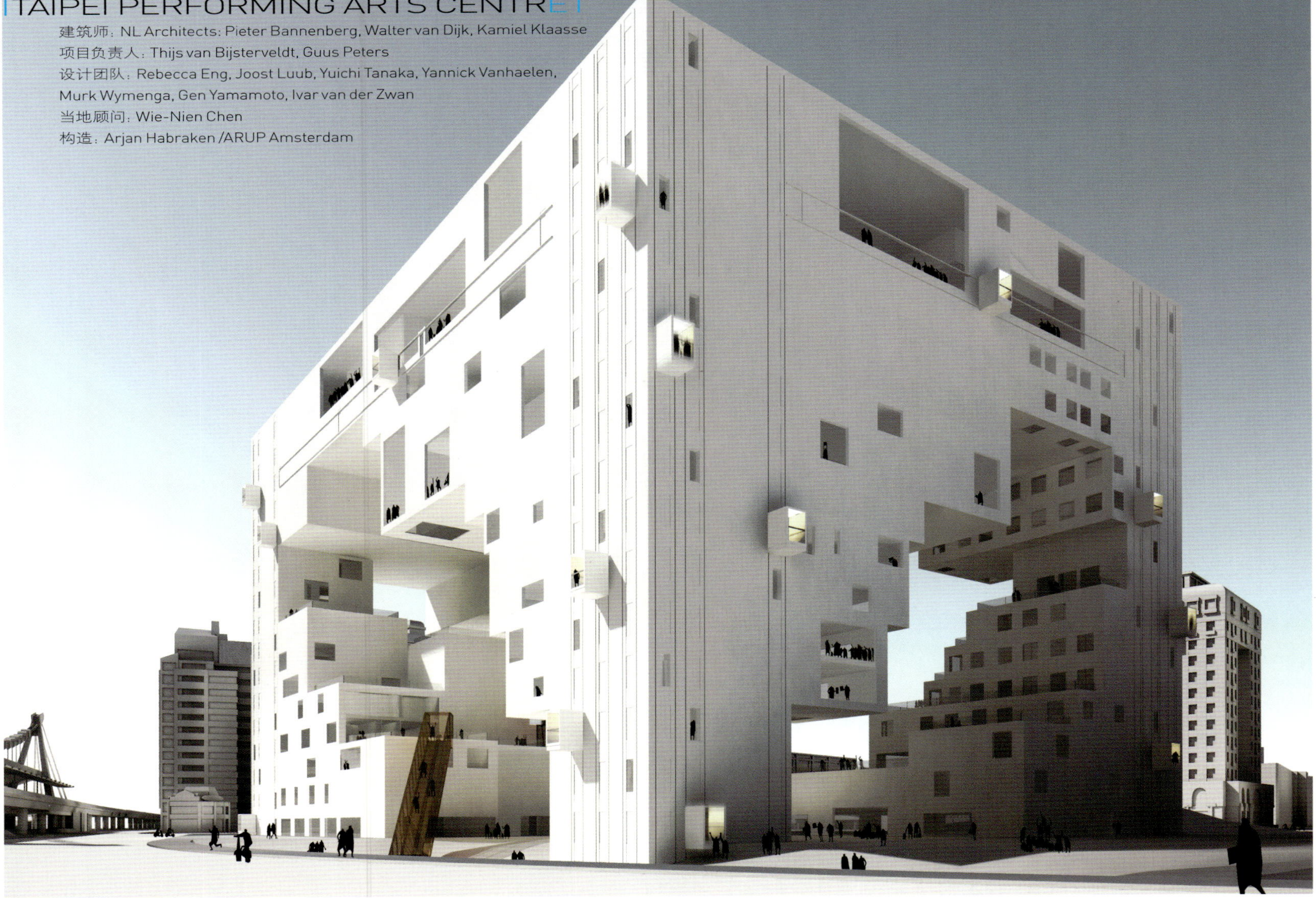

高雅文化与通俗文化之间的对立正在慢慢地消失。但是我们是否可以创造一处唤起所有人的热情的场所？这一场所能否将高雅与通俗融为一体？我们如何预想一座真正的公共建筑？

台北表演艺术中心计划建成所有人都能利用的建筑。中心的公共特质因为实质功能部分的立面得到了保证，其下形成了一座公共广场。广场从根本上来说成为了建筑的一部分：它被容纳在建筑之内。

建筑的尺寸为110m × 80m × 64m。你可以把它当成一张“桌子”：四条“桌腿”支撑着一个容纳三层楼的“桌面”。在建筑的内部是被抬高了的城市，一处空中浏览空间。此处将是文化设施的领域：多媒体图书馆、音乐商店、展廊、大厅、酒吧、餐厅和俱乐部。具有不同功能的阳台和露台为空间注入了活力。有时候这些空间是向公众开放的，有时候却是封闭的专享空间。

The dichotomy between high and low culture is disappearing. But can we create an environment that is inspiring for everyone? Is it possible to be elitist and populist at the same time? How can we envision a truly public building?

The Taipei Performing Arts Centre aspires to become accessible for everybody. The public character of the Centre is guaranteed by the elevation of a substantial part of its program, creating a public square underneath it. As such the square fundamentally becomes part of the building: it is included inside it.

The block measures 110 x 80 x 64 meters. It could be considered as a "table": Four "legs" support as a "tabletop" that accommodates 3 stories. Inside you'll find an elevated fragment of the city, a public "browsing space" in the sky. This will be the domain for cultural facilities: the multi media library, music stores, galleries, lobbies, bars, restaurants and clubs.

Balconies and terraces with different programs activate the space. Sometimes they are open and public; sometimes exclusive or intimate.

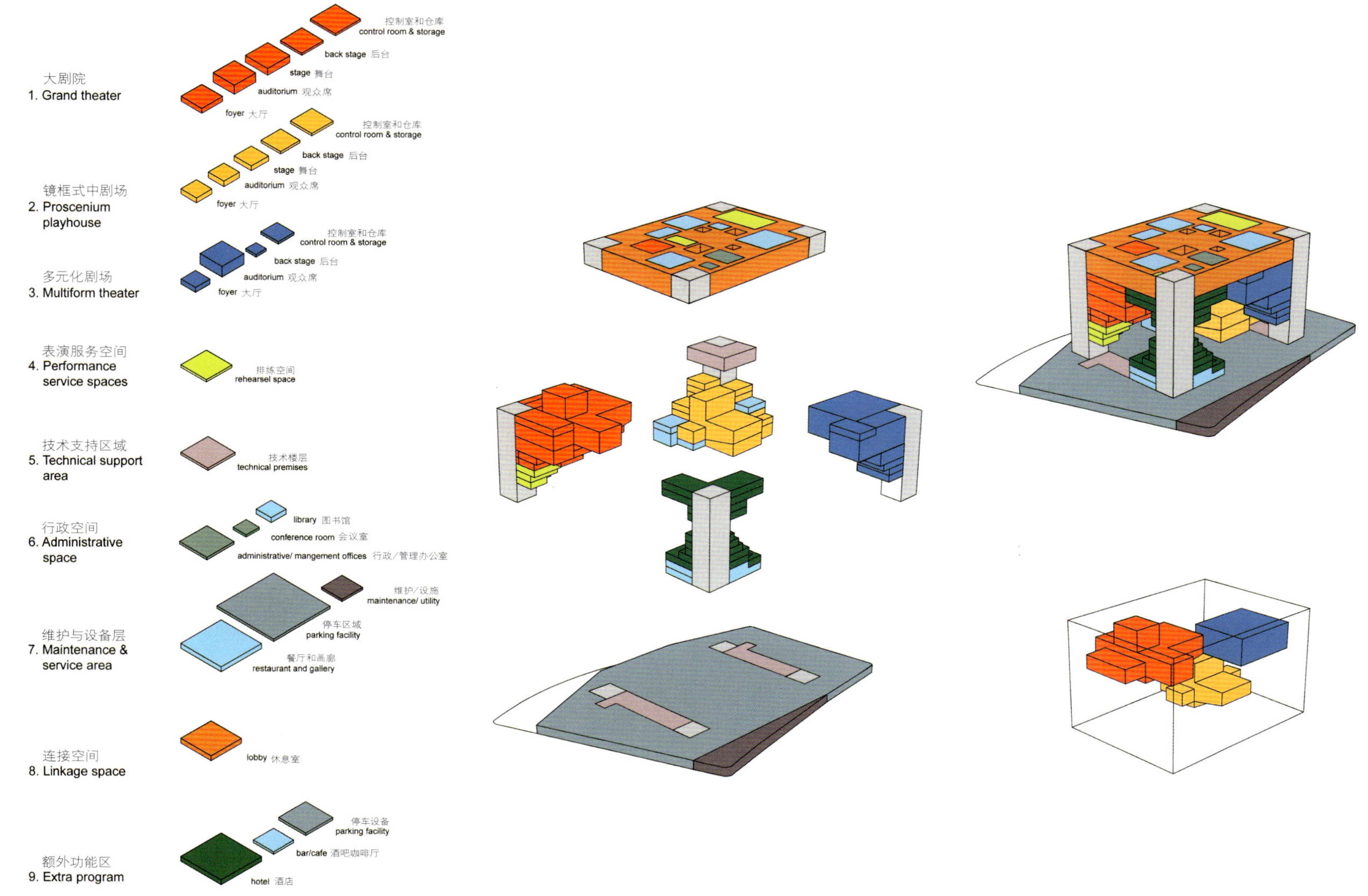

大剧院
1. Grand theater

控制室和仓库 control room & storage
back stage 后台
stage 舞台
auditorium 观众席
foyer 大厅

镜框式中剧场
2. Proscenium playhouse

控制室和仓库 control room & storage
back stage 后台
stage 舞台
auditorium 观众席
foyer 大厅

多元化剧场
3. Multiform theater

控制室和仓库 control room & storage
back stage 后台
auditorium 观众席
foyer 大厅

表演服务空间
4. Performance service spaces

排练空间 rehearsel space

技术支持区域
5. Technical support area

技术楼层 technical premises

行政空间
6. Administrative space

library 图书馆
conference room 会议室
administrative/ mangement offices 行政／管理办公室

维护与设备层
7. Maintenance & service area

维护／设施 maintenance/ utility
停车区域 parking facility
餐厅和画廊 restaurant and gallery

连接空间
8. Linkage space

lobby 休息室

额外功能区
9. Extra program

停车设备 parking facility
bar/cafe 酒吧咖啡厅
hotel 酒店

功能图
PROGRAM

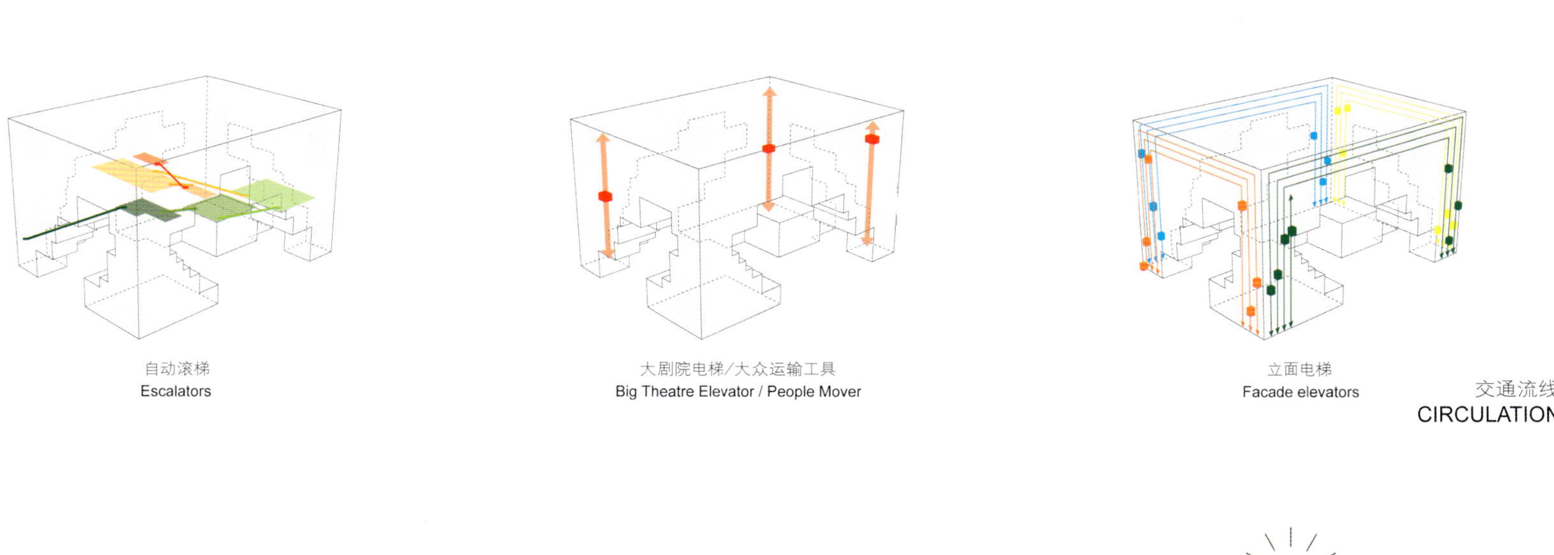

自动滚梯
Escalators

大剧院电梯／大众运输工具
Big Theatre Elevator / People Mover

立面电梯
Facade elevators

交通流线
CIRCULATION

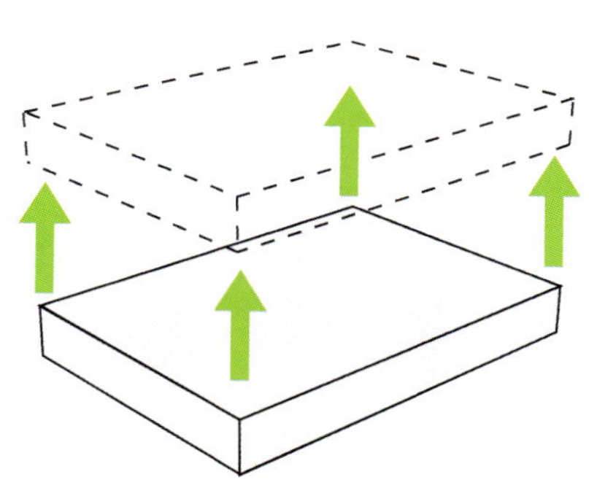

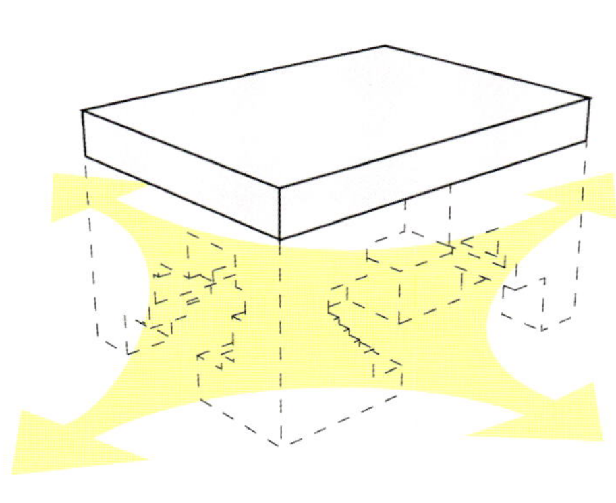

抬升的功能空间为城市创造了新的遮蔽广场，向市民提供了遮阳避雨的去处。
Elevated program creates a new covered square for the city that provides shade and shelter

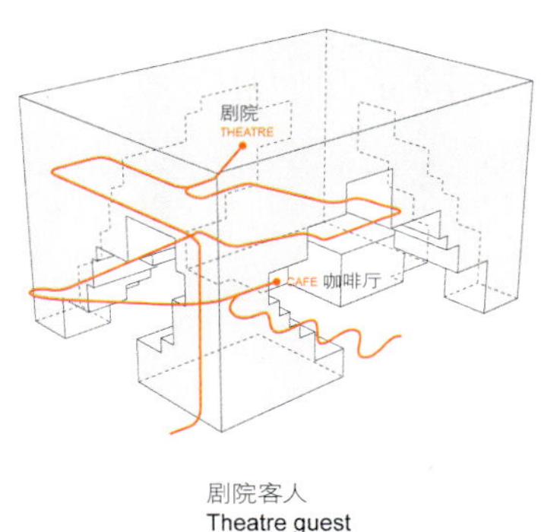

剧院客人
Theatre guest

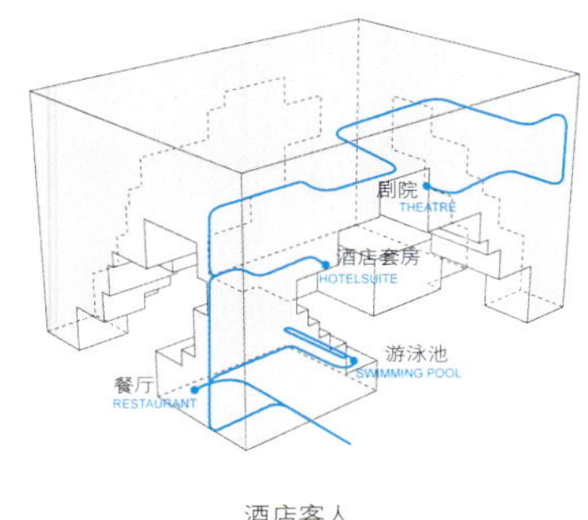

酒店客人
Hotel guest

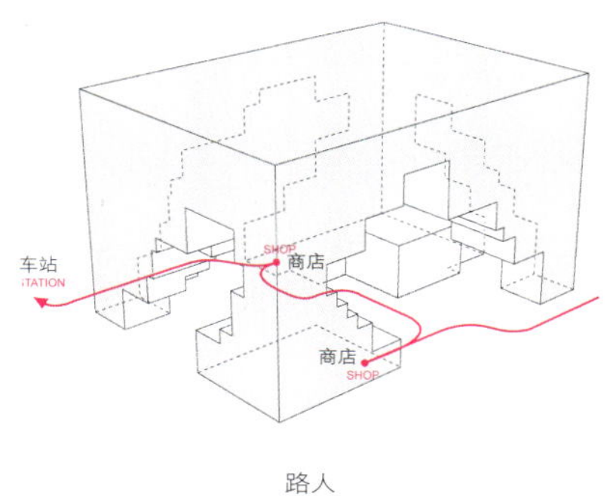

路人
Passer by

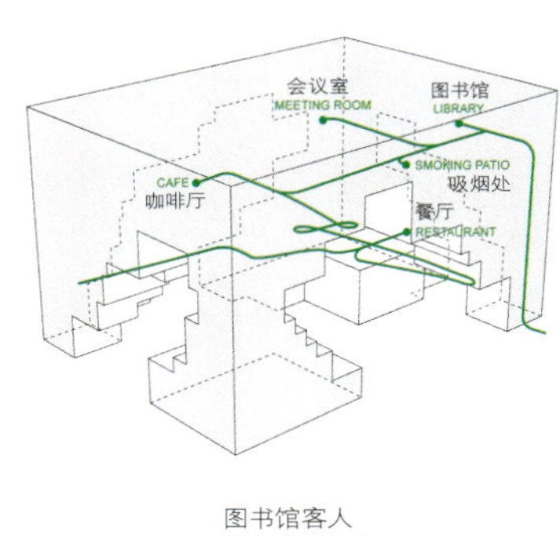

图书馆客人
Library guest

游客可能的路线
POSSIBLE VISITOR SCENARIOS

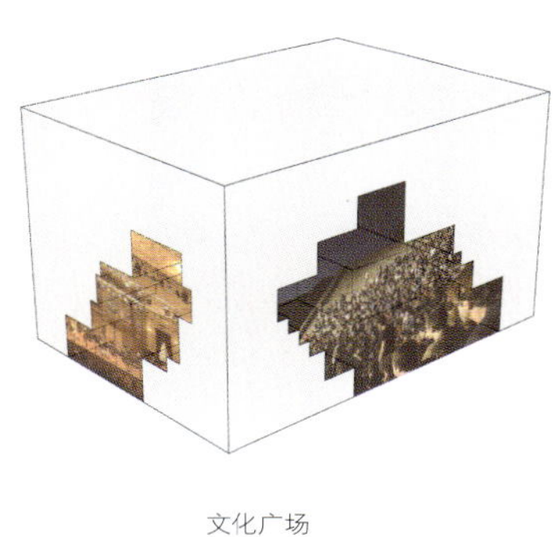

文化广场
Cultural Square

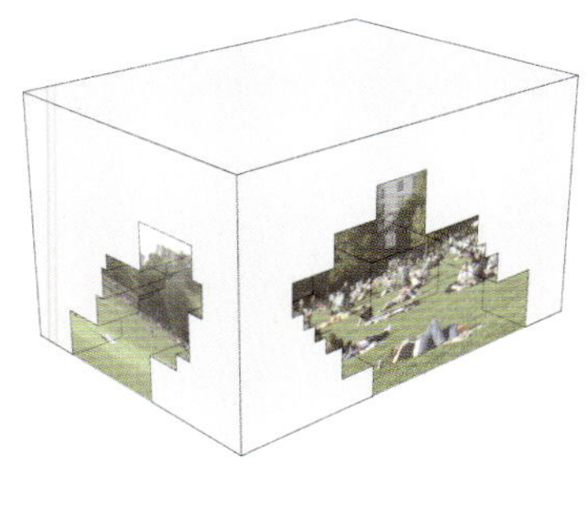

绿色广场
Green Square

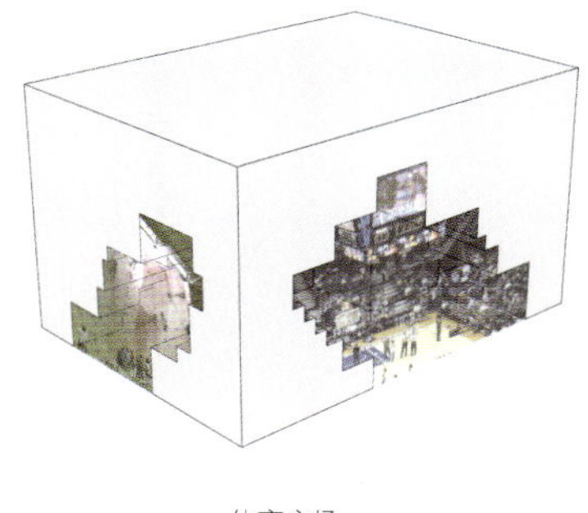

体育广场
Sport Square

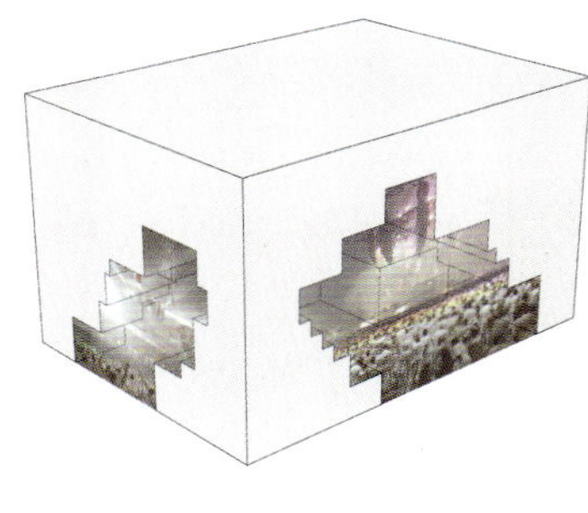

聚会广场
Party Square

预计的广场景观
POSSIBLE SQUARE SCENARIOS

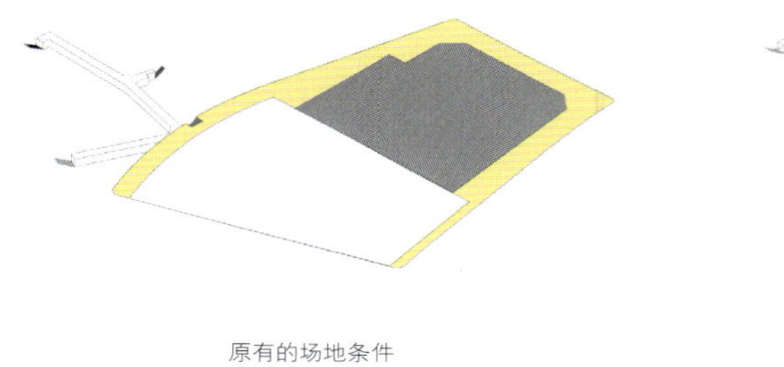

原有的场地条件
Existing Situation

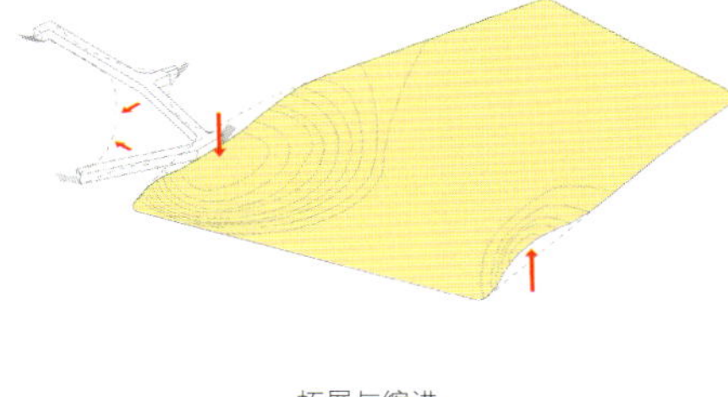

拓展与缩进
Push and Pull

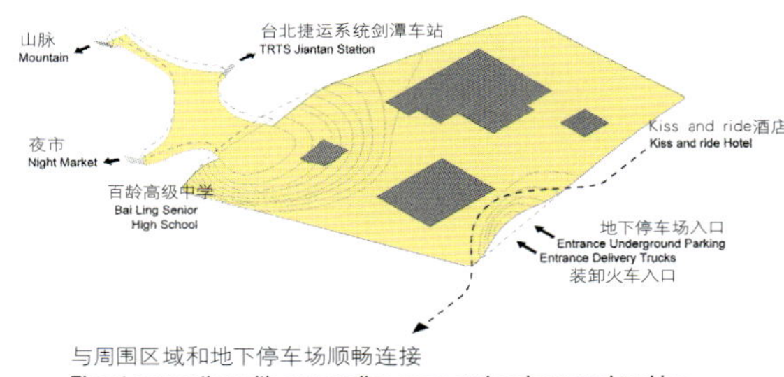

与周围区域和地下停车场顺畅连接
Fluent connection with surrounding areas and underground parking.

广场设计
SQUARE DESIGN

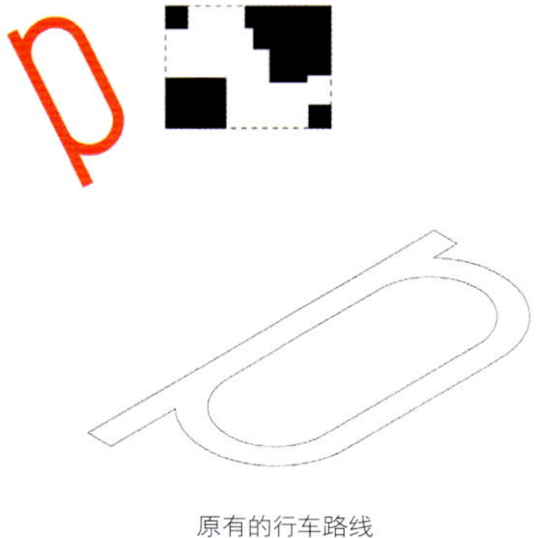

原有的行车路线
Existing Running Track

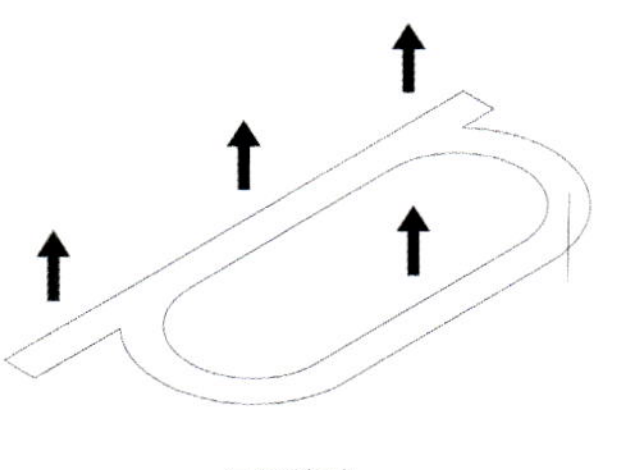

复制粘贴
Copy Paste

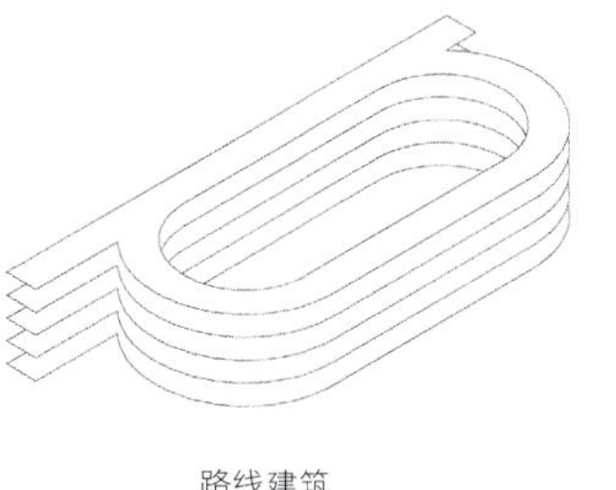

路线建筑
Track Building

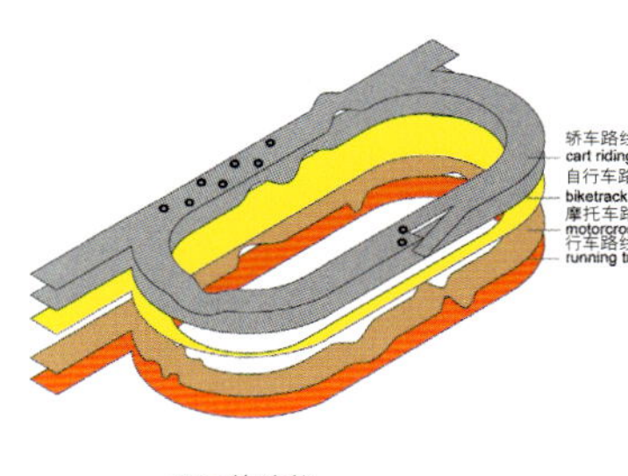

不同的功能
Differentiate program

规划：通过将原有的行车路线转换为一座路线建筑，场地周围的公共功能得到了加强
PROPOSAL: INTENSIFY PUBLIC PROGRAMS AROUND THE SITE BY TRANSFORMING EXISTING RUNNING TRACK INTO A TRACK BUILDING

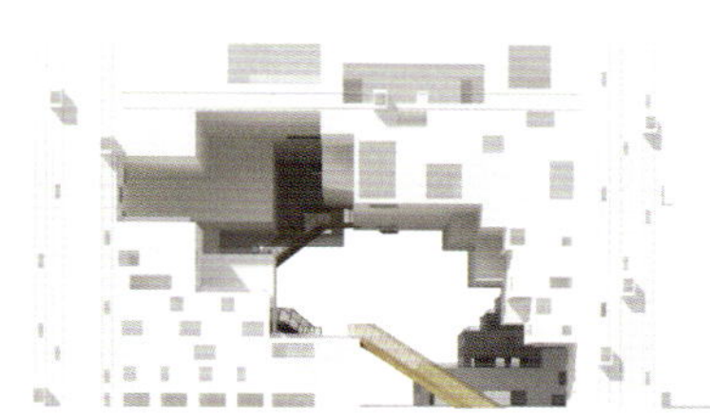

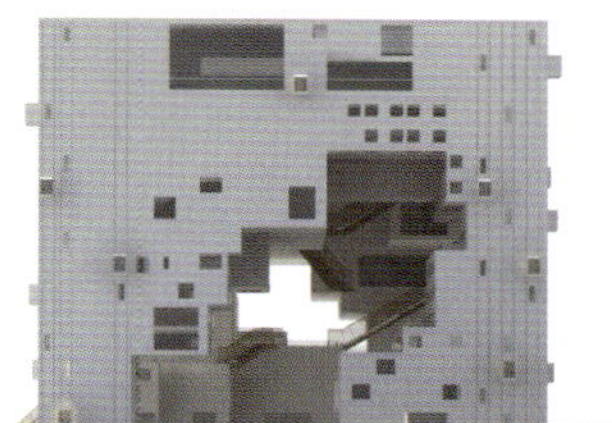

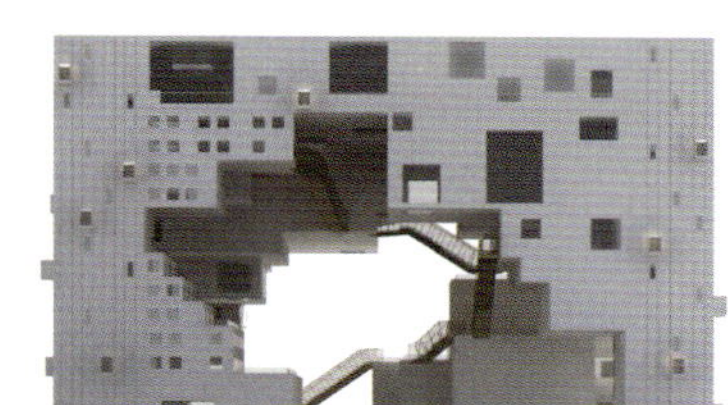

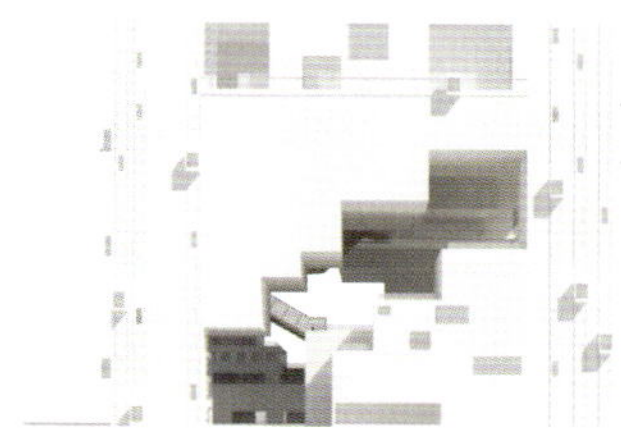

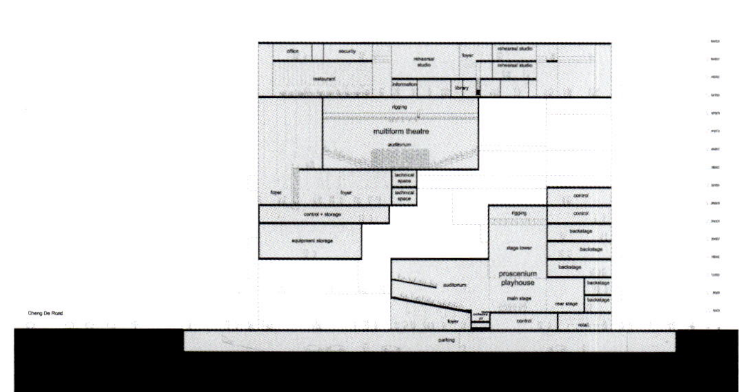
multiform theatre
proscenium playhouse
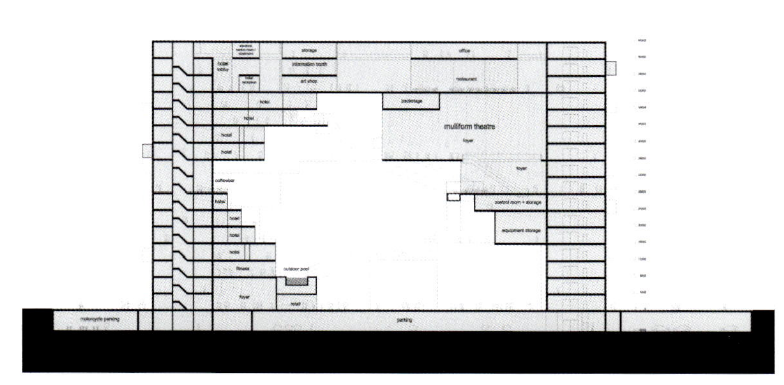
multiform theatre
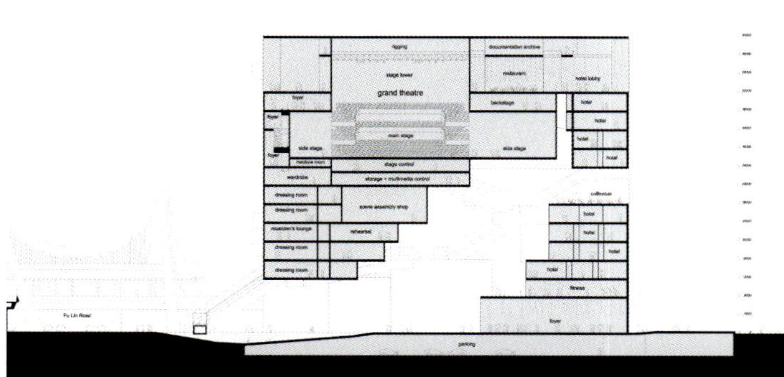
grand theatre
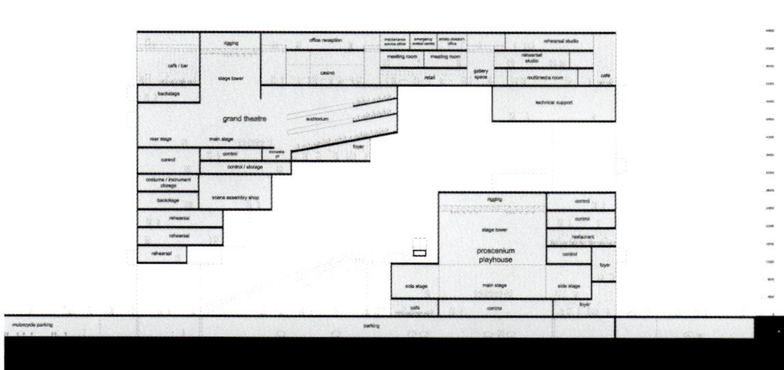
grand theatre
proscenium playhouse

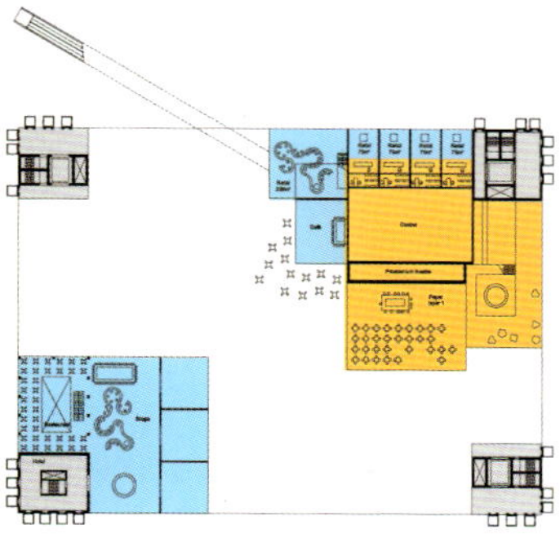

GROUND FLOOR
一层

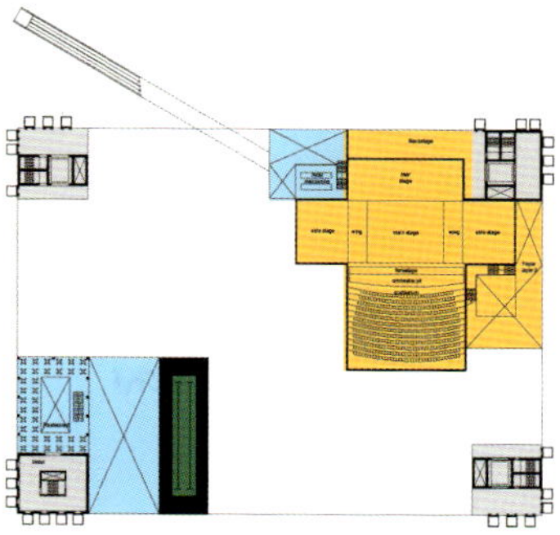

LEVEL 01
二层

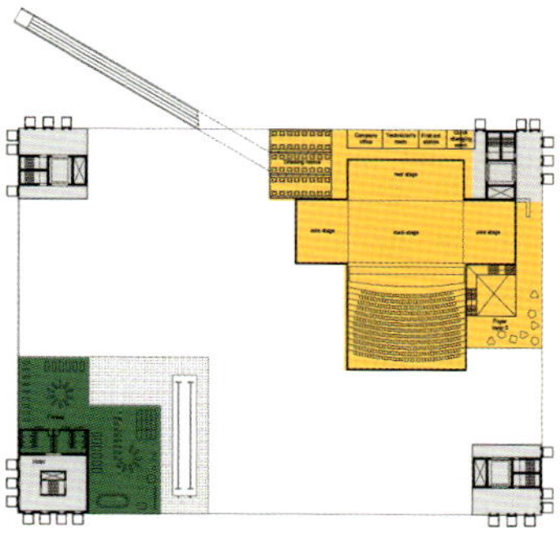

LEVEL 02
三层

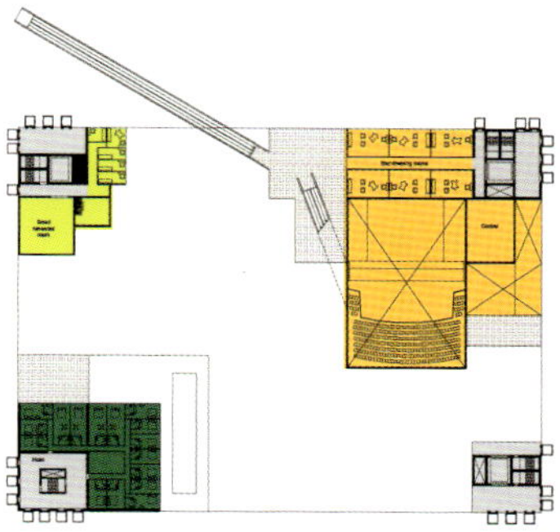

LEVEL 03
四层

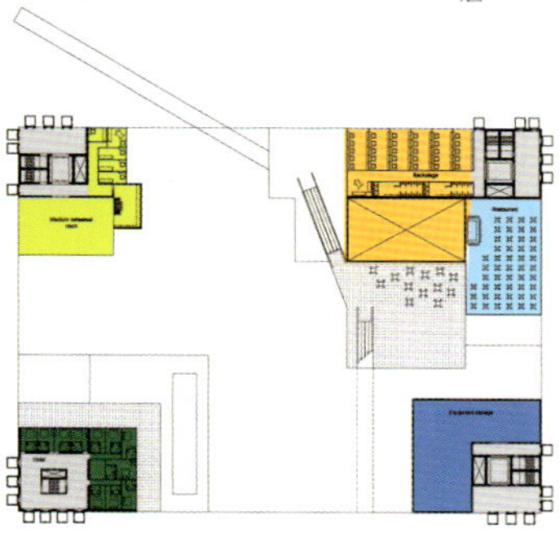

LEVEL 04
五层

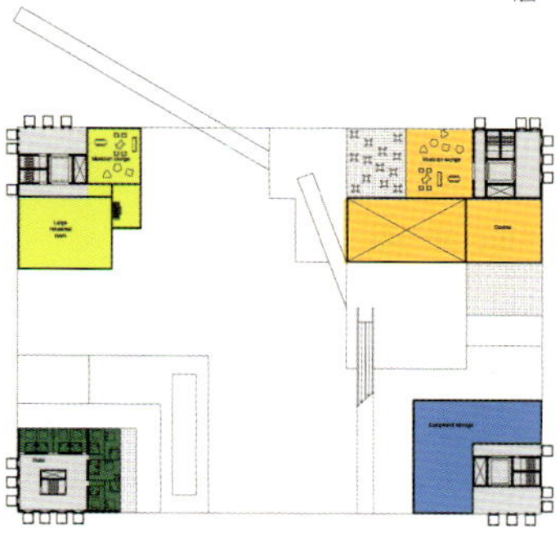

LEVEL 05
六层

LEVEL 06
七层

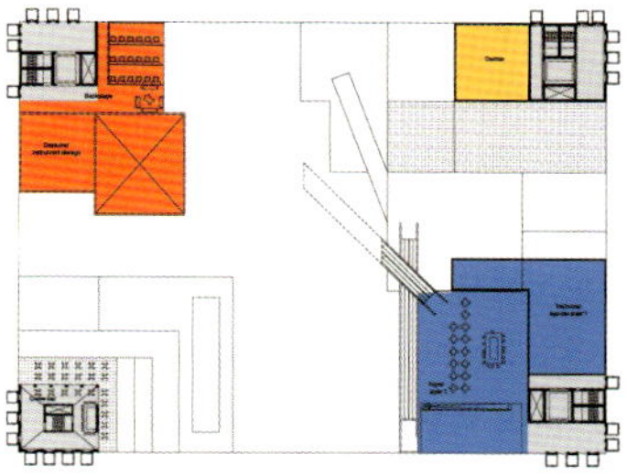

LEVEL 07
八层

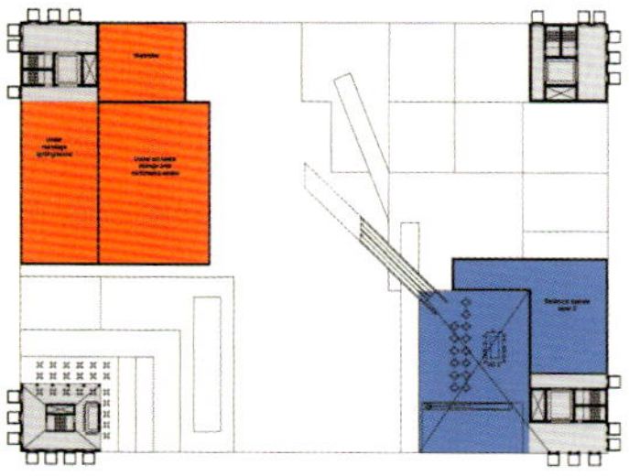

LEVEL 08
九层

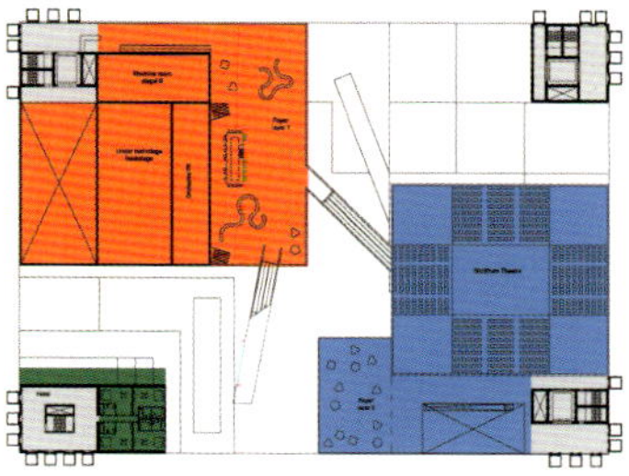

LEVEL 09
十层

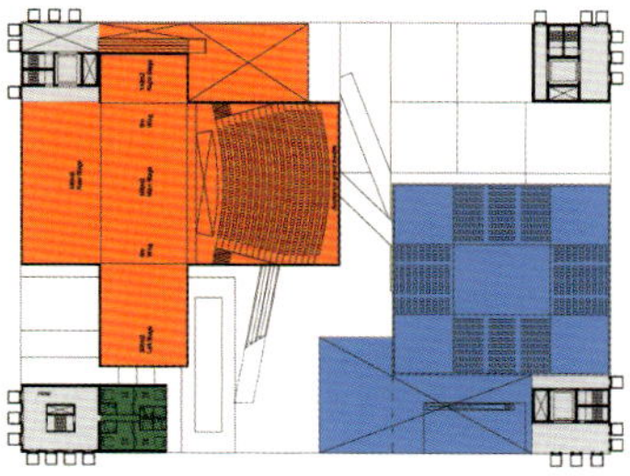

LEVEL 10
十一层

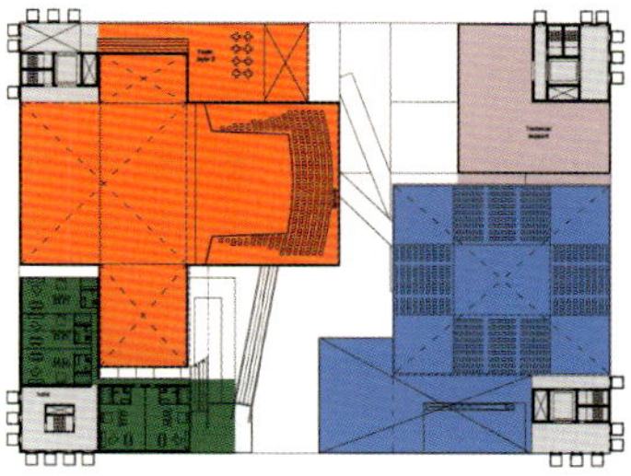

LEVEL 11
十二层

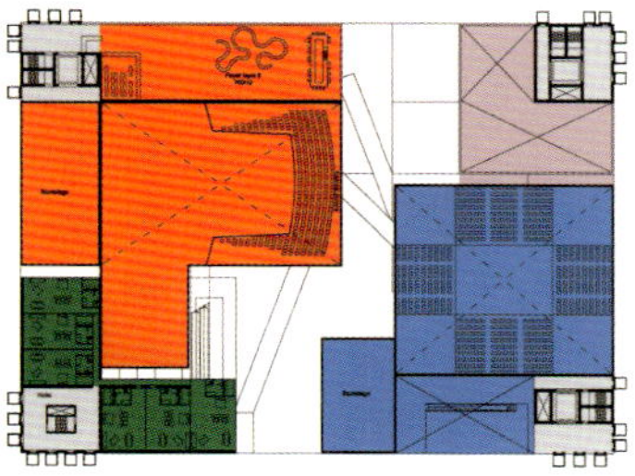

LEVEL 12
十三层

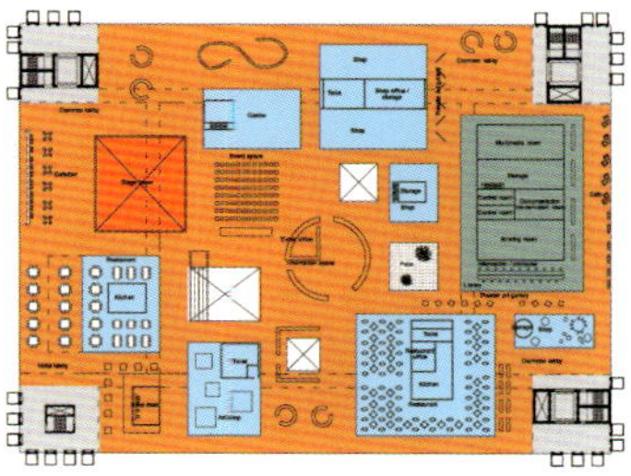

LEVEL 13
十四层

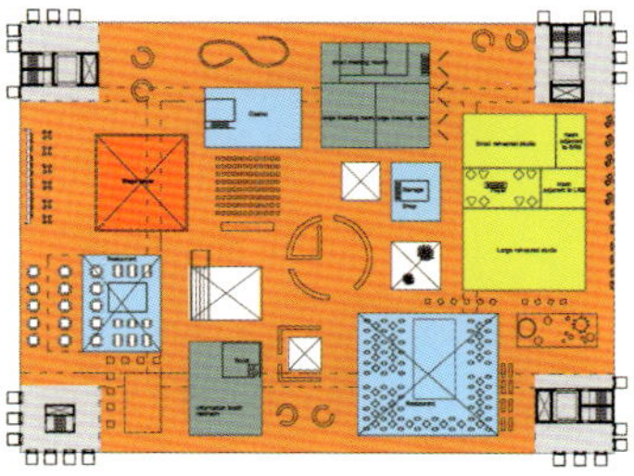

LEVEL 14
十五层

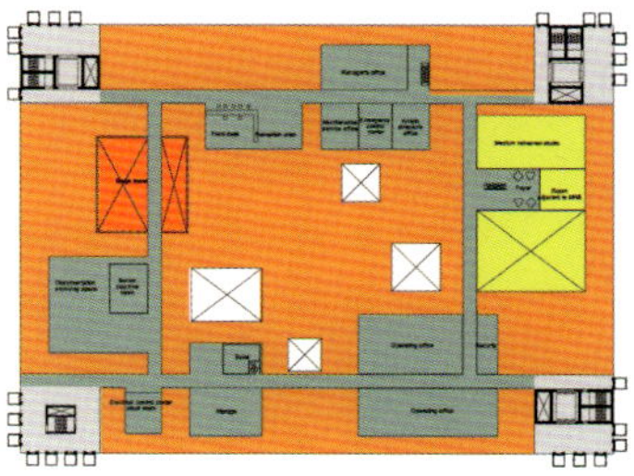

LEVEL 15
十六层

Deichmanske图书馆
THE DEICHMANSKE LIBRARY

建筑师：schmidt hammer lassen architects
甲方：Oslo Kommune and Hav Eiendomme
工程师：WSGreen Technologies
景观建筑师：schmidt hammer lassen architects

Deichmanske图书馆是一座方形的、具有雕塑特点的建筑。它的基座富于表现力，其上的楼层具有沉静的气质。新的图书馆比邻挪威首都奥斯陆水畔新建的歌剧院，建筑想借助优越的地理位置成为主要的集会场所。

规划强调建筑要有自己的特色，而不去与比邻的歌剧院争奇斗艳。因此，设计援用了垂直和水平的线条，不去打破歌剧院某些部分所具有的对角线特质。为了将市民的目光吸引到这座与歌剧院比肩的特色图书馆中来，设计师将建筑的西南角设置向内凹陷的三角形，形成了朝向水边的有遮挡户外空间。这一向内切割的手法创造了23m高的户外平台，平台下为一处类似圆形剧场的下沉空间。该场所将成为重要的集会场所，在此避风的游客将会欣赏到阳光、水景和歌剧院的景色。这使得建筑成为广场上举行的多种活动的背景。

图书馆的立面由透明玻璃和丝网印刷玻璃构成，使得图书馆内的使用者具有360° 的全景视野。游人也可以看到室内举行的各种活动。

通过对可持续性的整体研究，Deichmanske图书馆被设计成世界上第一座被动能源图书馆。它是在被动住宅基础上的进一步开发。图书馆的所有能耗都以可持续资源为基础。建筑将通过玻璃立面中的光电电池生成额外的能源，并通过附近的阿克尔河进行水力发电。项目的总体目标是建成一座舒适度高、运行费用低的碳中性建筑。

设计师在建筑的感性方面同样投入的大量的精力。图书馆开放、民主，具有可及性，充分考虑了人类福祉和社会责任。建筑的关键词就是“透明”和“开放”，因为它是作为为民众服务的建筑。此外，宽敞明亮的房间提供了欣赏周围美景的绝佳视野。

The Deichmanske Library is conceived as a cubic and sculptural building with an expressive base and calmer upper storeys. The new main library seeks to become a central meeting place given its privileged location as the neighbour of the much acclaimed new Opera House on the waterfront of Oslo, the capital of Norway.

The proposal places great emphasis on the fact that the building has its own distinctive identity without competing with the Opera House next door. Consequently, vertical and horizontal lines have been chosen so as not to disturb the diagonal lines that characterise parts of the opera house. To preserve the direct sightline from the city to the characteristic "shoulder" of the Opera House, a triangle has been incised into the south-west corner of the library, creating a covered outdoor space facing the water. This cut in the building creates an outdoor terrace up to 23 metres high, with an amphitheatre-like concave space below it. This space will become a pivotal meeting place, where visitors – sheltered from the wind – can enjoy the sun and the view of the water and the Opera House, making it a pleasant backdrop of the many activities taking place in the square.

The facade of the library will consist of see-through, silk-screen printed glass, which will offer openness and a 360 degree view from within the library, while the activities inside the library will be visible from the outside.

Deichmanske Library is designed to be the first energy positive library building in the world and relies on a holistic approach to sustainability. It is a further development of the passive house. Its entire consumption of energy will be based on sustainable sources. The Library will also produce an excess of energy by photovoltaic cells in the glass facade and hydroelectric power generated from the Akers River nearby. The overall goal is to develop a CO_2 neutral building with high comfort levels for the users and low operational costs.

Much effort has also been put into the sensuousness of the building. The emphasis is on human well-being and social responsibility, and the library is open, accessible and democratic. The main parameters of the library are transparency and openness, because it is a building meant for people. In addition, the splendid rooms offer a fantastic view of the surrounding environment.

丹麦城市媒体空间
DENMARK URBAN MEDIA SPACE

建筑师：schmidt hammer lassen architects
工程师：Alectia Consulting Engineers
甲方：The Municipality of Aarhus and Realdania

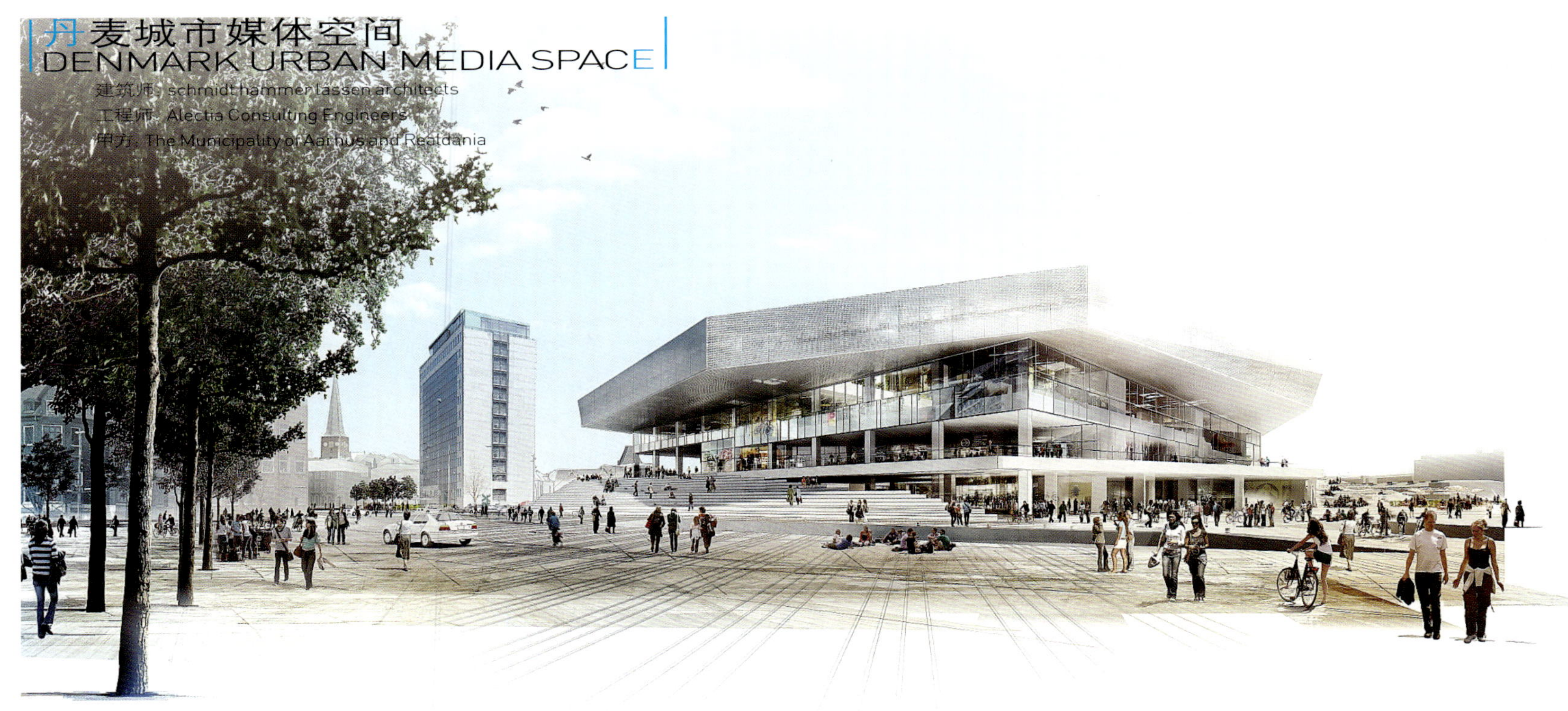

城市媒体中心将成为斯堪地维亚最大的公共图书馆，它代表着新型城市综合图书馆的诞生。该建筑同时具有多重功能。建筑位于奥尔胡斯河的河口处，建筑场地是奥尔胡斯城市中心最突出的场地之一。

城市媒体空间是城市区域规划宏图的组成部分。建筑师将建筑场地和城市的历史中心在视觉和地理上进行连接，使原先港口的货物码头重新焕发了生机。

设计的主导思想是一片有遮挡的城市空间。一个大型的七边形屋顶体量“盘旋”在玻璃棱柱体量上，后者位于一块冰花状的台阶上。台阶朝海边成扇形散开。“冰花”台阶形成了宽敞的台地，可以举行各种娱乐活动和户外活动。

七边形体量将包括媒体屋行政区和租赁用写字间。其下的建筑空间由于玻璃的应用具有了透明的特质，使得路人可以看到室内举行的各种活动，室内的用户可以拥有360°的观景视角。图书馆各层之间呈交错状态并分为多个区域，包括文献和媒体区、展览区、儿童剧院、互动活动区、公共活动区、咖啡馆和餐馆，这些功能空间构成了一条贯穿整个建筑的重要散步走廊。建筑的地下停车场将对公众开放。滨水区的部分交通将分流，经由建筑地下通过。为了促进公共交通的发达，新的电车将在建筑所在地设置站点。

Urban Media Space will be Scandinavia's largest public library and represents a new generation of modern hybrid libraries and thus the building contains multiple potentials. The building is situated at the mouth of the Aarhus River in one of the most prominent sites of the city centre of Aarhus.

Urban Media Space is part of the ambitious district plan and revitalises the former industrial cargo docks on the harbour front by connecting the area both visually and physically to the historic centre of the city.

The leading idea is a covered urban space. A large heptagonal slice hovers above a glazed prism, which is resting on a square of ice flake-shaped stairs fanning out to the edge of the sea. The ice flakes create wide plateaus and accommodate recreational activities and outdoor events.

The heptagon will contain the media house administration and offices for rent. The glass building below is transparent and allows passers-by visual access to the activities in the building while the users have a 360 degree panoramic view from the inside. The library contains several divisions in staggered levels that cover literature and media areas, exhibitions, children's theatre, interactive activities, public events, cafés and restaurants and hence, they form an eventful promenade through the building.Below ground the large parking area will be available to the whole city. Part of the traffic along the waterfront will run beneath the building. To boost public transportation the new tram will have a stop here.

牛津大学化学系建筑
OXFORD UNVERISITY CHEMISTRY DEPARTMENT

建筑师：schmidt hammer lassen architects

甲方：The University of Oxford Chemistry Research Lab 2

位于牛津科学区的第二化学实验室将会成为化学系新的展示先锋。结合了先进的实验室和教学区的第二化学实验室将会吸引全世界的科学家、学生和投资者。规划强调这次扩建的规模 并为第二化学实验室与其他学术和艺术团体的合作寻求支持。

建筑的形式简单而有冲击力：是对城市环境的明确呼应，并符合充分利用场地容纳能力的要求。建筑沿着长度方向在中心处被分隔开来，形成了一个贯穿整个高度的中庭，这样就可以欣赏到建筑各个部分的美景。建筑地上四层、地下两层，功能区清晰地分为研究实验室和办公室。

考虑到实验室和基本的设备核心筒的设置，建筑室内的结构设置非常简单。两个线性的主建筑体量每个都分成四个小的部分，这种结构在室外和室内设计中都起到了重要的作用。该建筑与校园区域中的城市环境形成了良好的呼应。

Situated in Oxford's Science Area, CRL 2 will be the new exposition front for the Department of Chemistry. Containing state-of-the-art labs and teaching spaces, CRL 2 will attract scientists, students and investors from all over the world. The proposal emphasises this ambitious outreach and seeks to support the synergy between CRL 2 and a wide professional and academic community.

The form of the building is simple and strong; an explicit response to its city context and the requirement to maximise the capacity of the site. The building is split in the centre along its longitudinal direction, creating an impressive atrium – rising throughout the building – so that you from the entrance areas have a magnificent view of the many components of the building. With four floors above ground and two below, the functions of the building are distinctly separated between research labs and offices.

A simple structure characterises the interior arrangement in consideration for the placement of the labs and essential service cores. The volume of each of the two linear main buildings is divided into four smaller sections and this structure creates a significant signature in the exterior as well as the interior design. The building corresponds with the surrounding urban spaces in the campus area.

南丹麦大学的工学院
THE TECHNICAL FACULTY OF UNIVERSITY OF SOUTHERN DENMARK

建筑师：C. F. Møller Architects
景观建筑师：Schønherr Landskab A/S
工程师：Moe & Brødsgaard A/S
规模：21.000 m²
图片：C. F. Møller Architects
地点：SDU, Odense, Denmark
甲方：University and Building Agency

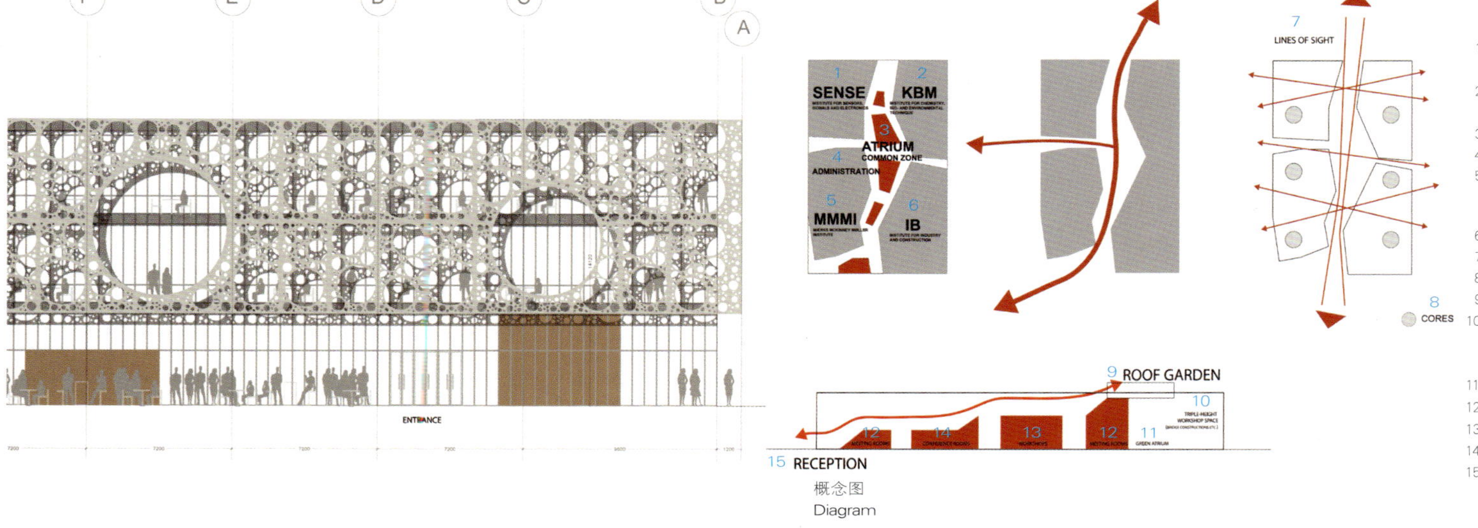

概念图
Diagram

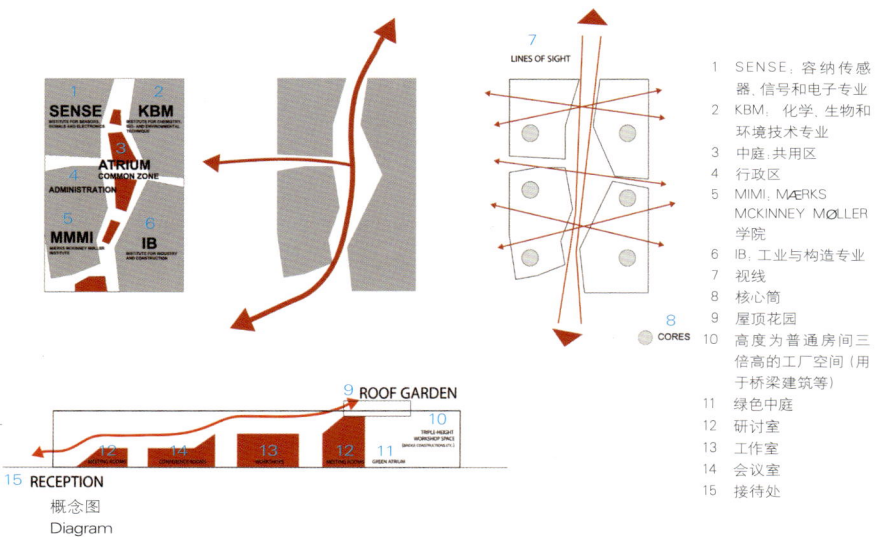

1　SENSE：容纳传感器，信号和电子专业
2　KBM：化学，生物和环境技术专业
3　中庭：共用区
4　行政区
5　MIMI，MÆRKS MCKINNEY MØLLER学院
6　IB：工业与构造专业
7　视线
8　核心筒
9　屋顶花园
10　高度为普通房间三倍高的工厂空间（用于桥梁建筑等）
11　绿色中庭
12　研讨室
13　工作室
14　会议室
15　接待处

工学院是位于欧登赛的南丹麦大学的一部分，形成了由四个不同机构共用的研究与教育环境。项目设计成由5个建筑构成的一个大的外围护结构，彼此之间用穿过房间中心的层高不同的桥梁、作为公共设施与会议室的“家具”连接，提供了通向屋顶花园/咖啡馆/休闲区的通道。多种连接使得阳台的设计更加流畅，公共的交流与知识的分享更加畅通。建筑被设计成一个玻璃房，外立面的材料因采光需求和朝向求而采用不同的材料。东西立面采用高雅的纤维增强混凝土屏，其下带有遮阳屏，并可以自然通风。南立面全部为玻璃，带有太阳能电池和与电池图案相同的遮阳构件。通过实体核心筒和推拉墙系统的组合，室内的布局灵活多变，可根据组群的规模进行适应性的二次调整。南丹麦大学工学院的节能等级符合BR95（丹麦建筑编码）的一级要求。这意味着建筑以最低的能耗达到良好的室内气候条件，并从生命周期的角度出发选用对环境影响最小的材料。

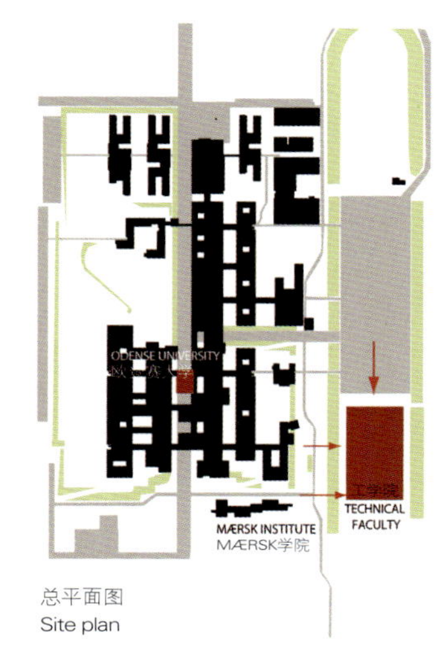

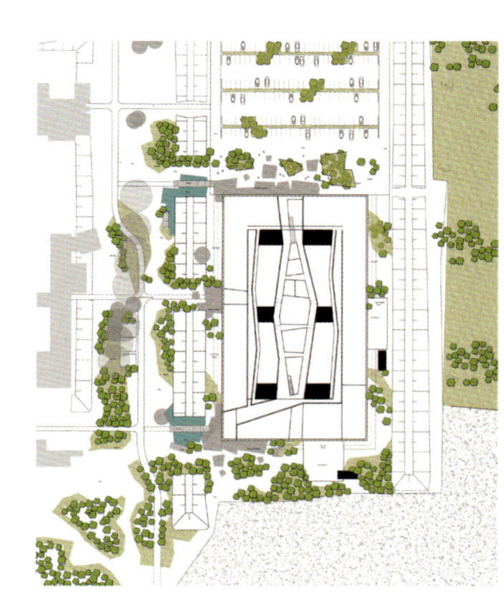

总平面图
Site plan

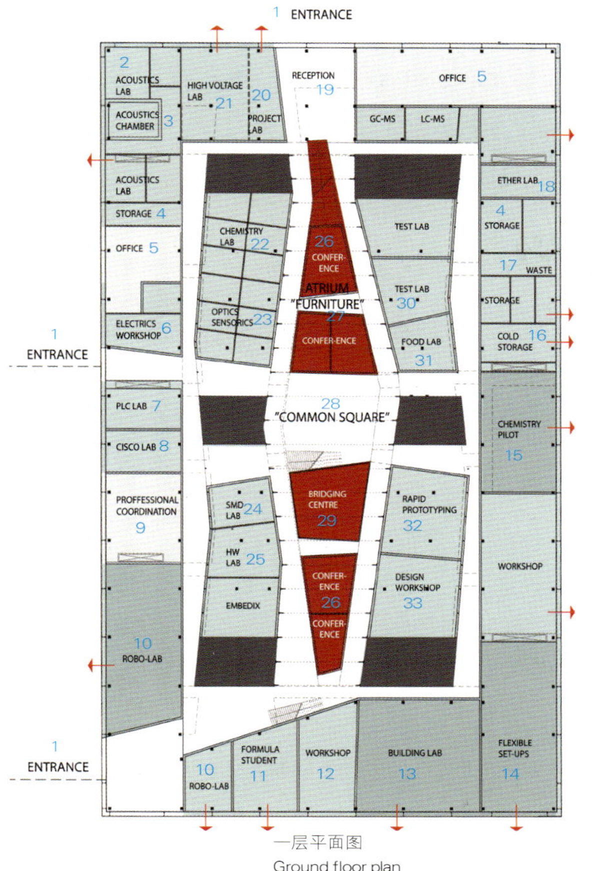

一层平面图

Ground floor plan

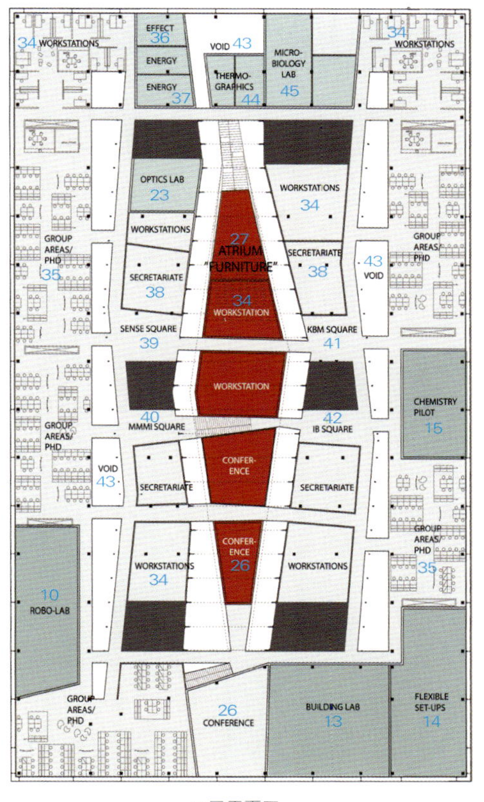

二层平面图

First floor plan

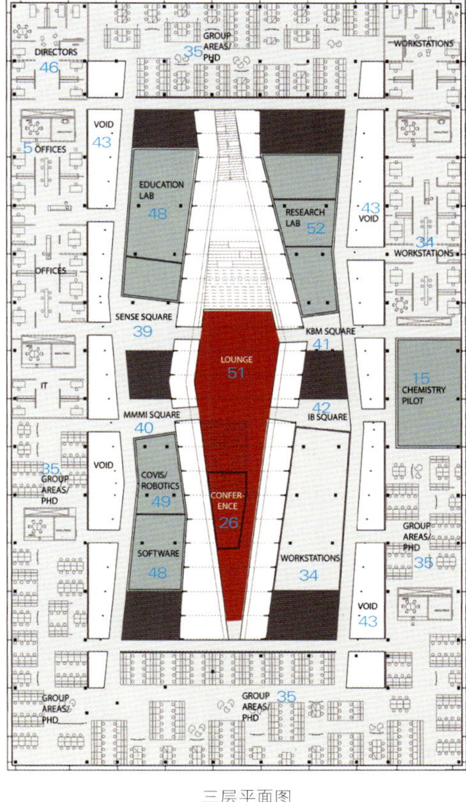

三层平面图

Second floor plan

1 入口
2 声学实验室
3 声学房间
4 储藏室
5 办公室
6 电气车间
7 PLC实验室
8 CISCO实验室
9 职业协调
10 机器人实验室
11 学生方程式
12 工作室
13 建筑实验室
14 灵活的组织结构
15 化学实验室
16 冷库
17 垃圾回收处
18 乙醚实验室
19 接待处
20 项目实验室
21 高压实验室
22 化学实验室
23 光学感应器
24 SMD实验室
25 HW实验室
26 会议室
27 中庭"家具"
28 公共广场
29 桥梁中心
30 测试实验室
31 食品实验室
32 快速原型
33 设计工作室
34 工作站
35 群组区域/博士
36 效果
37 能量
38 秘书处
39 SENSE广场
40 MMMI广场
41 KBM广场
42 IB广场
43 上空空间
44 温度记录
45 微生物学实验室
46 董事
47 信息技术
48 教育实验室
49 COVIS/机器人学
50 软件
51 休息室
52 研究实验室

剖面图
Sections

1 群组区域
2 实验室
3 天窗
4 学习室
5 工作室
6 休息室
7 会议室
8 圆形天窗
9 管道
10 北入口
11 接待处
12 休息室/咖啡厅
13 公共广场
14 会议空间/屋顶花园
15 中庭
16 南入口

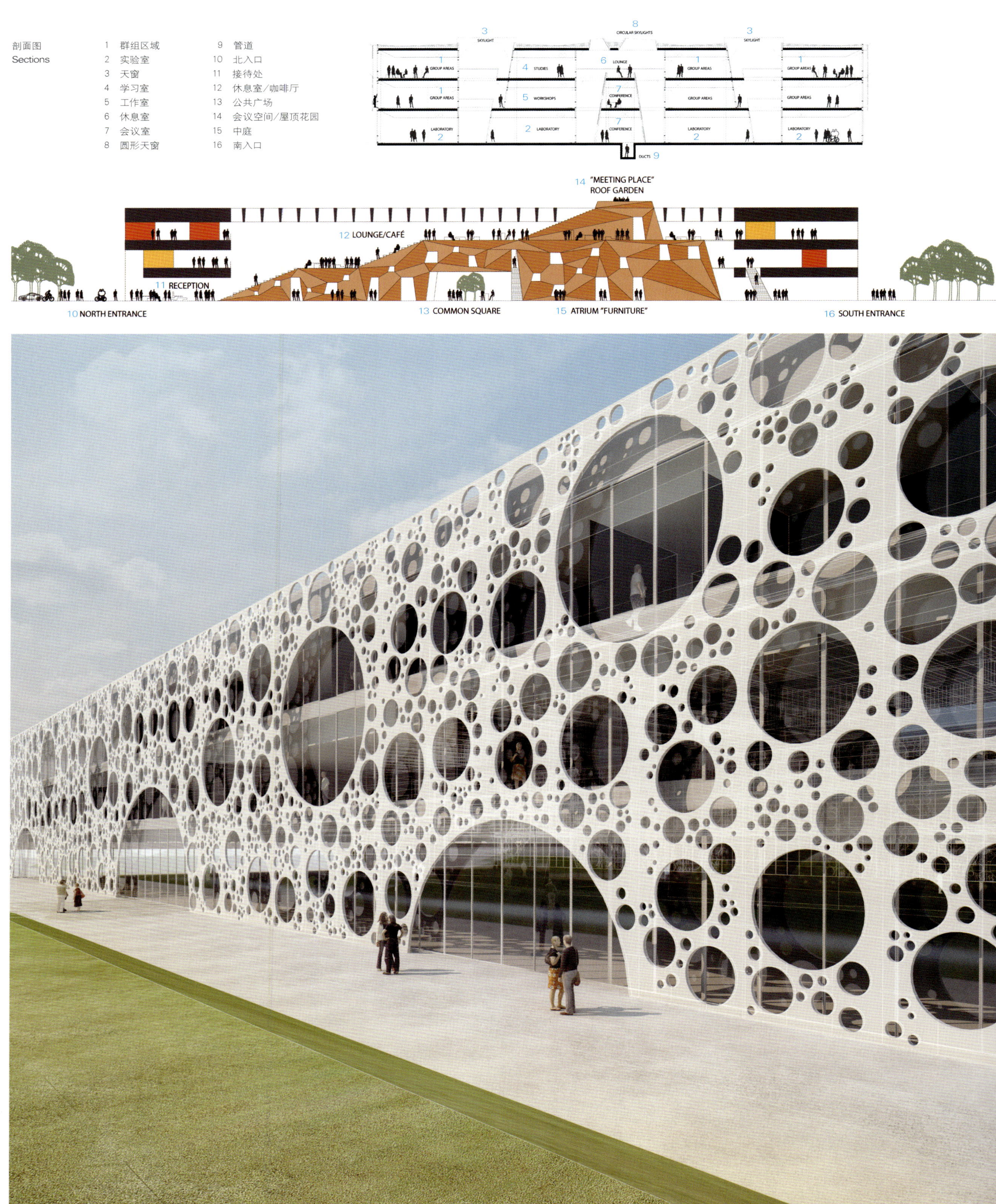

The Technical Faculty is part of the University of Southern Denmark in Odense, and constitutes a shared research and education environment for four different institutes. The building is designed as one big envelope consisting of 5 houses connected by bridges at multiple levels crossing the heart of the house, a "piece of furniture" containing common functions and meeting-rooms, and giving access to a roof garden/café/lounge area. The many connections allow for more fluid boundaries, and more community and knowledge sharing. The building is designed as a glass house with an external screen of varying materials, depending on exposure and orientation. An elegant screen of fibre-reinforced concrete is designed to the west and east, with an underlying solar screen and natural ventilation, whereas the south facade is fully glazed with solar cells and solar shading in a similar pattern.The interior layout creates great flexibility, by a combination of solid cores and sliding wall system for adaptable sub-divisions depending on group sizes. The Technical Faculty at the SDU is to meet the requirements for low energy class 1 according to BR95 (Danish building codes). This means minimal energy consumption, good indoor climate and use of materials with a low environmental impact in a life cycle perspective.

根特图书馆和多媒体中心
LIBRARY AND MULTIMEDIA CENTRE IN GHENT

建筑师：schmidt hammer lassen architects
甲方：CVBA Waalse Krook
工程师：Technum
当地合作建筑师：ELD partnership
其他合作者：Venac (acoustics)

尼尔 · 玻尔科学公园
NIEL BOHR SCIENCE PARK

建筑师：schmidt hammer lassen architects
甲方：The Danish University and Property Agency in cooperation with Copenhagen University
景观建筑师：Nørgaard & Holscher
工程师：Moe & Brødsgaard
顾问：Andy Horsewell
建筑师：DR CUH2A, Jørn Langvad Arkitekter

位于哥本哈根中心位置的尼尔 · 波尔科学公园是城市开发计划的一部分。开发以推进哥本哈根大学的科学和研究氛围。科学公园为人们提供了寻求答案、交换知识和分享经验的场所，新知识也因此得以发展。鉴于此，科学公园将对市民、学生、研究人员和商人开放，并为他们提供可遮风挡雨和私密安全的环境。科学公园将位于Nørre校园中。该校园中已经包括了其他的科学研究组织和国立大学医院。建筑建于穿过Nørre校园的Jagtvej街两侧的两处平台上。

建筑的分割激发了公共通道"兰布拉"的创造，该公共通道从Jagtvej的底部通过，连接起两处场地。两处建筑场地是城市中的焦点，而兰布拉是位于两处场地出入口空间一个独特和明显的特征，它使得校园与城市相融合，而这正是校园长久以来缺少的东西。

带图案的立面和屋顶勾勒出两座建筑的体量。屋顶和立面与朝向医科学生公寓和兰布拉公共通道的山墙形成鲜明的对与。建筑体量的中部稍稍弯曲，这样一来，它就从直角的平面中解放出来，通过朝向各个方向的通道将自身与周围环境融为一体。

Niels Bohr Science Park in central Copenhagen is part of an urban development aimed at boosting the scientific and research environment of Copenhagen University. Niels Bohr Science Park provides a framework where people can meet in search of answers, exchange of knowledge and sharing of experiences, so that new knowledge can be developed. As a result, Niels Bohr Science Park will be open, visible and accessible to the city, students, researchers and businesses while at the same time providing a sheltered, intimate and secure environment. Niels Bohr Science Park will

be situated at the Nørre Campus already hosting other science faculties and the National University Hospital. It will be built in two stages placed on each side of the Jagtvej street, which runs through Nørre Campus.

This division has sparked the creation of a public pathway – the "rambla"– that crosses under the Jagtvej to connect the two sites. The rambla is a distinct and easily perceptible feature anchoring the arrival and entrance areas of the two sites as visible fixpoints in the city. It also establishes coherence with the surrounding campus and cityscape, which is something the campus has long missed.

The two buildings are well defined volumes with patterned facades and roofs contrasted by significantly open gables facing the medical student residences and the public rambla. By a slight bend midway in their bodies, the buildings are liberated from the right-angled plan geometry, and they open themselves up to their surroundings with access from all sides.

克里斯蒂安桑德的歌剧和文化中心
KRISTIANSUND OPERA AND CULTURE CENTRE

建筑师：C. F. Møller Architects
地点：克里斯蒂安桑德城中心
规模：15400 m²
甲方：Kristiansund Kommune and FG Eiendom

克里斯蒂安桑德的歌剧和文化中心将会上演在挪威可追溯至1805年的最古老的歌剧，并将于每年举办歌剧周——挪威最大也是最全面的歌剧和音乐剧展示周，同时它也是斯堪地维亚半岛最大的歌剧节。

新歌剧院的设计愿景是在特隆赫姆西部努尔莫勒地区的首府克里斯蒂安桑德创造一个歌剧文化中心，使其成为整个区域的地标。除了歌剧中心外，新的文化中心还将容纳芭蕾舞中心、图书馆、学院中心、会议室、一家餐馆和一家咖啡厅。

C. F. Møller建筑师事务所将这个项目命名为Kulturkvartalet。就像它的名字一样，项目将不同的文化功能融入到动感和富有创造性的环境中，将这些功能空间围绕这主音乐厅布置。音乐厅的设计将提供顶级的音响效果和超级灵活的剧院设备。

Kristiansund Opera and Cultural Centre will house Norway's oldest opera, dating back to 1805, which every year arranges the Opera Festival Weeks - Norway's largest and most comprehensive presentation of opera and musical theatre and one of Scandinavia's largest opera festivals.

The vision of the new Opera House is to create a cultural power centre in Kristiansund, the capital of the region of Nordmøre, west of Trondheim – a landmark to the entire region. Apart from the opera the new cultural centre will house a ballet centre, library, college centre, conference rooms, a restaurant, and a café.

Kulturkvartalet, as the project made by C. F. Møller Architects is called, unites the different cultural functions in a dynamic and creative environment, centred round the main concert hall. The concert hall has been designed to offer sublime acoustics and super flexible theatre settings.

诺托登图书与文化馆
NOTODDEN BOOK AND CULTURE HOUSE

建筑师：WE architecture

设计团队：Marc Jay, Julie Schmidt-Nielsen, Nora Fossum, Søren Thiesen, Lena Reeh Rasmussen, Juan Olivara, Zsofia Horvath, Luis Gil, Edward Gentry Becker

甲方：Notodden Municipality

面积：4000 m²

项目地点：Notodden, Norway

诺托登市政厅决定在同一座建筑内设置多个文化功能，如咖啡厅、博物馆和一个图书馆，从而创造一座不同年龄段和不同兴趣爱好的人们可以见面与交流的建筑。建筑屋顶好似一个圆形剧场，围绕着装卸码头盘旋而设。码头本身容纳了一个漂浮舞台。这里就好像一个希腊剧院，周围的景观都变成了透视图。

Notodden Townhall has decided to join several cultural functions such as a cafe, a museum, and a library under one roof; this allows the possibility of creating a building where people of different ages and interests can meet and interact. The roof of the building is laid out as an amphitheatre which spirals around the old shipping dock, and the Dock itself houses a floating stage. Much like in a Greek theatre, the surrounding landscape becomes the scenography.

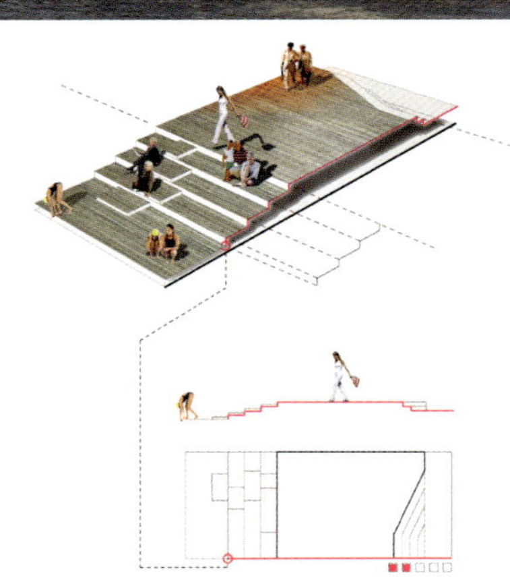

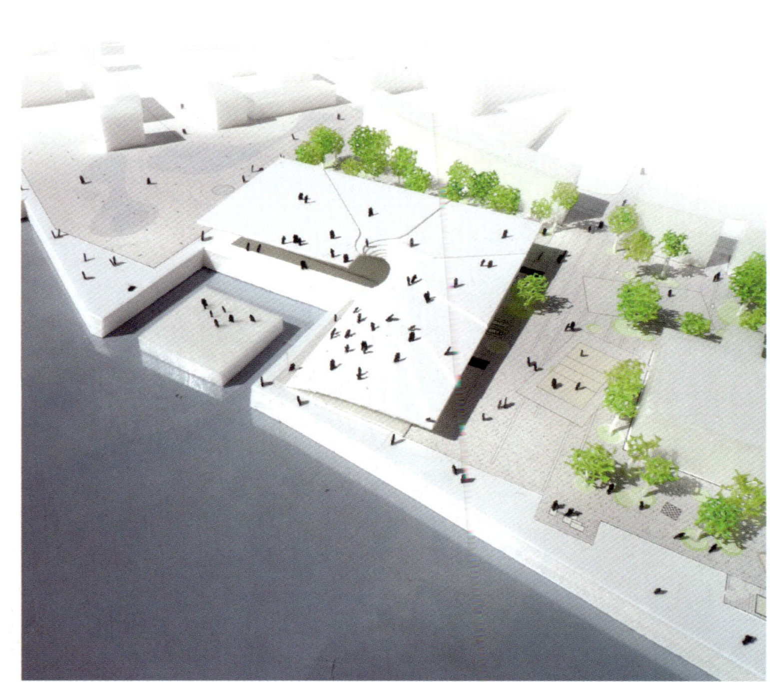

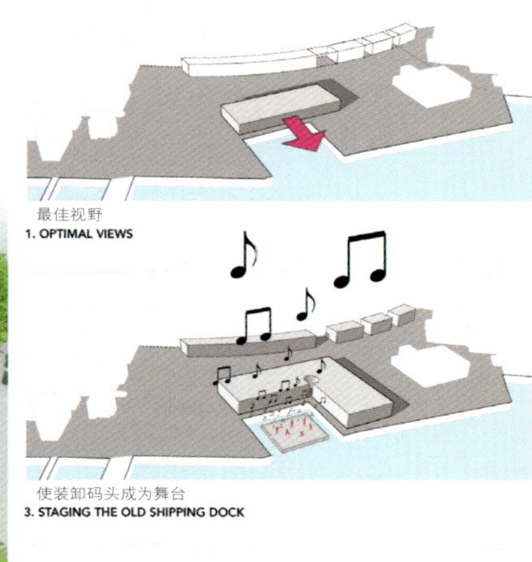

最佳视野
1. OPTIMAL VIEWS

使装卸码头成为舞台
3. STAGING THE OLD SHIPPING DOCK

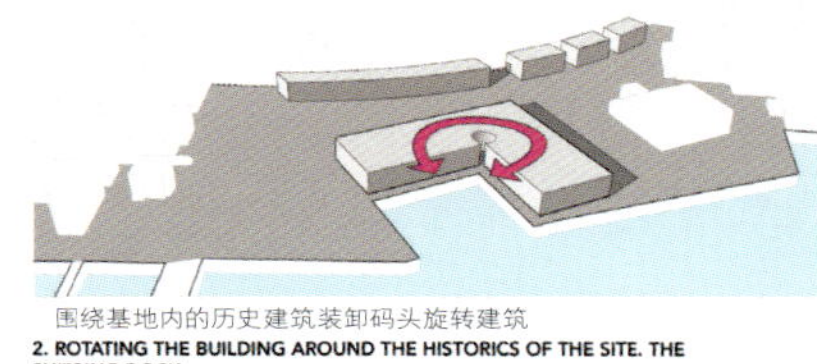

围绕基地内的历史建筑装卸码头旋转建筑
2. ROTATING THE BUILDING AROUND THE HISTORICS OF THE SITE. THE SHIPPING DOCK

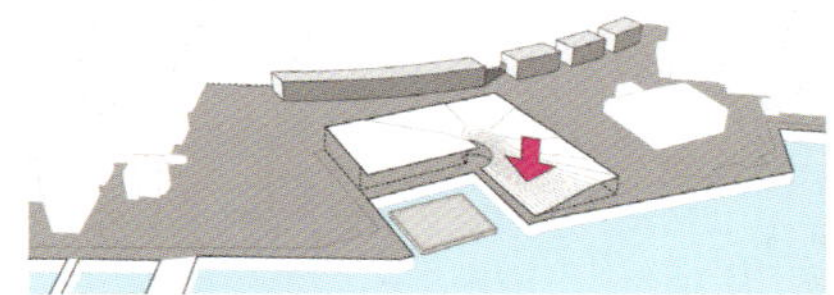

降低屋顶，使图书馆免受吵闹的舞台的影响
4. PUSHING DOWN THE ROOF TO PROTECT THE LIBRARY FROM THE NOISY STAGE

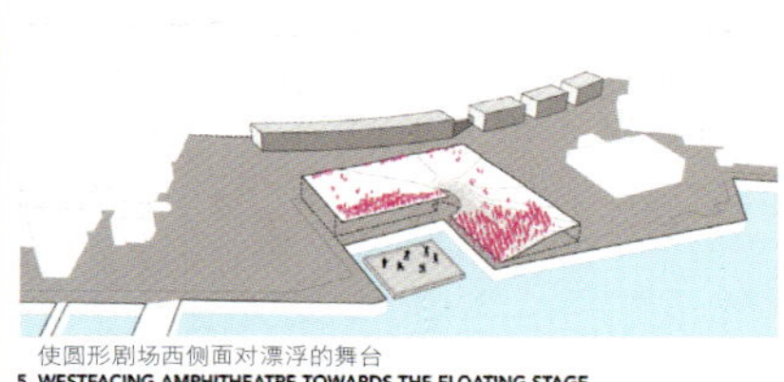

使圆形剧场西侧面对漂浮的舞台
5. WESTFACING AMPHITHEATRE TOWARDS THE FLOATING STAGE

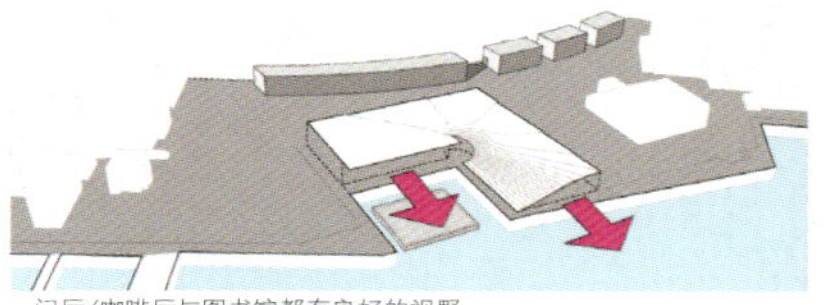

门厅/咖啡厅与图书馆都有良好的视野
6. GOOD VIEWS FROM BOTH FOYER /CAFE AND THE LIBRARY

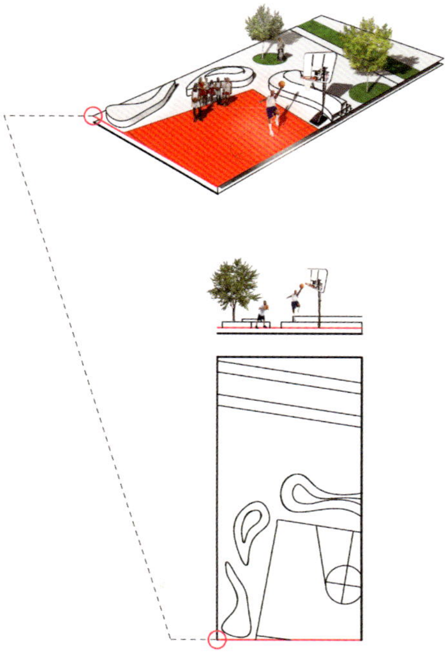

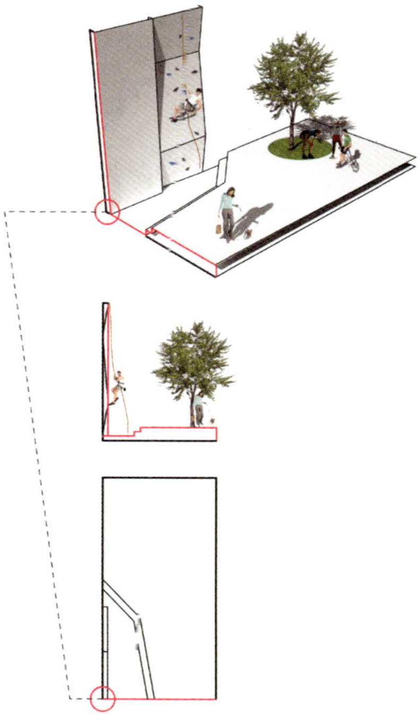

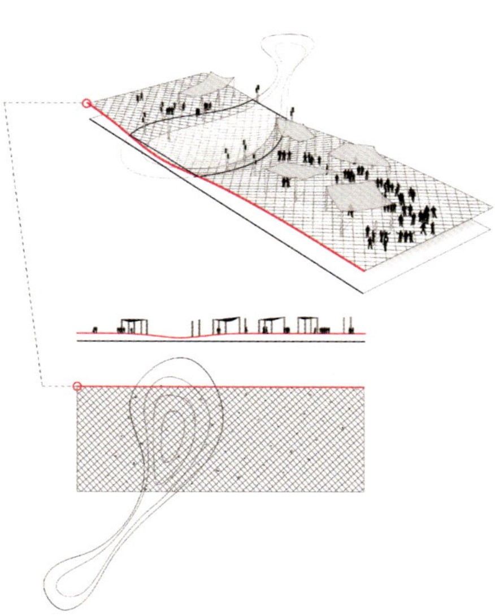

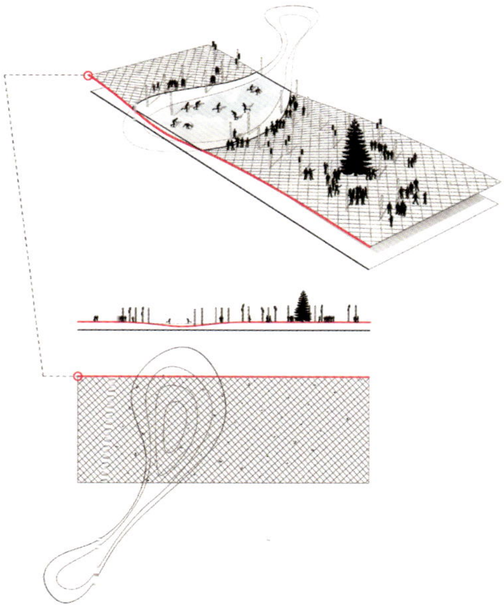

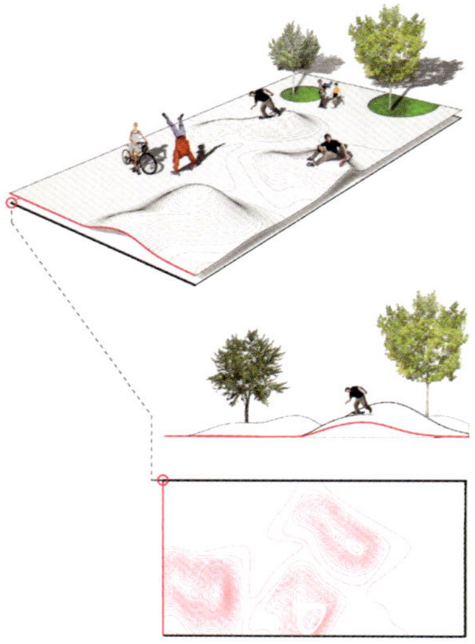

BILLIE HOLIDAY

Mariehøj 文化中心
MARIEHØJ CULTURCENTER

建筑师：WE architecture + Sophus Søbye Arkitekter

设计团队：Marc Jay, Julie Schmidt-Nielsen, Hermanus Neikamp, Lawrence Mahadoo, Lena Reeh Rasmussen, Jenny Selldén, Zsofia Horvath, Nora Fossum, Søren Thiesen, Luis Gill,

合作者：Sophus Søbye Arkitekter, MASU Planning, Øllgaard Consulting Engineer, Spangenberg & Madsen Consulting Engineer, Hausenberg

甲方：Rudersdahl Municipality

面积：8000 m²翻修，800 m²新建

项目地点：Holte, Denmark

获奖情况：竞赛一等奖

未来的Mariehøj文化中心将在周围的景观中勾勒出一幅清晰的美丽画面。新的门厅会呈现出新的面貌，吸引Ruderdah市的所有人来到这里，并成为举办各种活动的场所。

文化中心与绿色的景观融为一体，连接着到达区域，文化广场和Mariehøj魅力的后庭院，面向周围环境开敞，使室内的各种活动充满绿色的气息。

The future "Mariehøj culturcenter" draw a clear profile in the landscape. With a new foyer, the culture center will get a new face that invites all people in Ruderdahl's municipality and a heart that can bring together and highlight the many users and activities in the house.

The culture center merges together with the green landscape: it bridges the gap between the arrival area, the cultural plaza and beautiful backyard of Mariehøj, it opens up towards the surroundings and incorporates the green qualities to the activities in the house.

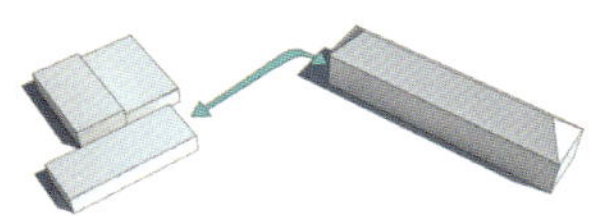

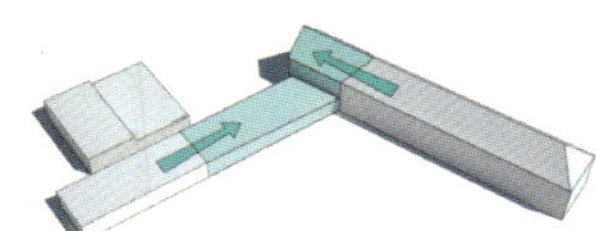

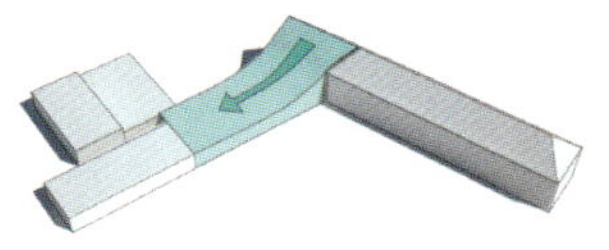

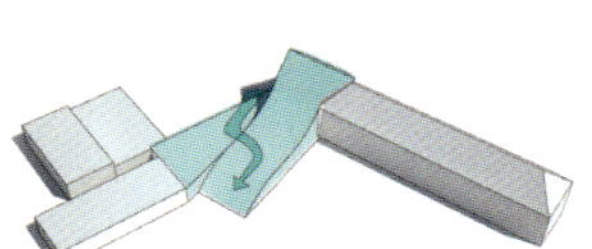

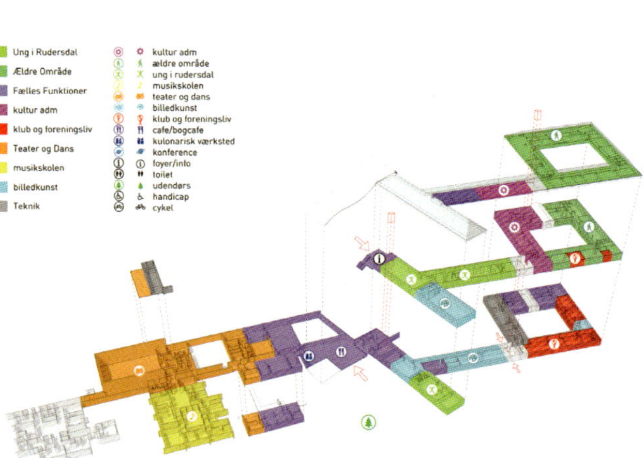

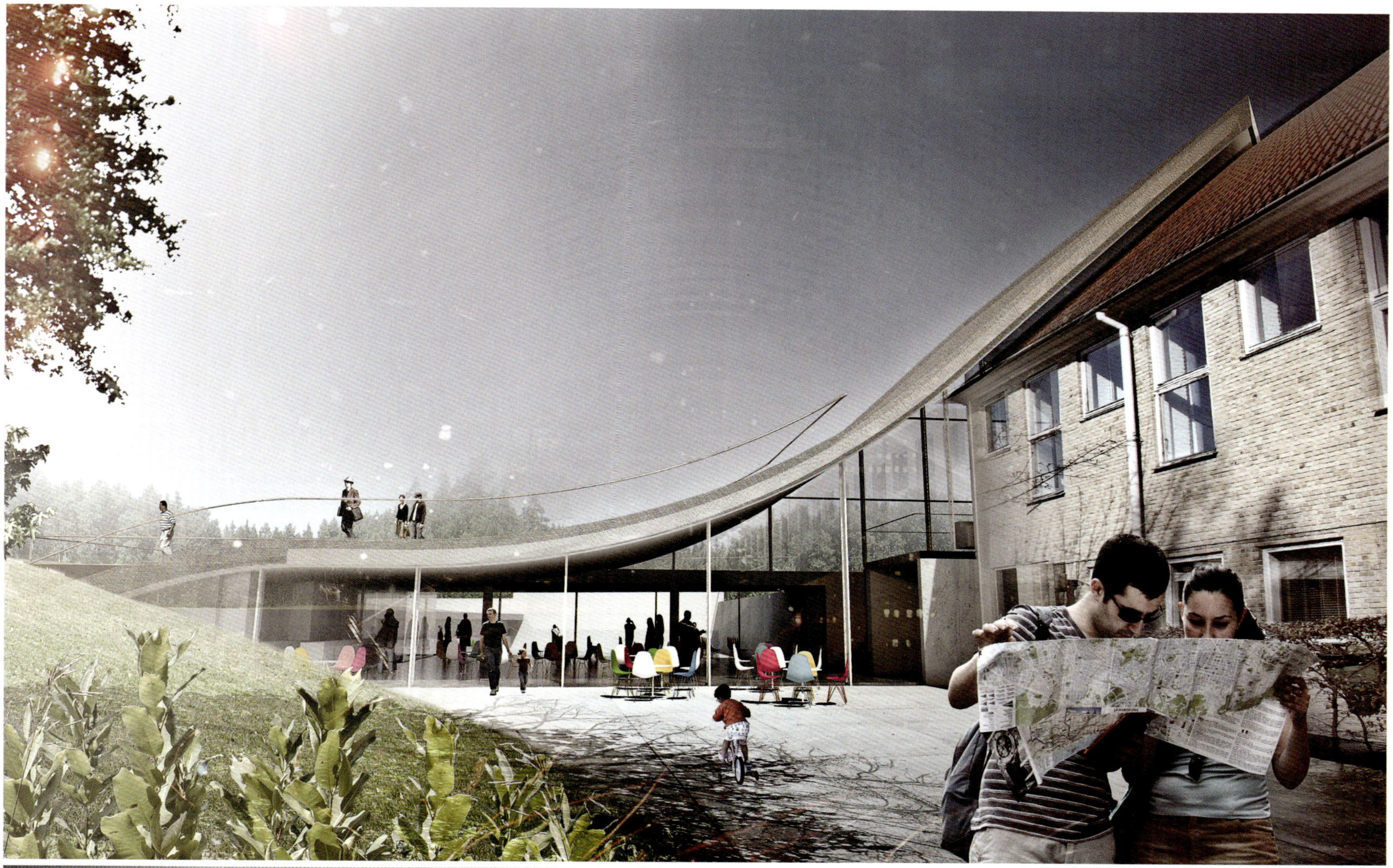

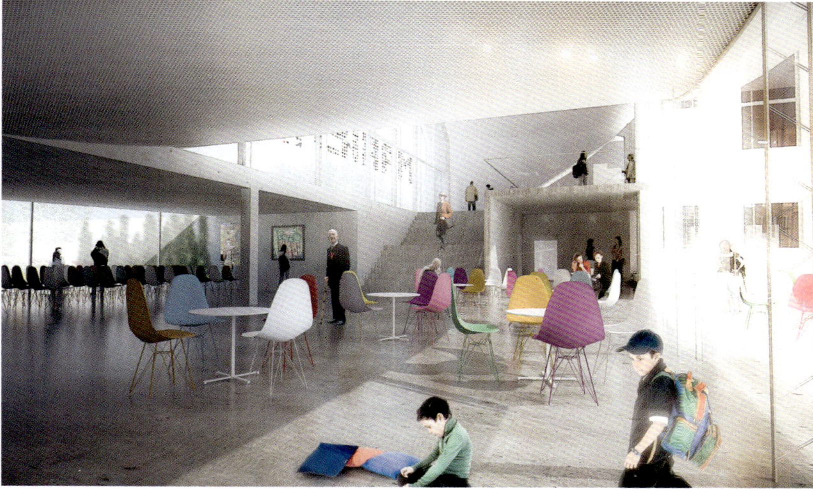

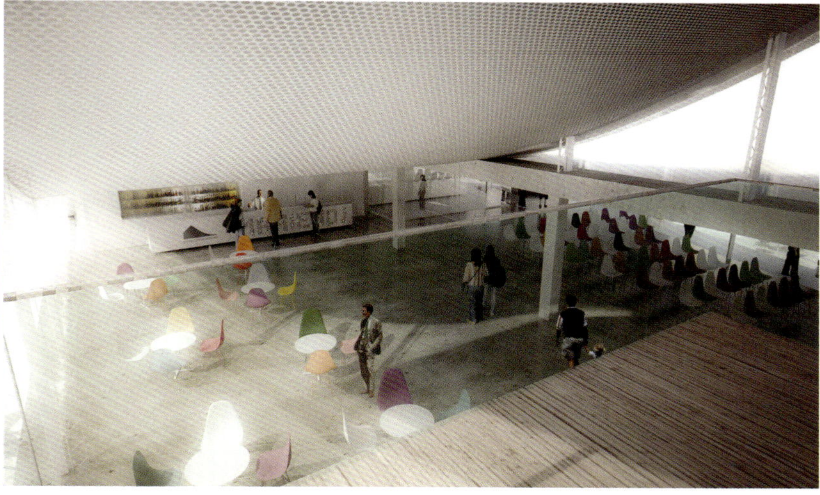

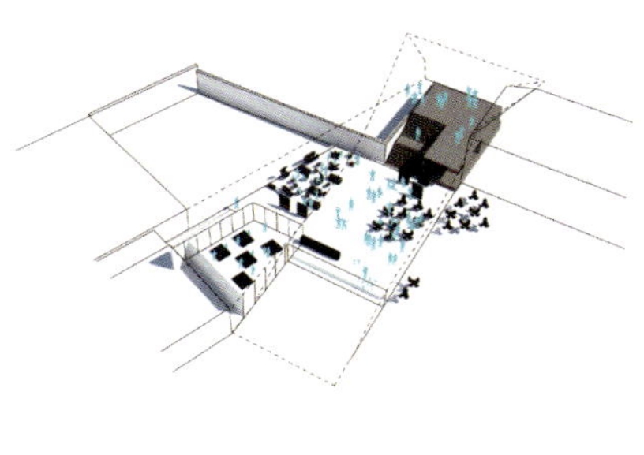

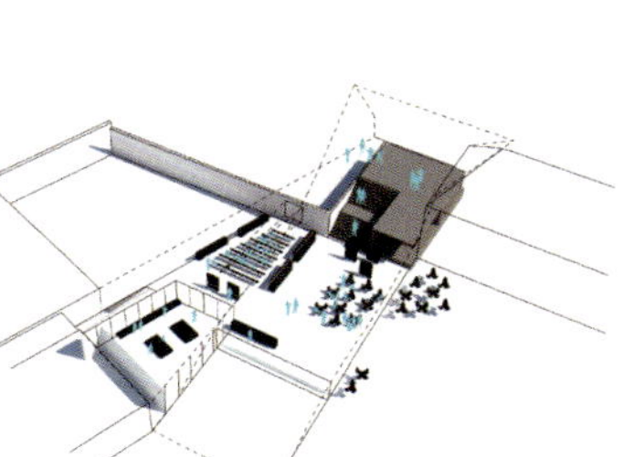

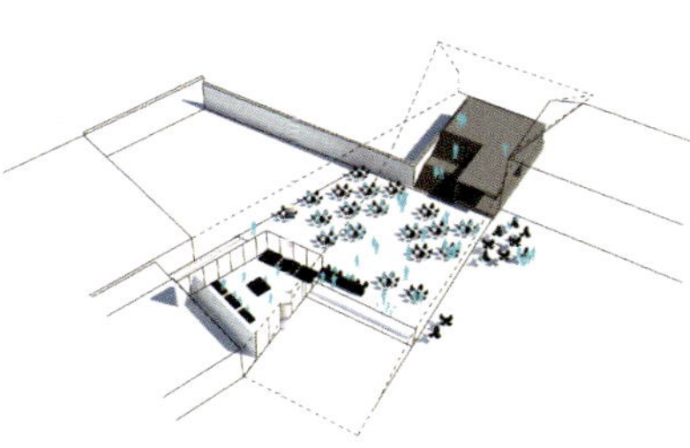

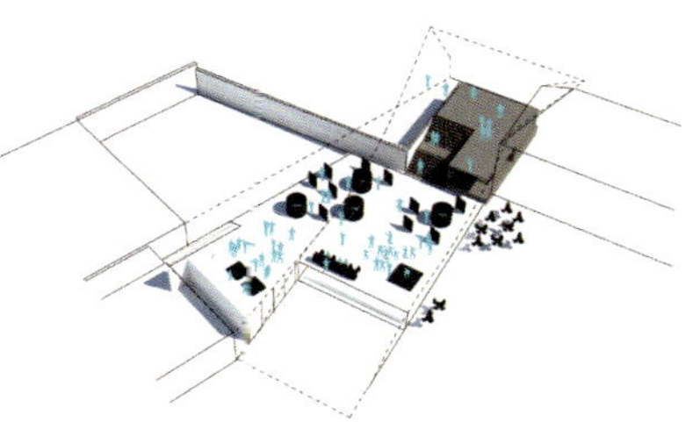

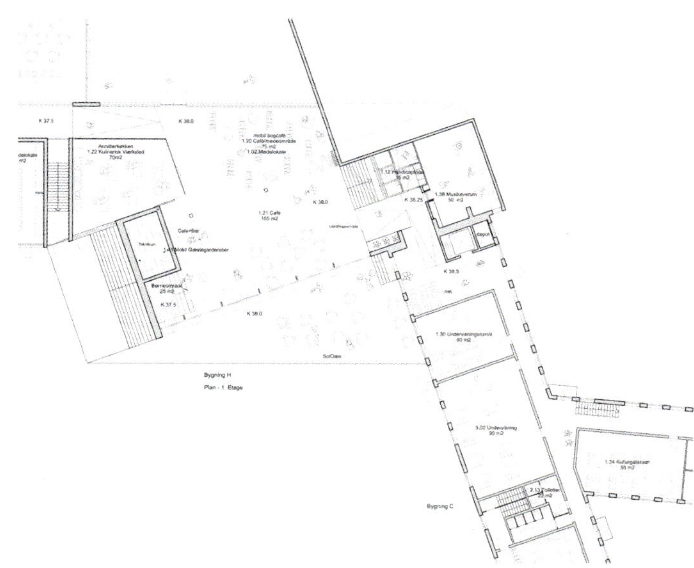

K 37.5
K 38.0
mobil bogcafe
1.20 Cafe/medieområde
75 m2
1.02 Mødelokale
Annettenkøkken
1.22 Kulinarisk Værksted
70m2
1.21 Cafe
155 m2
Cafe+Bar
1.38 Musikøverum
50 m2
K 38.25
depot
K 38.5
Børneområde
25 m2
K 37.5
K 38.0
1.30 Undervisningsrum
90 m2
SolDæk
Bygning H
Plan - 1. Etage
5.02 Undervisning
90 m2
1.24 Kulturgalleriet
58 m2
2.13 Toiletter
22 m2
Bygning C

亚眠大学礼堂与图书馆
AUDITORIUM AND LIBRARY FOR THE UNIVERSITY OF AMIENS

设计团队：SERERO Architectes
项目地点：Amiens (80), France
甲方：Rectorat de l'Académie d'Amiens
楼面面积：1424 m²
城市规划：David Serero, Yoichi Ozawa, Ran She, Amadine Quillent, Fabrice Zaini
工程师：BETOM Ingénierie
获奖情况：竞赛一等奖

新建的图书馆与礼堂坐落在亚眠大学的中心，是师生们交流与研究科学问题的中心场所，也是密集的校园社交生活的催化剂。本案的图书馆和礼堂两座实体建筑与大学中心花园位于同一高度。礼堂的地面随着基地的自然景观而倾斜。在两座建筑之间是一个接待大厅，通过长长的石头台阶将基地较低的楼层与花园连接在一起。

Located in the heart of Amiens University, the new library and auditorium constitute a central place for exchanges and meeting around technologies as well as the catalyst of an intense campus social life.

The project sets the two entities, library and auditorium, at the same level with the central garden of the University. The auditorium floor is sloped and follows the natural landscape of the site. In between these two volumes, the reception hall connects the lower level of the site to the garden by long steps of stone.

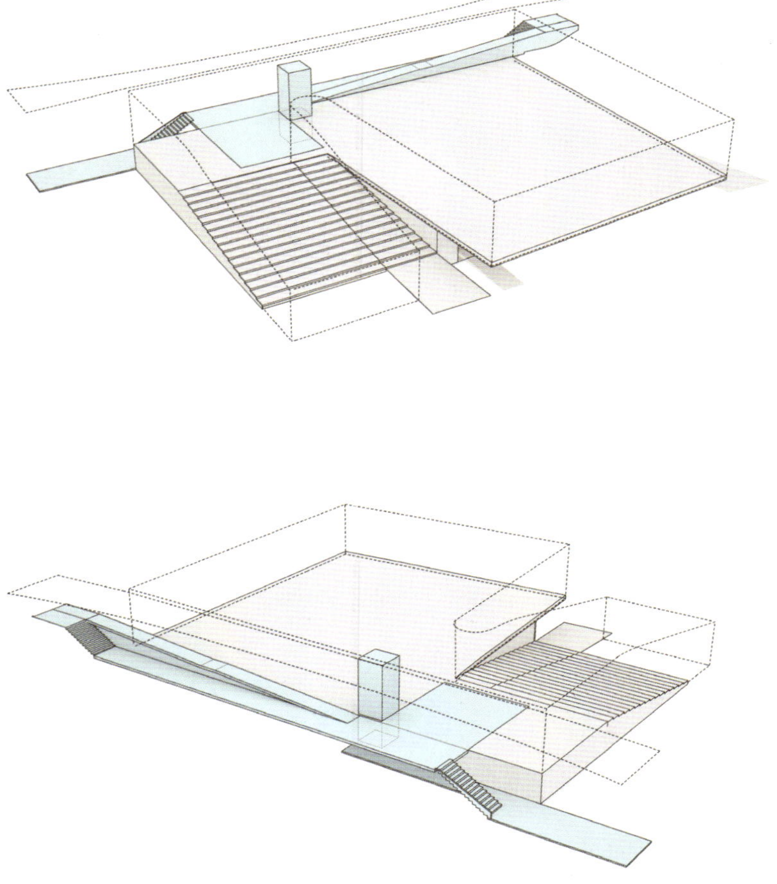

轴测图
Axonometry

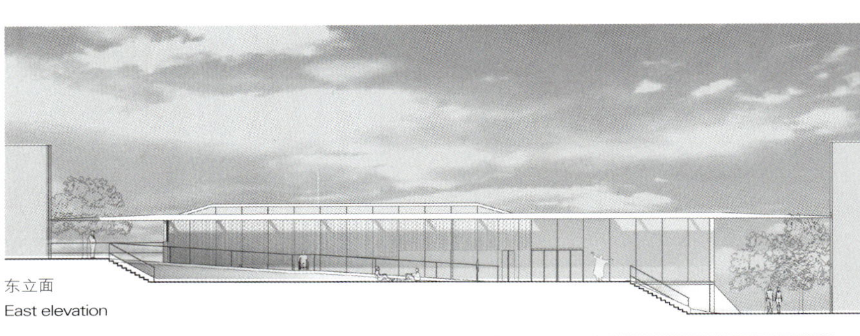

东立面
East elevation

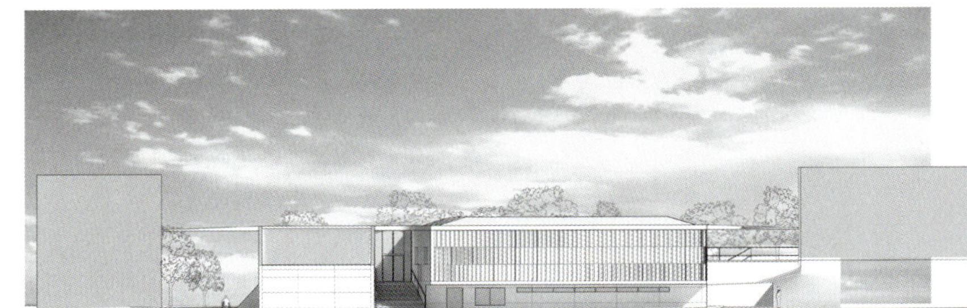

西立面
West elevation

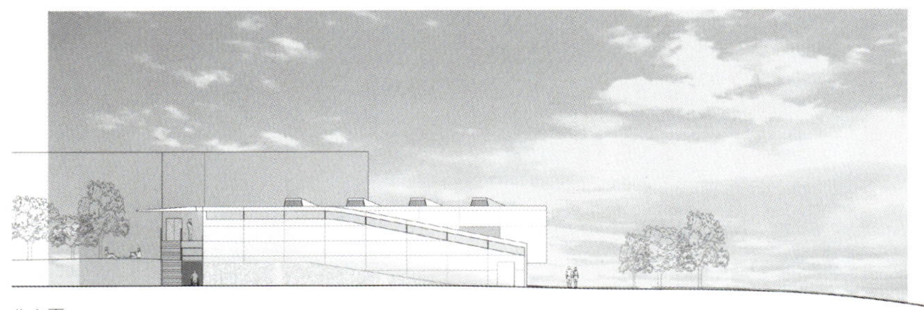

北立面
North elevation

BIBLIOTHEQUE
AMPHITHEATRE

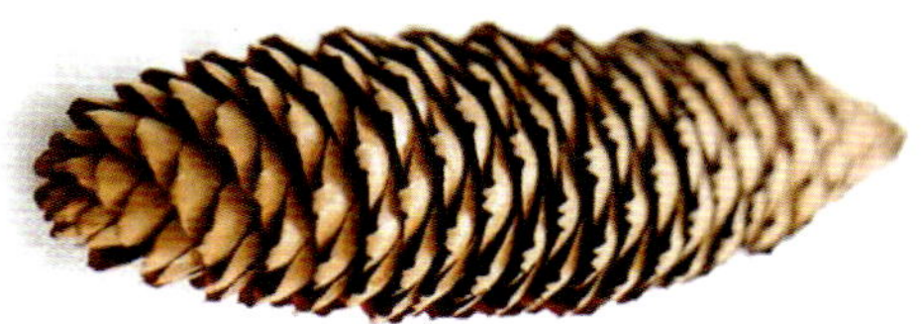

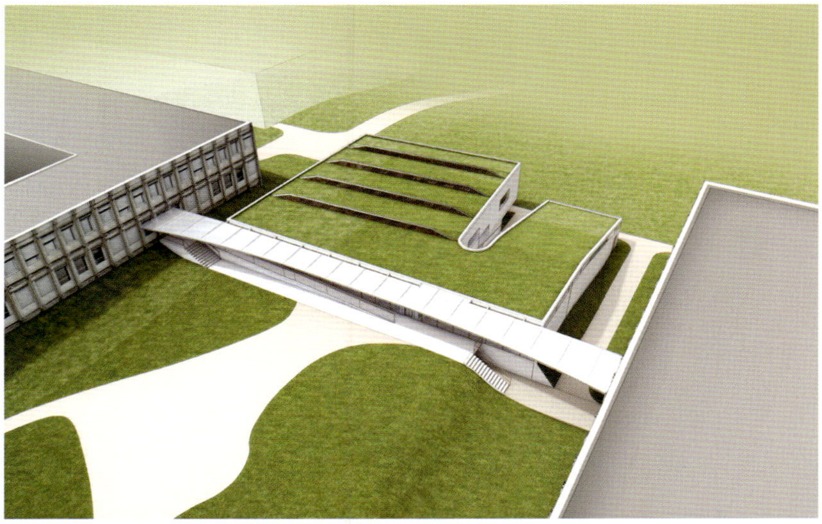

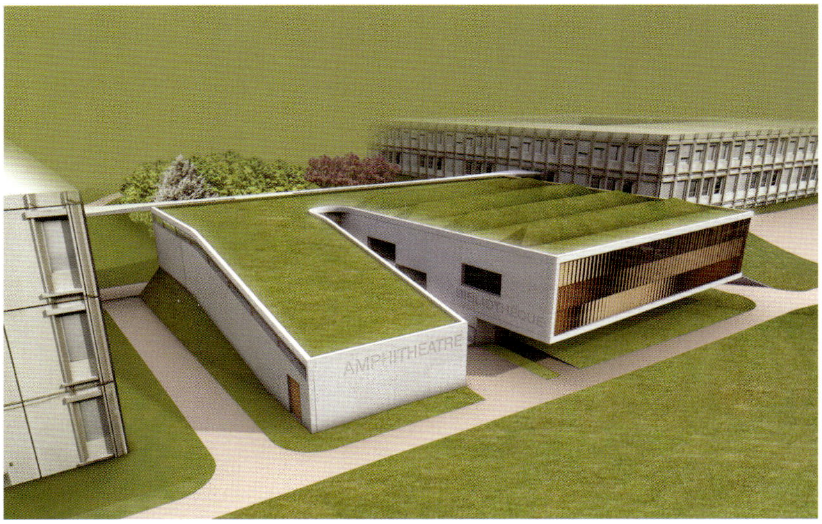

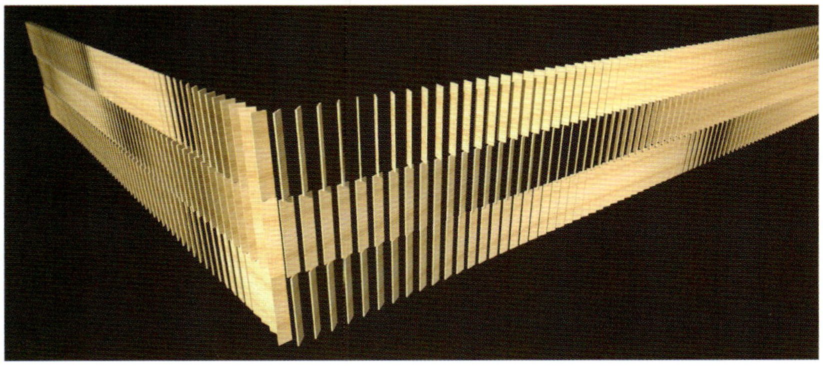

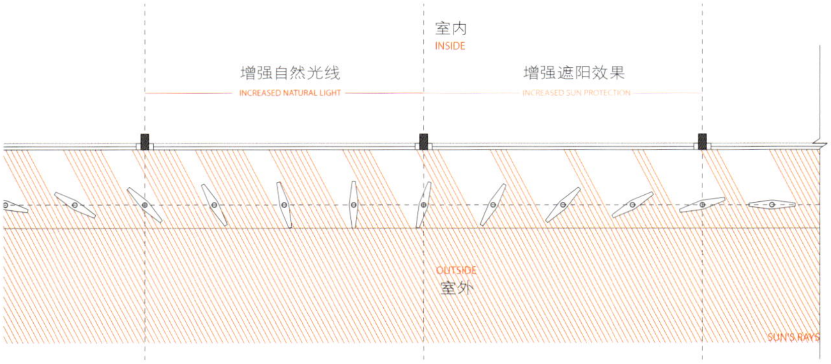

室内
INSIDE
增强自然光线
INCREASED NATURAL LIGHT
增强遮阳效果
INCREASED SUN PROTECTION
OUTSIDE
室外
SUN'S RAYS

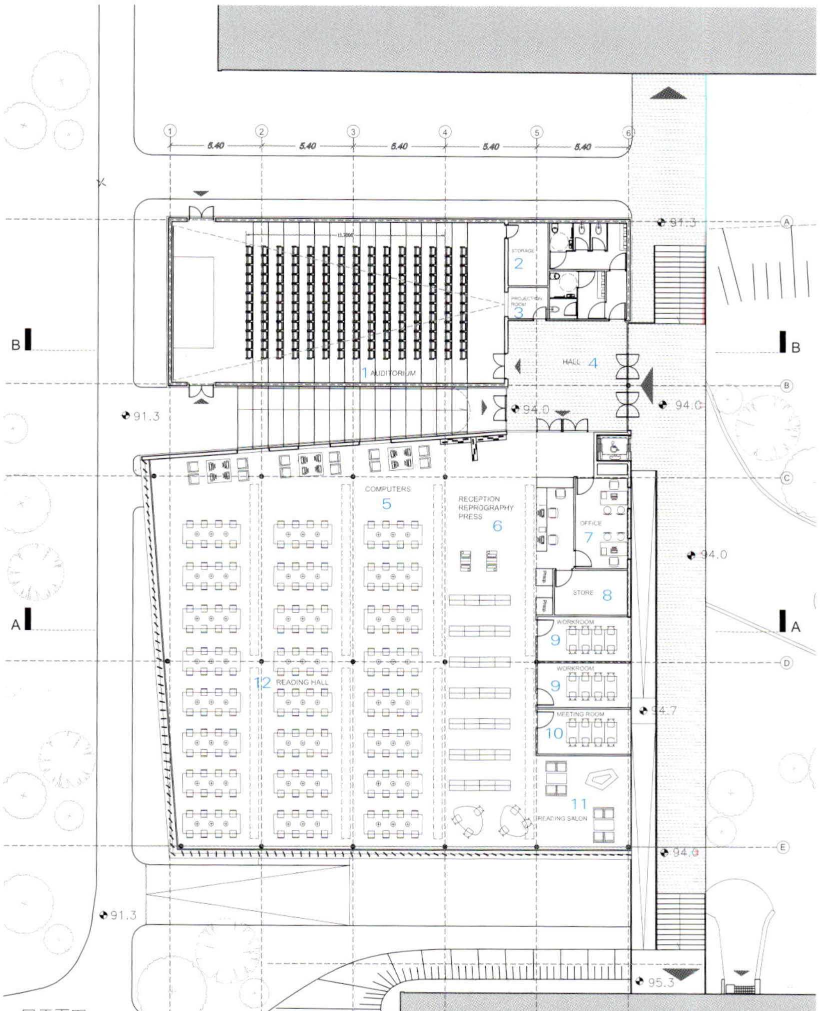

一层平面图
Ground floor plan

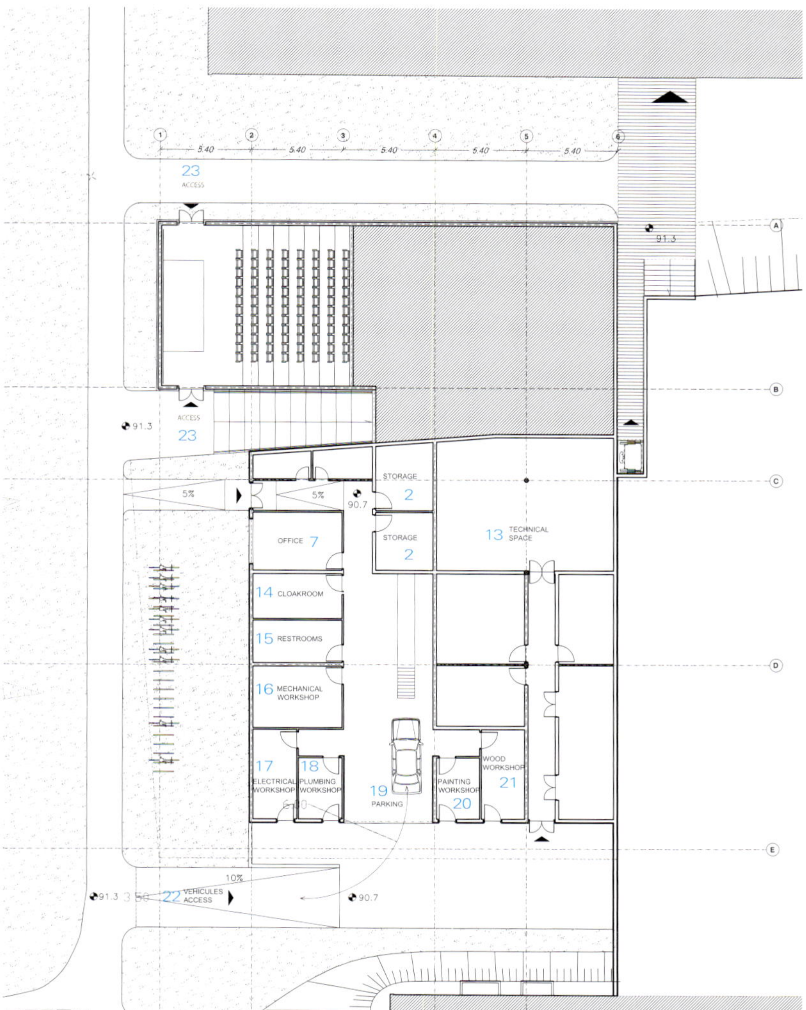

二层平面图
First floor plan

1 礼堂
2 储存室
3 放映室
4 大厅
5 计算机
6 复印室
7 办公室
8 商店
9 工作室
10 会议室
11 阅读沙龙
12 阅览大厅
13 技术设备空间
14 衣帽间
15 卫生间
16 机械工作室
17 电力工作室
18 管道工作室
19 停车场
20 绘画工作室
21 木制品制作工作室
22 机动车入口
23 入口

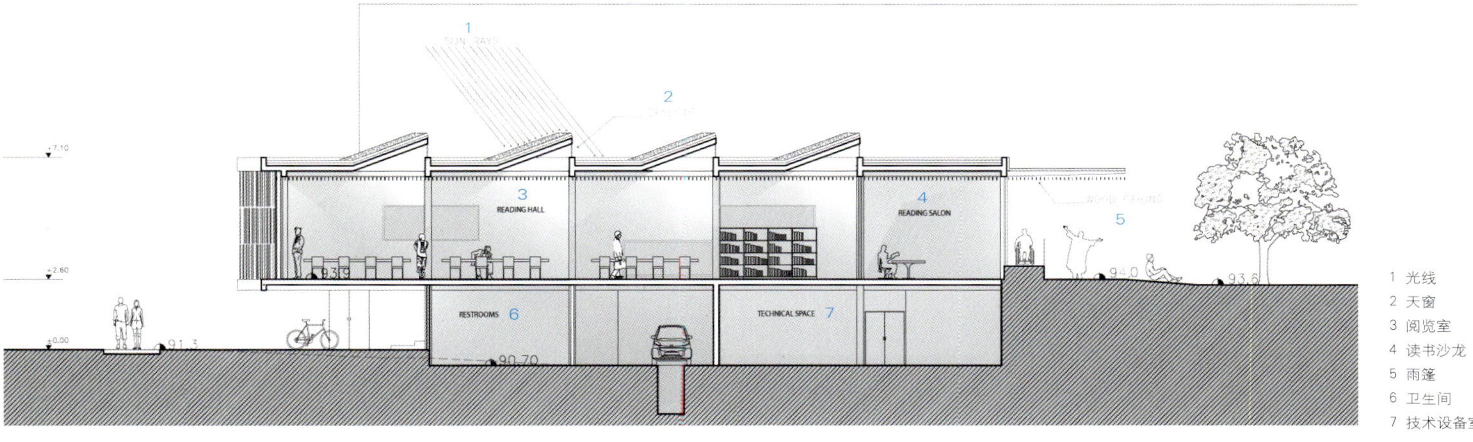

1 光线
2 天窗
3 阅览室
4 读书沙龙
5 雨篷
6 卫生间
7 技术设备室

剖面图
Sections

英国伦敦布伦特市民中心
BRENT CIVIC CENTRE, LONDON, UK

建筑师：Hopkins
顾问：Turner & Townsend（成本管理）
Scott Wilson（结构、机电与环境工程）
Cordless Consultant（IT）
AECOM（防火顾问）
Ann Sawyer（通道设计顾问）
图片提供：Hopkins

具有标志性意义的市民中心大楼建在温布里的重建区，位于国际知名的温布里体育场和温布里体育场广场的对面。设计团队将为布伦特市民中心提供一个量身定制的解决方案，符合布伦特理事会对该方案的设计要求，这里实际上就是为布伦特的各个社区提供服务的中心。设计特色如下：它能够体现出布伦特理事会的领导作用，并为温布里总体规划区正在进行改造的项目设立全新的基准。

The landmark Civic Centre building is designed within the Wembley regeneration area, opposite the internationally renowned Wembley Arena and Wembley Arena Square. The design team will provide a tailored solution that addresses Brent Council's strategic brief for a Civic Centre that is in essence a hub for delivering services to Brent's diverse communities. The quality of the design will be such that it demonstrates Brent Council's civic leadership role and sets a new benchmark for the ongoing regeneration within the Wembley master plan area.

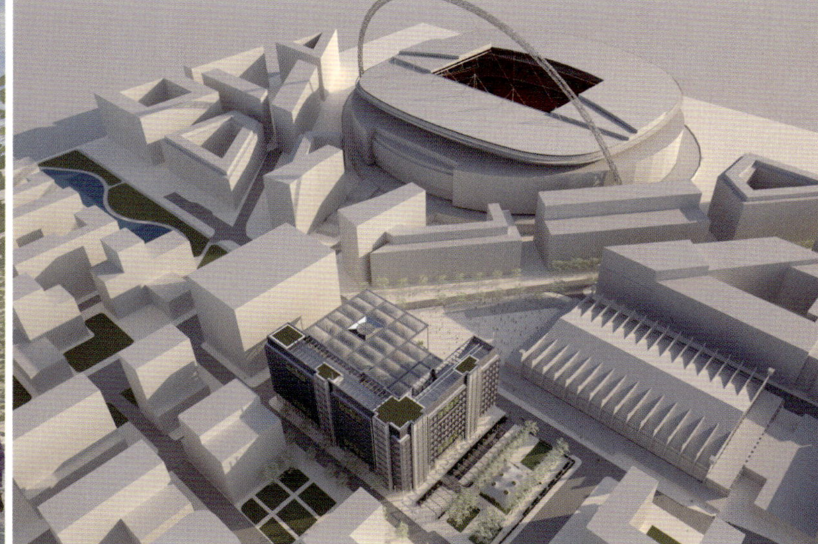

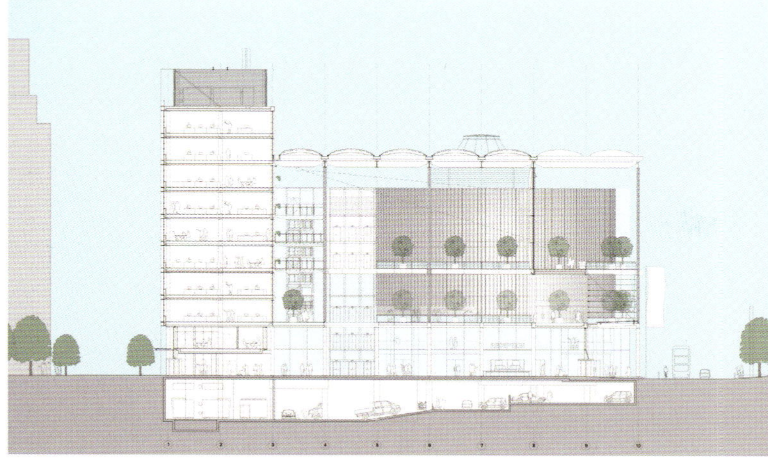

入口大厅横剖面
Cross Section through Entrance Foyer

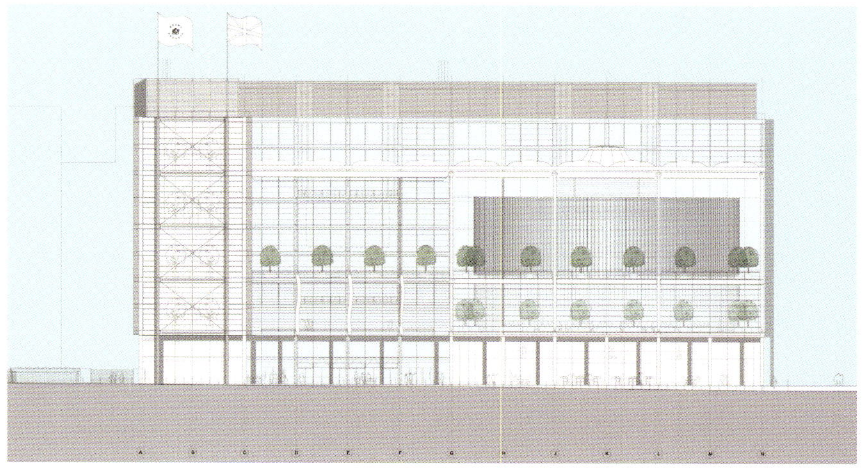

温布里活动广场立面
Wembley Arena Square Elevation

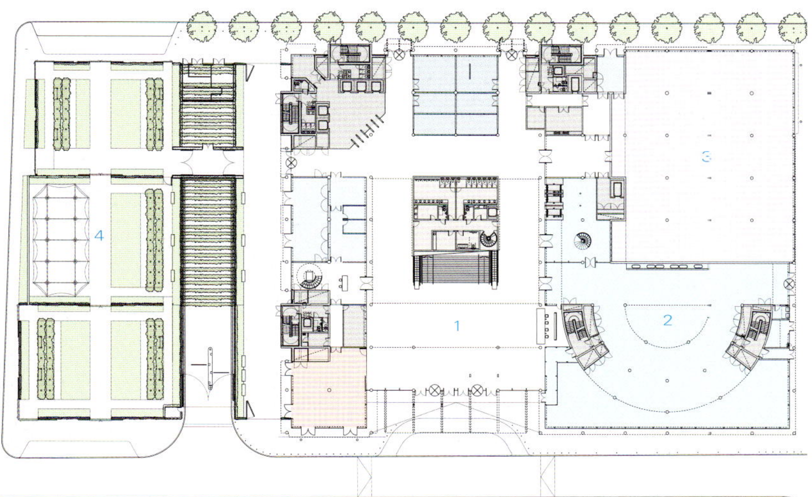

底层
Ground Floor Plan

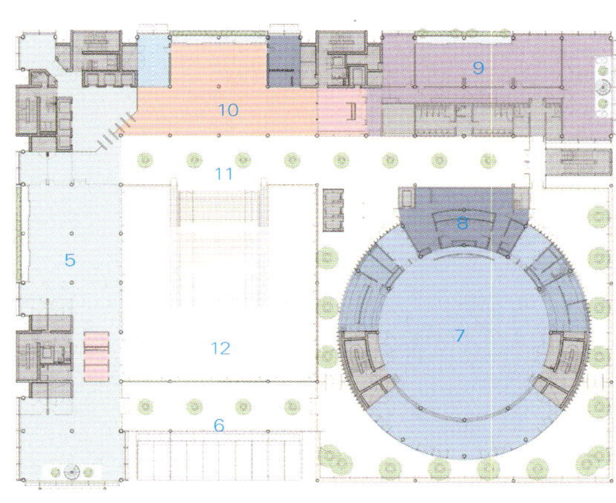

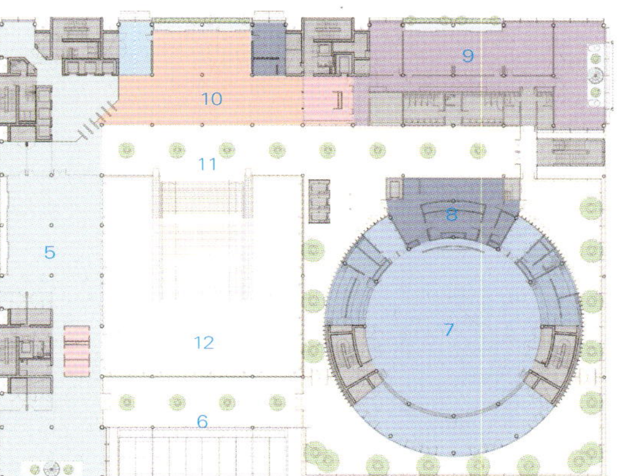

二层
First Floor Plan

1 入口大厅
2 图书馆
3 零售区
4 婚礼花园
5 管理办公区
6 架桥
7 社区大堂
8 餐厅
9 培训
10 咖啡厅
11 画廊
12 中庭

1 Entrance Foyer
2 Library
3 Retail
4 Wedding Garden
5 Administration
6 Bridge
7 Community Hall
8 Catering
9 Training
10 Cafe
11 Gallery
12 Atrium

歌剧与文化馆
OPERA AND CULTURE HOUSE

建筑师：Brisac Gonzalez + Space Group

设计团队：

Brisac Gonzalez

Edgar Gonzalez, Cécile Brisac, João Baptista, Eleftherios Ambatzis, Husain Jaorawala, Miguel Goncalves, Gordon Swapp, Franck Lebouc-Mazé

Space Group

Gary Bates, Gro Bonesmo, Adam Kurdahl, Wenche Andreassen, Gesine Gummi, Ingjerd Sandven Kleivan, Naofumi Namba, Jens Noach, Erich Gerlach, Sassi Heiskanen, Tudor Vlasceanu, Jens Niehues, Rebekah Schaberg

甲方：Kristiansund Kommune

项目地点：Kristiansund, Norway

面积：15 000 m²

新的歌剧与文化馆由600座的礼堂、交响乐厅、芭蕾舞中心、餐厅、图书馆、学校、会议大厅、文化设施和青年中心组成。基地上原来有两座重要的建筑，一座是19世纪的学校建筑，另一座是20世纪初的下议院大楼。如果在它们中间简单地插入第三座较大的建筑将会破坏它们的整体性。最终的设计策略是创造一个由三座不同的建筑组成的能够相互渗透文化的综合体，这三座建筑相互独立却又彼此相连。原来的两座建筑将会被彻底翻修，以举办各种新活动。一座丝带一样的玻璃连接桥同时还被用作展览画廊，它连接着下议院大楼的图书馆和新建筑的主门厅。学校建筑中将容纳青年中心。一条地下通道将学校建筑与新礼堂的舞台层连接起来。

The New Opera and Culture House comprises a 600 seat auditorium, symphony, ballet centre, restaurant, library, school, conference hall, cultural facilities and youth centre. There are already two significant buildings on the site – a 19th century school building and an early 20th century Folkets Hus (People's House). To engulf them by a third larger building would diminish their integrity. The design strategy creates a porous cultural compound of three very different free standing buildings that are autonomous yet connected. The two existing buildings will be fully refurbished and filled with a host of new activities. A ribbon-like glass bridge that doubles as exhibition gallery connects the library in the Folkets Hus to the main foyer of the new building. The school building will contain youth cultural activities. An underground passage will link it to the stage level of the new auditcrium.

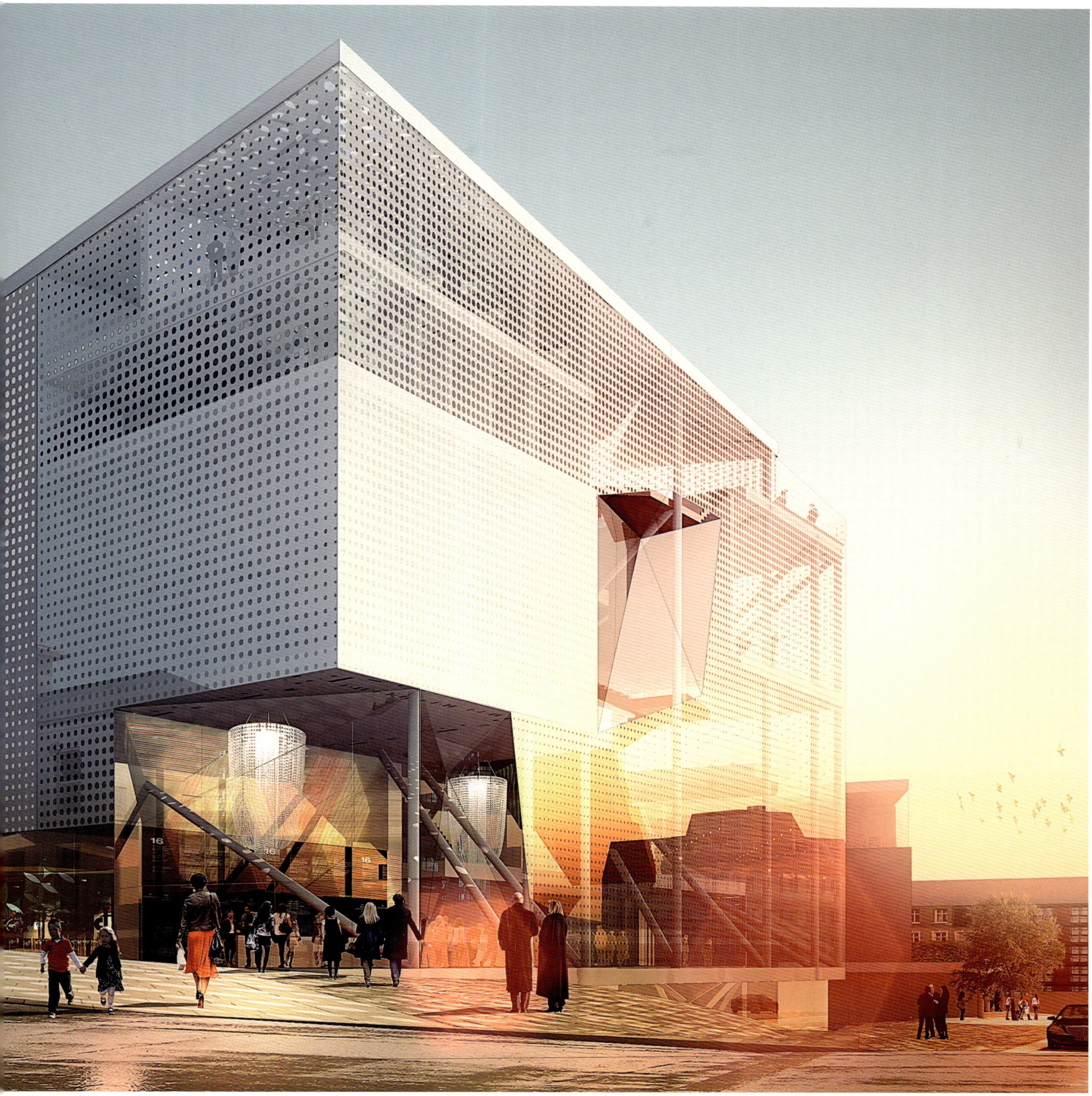

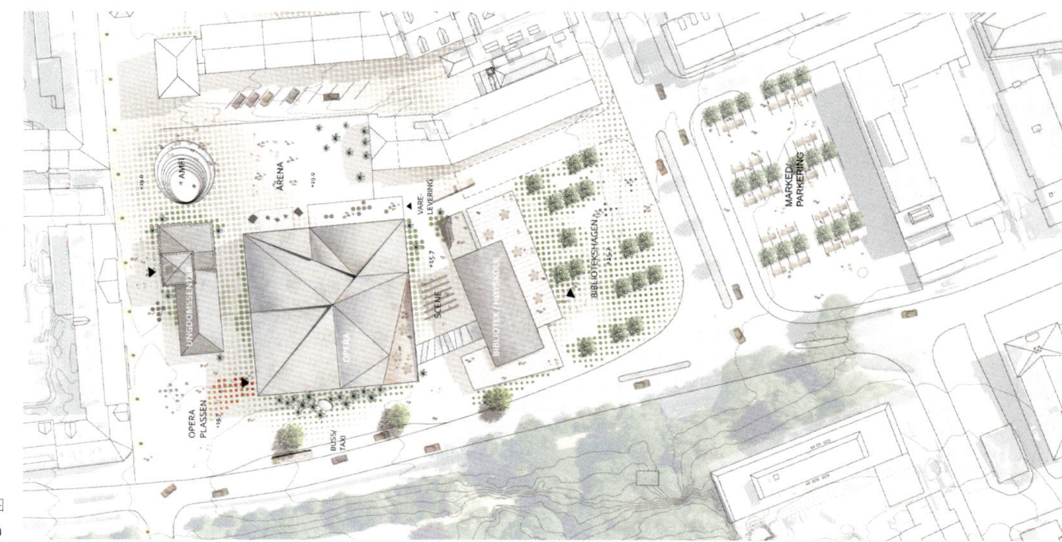

总平面图
Site plan

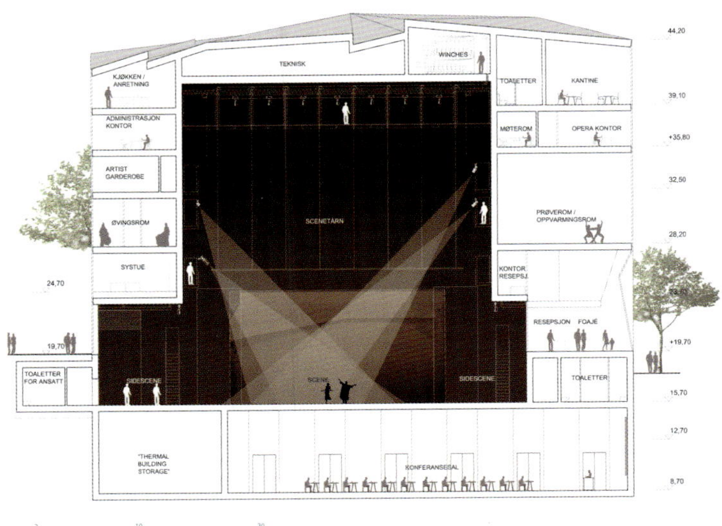

BB剖面图
Section BB

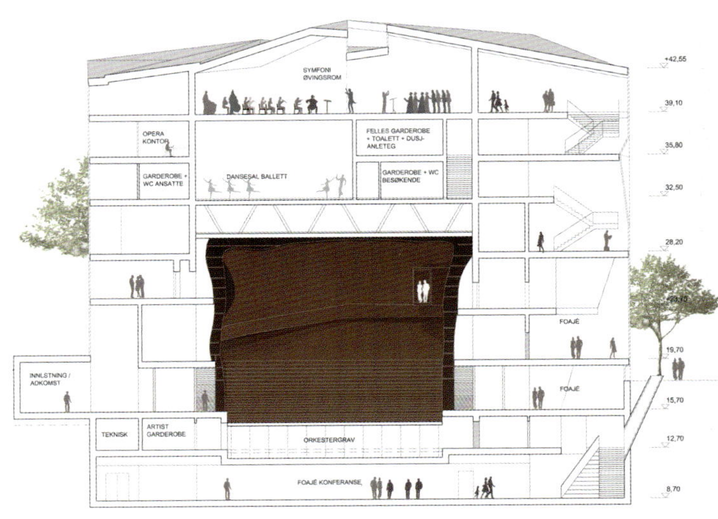

CC剖面图
Section CC

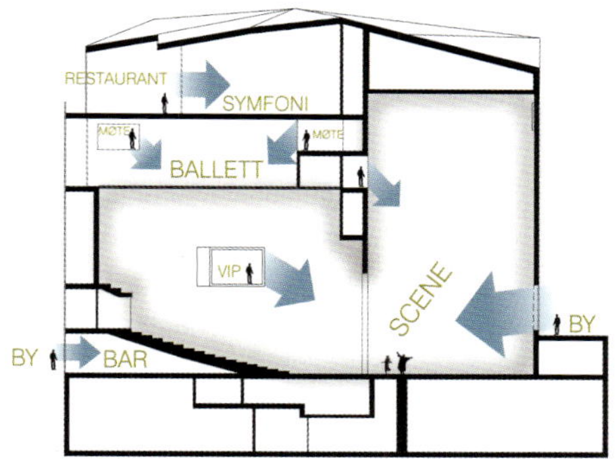

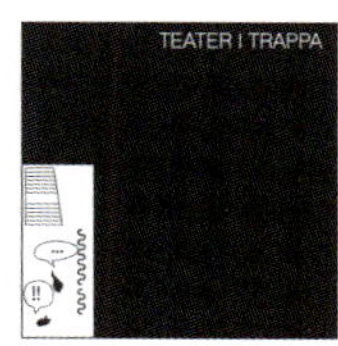

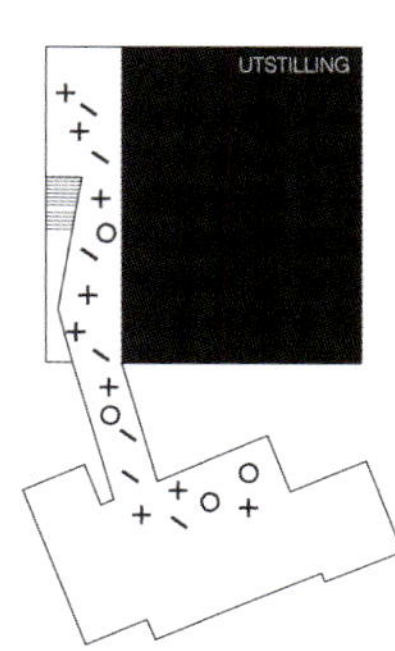

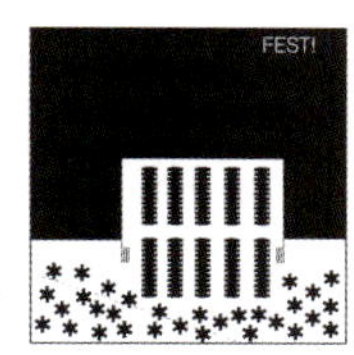

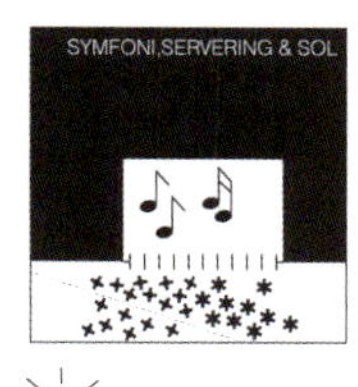

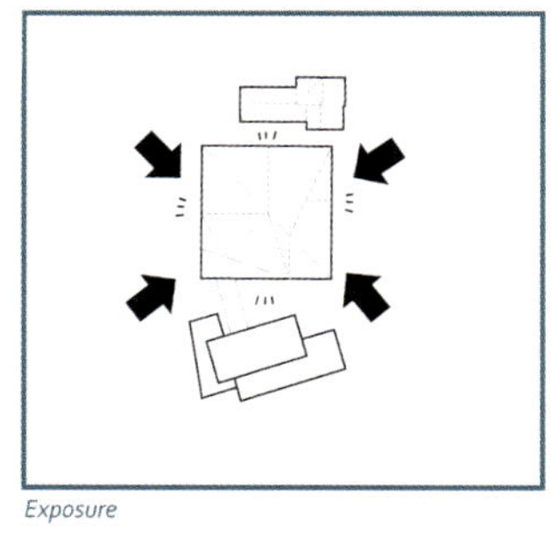
Exposure

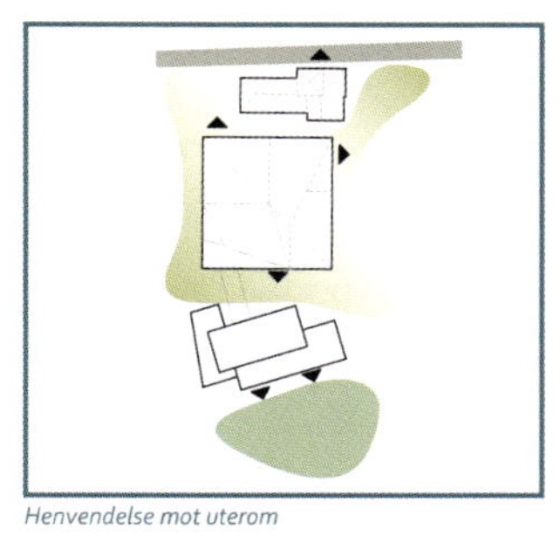
Henvendelse mot uterom

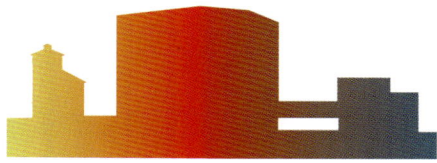

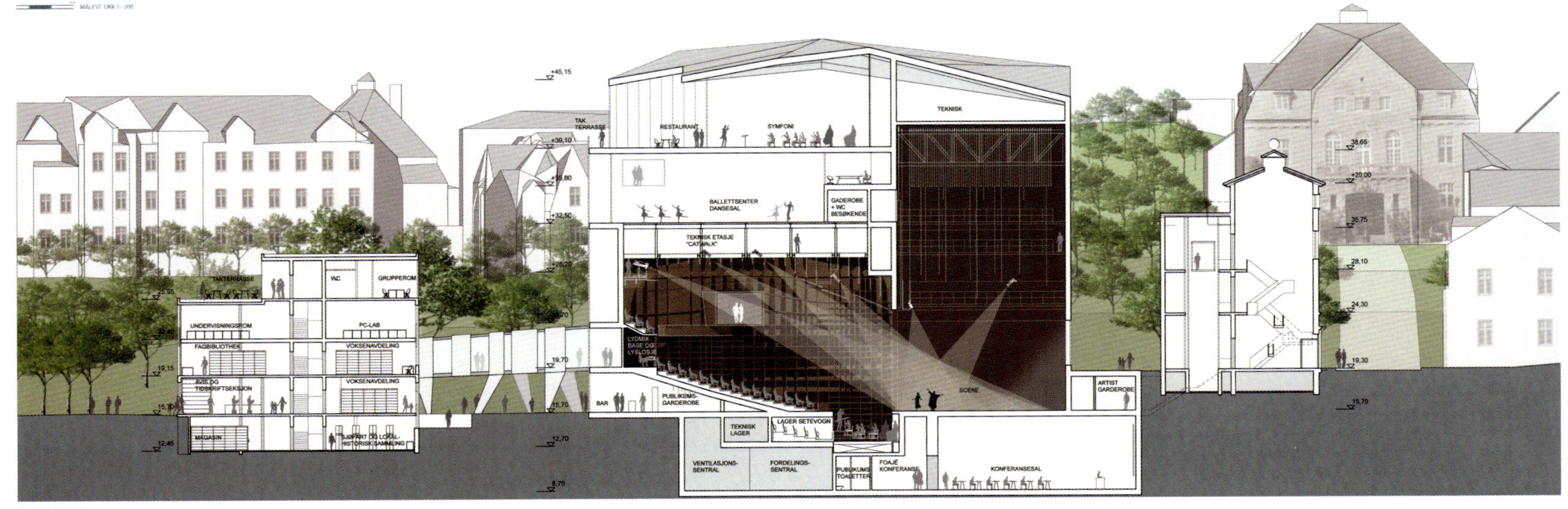

AA剖面图
Section AA

南立面
South elevation

北立面
North elevation

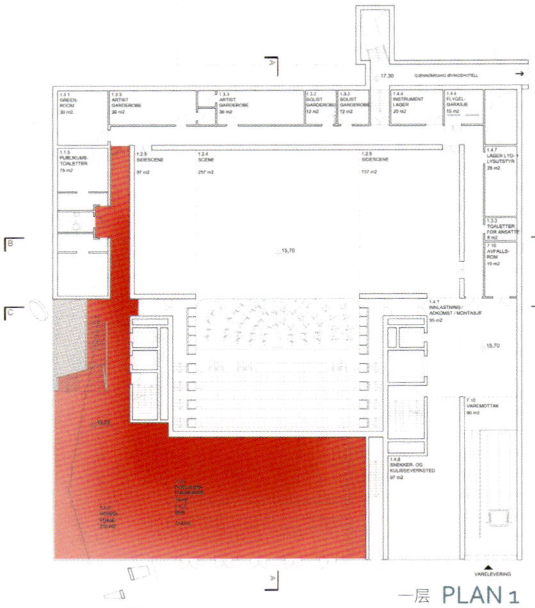

一层 PLAN 1

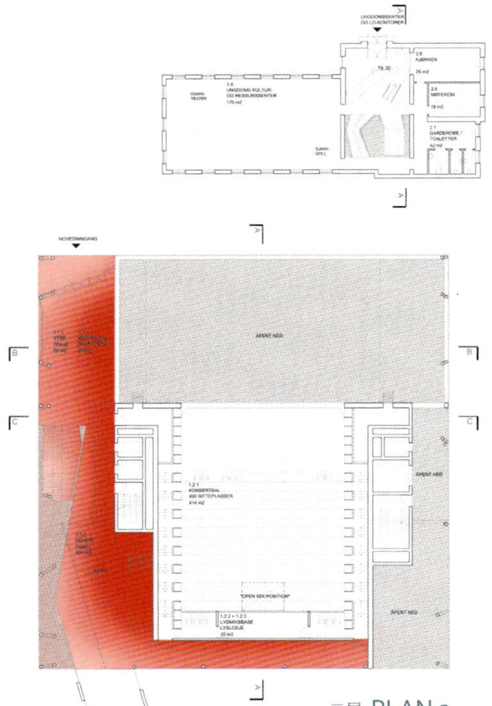

二层 PLAN 2

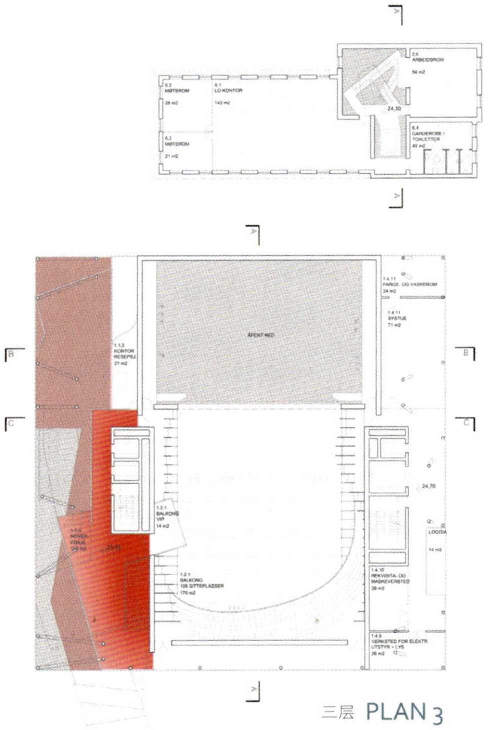

三层 PLAN 3

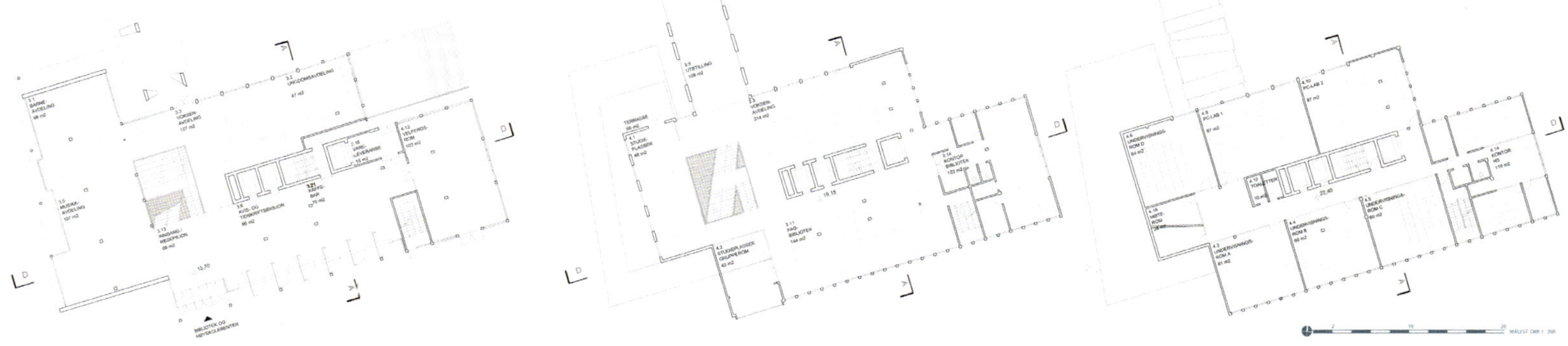

东立面
East elevation

西立面
West elevation

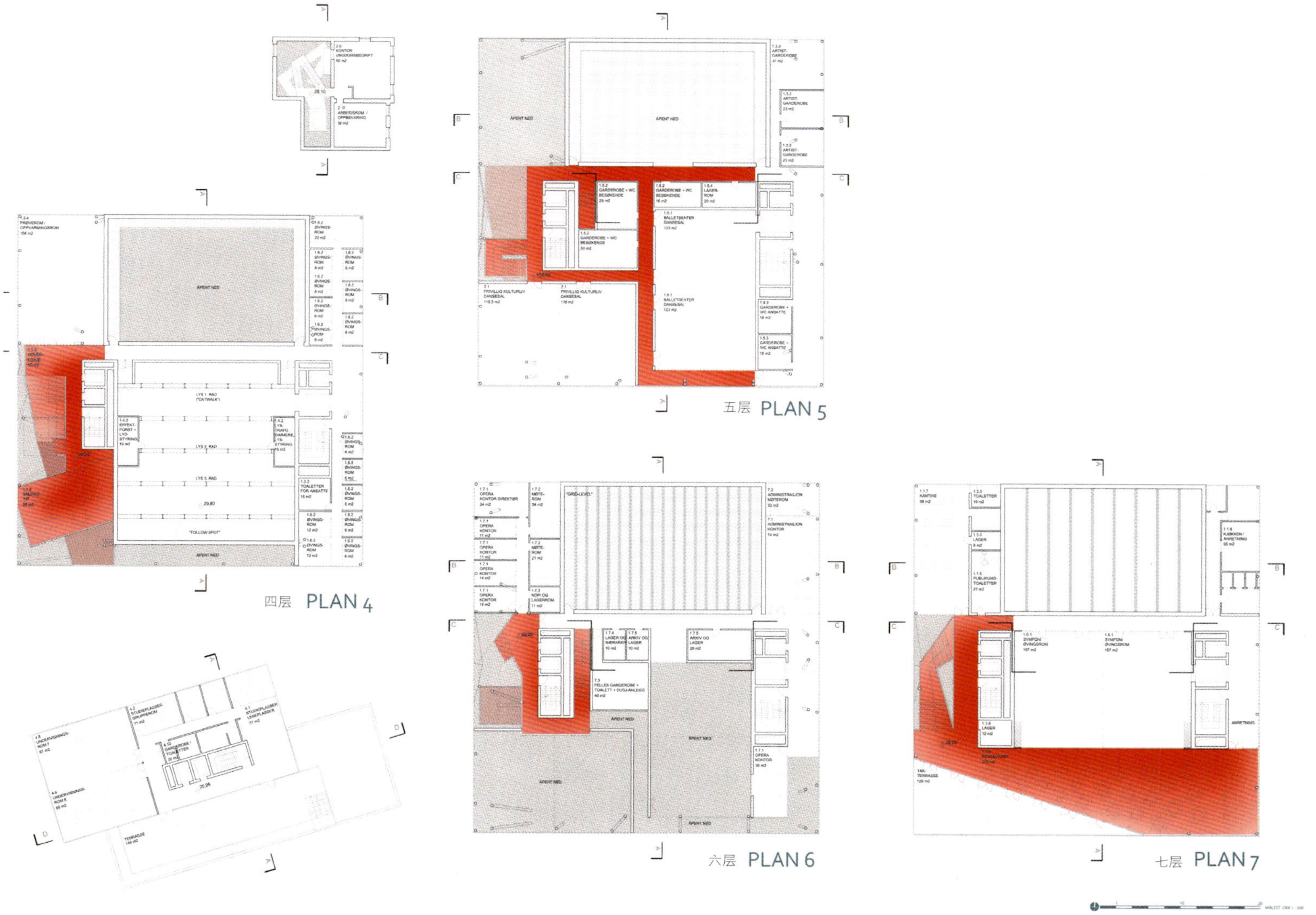

四层 PLAN 4

五层 PLAN 5

六层 PLAN 6

七层 PLAN 7

杜伊斯堡—埃森大学图书馆与研究中心
UNIVERSITY LIBRARY + RESEARCH CENTRE, DUISBURG – ESSEN

建筑师：Wiel Arets Architects
项目团队：Wiel Arets, Bettina Kraus, Natalie Gagro, Jochem Homminga, Julius Klatte, Joana Varela
甲方：University of Duisberg
项目地址：Duisberg, Germany
建筑面积：15 820 m²
获奖情况：竞赛四等奖

本案两座新建的紧凑型建筑坐落于城市中心和大学城交界处，在所处环境中呈现出不同的规模和空间结构。建筑体量清晰的垂直与水平姿态彰显出建筑独特的图像式外观，在周围环境中脱颖而出。

为了强调建筑的公众辐射性以及图书馆的功能，较高的一座建筑通过两种不同的亮度进行了细分。23个立方体结构使整栋楼充满变化，这样的层次无法一眼识破。

较低的研究中心大楼有八个天井，分别为不同的工作单元提供空间，构成相邻的区域。平面网状路径使我们联想起街道的分布，而不是常规办公室的布局。

图书馆立面图一
LB elevation 1

图书馆立面图二
LB elevation 2

研究中心立面图一
RB elevation 1

研究中心立面图二
RB elevation 2

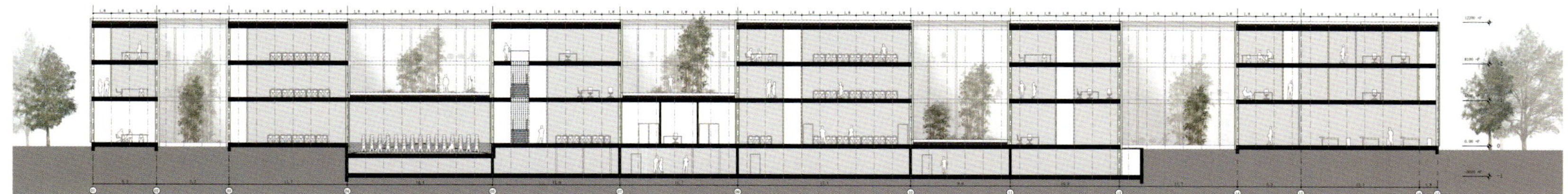
研究中心剖面图一
RB section 1

Positioned at the interface between city and university centre two new compact solitaires establish an independent scale and spatial structure within their heterogonous proximity. The outspoken vertical and horizontal gesture of the volumes results in an iconographic complex whose presence goes far beyond the immediate context.
In order to strengthen the public radiation and the action at the library, two different magnitudes subdivide the taller building. Twenty-three cubes form a conjugated aggregate of spaces, whose levelling can only be recognized on second sight.
The low research centre building is organized by eight patios, which provide identity and territory to the work units, creating a neighborhood. The horizontal network of routes reminds us of a street map, rather than a regular office layout.

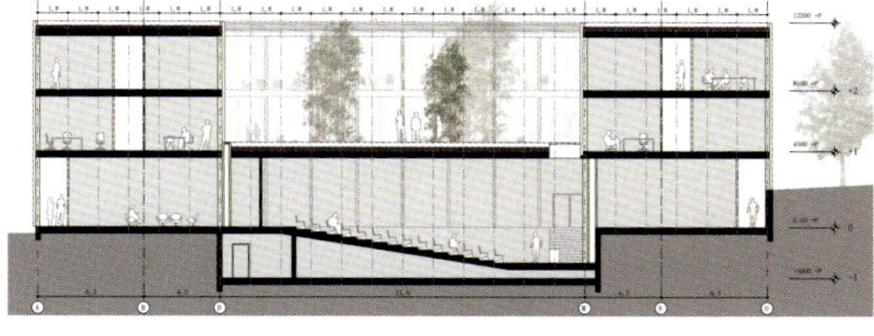
研究中心剖面图二
RB section 2

总平面图
Site plan

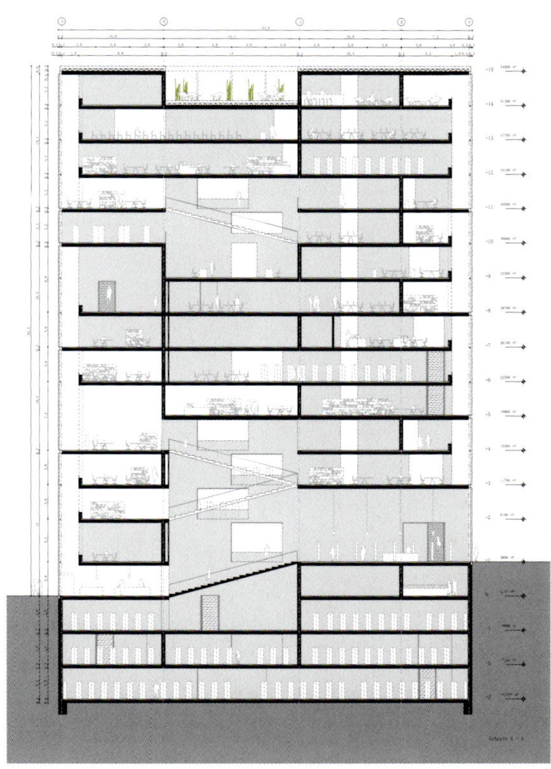
图书馆剖面图
LB section

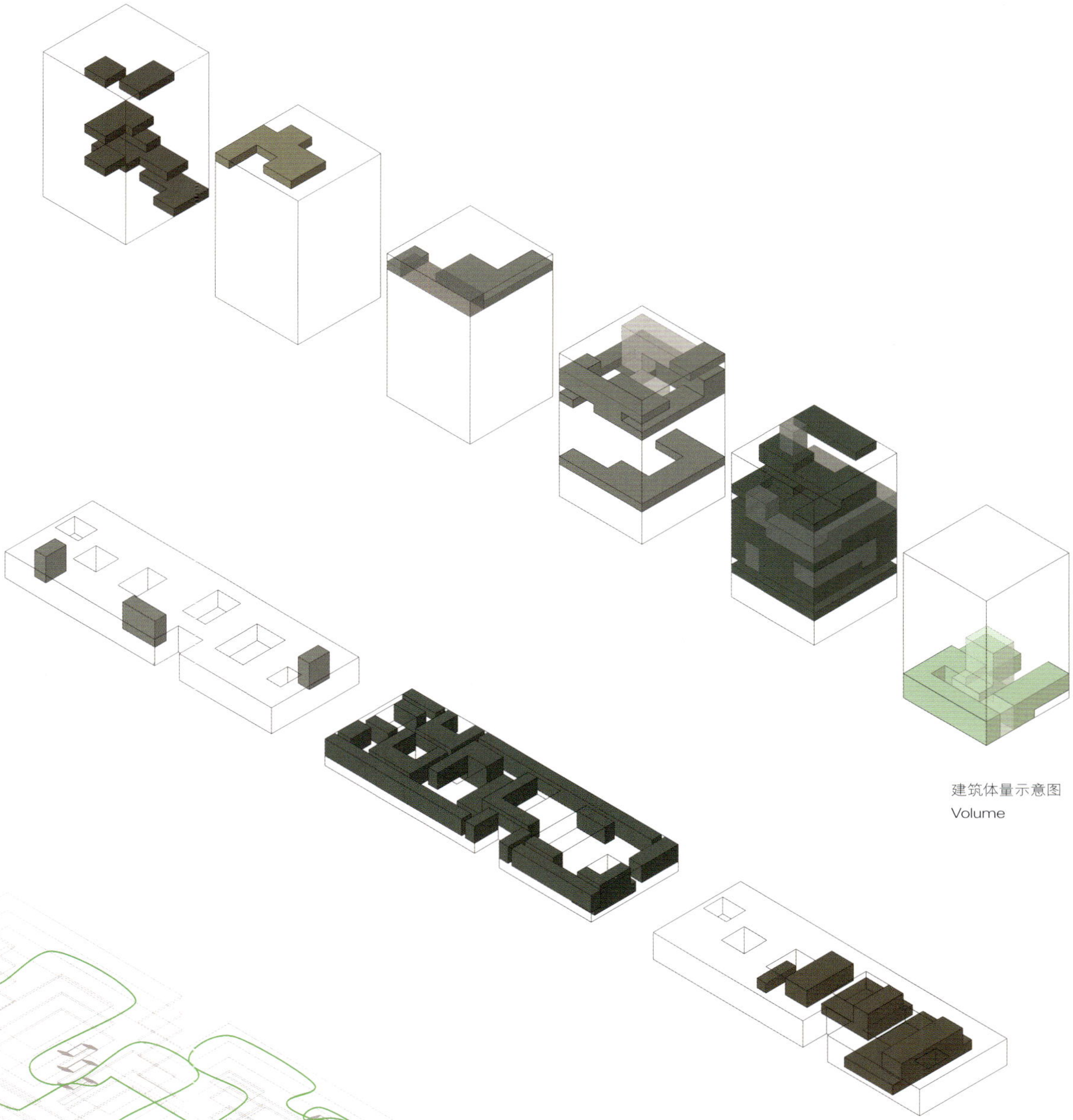

建筑体量示意图
Volume

办公布局示意图
Office

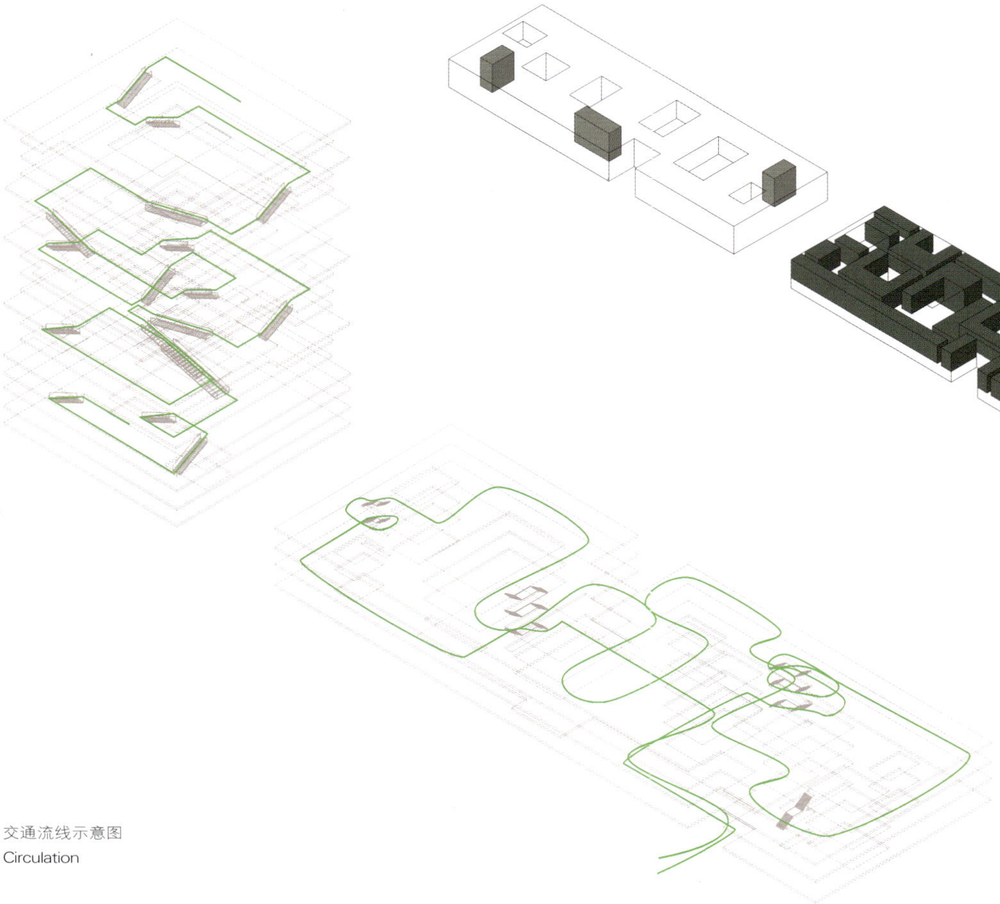

交通流线示意图
Circulation

Saint-Hilaire du-Harcouët媒体中心
SAINT-HILAIRE DU-HARCOUËT MEDIA CENTRE

设计团队：SERERO ARCHITECTS (David Serero, Fanny Lenoble, Yoichi Ozawa, Amandine Quillent

工程师：Beterem Ingénierie

甲方：City of Saint-Hilaire-du-Harcouët

楼面面积：800 m²

项目组成：带门厅的300座表演大厅，四间艺术与音乐工作室，一家餐厅，一个排练大厅，行政管理办公室和相关技术设备空间

项目地点：Saint-Hilaire- (50), France

获奖情况：竞赛一等奖

Saint-Hilaire-Du-Harcouët的新媒体中心坐落在特殊的历史及城市环境中。设计师将其看作是周围城市地形的延伸，并将其设计得尽量低，还有一部分设计在地面以下，以此来表达对历史环境的敬意。媒体中心的南面是一个长长的窗户，可以看到外面的景观、教堂和周围的建筑。自然光线在媒体中心扮演着最重要的角色，也是整个空间设计的中心。

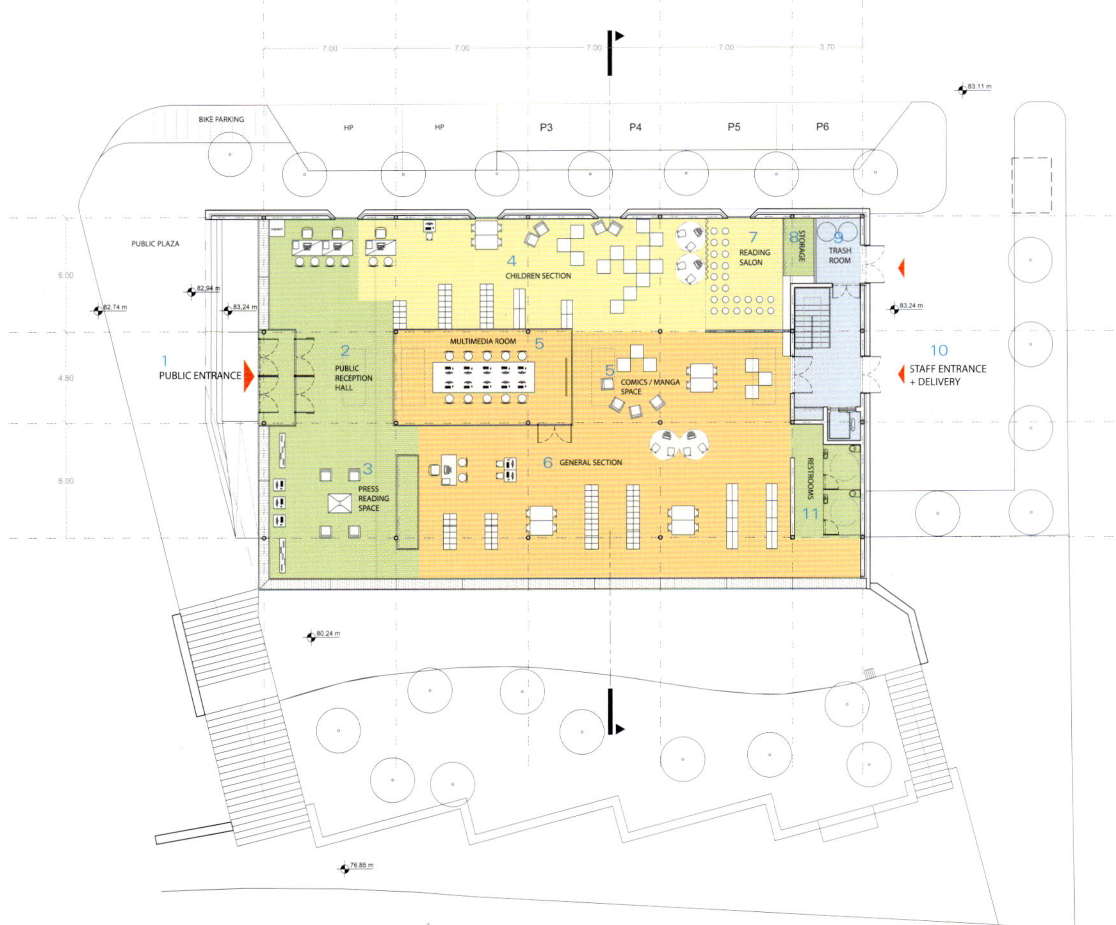

平面图

Floor plan

1　公共入口
2　公共接待大厅
3　媒体信息阅读大厅
4　儿童区
5　多媒体室
6　成人区
7　阅读沙龙
8　储存室
9　垃圾存放处
10　员工与送货入口
11　卫生间

Saint-Hilaire-Du-Harcouët's new Media Centre sets in an exceptional historical and urban context. The architects conceived it as an extension of the surrounding urban morphology. We designed the building as low as possible to respect the historical context and partially integrated it in the ground. South side, the media center is conceived as a long window giving an open view on the landscape, on the church and the buildings around. The media centre is a place where the natural light has an essential role and is placed in the centre of the spaces designed.

大教堂圆花饰图例
EXAMPLE OF CATHEDRAL ROSETTES

当地石墙
LOCAL STONE WALL

教堂窗户
WINDOWS OF THE CHURCH

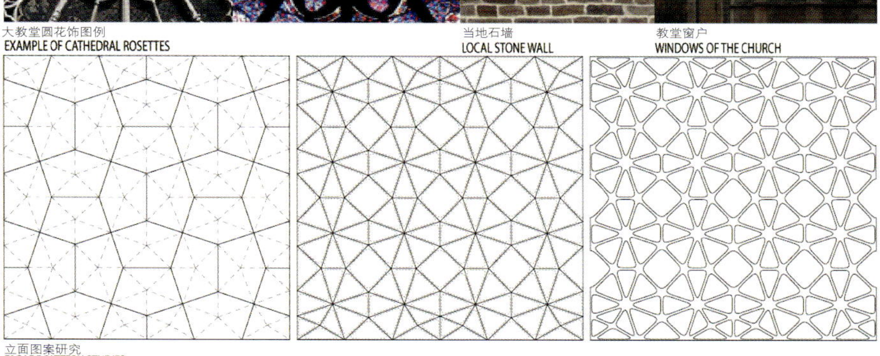

立面图案研究
FACADE PATTERN STUDIES

1　混凝土立面上覆盖石板
2　儿童区
3　图书储存室
4　绿色屋顶
5　成人区
6　办公室
7　带通风的双层玻璃立面系统
8　遮阳铜网
9　计算机控制的卷帘
10　露台
11　金属遮阳帘
12　冬至
13　夏至
14　采光天窗

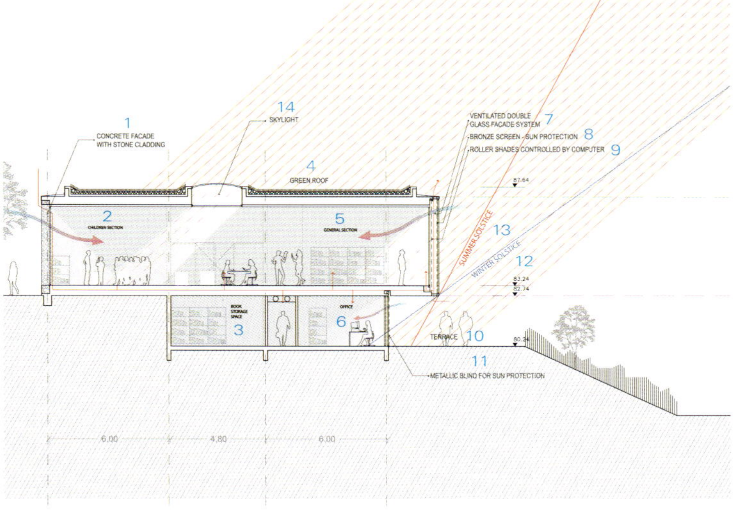

剖面图 Section

悉尼理工大学入口建筑
UTS PODIUM

建筑师：Lacoste+Stevenson in association with DJRD and 6o
工程师：ARUP
甲方：UTS (University of Technology Sydney)
项目地点：Broadway, Sydney
用途：大学入口建筑、600座讲演厅、多媒体电影院、展览空间、书店、咖啡厅
基地面积：2530 m²
建筑面积：14 500 m²
建筑规模：地上二层，地下四层
效果图制作：Thierry Lacoste and Ivolve
获奖情况：竞赛一等奖

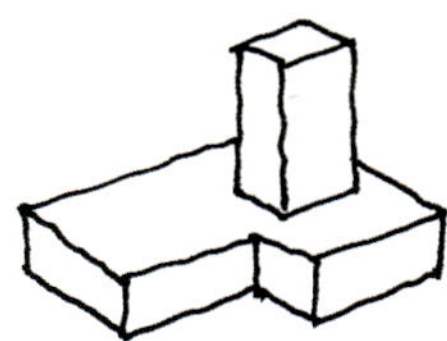

原有建筑
existing

拆除立面
demolish the facades

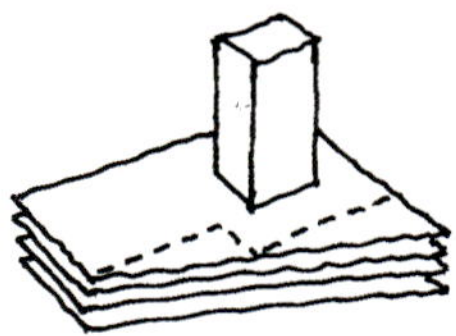

延伸楼板
extend the floors

用新立面包围
wrap around a new facade

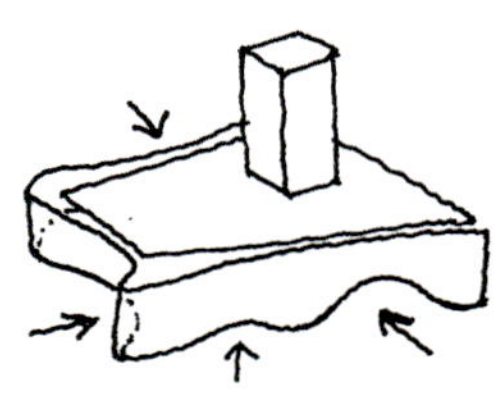

弯曲立面设置入口
bend it to accomodate entries

建筑形式生成过程
Form generation

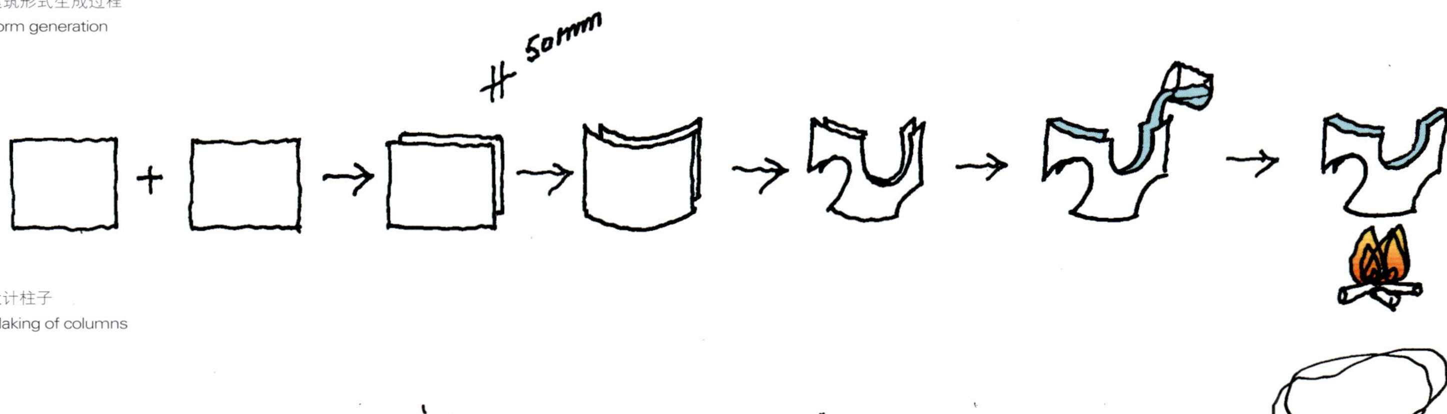

设计柱子
Making of columns

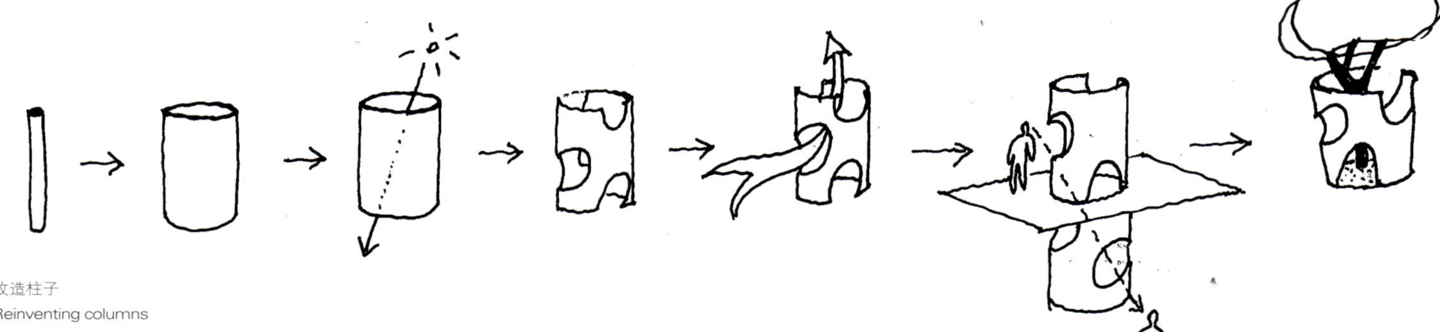

改造柱子
Reinventing columns

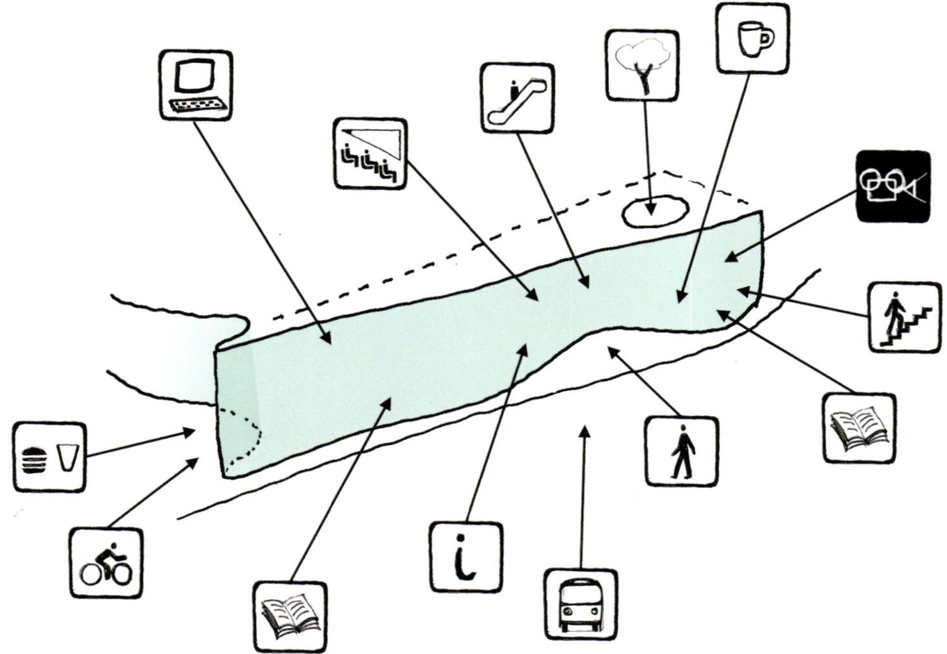

图示说明
Pictograms

本案扩建建筑将为悉尼理工大学提供一个新的主入口。与野兽派的塔楼建筑不同，这座新的建筑则摆出开放与欢迎的姿态。它是一座波浪形的半透明建筑，保卫着塔楼的基座。建筑的表面是曲面玻璃，并带有森林图案的白色釉料，表现出柔和的特质，就像窗帘一样柔软。光滑的表面大大地向内倾斜，标志着Broadway上的入口。

This extension will provide a new main entrance to the university. In contrast to the Brutalist architecture of the Tower, the new podium is open and welcoming. It intervenes as an undulating semi-transparent building wrapping the base of the tower. This is achieved with curved glass and a white frit in the pattern of a forest. It appears soft and almost pliable like a curtain; its smooth surface blowing inward dramatically to mark the main entrance on Broadway.

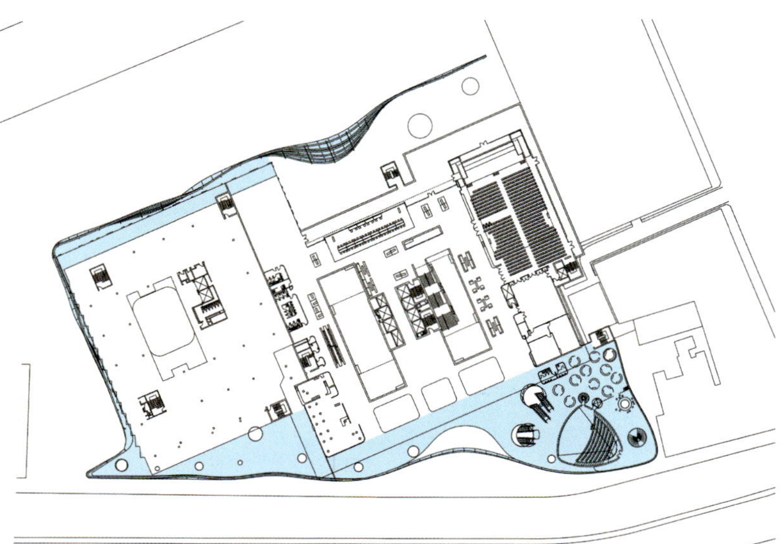

五层平面图
Level 5

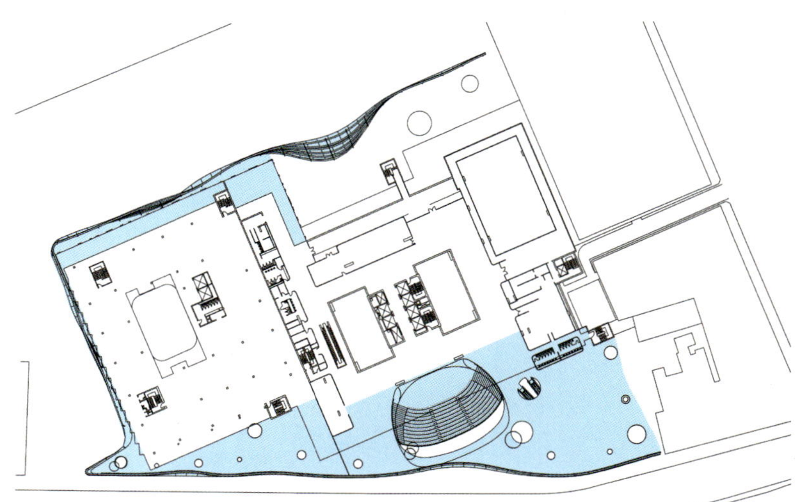

六层平面图
Level 6

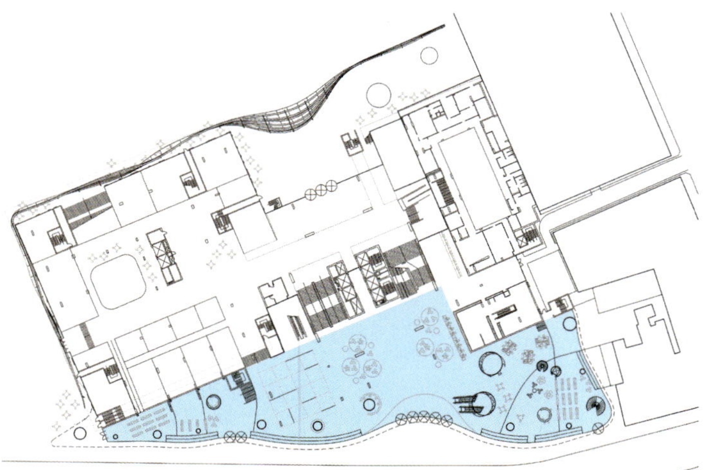

三层和四层平面图
Level 3 and 4

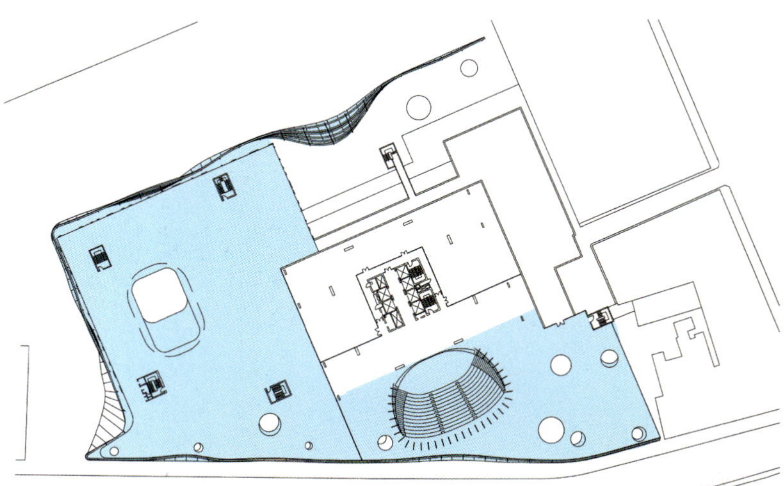

七层平面图
Level 7

1 塔楼建筑表现了力量与直接，是对其所处时期建筑哲学的一种纪念。
与塔楼建筑不同，扩建的入口建筑几乎是无形的，好像是塔楼建筑的面纱。它柔软，波浪起伏，又显轻盈，是建于20世纪70年代的塔楼一层温柔的屏障。
新的入口建筑将成为当代建筑的典范，促进悉尼理工大学原有建成环境的革新。

2 悉尼理工大学入口建筑所使用的材料与塔楼建筑截然不同，使大学的边缘更加柔和。它轻盈、透明，与塔楼建筑内封闭的教学管理楼层形成鲜明的对比。

设计概念
Design concept

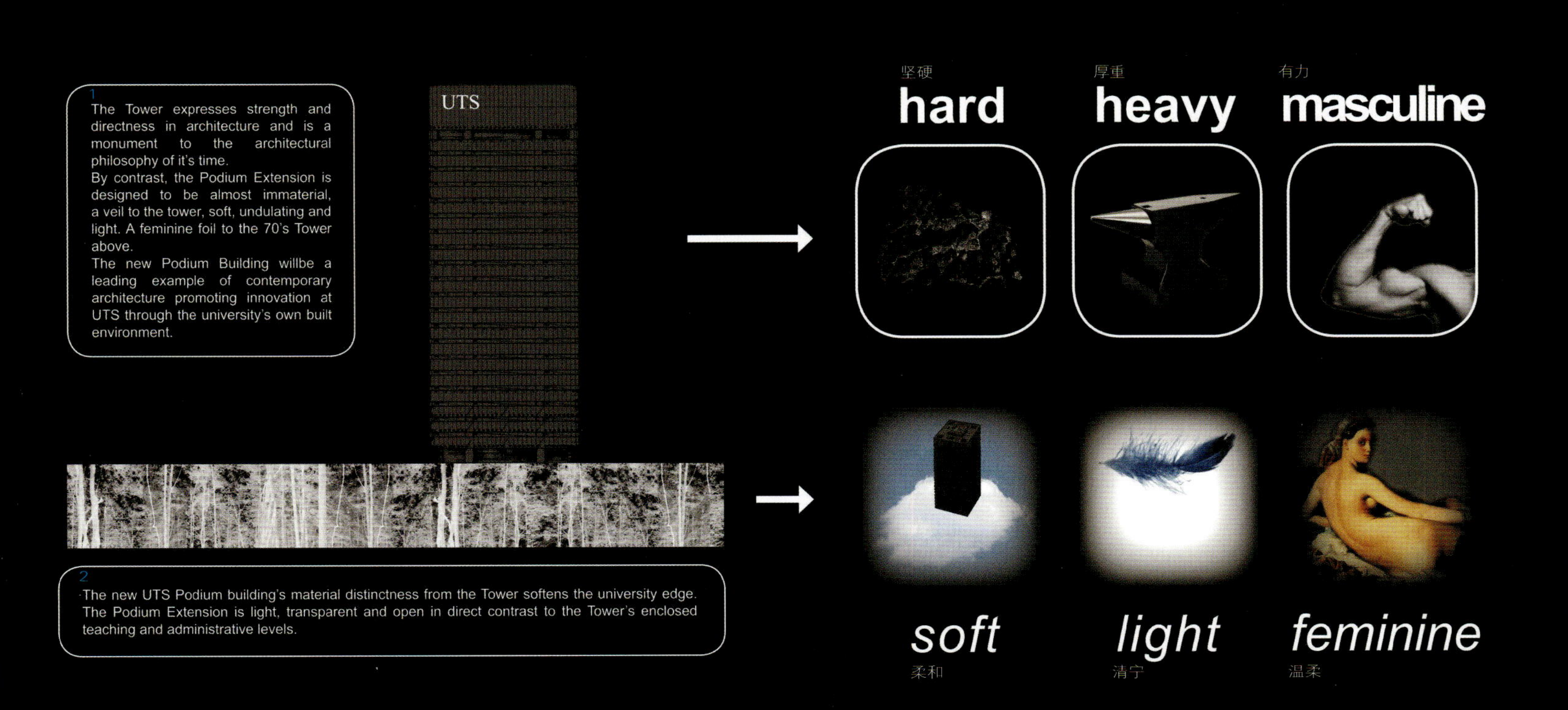

1
The Tower expresses strength and directness in architecture and is a monument to the architectural philosophy of it's time.
By contrast, the Podium Extension is designed to be almost immaterial, a veil to the tower, soft, undulating and light. A feminine foil to the 70's Tower above.
The new Podium Building willbe a leading example of contemporary architecture promoting innovation at UTS through the university's own built environment.
UTS
坚硬
hard
厚重
heavy
有力
masculine
2
The new UTS Podium building's material distinctness from the Tower softens the university edge. The Podium Extension is light, transparent and open in direct contrast to the Tower's enclosed teaching and administrative levels.
soft
柔和
light
清宁
feminine
温柔

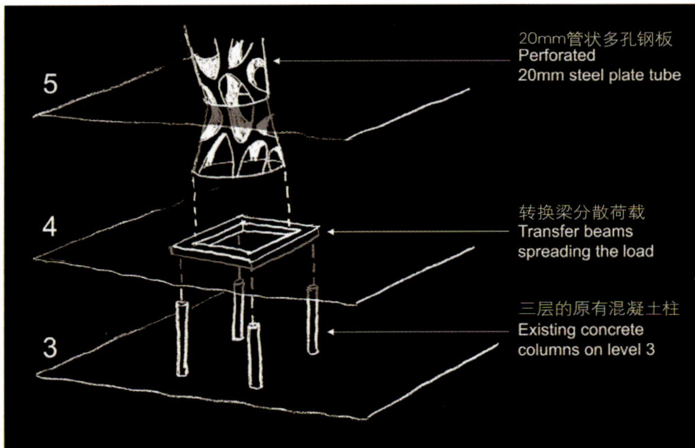

结构原理

Structural principal

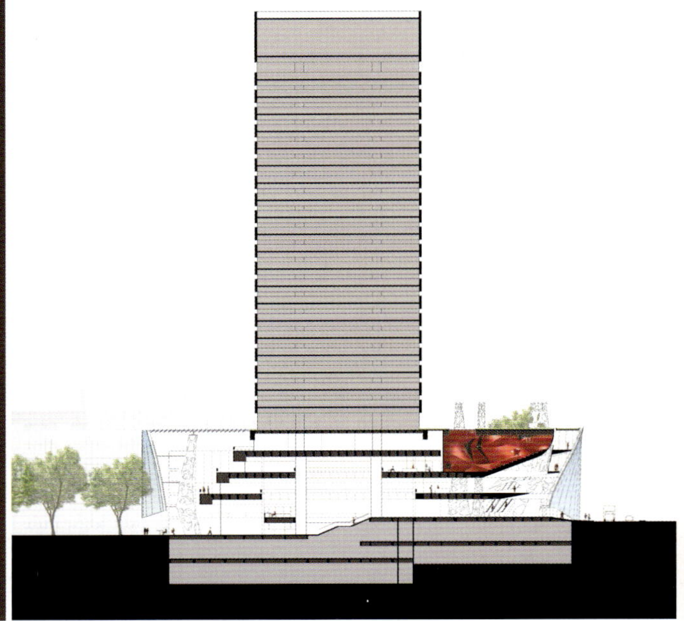

剖面图

Section

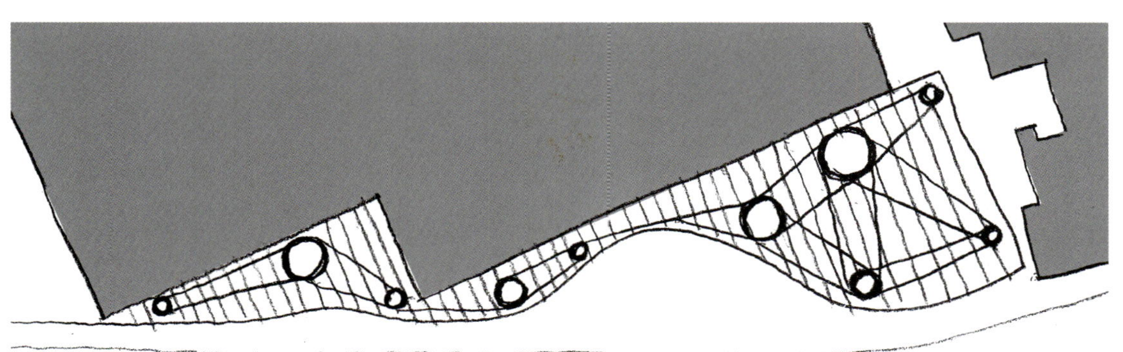

结构框架

Structural framing

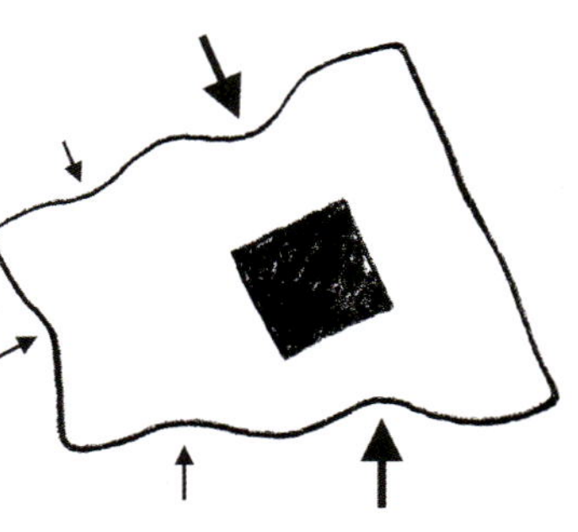

入口示意图

Entries

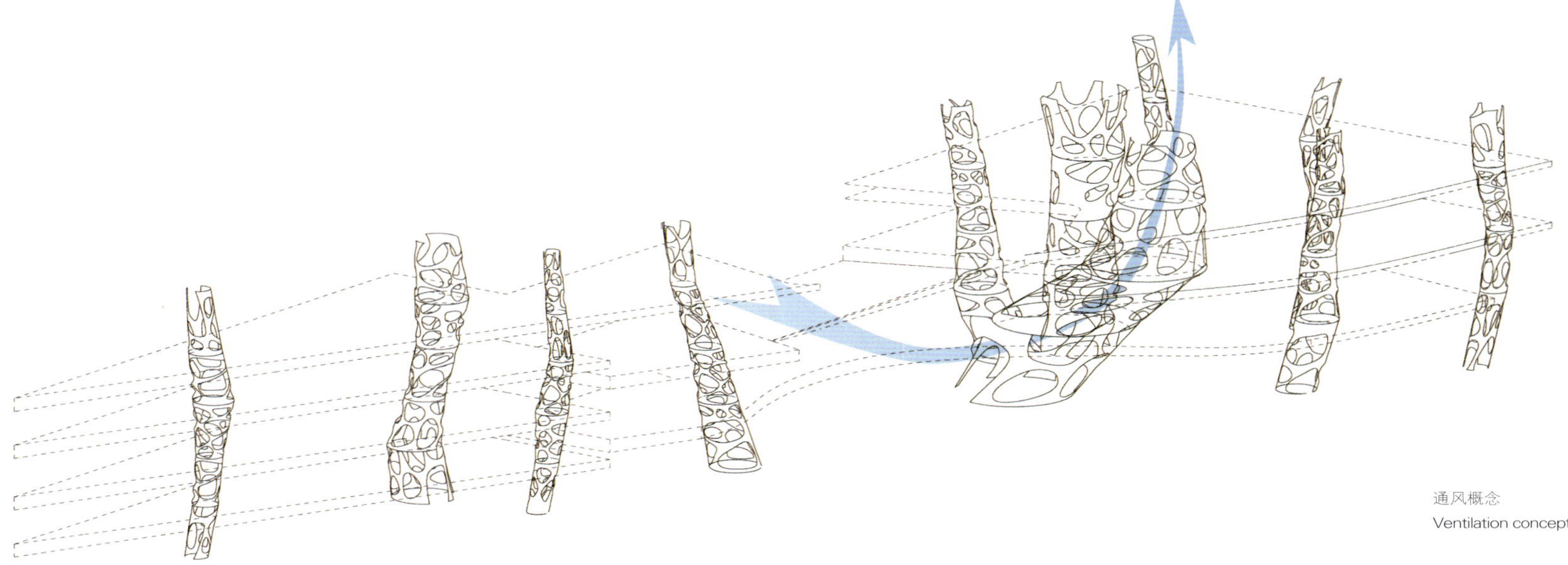

通风概念

Ventilation concept

纳维达斯公园
NAVITAS PARK

建筑师：schmidt hammer lassen architects
甲方：Ingeniørhøjskolen Aarhus, Maskinmesterskolen Aarhus, Incuba Science Park and the Municipality of Aarhus
承包商：A. Enggaard A/S, Per Aarsleff A/S (DK)
工程师：Rambøll A/S (DK)
其他顾问：Lund+Slaatto Arkitekter (NO), Werner Sobek Green Technologies (GER)

纳维达斯公园是奥尔胡斯工程学院、奥尔胡斯船舶工程学校和Incuba科学园合并后的新园区。建筑位于丹麦第二大城市奥尔胡斯市的中心海港地区。

建筑是对场地南部要塞和新的多媒体中心图书馆——城市媒体空间的补充。纳维达斯公园是一座从远处就能看得到的醒目建筑，以其巨大的体量对内港的总体规划作出了重要的贡献。港口建筑要具有大尺度、简洁和雕塑感的特质，纳维达斯公园符合上述所有要求。建筑将为学生和散步道上的路人提供活动和休闲的场所，这些场所非常人性化。巨大挑出部分在形成遮阳和建筑切口，这些切口使得日光可以深入到建筑室内的活动场所，从而使建筑具有雕塑般的简洁特质。

纳维达斯公园是奥尔胡斯海滨升级建设中最后一座待建的建筑。园区就像是一座轮廓分明岛屿，它可作为独立的单元使用，同时又与重建的城市肌理顺畅地融合。

建筑围绕着中心中庭布置。中庭上方的大型天窗和引入日光的切口在各个功能区之间形成了视觉连接，同时形成了愉悦放松的室内环境。会议室和教室中透明和半透明玻璃的交替使用使得走廊充满了活力和开放的感觉。

屋顶上的“能源公园”由一系列的太阳能板、风力涡轮机、景天属绿色植物区和白色的二氧化碳反射屋顶构成。反射屋顶安置在围绕着采光井和天窗的格栅上。对于学生来说，学校同样拥有一个能源操场——屋顶的太阳能实验室，此处拥有欣赏城市风光的最佳视野。实验室周围将修建一座大型的露台，学生可以在此处研究与可再生能源相关的最新技术。

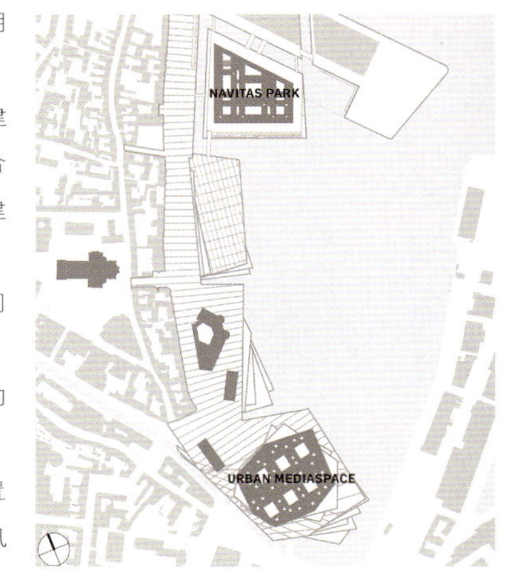

Navitas Park is a campus that gathers the Aarhus Engineering School, the Aarhus School of Marine Engineering and Incuba Science Park under one roof. It will be situated in the central harbour area of Aarhus – Denmark's second largest city.

The building is the northern complement to the southern bastion and the new multimedia library, Urban Mediaspace. Navitas Park is a building, which can be seen from a far distance, and with its size it will be an important contribution to the inner harbour masterplan. The architecture of the harbour is planned to be large scale, and the simple, sculptural form of Navitas Park complies with this idea. However, it will provide niches of activity and leisure for both students and passers-by on the esplanade, which are adjusted to the human scale. The large cantilevers which will create distinctive shadows and the incisions which will let daylight far into the niches give the building a sculptural simplicity.

Navitas Park is the last missing piece in the upgrading of the Aarhus waterfront. The campus is like an island with four clear sides, which is treated as an independent unit, and it integrates freely and smoothly into the regeneration fabric.

The building is organised around a central atrium. The large skylight above the atrium and the daylight-generating incisions create visual connections between the functions and, at the same time, offer a pleasant and relaxed interior. The alternating use of opaque and transparent walls in the seminar rooms and classrooms makes the hallways vibrant and open.

On the rooftop, an energy park will be established with a blend of solar panels, wind turbines, green areas of sedum grass and white CO_2-reflecting roof surfaces placed in a grid around the light shafts and the skylight. For the students, the school will also have an energy playground – the Solar Lab – on the roof with the best view in town. A large terrace will be built around the lab, where the students can test their research in the latest technology regarding renewable energy sources.

曼哈顿的现代艺术博物馆大厦
MOMA TOWER AT 53 W 53, NEW YORK

建筑师：Axis Mundi
概念设计与总建筑师：John Beckmann
项目团队：John Beckmann, CarloMaria Ciampoli, James Coleman [LAN],
Nick Messerlian, Margaret Janik, Pauline Marie d'Avigneau, Taina Pichon
参数模型：CarloMaria Ciampoli, James Coleman [LAN]
效果图：Orchid 3D
图表制作：Michael Wartella

随着城市进入后期繁荣时代，纽约的高楼建筑都变得相差无几，建筑师约翰·贝克曼认为该好好考虑这个问题了。贝克曼说："我们寻找到一种途径来展现多样性，而不是将多用途高楼的丰富价值掩盖起来。"公司采用了计算机参数模型软件对各种可能性都进行了检测。经过这些迭代过程，贝克曼及其公司同事都建议采用新的方式来构建和展示高楼，即垂直社群的形式。贝克曼说："以前，高楼大厦的建筑轮廓线给人的感觉很单一，而现在，这些建筑将城市变得更加多样化、综合化，也更加符合环艺要求。"

As the city takes stock in a post-boom era, architect John Beckmann sees this is the time to rethink the tall buildings that have become synonymous with New York City's identity. "Instead of disguising the rich potential of towers that have a mix of uses, we looked for a way to express that diversity," Beckmann explained. The firm used parametric computer-modeling software to test a wide range of possibilities. Out of this iterative process, Beckmann and his firm, proposes a new way to organize and express tall buildings: the Vertical Neighborhood. "A more diverse, complex, heterogeneous, and environmentally minded city need no longer be represented on its skyline by one-note architecture that makes a singular visual image and little else," explained Beckmann.

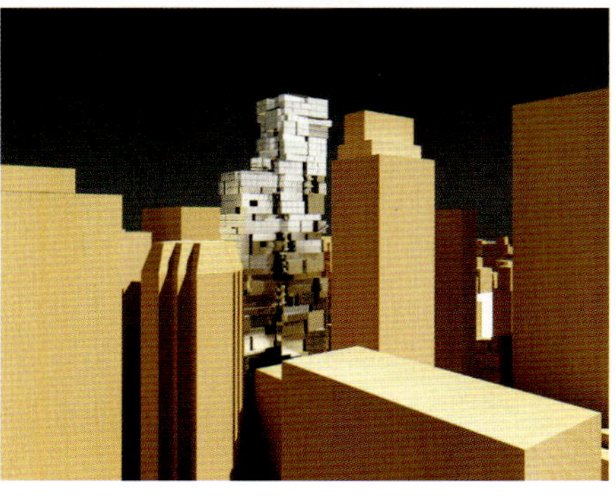

设计示意图
PROGRAM

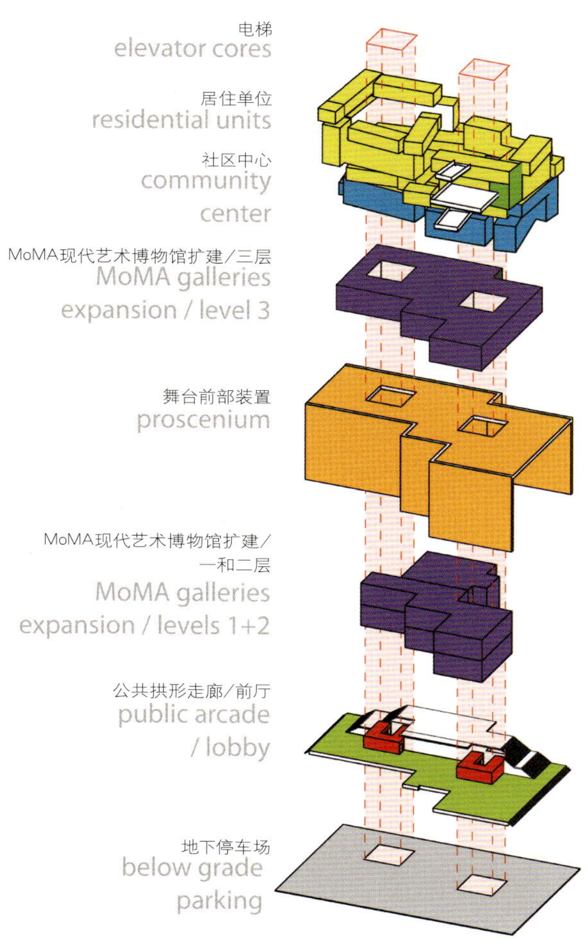
电梯
elevator cores
居住单位
residential units
社区中心
community center
MoMA现代艺术博物馆扩建/三层
MoMA galleries expansion / level 3
舞台前部装置
proscenium
MoMA现代艺术博物馆扩建/一和二层
MoMA galleries expansion / levels 1+2
公共拱形走廊/前厅
public arcade / lobby
地下停车场
below grade parking

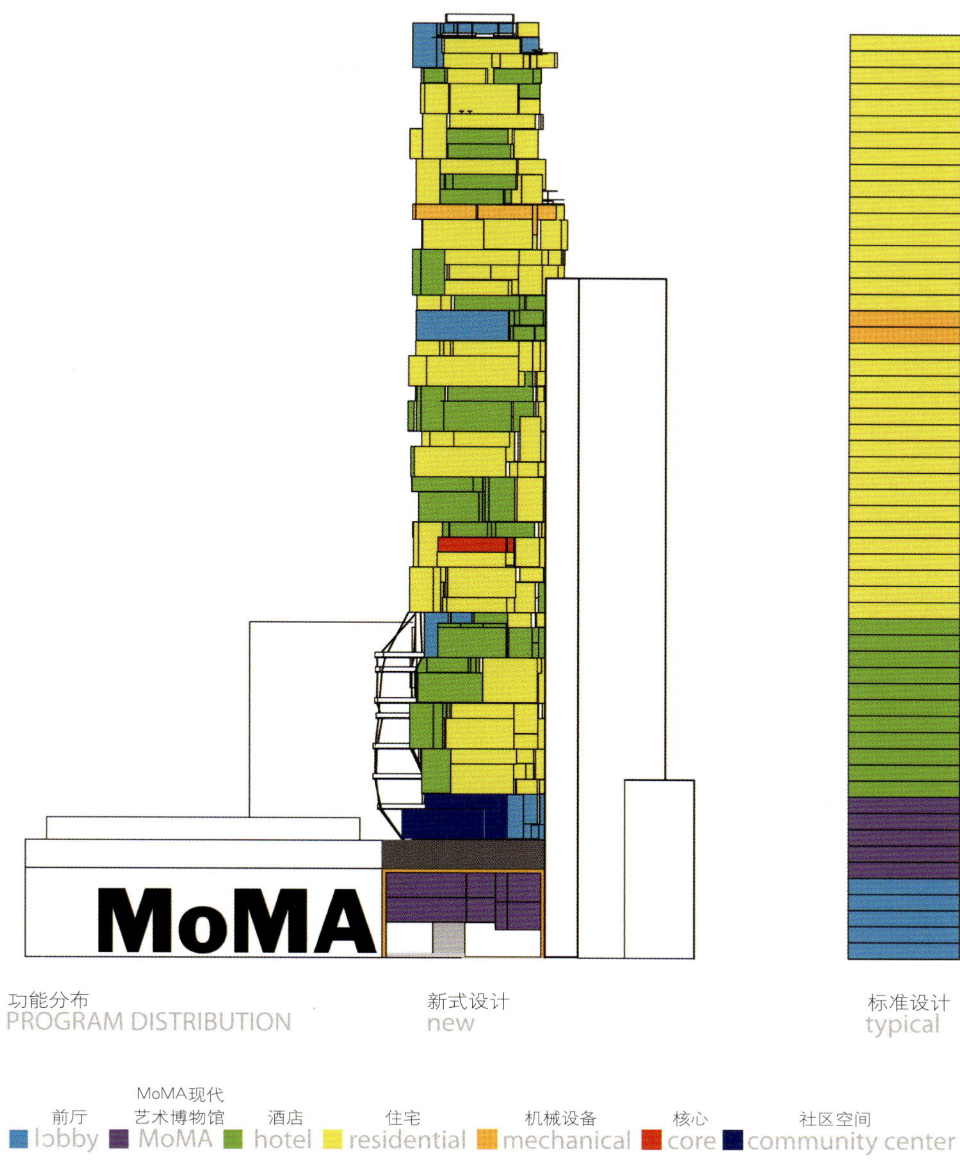
MoMA
功能分布
PROGRAM DISTRIBUTION
新式设计
new
标准设计
typical
前厅 lobby
MoMA现代艺术博物馆 MoMA
酒店 hotel
住宅 residential
机械设备 mechanical
核心 core
社区空间 community center

6th avenue
53rd street

54th street
6th avenue

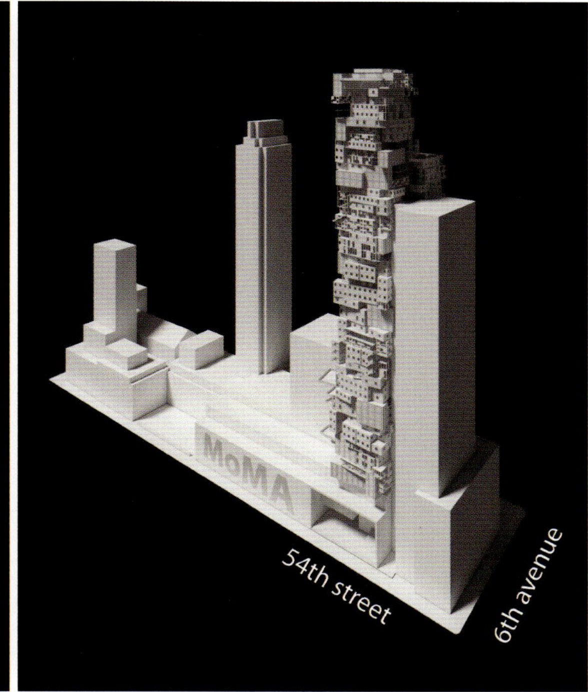
54th street
6th avenue

RESIDENTIAL UNIT SIZE 居住单位规模

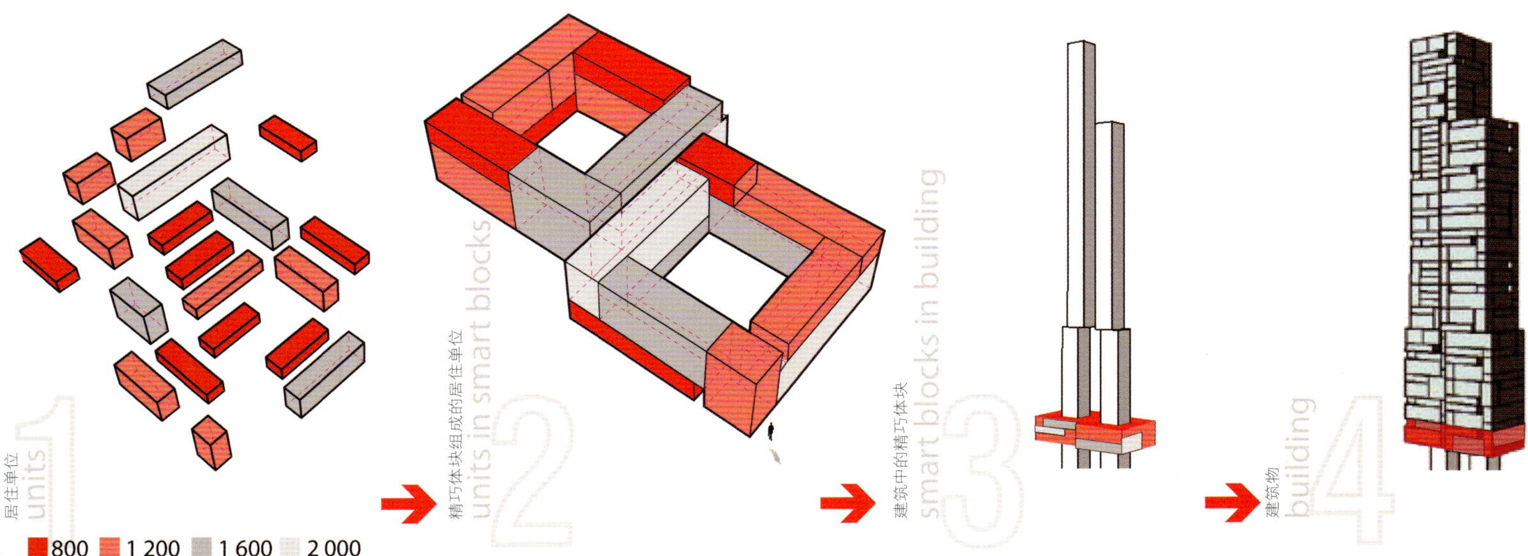

居住单位可水平或竖直结合，形成精巧的建筑体块

UNITS CAN BE COMBINED VERTICALLY OR HORIZONTALLY TO CREATE SMART BLOCKS

DENSITY CONTROL 密度控制

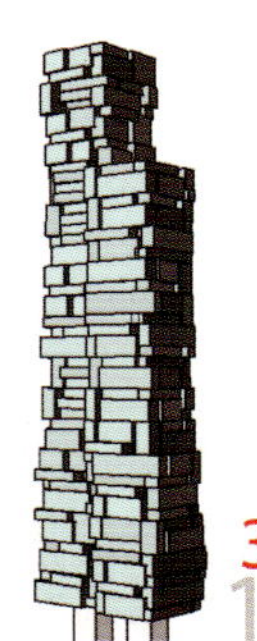

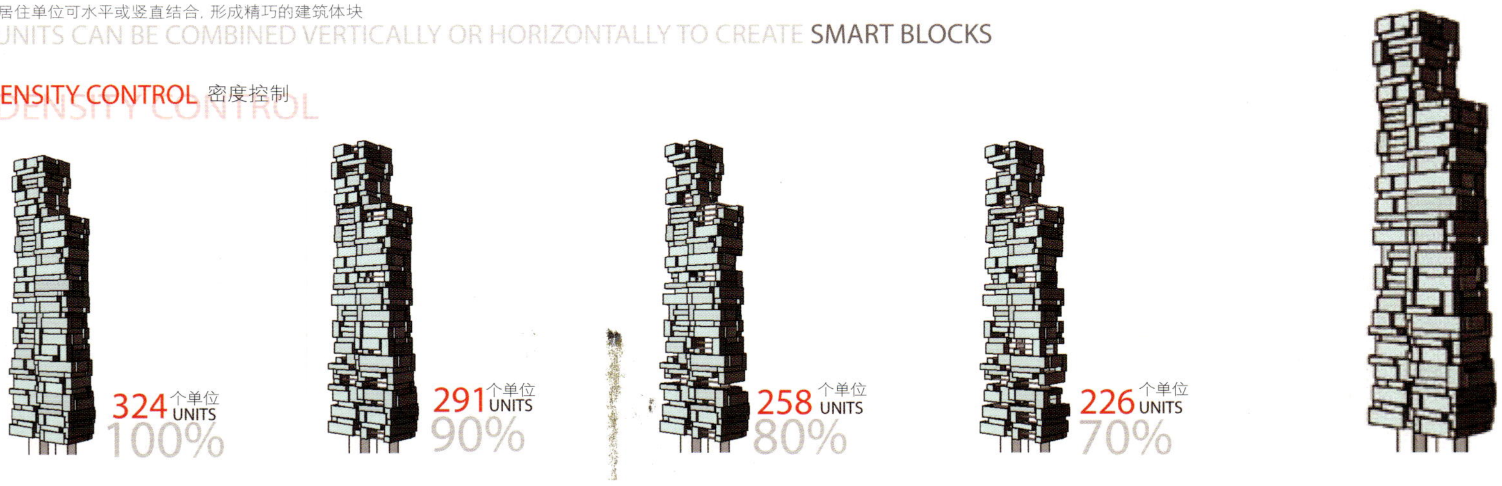

324 个单位 UNITS
100%

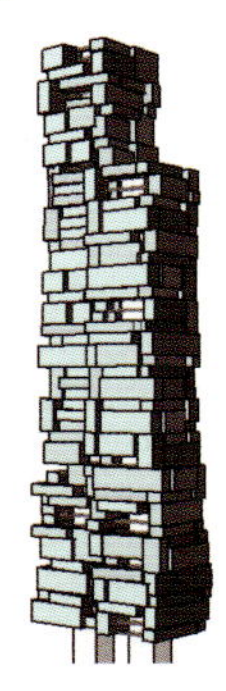

291 个单位 UNITS
90%

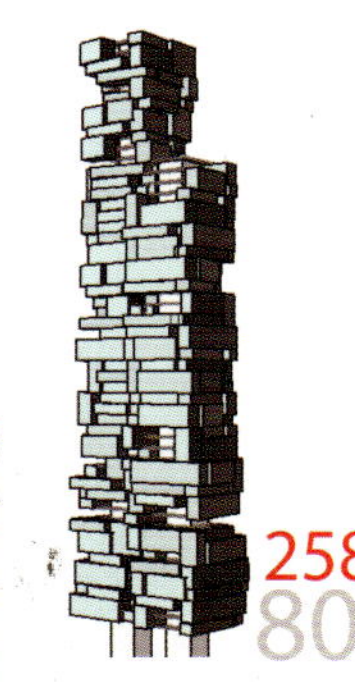

258 个单位 UNITS
80%

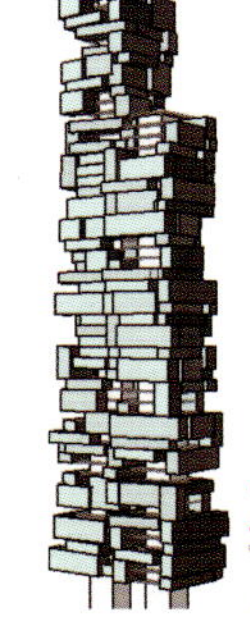

226 个单位 UNITS
70%

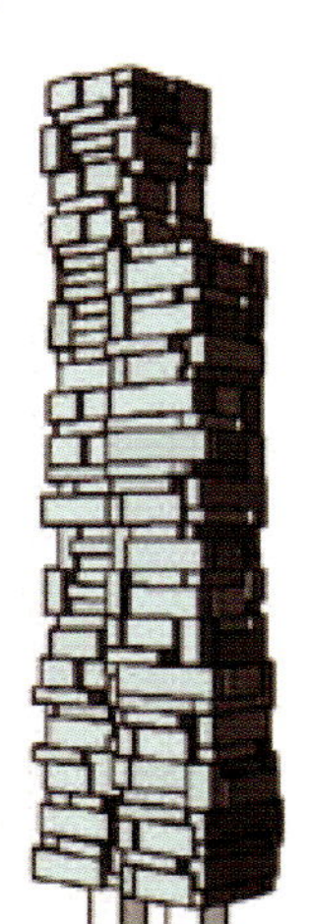

INTERSTITIAL GREEN SPACES 点缀在建筑中的绿色空间

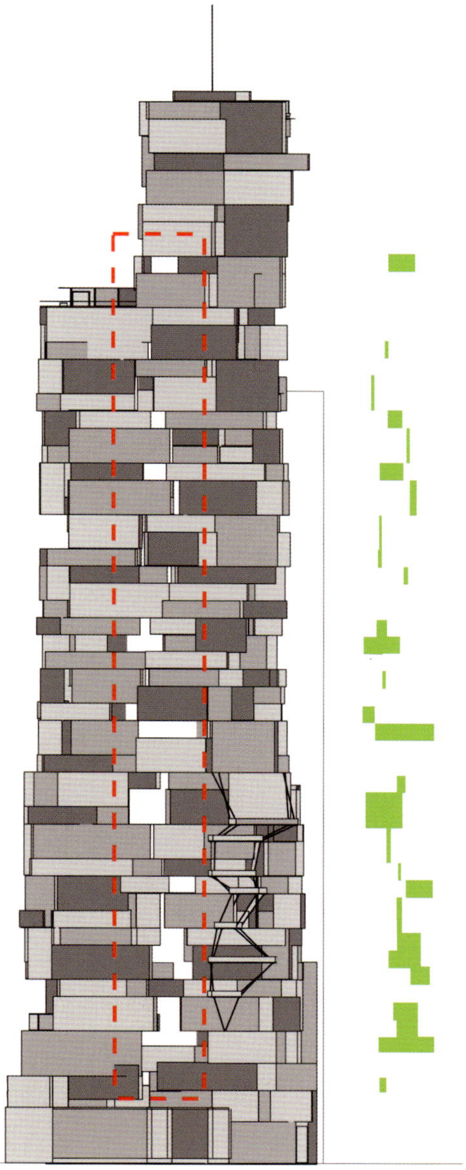

814'
672'
596'
588'
496'
491'

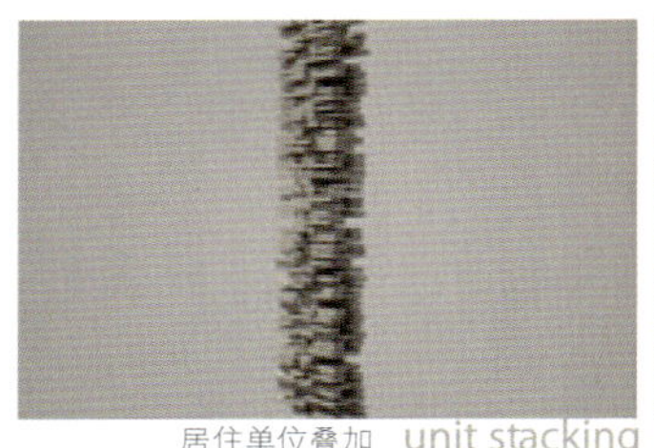
居住单位叠加 unit stacking

层次分明的立面 scaled facade

引人注目的大厦 attractor tower

自由把握密度 density randomizer

偶尔有居住单位转动方向 random unit rotation

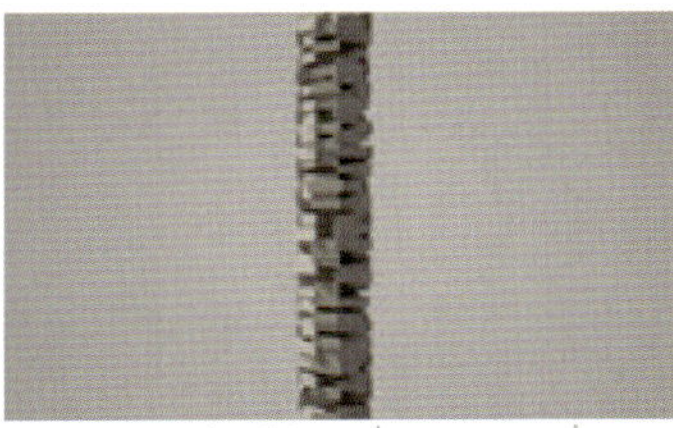
六角形核心 hexagonal core

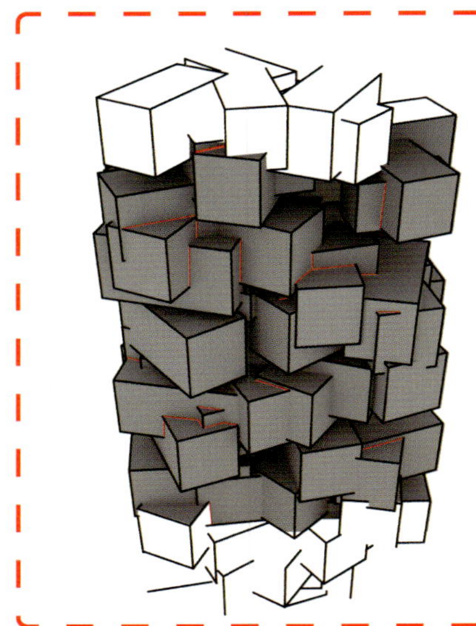

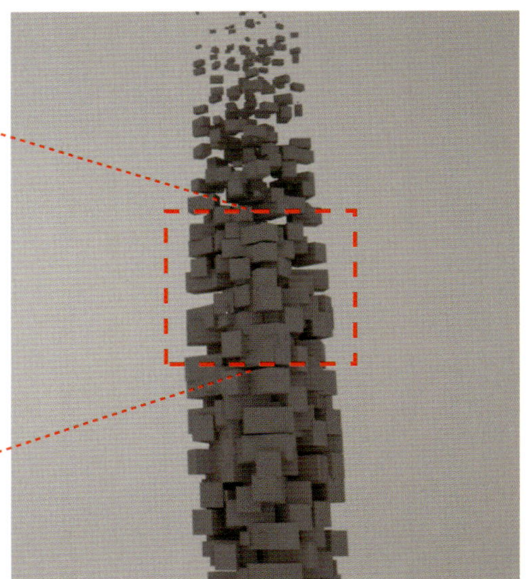

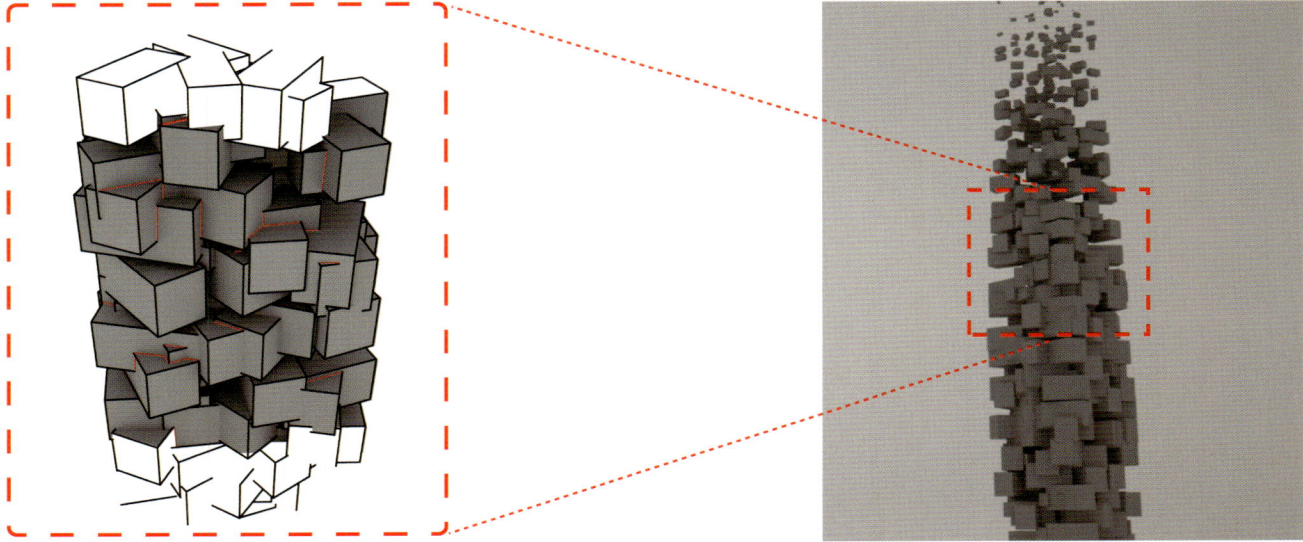

一层平面
GROUND FLOOR PLAN

西54大街
West 54th Street

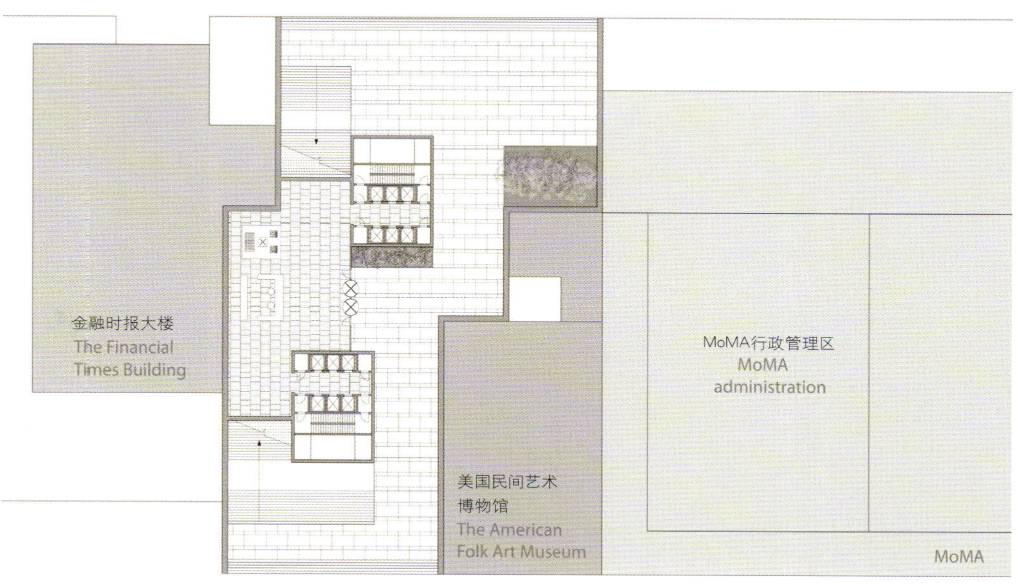

西53大街
West 53rd Street

海牙音乐与舞蹈中心
DANCE AND MUSIC CENTRE, THE HAGUE

建筑师：Mecanoo architecten, Delft, The Netherlands
结构工程师：Corsmit Raadgevend Ingenieurs, Rijswijk
声学、建筑物理、能源与可持续顾问：Peutz b.v., Zoetermeer
剧院顾问：Theateradvies bv, Amsterdam
甲方：City of the Hague
建筑地点：The Hague
用途：52 000 m^2剧院与音乐学校综合建筑，
其中有音乐厅（1500座）、音乐剧大厅（1000座）、
小音乐厅（500座）、小剧院（350座）、
门厅、办公室、餐厅、商店、250位停车场与自行车停放处

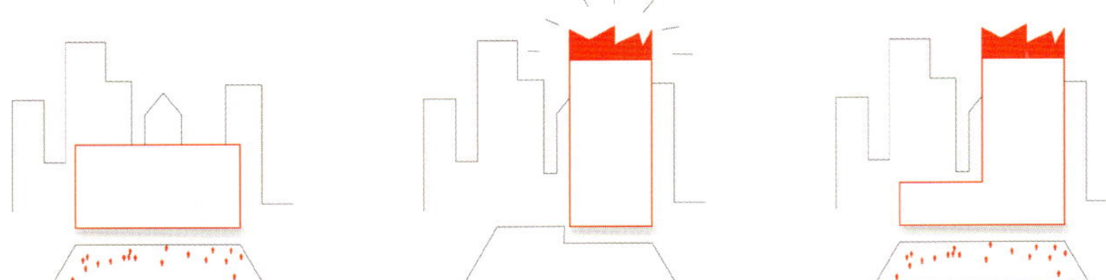

音乐与舞蹈中心像一个文化图标矗立在海牙市内，坐落在一个连接着架高公园的基座上。剧院与音乐学校被抬离地面，创造了一个透明的基座，基座内容纳了餐厅、商店以及广场上一个两层的入口大厅。建筑就好像一个巨大的展示窗，展示着它所有的文化功能。它不是一个封闭的剧院盒子，而是一个开放的、充满活力的建筑，在Spui广场上一眼就能看到。到了晚上，剧院内灯火通明，吸引着远处的观众来到这里。

The tower stands as a cultural beacon in the city and rests on a plinth with an elevated park. The theatres and the Conservatory are raised, creating a transparent plinth with restaurants, shops and a two storey entrance hall on the square. Like a grand showcase, the building exhibits all of its cultural functions. This is not a closed theatre box, but an open and inviting building full of vitality; palpable and visible from the Spui Square. From afar the building's theatrical lighting at night draws in the audience.

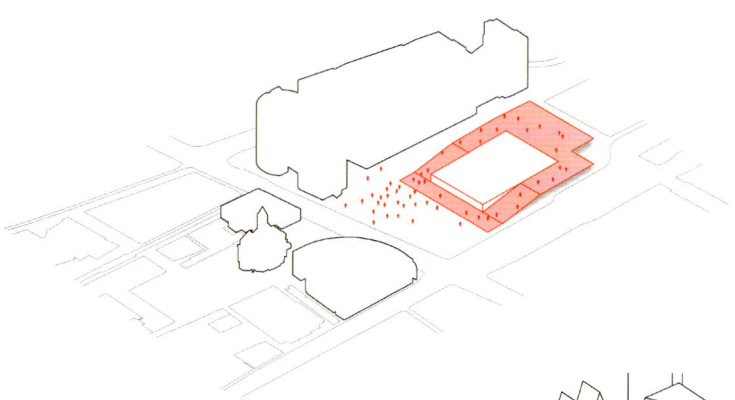

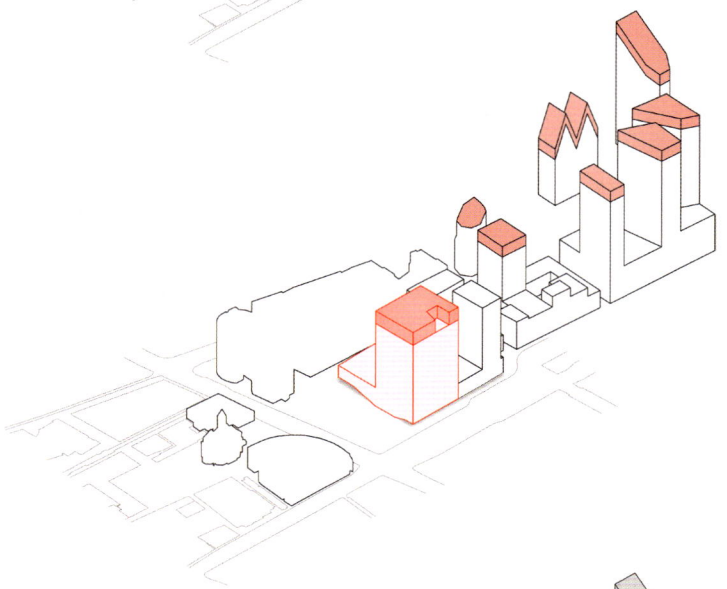

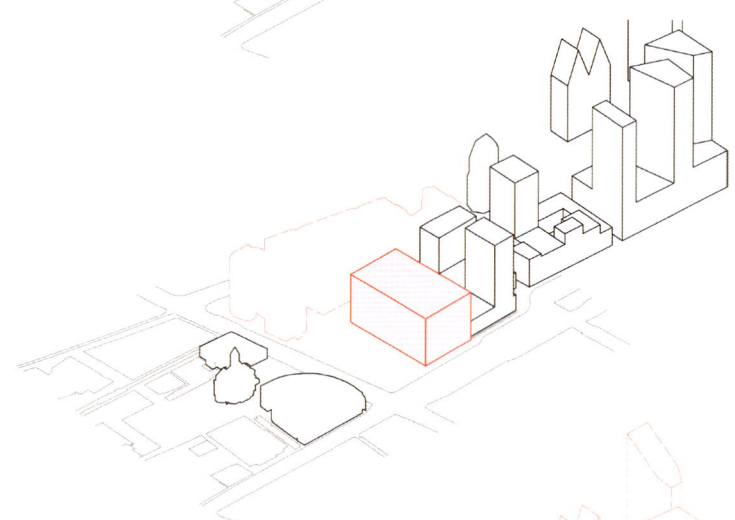

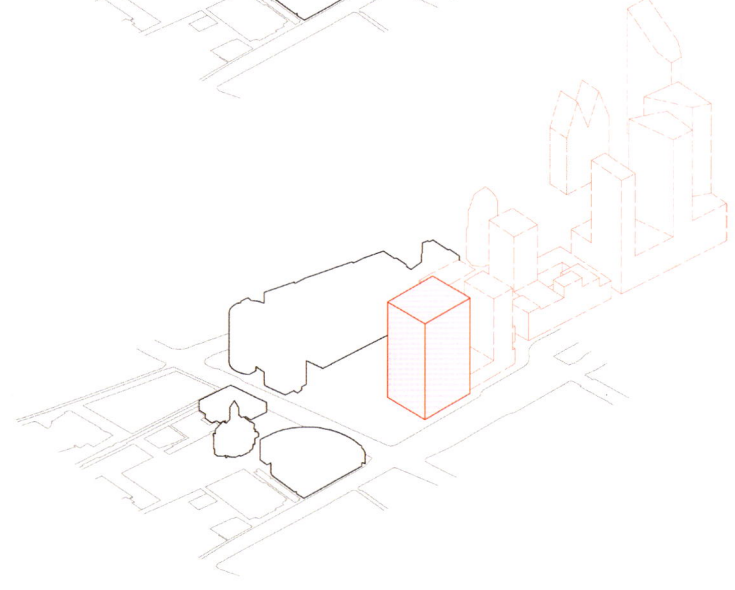

捷克罗兹托基的一所小学
ELEMENTARY SCHOOL IN ROZTOKY, CZECH REPUBLIC

建筑师：Marek Chalupa, Stepan Chalupa, Tomas Havlicek, Michal Rosicky/Chalupa architekti
甲方：Municipal Office of Roztoky, mayor Olga Vavrinova
效果图：Vit Musil, Radim Petruska / miss3 s.r.o.

该学校的校园扩建由"A""B""C"三栋楼组成。在现有"A"栋教学楼的基础上进行扩建，与现有单边走廊的镜像对称。现有的侧楼与新设计的平行花园环绕着新建的室内大厅，组成了一个矩形的环，这已成为学校的新核心区域。"A"楼体积小巧，外立面面积也小，带有一个开放的室内大厅，大大改善了建筑的热工性能及热学技术要求。还有一幢新设计的"C"楼，跨度较大，配有体育馆及图书馆，后者面向学校以及公众开放。玻璃高架桥把"C"楼与现有的"B"楼连接到了教学楼，"B"楼中有俱乐部与食堂。

The design of an extension of the school campus area consists of three individual buildings "A", "B", "C". The extension of an existing educational building "A" is designed as a mirrored copy of an existing single loaded corridor. The existing wing and a newly designed parallel garden wing are connected into a rectangular ring around a newly created inner hall, which has become a new core of the school. A compact volume of building "A" with its small area of external facade and an open space of the inner hall are the benefits that significantly improve the thermal technical characteristics of the building and its thermal technical demands. There is also a newly designed the individual large-span building "C" with school gymnasium and library for the school and public. This completely new object"C", as well as the existing individual building "B" with a school club and a school canteen are linked by glazed fly-over crossings to the educational building.

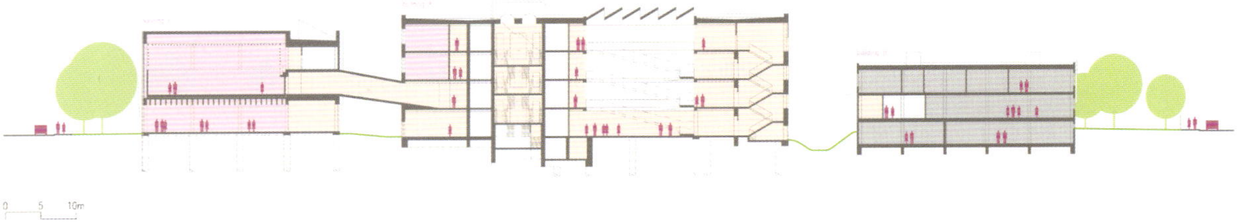

纵剖面
Longitudinal section

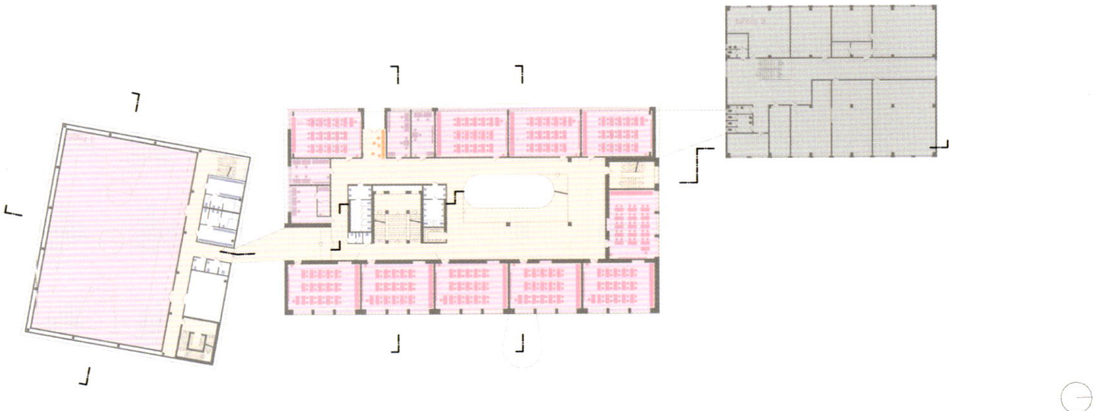

标准层
Typical floor

底层
Ground floor

现有学校
Existing school

扩建设计
Design of the extension

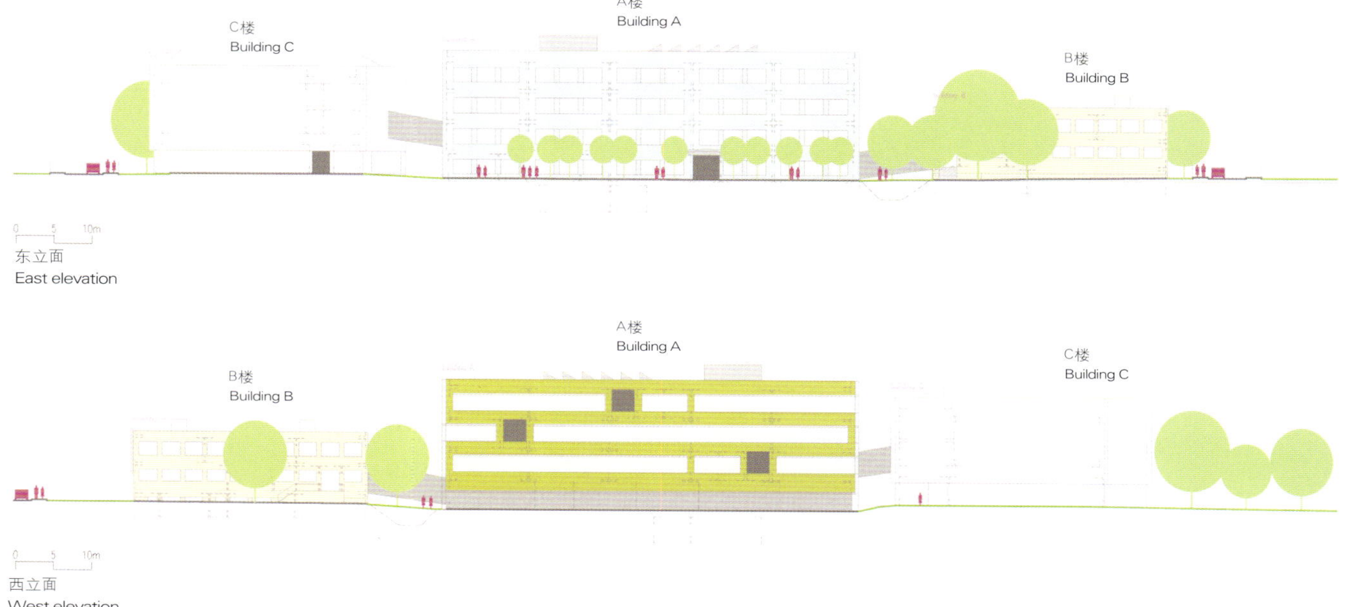

东立面
East elevation

西立面
West elevation

100 Global Architectural Schemes

国际最IN建筑设计

100

商业办公

2 COMMERCE AND OFFICE

日出塔楼
SUNRISE TOWER

甲方：Sunrise Berhad
建筑师：Zaha Hadid Architects
设计：Zaha Hadid with Patrik Schumacher
工程师：Buro Happold

扎哈·哈迪德建筑师事务所为日出塔楼所做的设计以多种方式与城市相融合。通过在不同层面开发协同效应并将建筑融入原有的城市肌理中，设计创造出一座与周围设施相交接的服务平台。建筑规划将多种功能融为一体，将其自身与传统的塔楼和裙楼类型相区别。得益于精细的景观设计，塔楼与地面交织在一起，将场地上不同的区域扩展并连接起来，使新的人形步道和内部的道路系统融为一体，并为新开发的设施规划出肌理。

建筑包括五种功能：住宅、酒店、写字间、零售商店和停车场。为了创造出流畅的设计，各个部分之间的连通性对于项目来说是非常重要的，这样就可以将各个部分的规模与不同的功能整合在一起，形成连贯的规划。建筑的各个功能区分层设置，或垂直分布或在塔楼的两个分支处平行分布。

Zaha Hadid Architects' design for Sunrise Tower engages with the city in multiple ways. By exploring potential synergies at different levels and anchoring itself to the existing urban fabric, it creates a platform of services that engage with neighbouring developments. The scheme merges all programmes into one building, distancing itself from the traditional tower and podium typology. Through a detailed landscape strategy the design interweaves tower and ground, extending and connecting the different parts of the site, integrating the new pedestrian routes and internal road system, structuring the fabric of the new development.

The design houses 5 different programmatic components: residential, hotel, offices, retail and parking. Connectivity between these parts becomes central to the project in order to produce an articulated design that encompasses both the scale and the different qualities of each of the parts, fusing them into a coherent scheme. The program is stratified, stacking one function over the other, or carrying them in parallel when the tower branches.

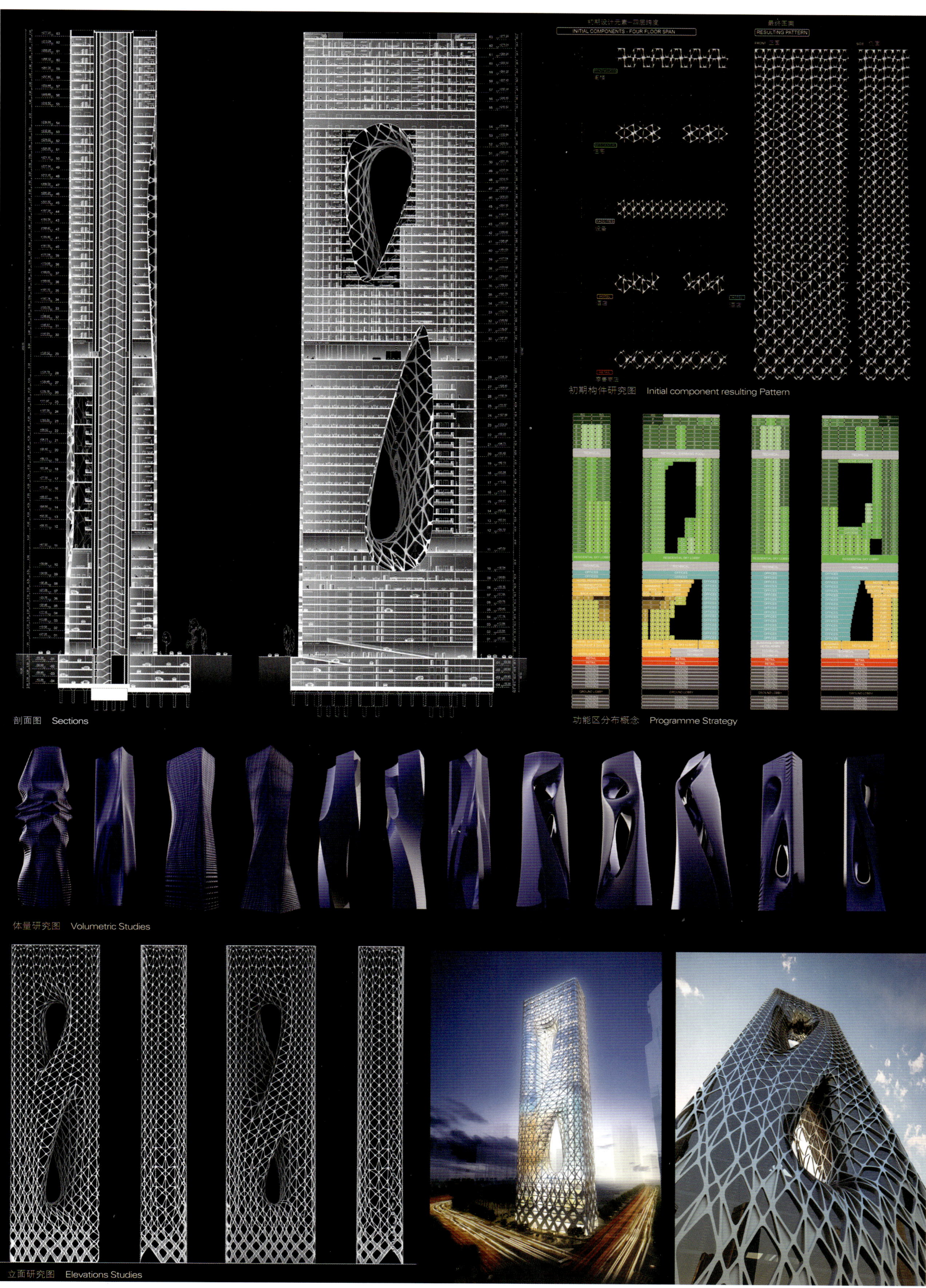
初期设计元素—四层跨度
INITIAL COMPONENTS - FOUR FLOOR SPAN
最终图案
RESULTING PATTERN
初期构件研究图 Initial component resulting Pattern
剖面图 Sections
功能区分布概念 Programme Strategy
体量研究图 Volumetric Studies
立面研究图 Elevations Studies

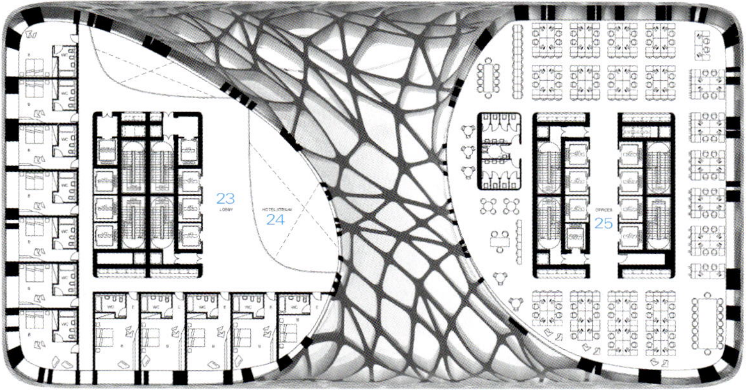

十二层平面图
酒店房间
Plan 12 hotel room/typical offices

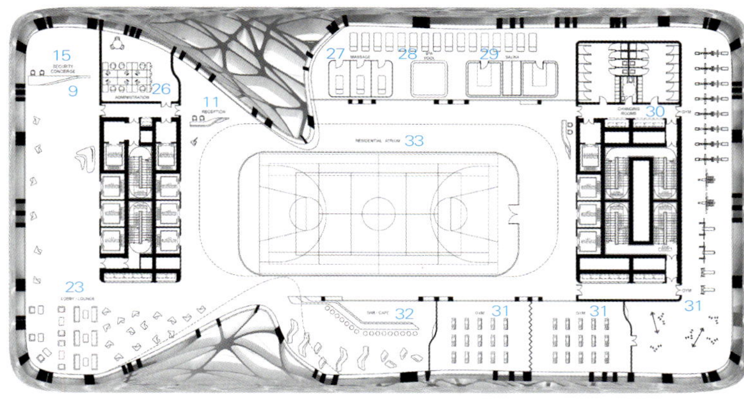

二十九层平面
住宅空中走廊-水疗馆/健康会所
Plan 29 residential sky lobby spa / health club

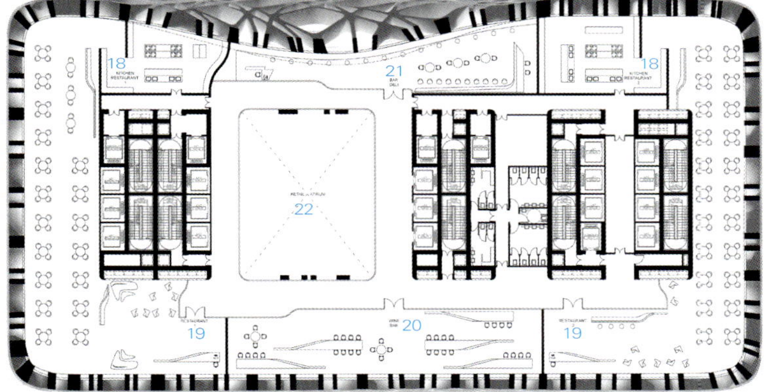

九层平面图
零售商店
Plan 9 retail

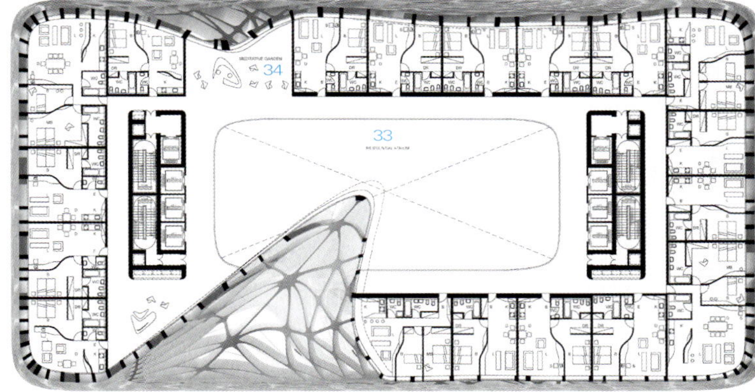

三十三层平面图
Plan 33 typical residential

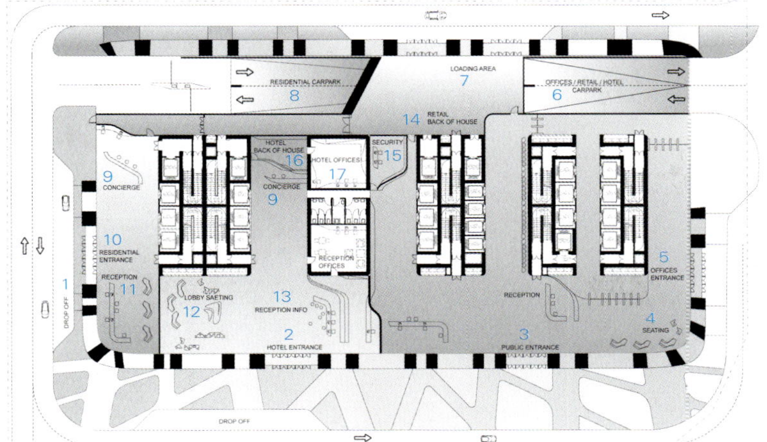

一层平面图
Ground floor plan

1　下车处
2　酒店入口
3　公共入口
4　座位区
5　写字间入口
6　写字间／零售商店／酒店用停车场
7　装卸区
8　住宅用停车场
9　门房
10　住宅入口
11　接待处
12　大厅座位区
13　接待咨询处
14　零售商店后台
15　保安处
16　酒店后台
17　办公室
18　餐厅厨房
19　餐厅
20　酒吧
21　熟食店
22　零售商店中庭
23　大厅
24　酒店中庭
25　办公室
26　行政管理处
27　按摩室
28　水疗泳池
29　桑拿室
30　更衣室
31　健身房
32　酒吧／咖啡厅
33　住宅中庭
34　冥想花园

斯德哥尔摩的大陆大厦
CONTINENTAL TOWER, STOCKHOLM

建筑师：C. F. Møller Architects
景观建筑师：C. F. Møller Architects
图片：C. F. Møller Architects
地点：Centralstationen, Stockholm, Sweden
甲方：FOLKSAM in co-operation with: Citybanan, Scandic Hotels, Stockholms Stadsbyggnadskontor, and Stockholms Exploateringskontor

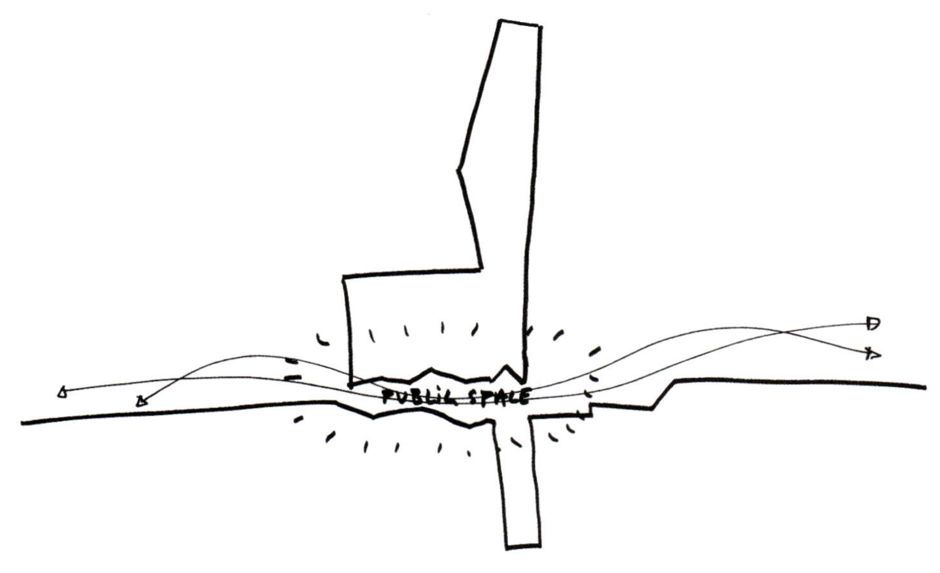
PUBLIC SPACE

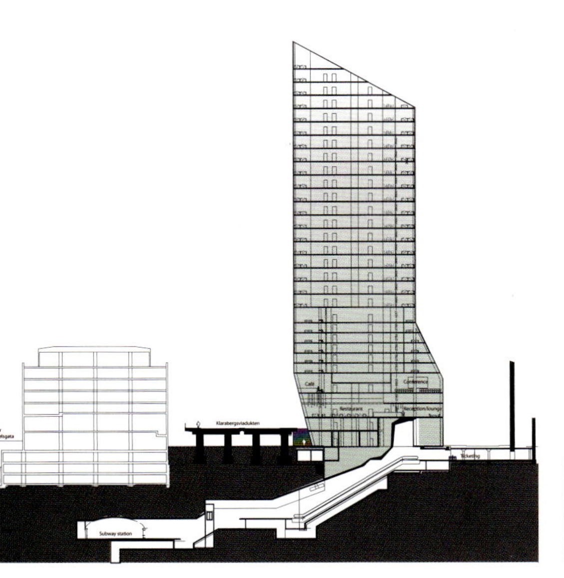

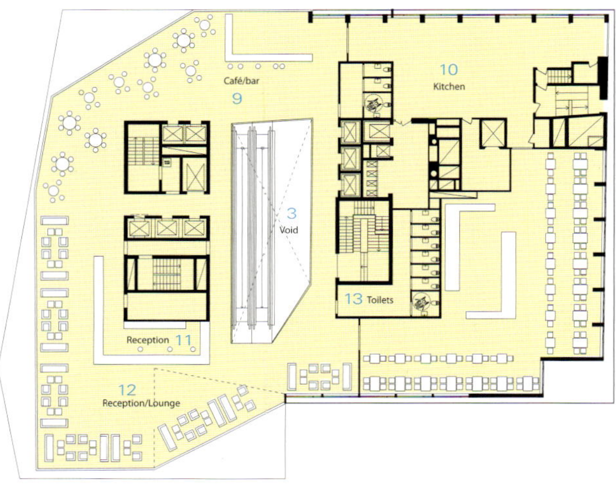

5层平面图
Fourth floor plan

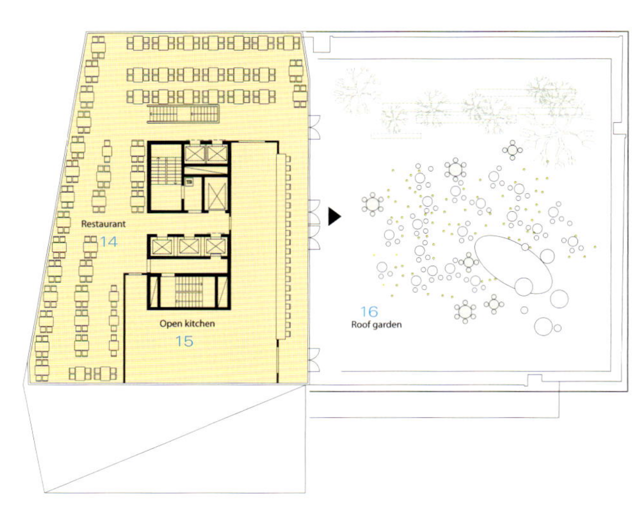

15层平面图
Fourteenth floor plan

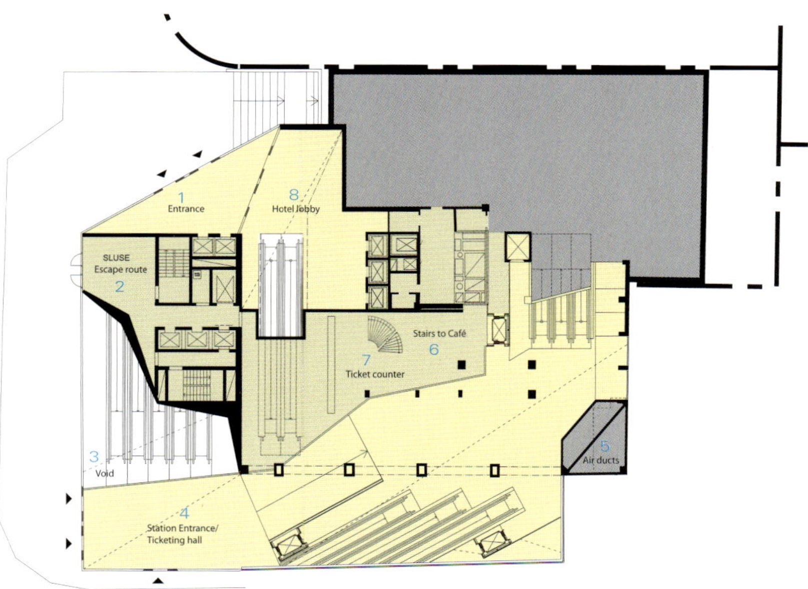

3层平面图
Second floor plan

11层（标准层）平面图
Tenth floor standard

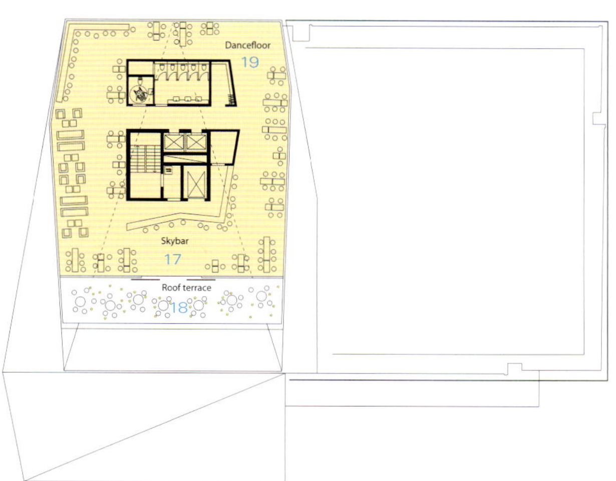

29层平面图/空中酒吧
Twenty - eighth floor plan/sky bar

1 入口
2 逃生通道
3 上空空间
4 车站入口/售票大厅
5 通风管道
6 通向咖啡馆的楼梯
7 售票柜台
8 酒店大堂
9 咖啡馆/酒吧
10 厨房
11 接待处
12 接待处/休息室
13 卫生间
14 餐厅
15 开放式厨房
16 屋顶花园
17 空中酒吧
18 屋顶露台
19 舞池

建于斯德哥尔摩的塔式大楼计划与城市铁道线路的扩建部分相连接。这座100m高、约30层的建筑将包括酒店和办公室，可能的话还包括住宅。建筑将配备一个通往与街道齐平的新地铁站的入口。新地铁站就在原来的Centralen站面。建筑将成为斯德哥尔摩新地铁的标志。

The tower block construction in Stockholm has been planned in connection with the extension of the City Line rail link. The complex of 100 metres, the equivalent of approximately 30 stories, will include a hotel, offices and possibly housing. The building will be equipped with an entrance to a new station at street level directly opposite the city's old main railway station, Centralen, and will therefore be a landmark for Stockholm's new metro.

曼哈顿下城区的埃德加街大厦
EDGAR STREET TOWERS, LOWER MANHATTAN

建筑师：IwamotoScott

主持建筑师：Lisa Iwamoto & Craig Scott

项目团队：Ryan Golenberg, Stephanie Lin, John Kim, Blake Altshuler

图片提供：IwamotoScott（除了P182页大图由Transparent House提供）

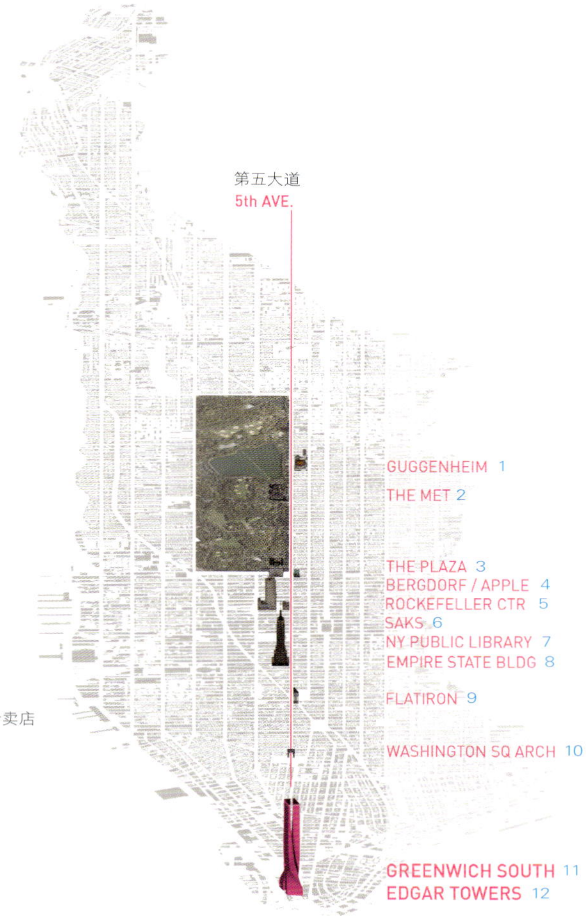

1 古根海姆博物馆
2 纽约大都会艺术博物馆
3 曼哈顿广场酒店
4 波道夫·古德曼／苹果专卖店
5 洛克菲勒中心
6 SAKS第五大道百货
7 纽约公共图书馆
8 帝国大厦
9 熨斗大厦
10 华盛顿公园拱门
11 格林威治村＆苏活区
12 埃德加街大厦

第五大道
5th AVE.

GUGGENHEIM 1
THE MET 2
THE PLAZA 3
BERGDORF / APPLE 4
ROCKEFELLER CTR 5
SAKS 6
NY PUBLIC LIBRARY 7
EMPIRE STATE BLDG 8
FLATIRON 9
WASHINGTON SQ ARCH 10
GREENWICH SOUTH 11
EDGAR TOWERS 12

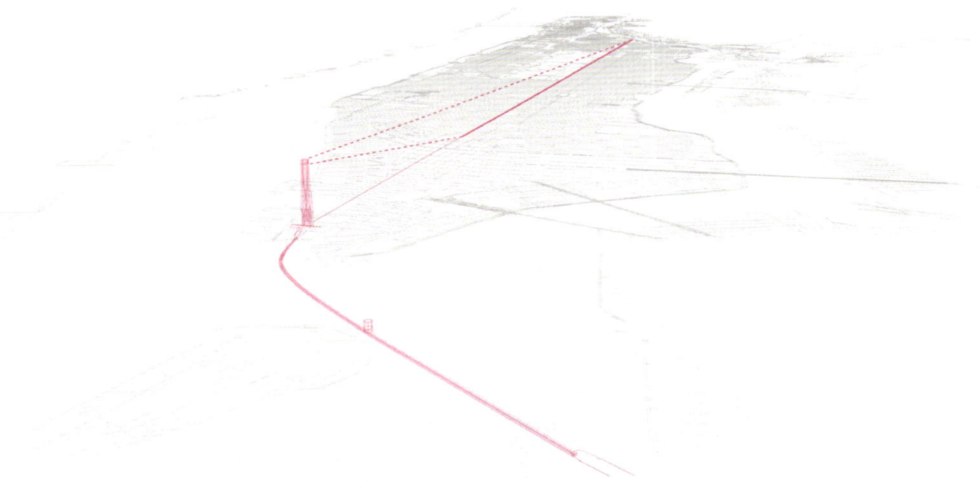

该大厦的设计目的在于让埃德加街成为东西部地区的公共道路，重新连接格林威治和华盛顿大街。这条通道的空间通过建筑的旋转上升，在中间楼层收拢，从而得到更大面积的楼板，并最终到达屋顶的空中大厅和公共空间。这座大厦的屋顶空间与曼哈顿大街网格成一条直线，正位于第五大道的轴线上。从宏观角度来说，埃德加街大厦充分利用了建筑基地所提供的清晰度和显著性，将以其充满活力的形式在这座城市中为来来往往的人们担当起城市地标的重任。

Twisting SkyVoid
from Edgar Street to sky
1

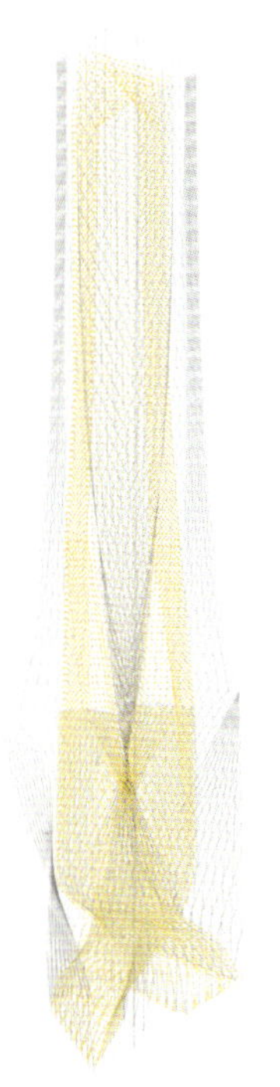

Fiberoptic Daylight Mesh
w/ glazing wireframe & atrium
2

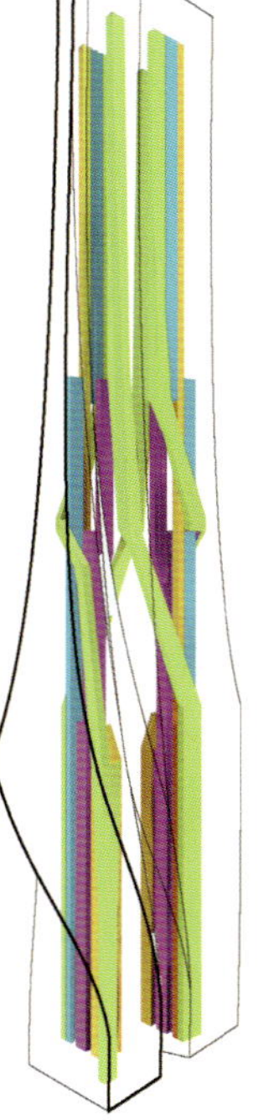

Infra-Structure
cores as interior structure
3

Programmatic Zoning
Floorplates and sky lobbies
4

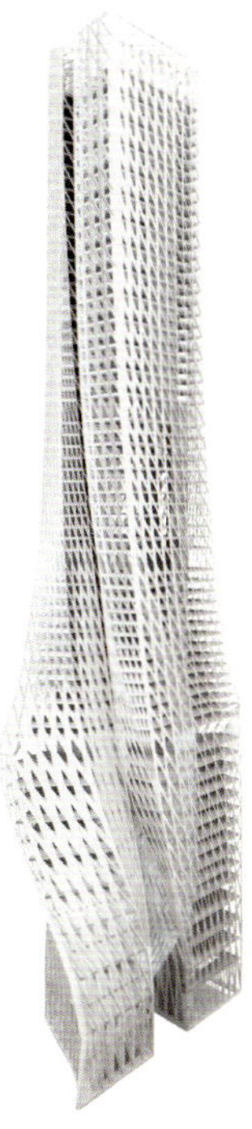

Structural Skin
modulated exoskeleton
5

1 扭曲的采光天窗上空空间
从地面直达天庭
2 光纤遮阳网
安装有玻璃窗、线框与中庭
3 基础设施
作为内部结构的核心
4 设计分区
楼板与采光门厅
5 结构表皮
适应性外部骨架

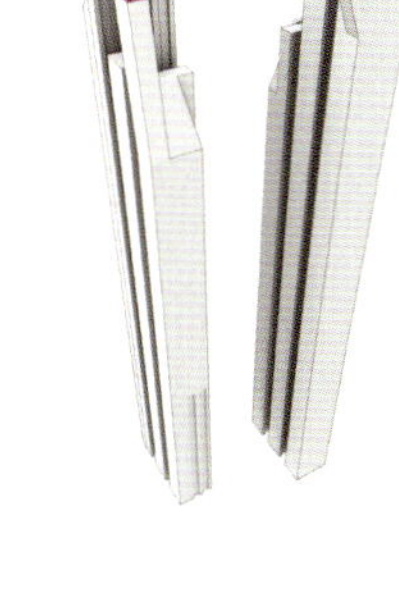

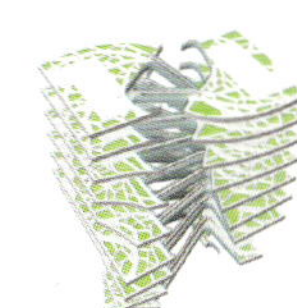

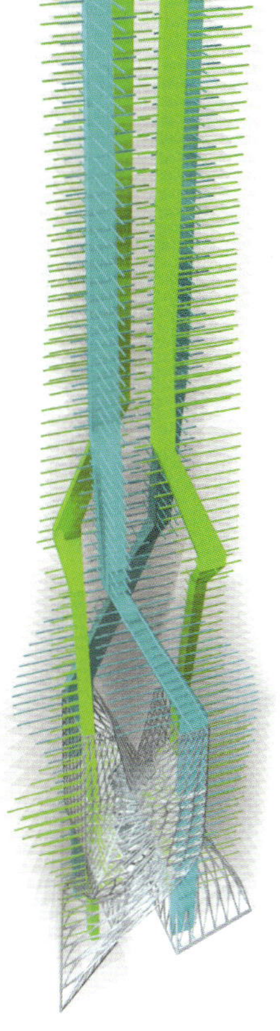

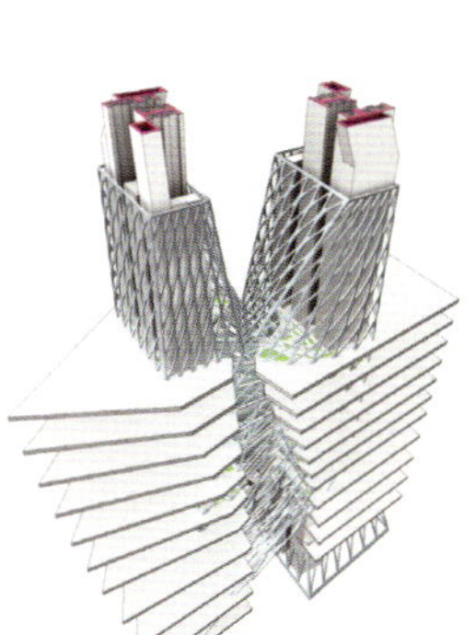

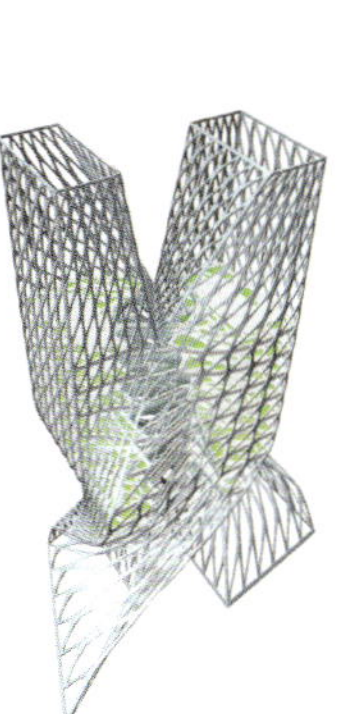

中庭的生物过滤
Atrium biofiltration

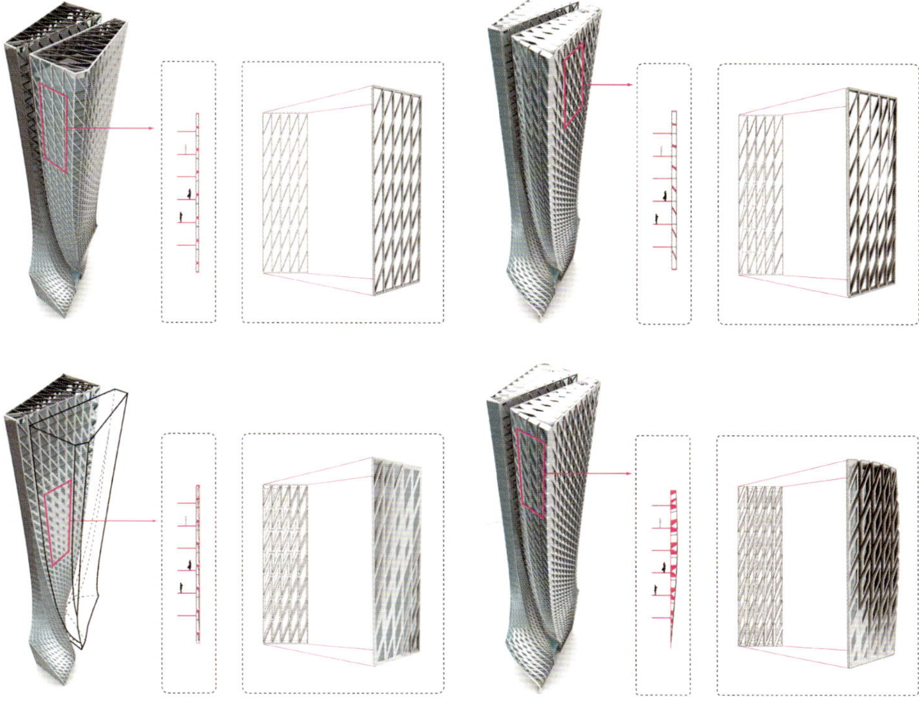

适应性表皮
Modulated skin

The towers' design seeks to reinstate Edgar Street as an east-west public way, reconnecting Greenwich and Washington streets. The space of this passageway through the building twists upwards, pinching at the mid level to allow for larger floor plates, and culminating at a rooftop sky lobby and civic space. This space at the towers' crown is aligned with the primary Manhattan street grid to the north, directly on axis with 5th Avenue. On a macro scale, Edgar Street Towers takes advantage of the visibility and prominence offered by its site, where its dynamic form acts as a civic landmark and beacon for those coming to and leaving the city.

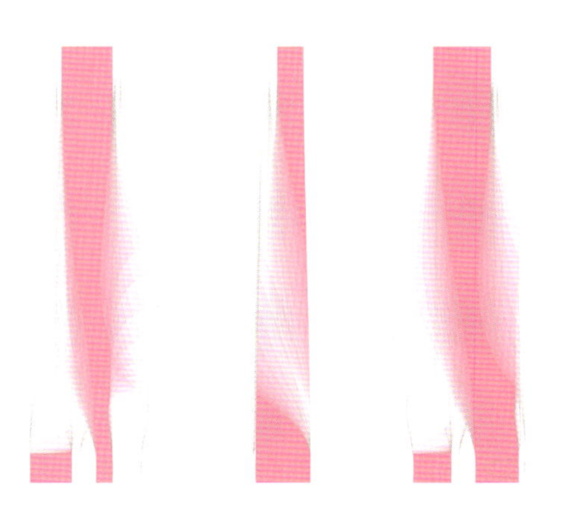

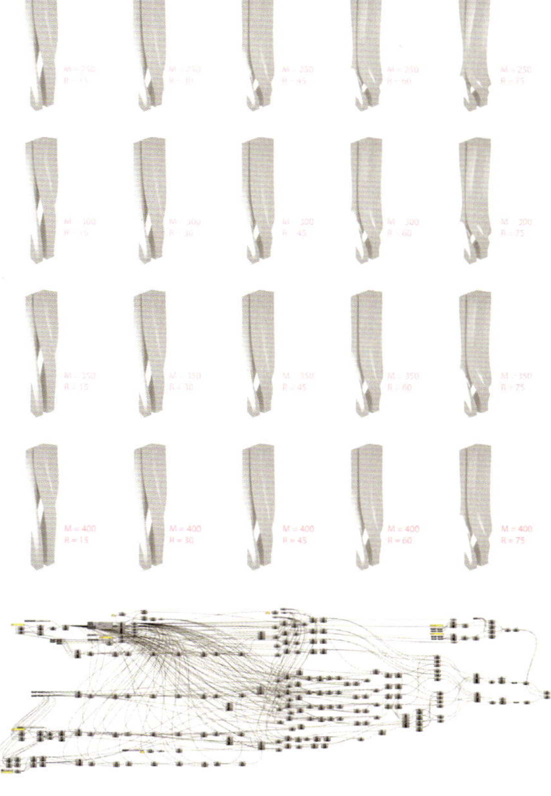

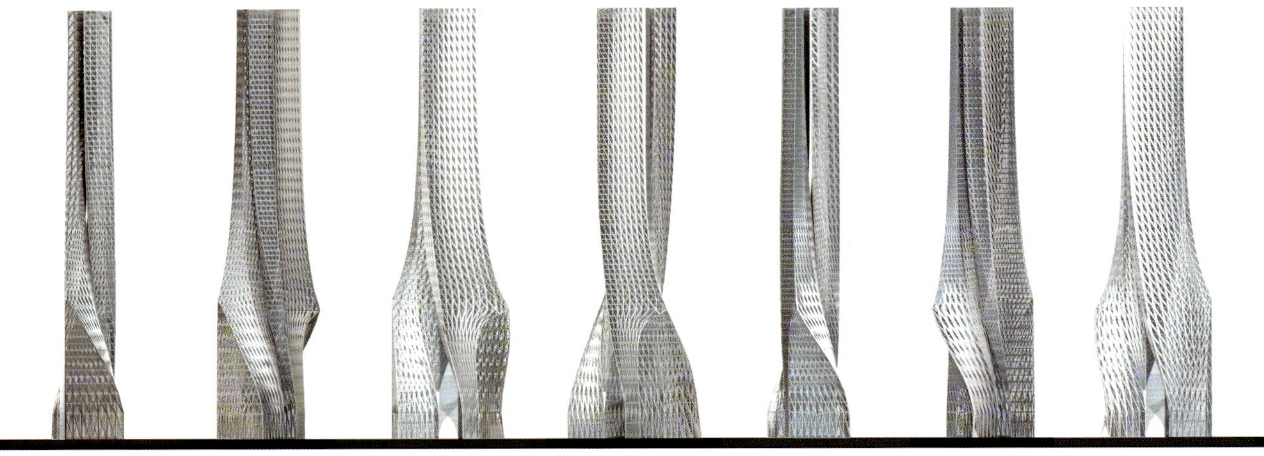

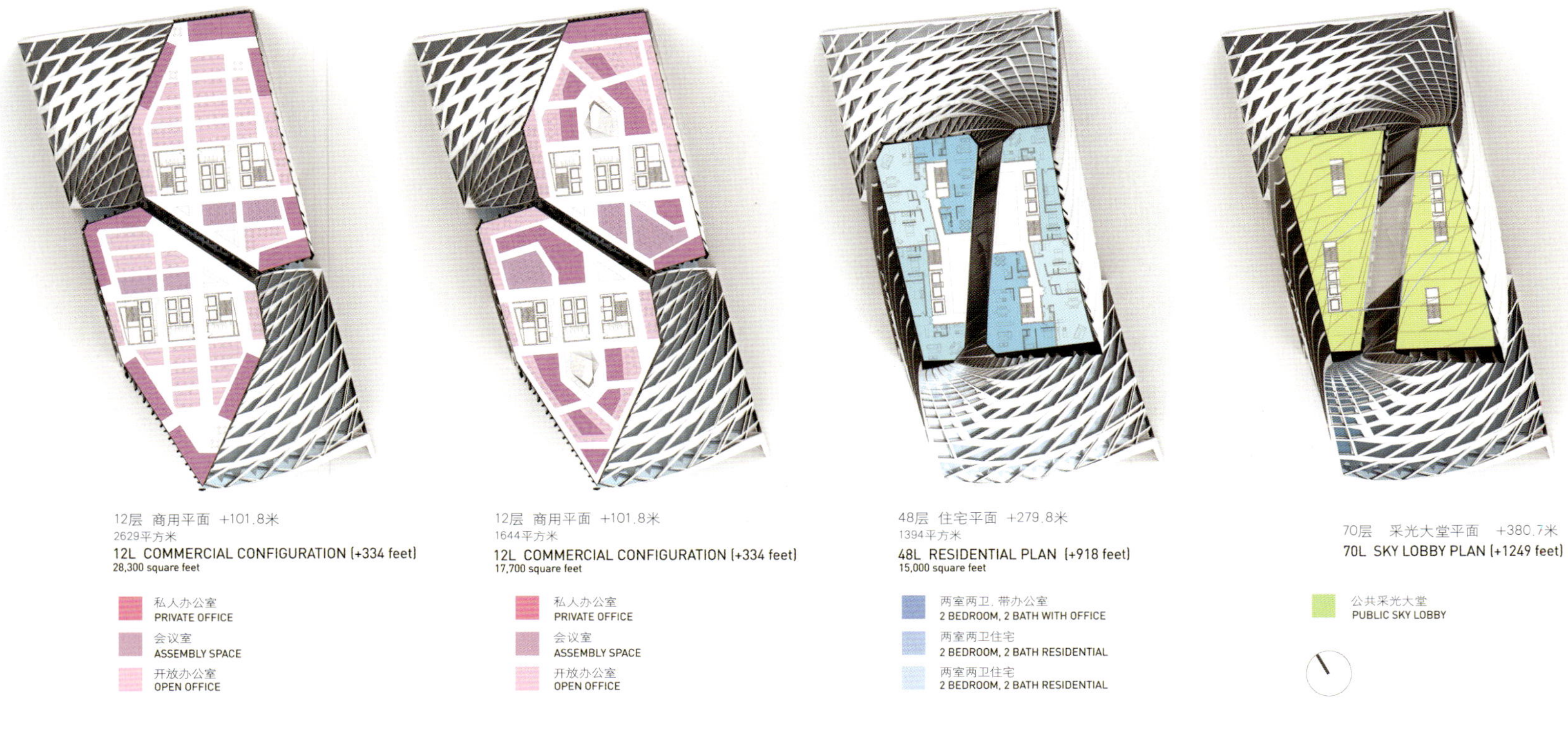

12层 商用平面 +101.8米
2629平方米
12L COMMERCIAL CONFIGURATION (+334 feet)
28,300 square feet
私人办公室
PRIVATE OFFICE
会议室
ASSEMBLY SPACE
开放办公室
OPEN OFFICE
12层 商用平面 +101.8米
1644平方米
12L COMMERCIAL CONFIGURATION (+334 feet)
17,700 square feet
私人办公室
PRIVATE OFFICE
会议室
ASSEMBLY SPACE
开放办公室
OPEN OFFICE
48层 住宅平面 +279.8米
1394平方米
48L RESIDENTIAL PLAN (+918 feet)
15,000 square feet
两室两卫，带办公室
2 BEDROOM, 2 BATH WITH OFFICE
两室两卫住宅
2 BEDROOM, 2 BATH RESIDENTIAL
两室两卫住宅
2 BEDROOM, 2 BATH RESIDENTIAL
70层 采光大堂平面 +380.7米
70L SKY LOBBY PLAN (+1249 feet)
公共采光大堂
PUBLIC SKY LOBBY

楼层平面图
Floor plans

法国巴黎的修道院广场大厦
HERMITAGE PLAZA IN PARIS, FRANCE

建筑师：Foster+Partners

甲方：Hermitage

顾问：Leslie E. Robertson Associates and Terrell, Michel Desvigne, Arup, Assistance Prevention Expertise, Bureau Veritas, DSA Engineering Ltd, Emmer Pfenninger Partner AG, Systematica, Turner International LLC

修道院广场大厦将为法国库尔贝伏瓦地区的拉德芳斯东部创造一个新型社区，该地区的中心沿着塞纳河边设有咖啡店、商店及阳光明媚的公共广场。设计包括两座323m高的建筑，这个新兴城市建筑群将在巴黎的天际线上创造一道与众不同的风景。在与其他建筑公司及事务所的紧密合作之下，该项目将通过创造可持续发展和高密度的社区，为拉德芳斯东部地区注入新鲜活力。两座大楼将容纳酒店、spa、全景公寓、办公室、酒店式公寓，以及基座中的商店。

Hermitage Plaza will create a new community to the east of La Défense, in Courbevoie, that extends down to the river Seine with cafés, shops and a sunny public plaza at its heart. The design incorporates two 323-metre-high buildings, which will establish a distinctive symbol for this new urban destination on the Paris skyline. The result of a close collaboration with other studios and firms, the project is intended to inject life into the area east of La Défense by creating a sustainable, high-density community. The two towers accommodate a hotel, spa, panoramic apartments, offices and serviced apartments, as well as shops at the base.

印度泰姬陵大地酒店
TAJ LANDS END, INDIA

建筑师：Yazdani Studio of Cannon Design

设计团队：Mehrdad Yazdani（设计主管）；Mark Erdly, AIA（项目主管）；Philip Ra, AIA（高级设计师）；Tarun Kumar, AIA（项目经理）；Yan Krymsky（设计师）；Mimi Lam, LEED AP; Jayashri Deshmukh, OAA; Kiduck Kim, LEED AP; Jenny Tse, LEED AP; Wai Yim; Hyun Jin; Dorin Popa

酒店外形上的"波浪"以连绵不断的姿态完美地涵盖了店内众多的公共和私人空间，立即在孟买直线型的天际线上刻画出与众不同的风景。建筑西边和南边的波谷部分是阶地状的瀑布水景和平台，这使得酒店的户外运动项目空间延伸到了海滩，并将其连接到更远处的海滨。前厅、宴会空间、餐厅以及零售区域都位于宽阔的基座中，围绕着中间的上升中庭，中庭靠采光天窗照明。波浪的弧线向着基地的北部逐渐升高，其中的客房可欣赏到充满活力的城市景色和海景，同时也保护了当地的景色。

In a single sweeping gesture, the "wave" gathers the multitude of public and private spaces within the new hotel into a seamless composition, immediately iconic against the rectilinear skyline of Mumbai. At the trough of the wave along western and southern edges of the building is a cascade of terraces and decks, which extend the outdoor event spaces of the hotel into the foreshore area and connect them to the sea's edge beyond. Within its generous base, lobbies, banquet spaces, restaurants and retail functions surround a rising sky lit atrium. The arc of the wave rises towards the north end of the site, offering the rooms within dramatic city and sea views, and also preserving views from Taj Lands End.

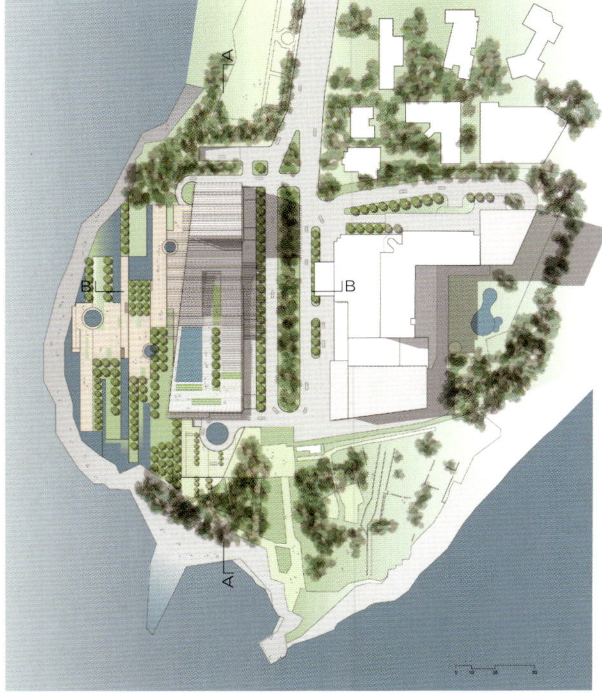

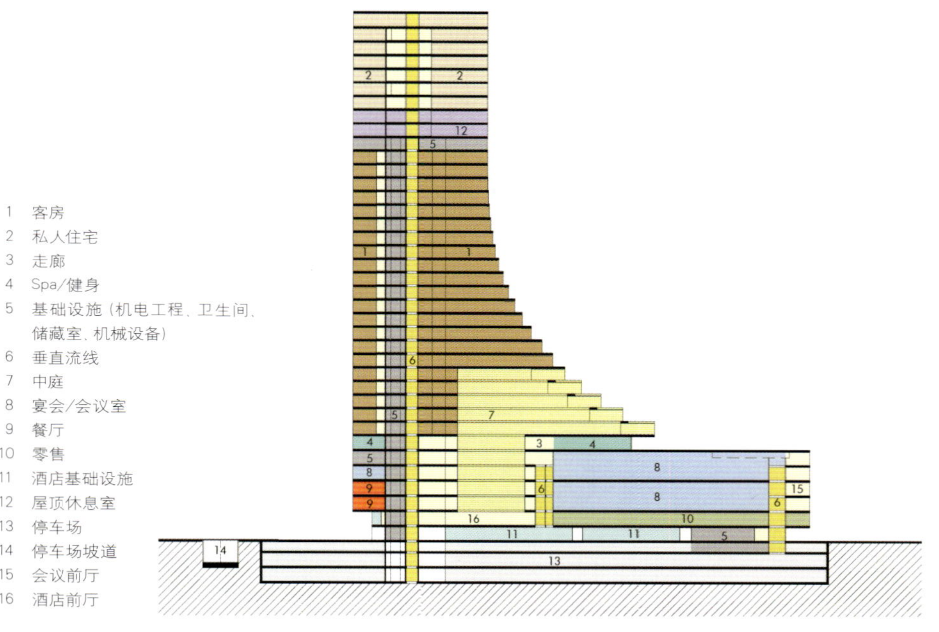

AA剖面

Section AA

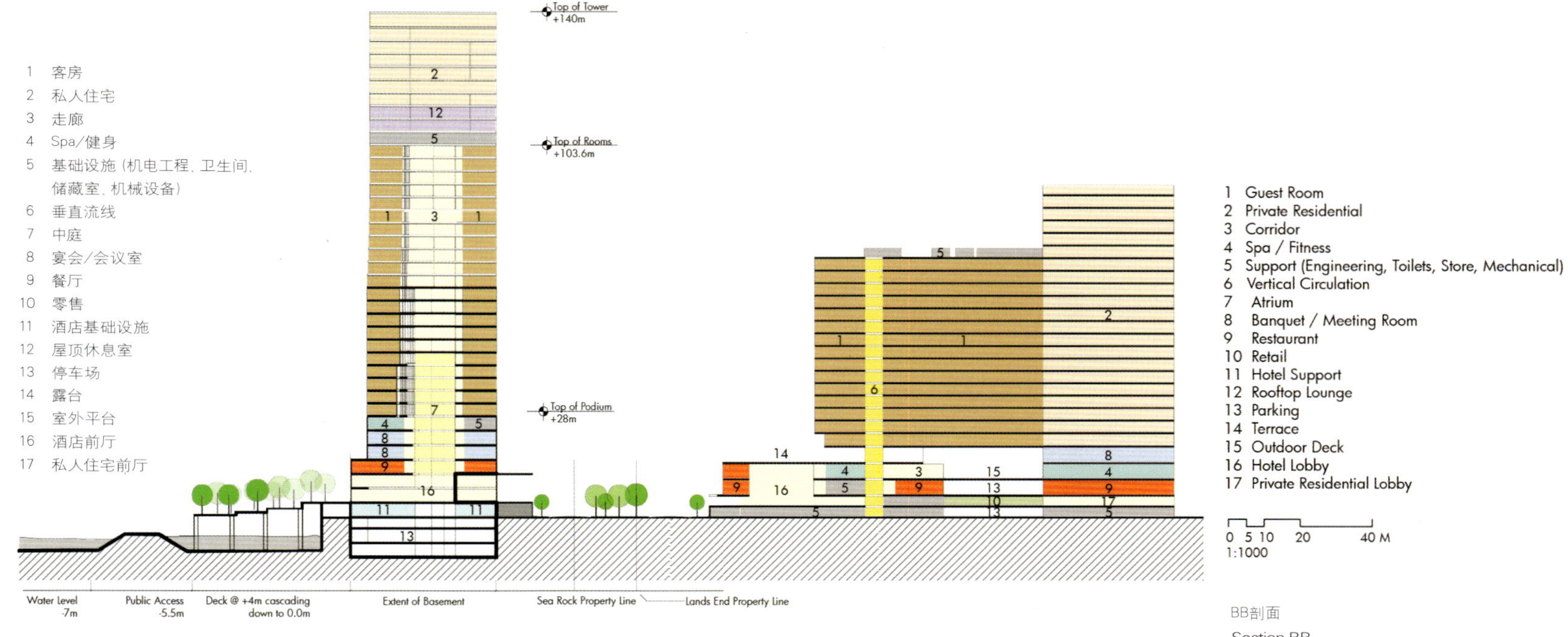

BB剖面

Section BB

H.A.N.D.—祝您愉快！
HAVE A NICE DAY - H.A.N.D.

建筑师：We Are You (Per Kaatman, Johan Zetterholm, Emil Lundh, Jonas Tjader)

"祝您愉快！"大厦主要由两部分组成：一部分是对公众开放的公共空间，另一部分是学生公寓。与其做一个楼层间彼此隔离的传统学生公寓项目，不如建一个从各楼层贯通的单一空间，这种"垂直客厅"将使空间更具多样性，并强调了楼层间的连接。为了给庞大的"垂直客厅"提供更多空间，只能尽量减少私人卧室的大小。公共空间有电脑、集体活动房间、健身室以及游泳池。这些公众场所免费对公众开放，并作为社区街头生活的重要组成部分。

建筑纵览

BUILDING OVERVIEW

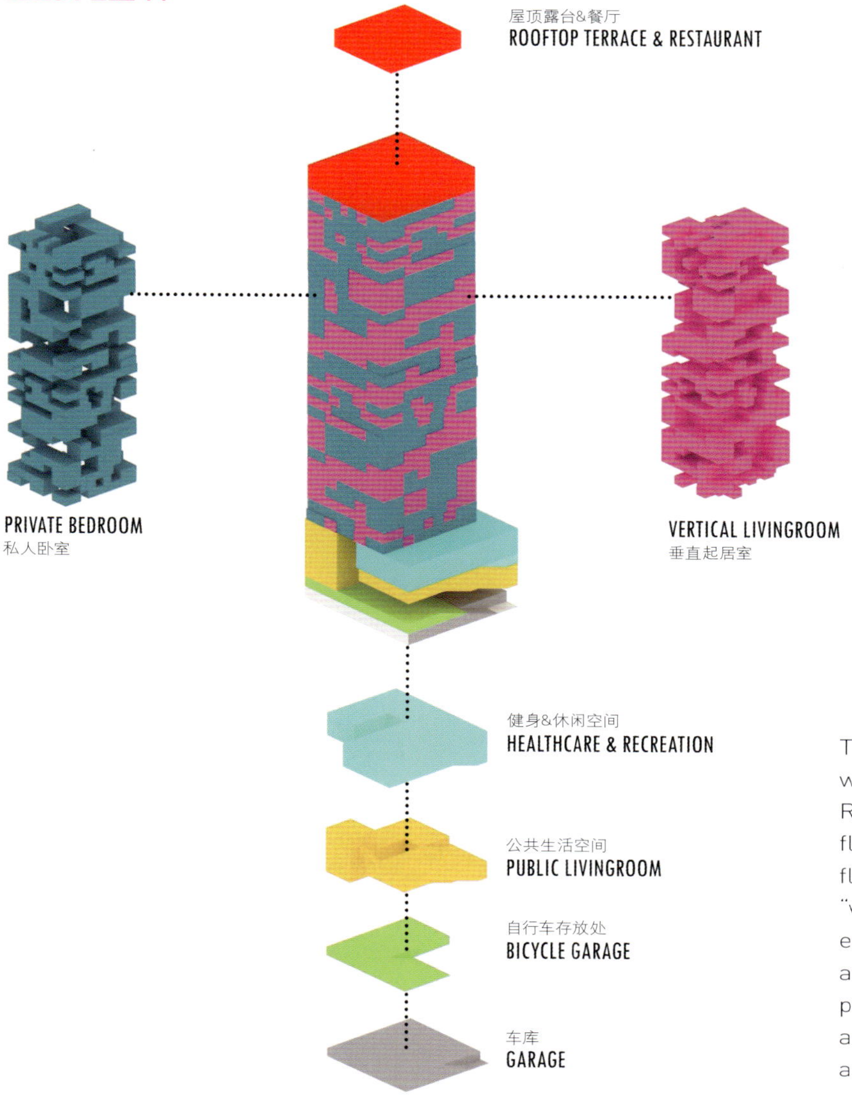

屋顶露台&餐厅
ROOFTOP TERRACE & RESTAURANT
PRIVATE BEDROOM
私人卧室
VERTICAL LIVINGROOM
垂直起居室
健身&休闲空间
HEALTHCARE & RECREATION
公共生活空间
PUBLIC LIVINGROOM
自行车存放处
BICYCLE GARAGE
车库
GARAGE

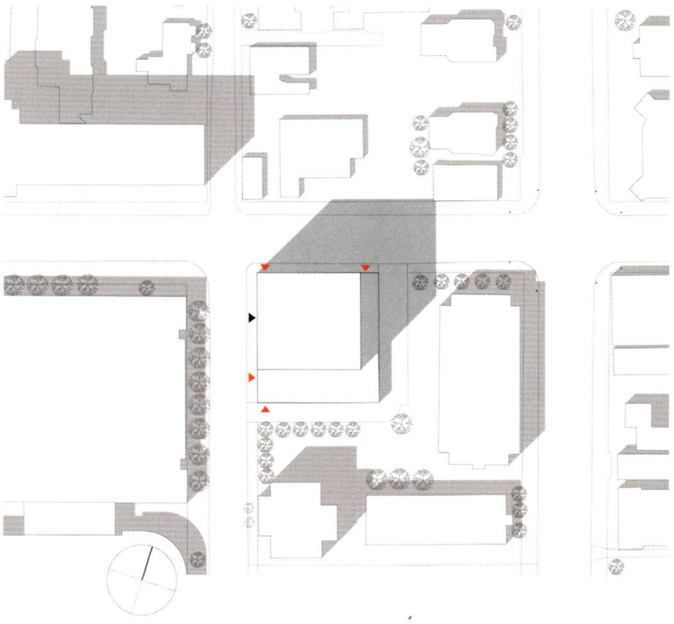

The Have a Nice Day building consists of mainly two parts: a public part with access for everyone and a student housing part for the students. Rather than making a conventional student housing project with the floors isolated from each other we have chosen to create a single space flowing from floor to floor throughout the student housing part, the "vertical living room". This creates a diversity in the spaces provided and emphasizes the connections between floors. The private sleeping units are minimized in order to give space to the large vertical living room. The public ground floors hold computers, rooms for group activities, gym, and a swimming pool. These are public spaces free for everyone to use and will serve as an important part of the neighborhoods' street life.

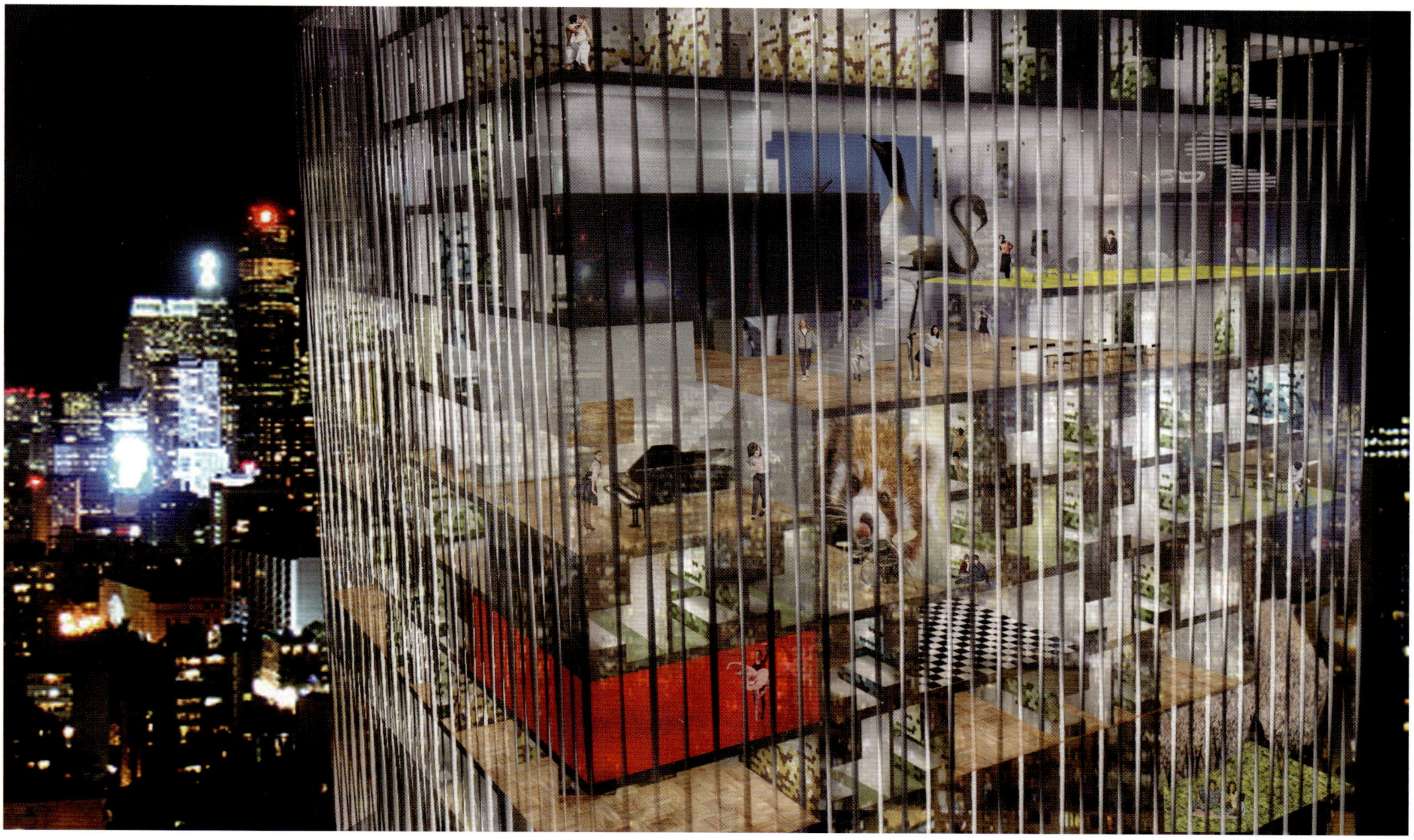

结构

STRUCTURE

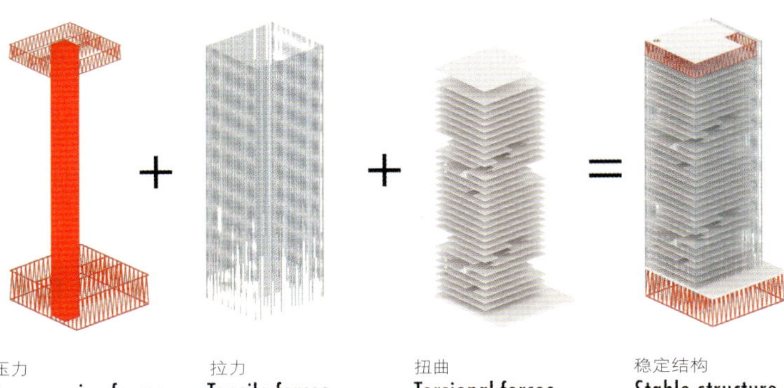

压力
Compressive forces

拉力
Tensile forces

扭曲
Torsional forces

稳定结构
Stable structure

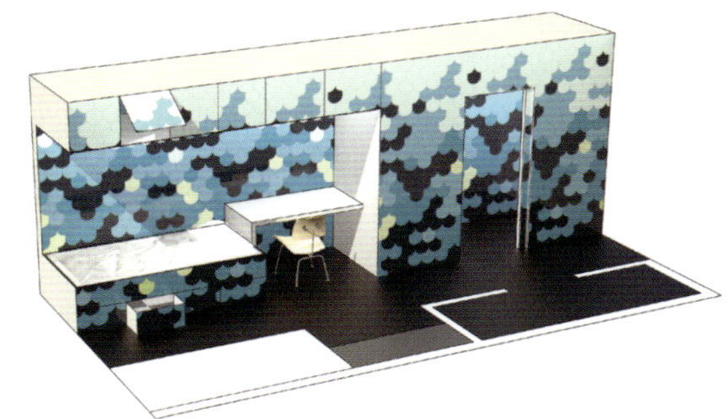

垂直校园

THE VERTICAL CAMPUS

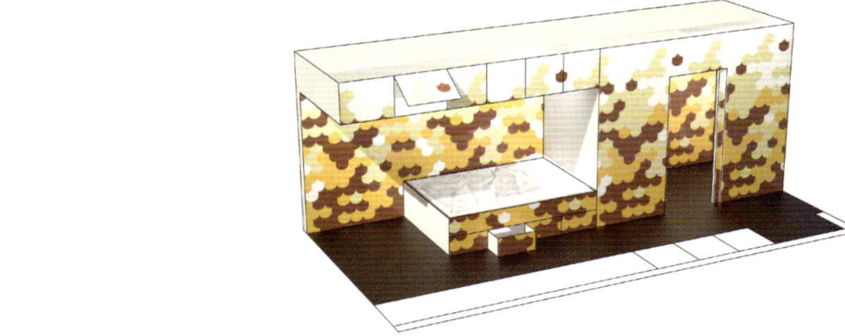

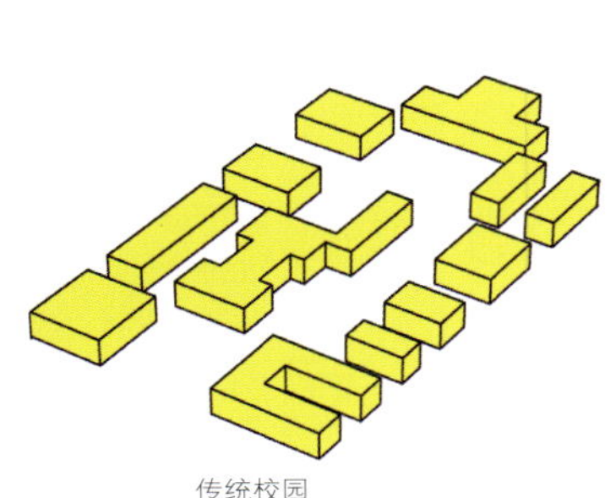

传统校园
Traditional campus

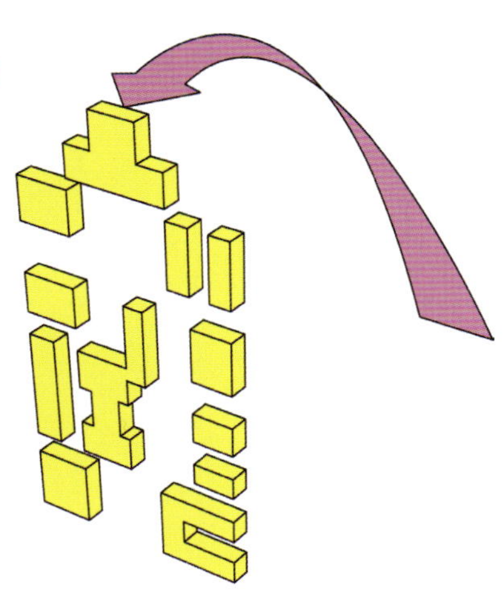

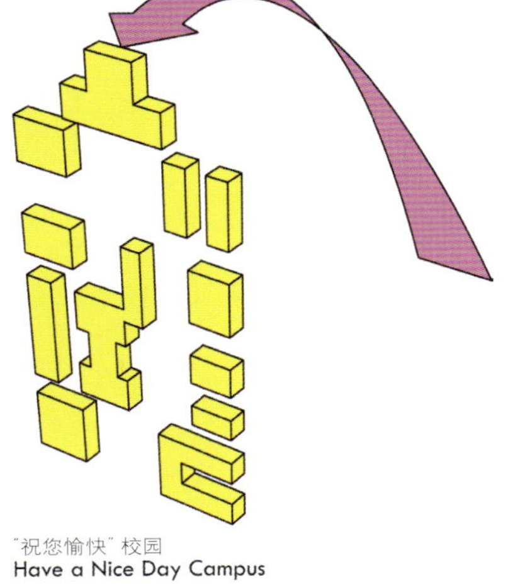

"祝您愉快"校园
Have a Nice Day Campus

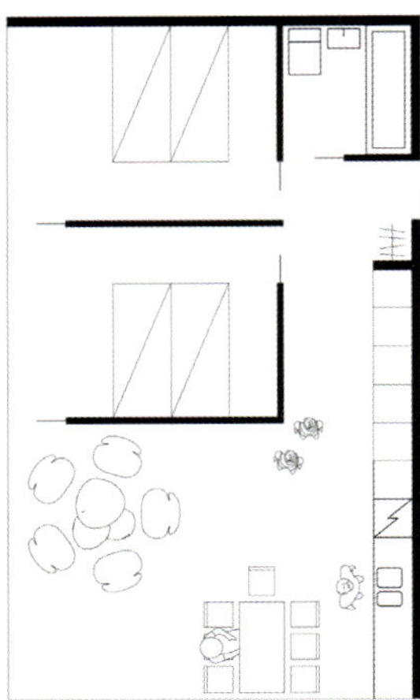

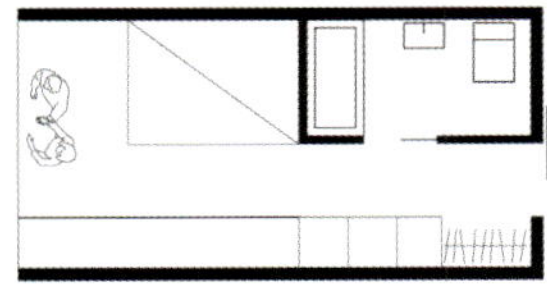

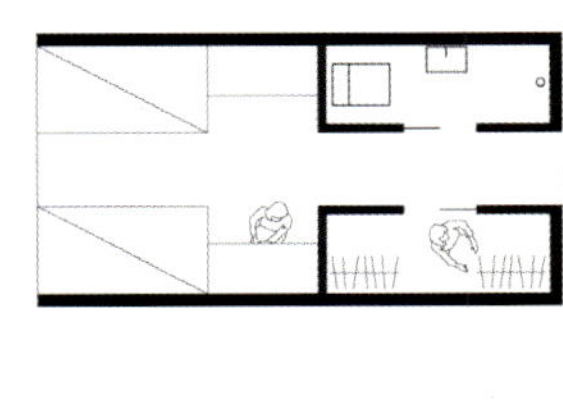

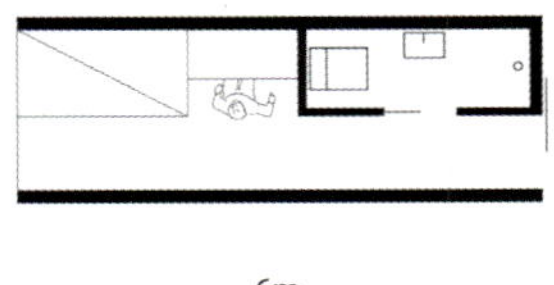

各种户型
Various plans of rooms

地下一层，自行车存放处
floor-1, Bicycle garage

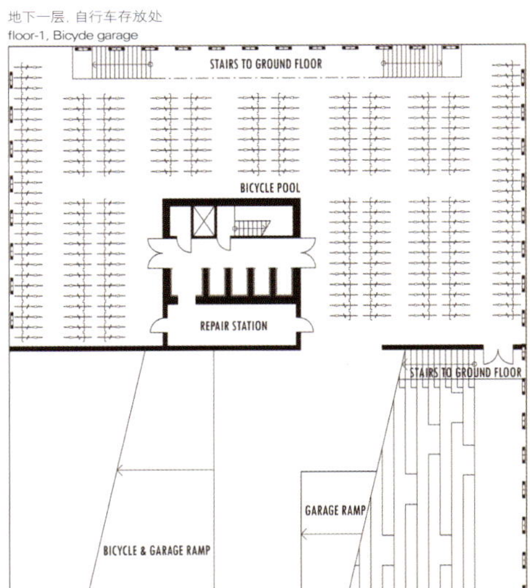

一层，公共生活空间
Floor 0, The public livingroom

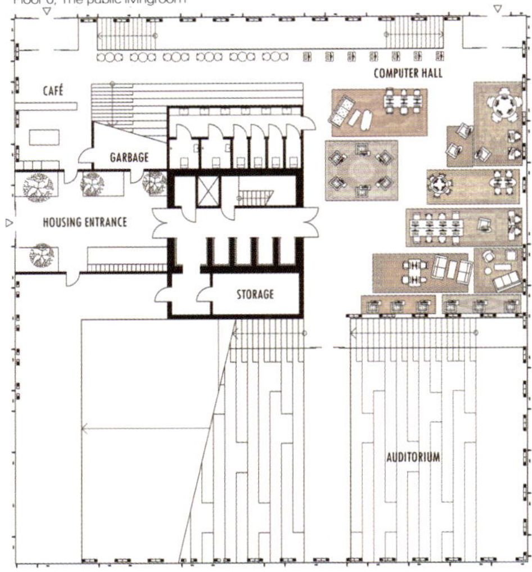

二层，公共生活空间
Floor 1, The public livingroom

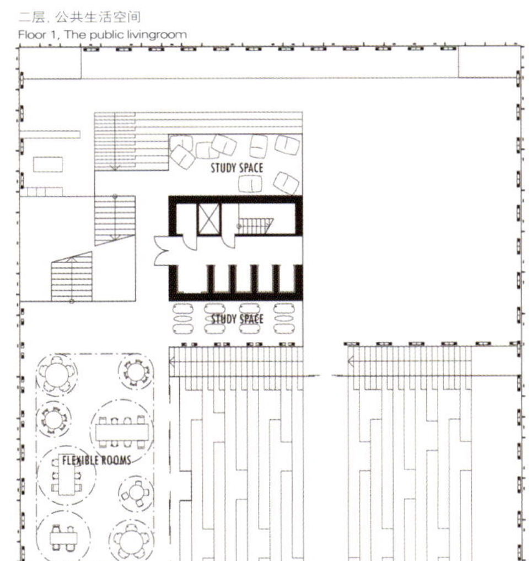

三层，运动及沐浴设施
Floor 2, Sports and bathing facilities

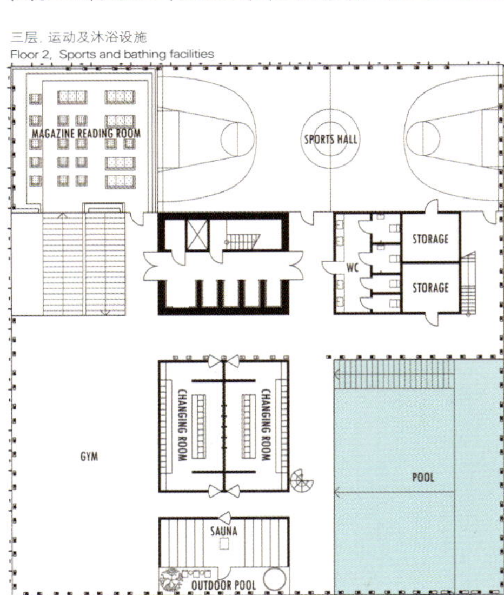

SWAP
YOUR
CLOTHES

奥斯陆的水晶大楼
CRYSTALCLEAR, OSLO

建筑师：C. F. Møller Architects in collaboration with Kristin Jarmund Arkitekter
景观建筑师：C. F. Møller Architects
工程师：ATKINS, Erichsen & Horgen AS
图片：MIR
地点：Central Station, Oslo, Norway
甲方：KLP Eiendom AS
获奖情况：建筑竞赛一等奖

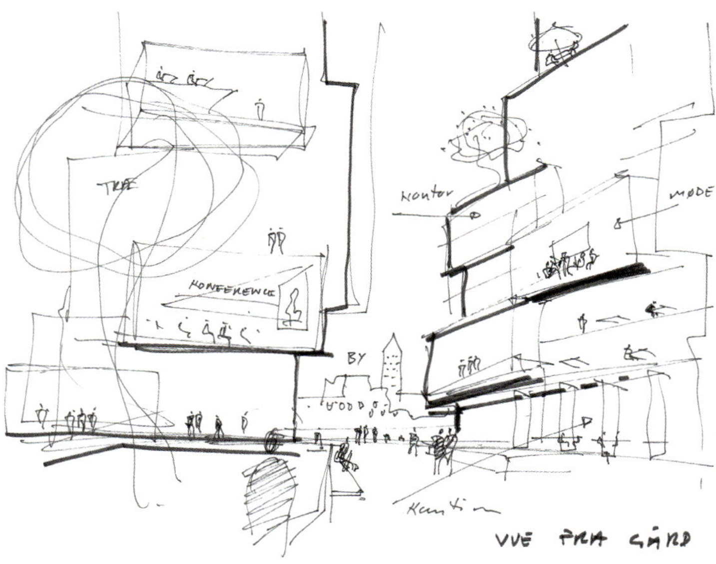

在挪威奥斯陆中心的最重要的交通中心矗立着一座高层建筑。这座建筑有着观看水景和远眺峡湾的奇妙视野。设计师的理念是用这些塔楼创造一组地标性的雕塑，但仍然与周围非高层的城市肌理相融合。三座塔楼的高度分别为110m、65m和55m。它们沿着场地的边缘布置，其中最高的一座与附近原有的奥斯陆广场和Postgirobygget塔楼保持一齐，而其余两座较低的建筑形成了与城市的连接。

A high-rise development, located at Norway's most important traffic hub in central Oslo, and with fantastic views of the waterfront and fjord-landscape beyond. The idea is to create a landmark sculptural ensemble of towers, yet observe the harmony with the surrounding, low-rise urban fabric of the capital. The three towers of approx. 110, 65 and 55 m height, are arranged along the edges of the site, and the tallest tower is aligned with the existing nearby Oslo Plaza and Postgirobygget towers, while the lower buildings form the link to the city.

Radisson SAS

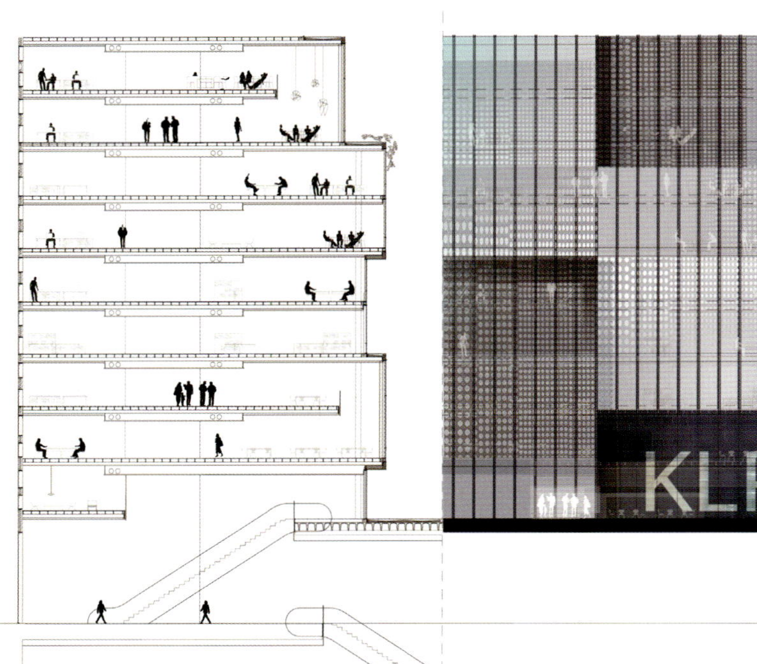

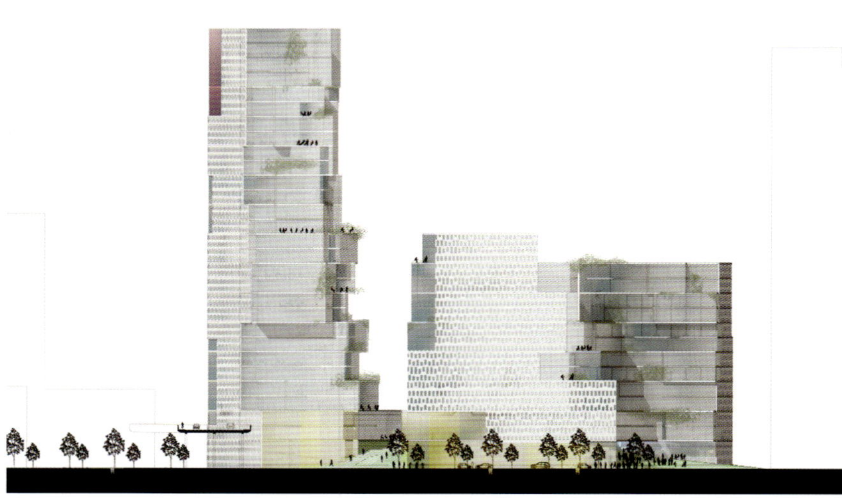

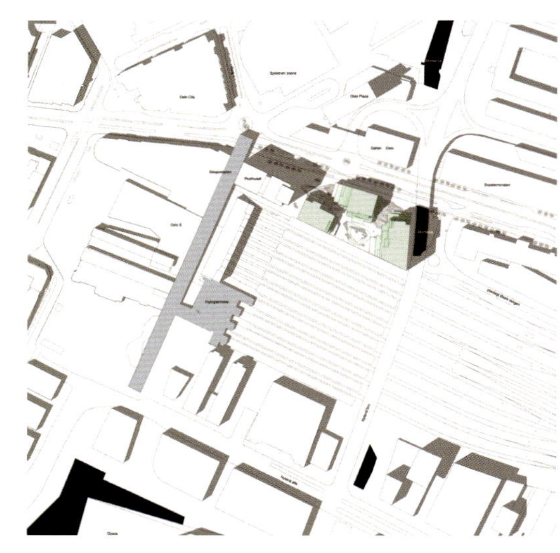

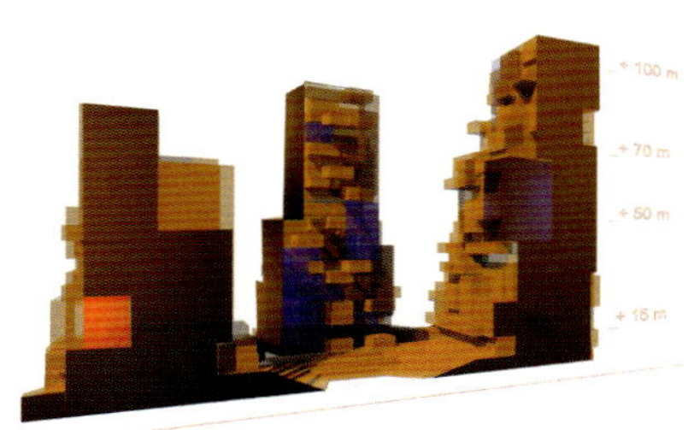

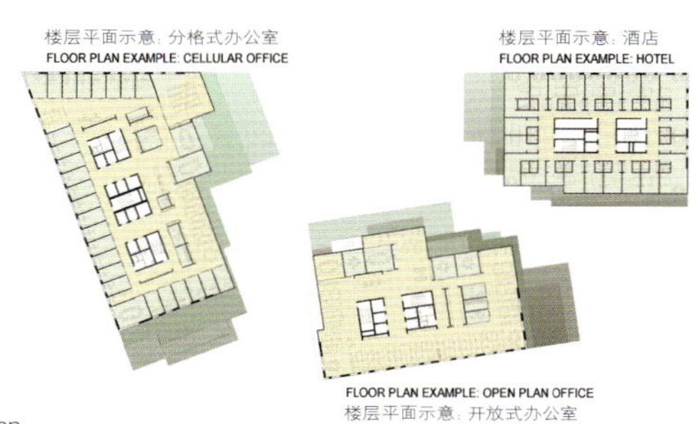

十一层平面
Tenth floor plan

1 商店
2 林荫路
3 大厅
4 露台
5 咖啡馆
6 办公室
7 酒店前台
8 餐厅
9 城市花园

9 URBAN GARDENS
RESTAURANT 8
5 CAFE
7 HOTEL RECEPTION
5 CAFE
4 TERRACES
6 OFFICE
TERRACES 4
6 OFFICE

三层平面
Second floor plan

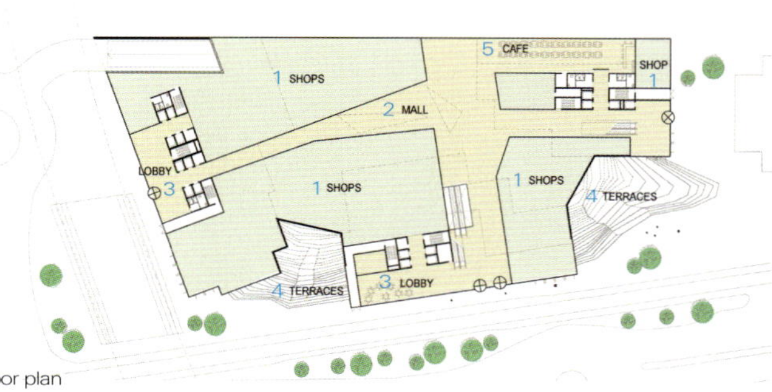

一层平面
Ground floor plan

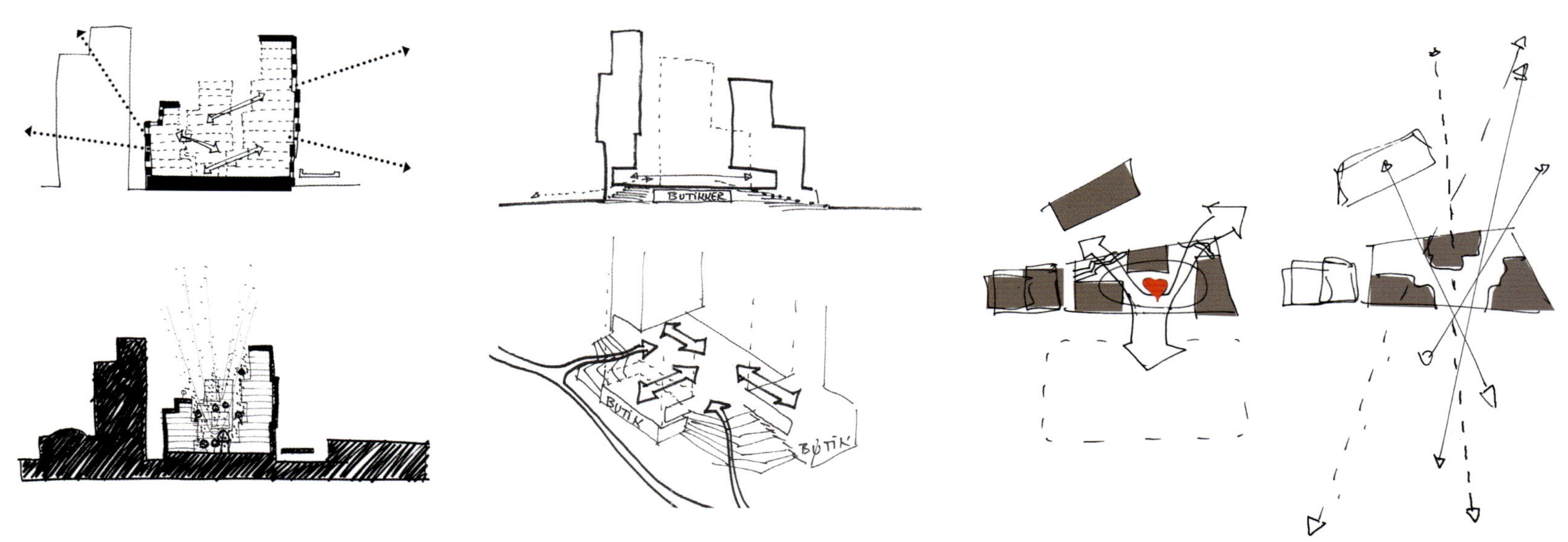
BUTIKKER
BUTIK
BUTIK

法兰克福君悦酒店
GRAND HYATT HOTEL, FRANKFURT AM MAIN

建筑师：UNStudio

Ben van Berkel, Caroline Bos, Astrid Piber with Arjan Dingsté, Marc Herschel, Marianthi Tatari, Steffen Riegas, Joerg Lonkwitz, Jesper Christensen, Junseung Woo, Peter Irmscher, Beatriz Zorzo, Leon Bloemendaal, Patrik Noome

甲方：Vivico Real Estate, Frankfurt am Main

项目地点：Frankfurt am Main, Germany

用途：酒店，顶层为餐厅和酒吧

占地面积：4400 m^2

总楼面面积：54 500 m^2

建筑规模：地上31层，地下2层

结构：混凝土

最大高度：113.30 m

停车位：187

君悦酒店大楼的设计突出强调了法兰克福的国际多样性。"大楼从不同的角度看起来各不相同，从一个角度看起来像针一样细，换一个角度就显得有力而端正，再从另一个角度看又似乎因带有一点扭曲而显得复杂。"

这座新的大楼将容纳一个五星级酒店，包括405个客房和套房、一个舞厅、一个spa室、多个餐厅、一个大厅酒吧和一个最顶层的公共天空休息区，还可设置一个相邻的会议中心。酒店大楼共30层，大约110m高，酒店大楼加上基座的面积共54 562.70m^2。

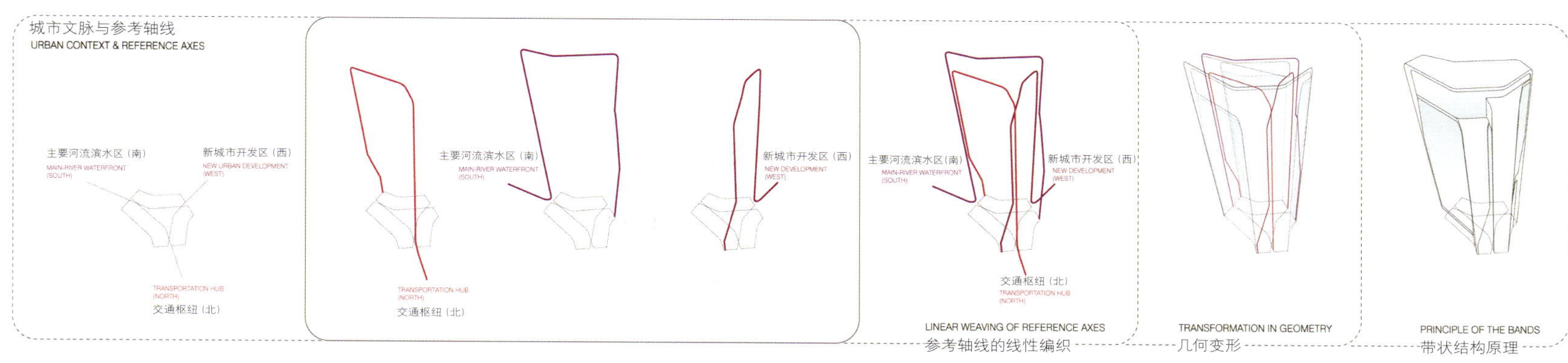

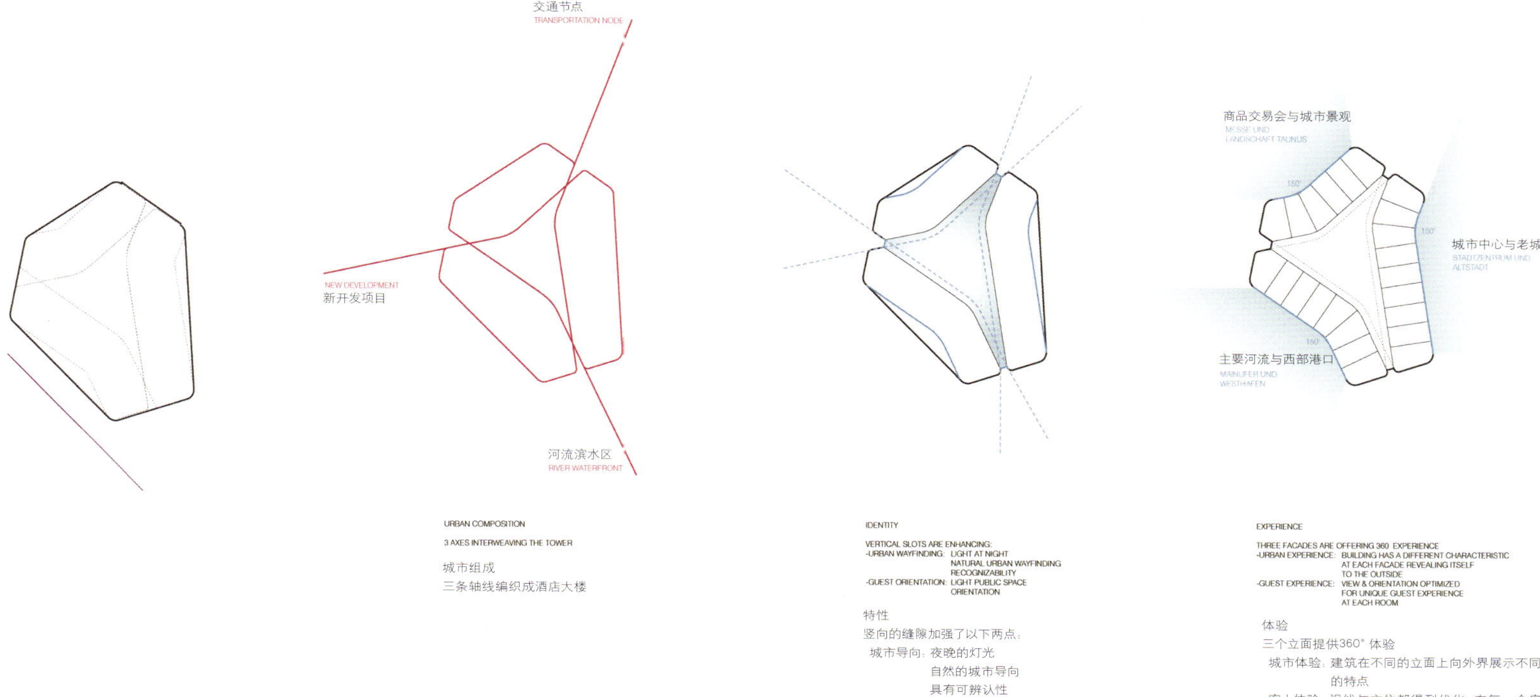

URBAN COMPOSITION
3 AXES INTERWEAVING THE TOWER
城市组成
三条轴线编织成酒店大楼
IDENTITY
VERTICAL SLOTS ARE ENHANCING:
-URBAN WAYFINDING: LIGHT AT NIGHT
NATURAL URBAN WAYFINDING
RECOGNIZABILITY
-GUEST ORIENTATION: LIGHT PUBLIC SPACE
ORIENTATION
特性
竖向的缝隙加强了以下两点：
城市导向：夜晚的灯光
自然的城市导向
具有可辨认性
客人定位：带有照明的公共空间
定位
EXPERIENCE
THREE FACADES ARE OFFERING 360 EXPERIENCE
-URBAN EXPERIENCE: BUILDING HAS A DIFFERENT CHARACTERISTIC
AT EACH FACADE REVEALING ITSELF
TO THE OUTSIDE
-GUEST EXPERIENCE: VIEW & ORIENTATION OPTIMIZED
FOR UNIQUE GUEST EXPERIENCE
AT EACH ROOM
体验
三个立面提供360°体验
城市体验：建筑在不同的立面上向外界展示不同的特点
客人体验：视线与方位都得到优化，在每一个房间内都有独特的体验

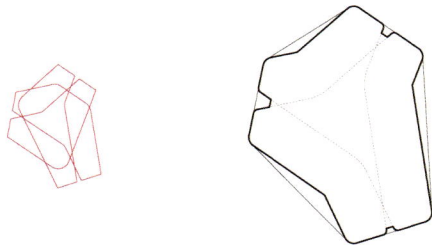

对于法兰克福的居民和游客来说大楼有三个同样重要的侧面
THE TOWER HAS 3/6 EQUALLY IMPORTANT SIDES FOR THE RESIDENTS AND VISITORS OF FRANKFURT

The design for the Grand Hyatt tower celebrates and highlights the cosmopolitan character and diversity of Frankfurt. "The tower can be perceived differently from each perspective; it appears needle-thin from one spot, strong and straight from another, and complex with a slight twist from yet another."

The new tower will house a 5-star-plus hotel with 405 rooms and suites, a ballroom, spa, various restaurants, a lobby bar and a public Sky Lounge on the top floor, and the possibility of an adjacent congress centre. The hotel tower consists of 30 floors at a height of approximately 110 meters and an area of 54,562.70 m^2 for hotel tower + plinth.

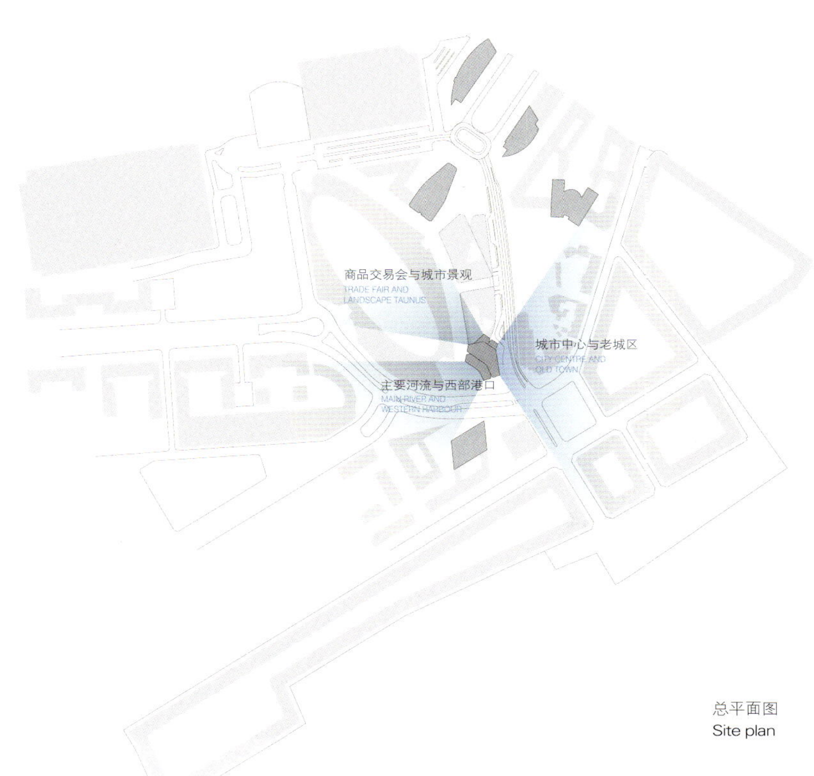

总平面图
Site plan

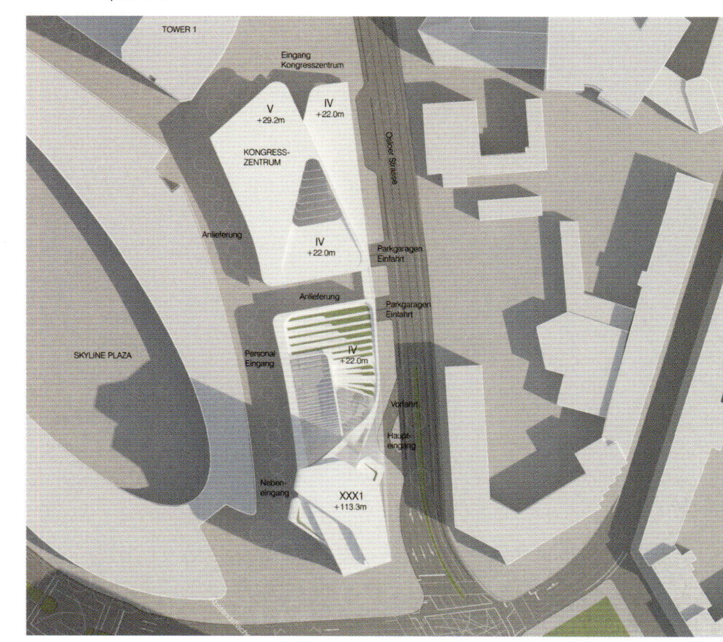

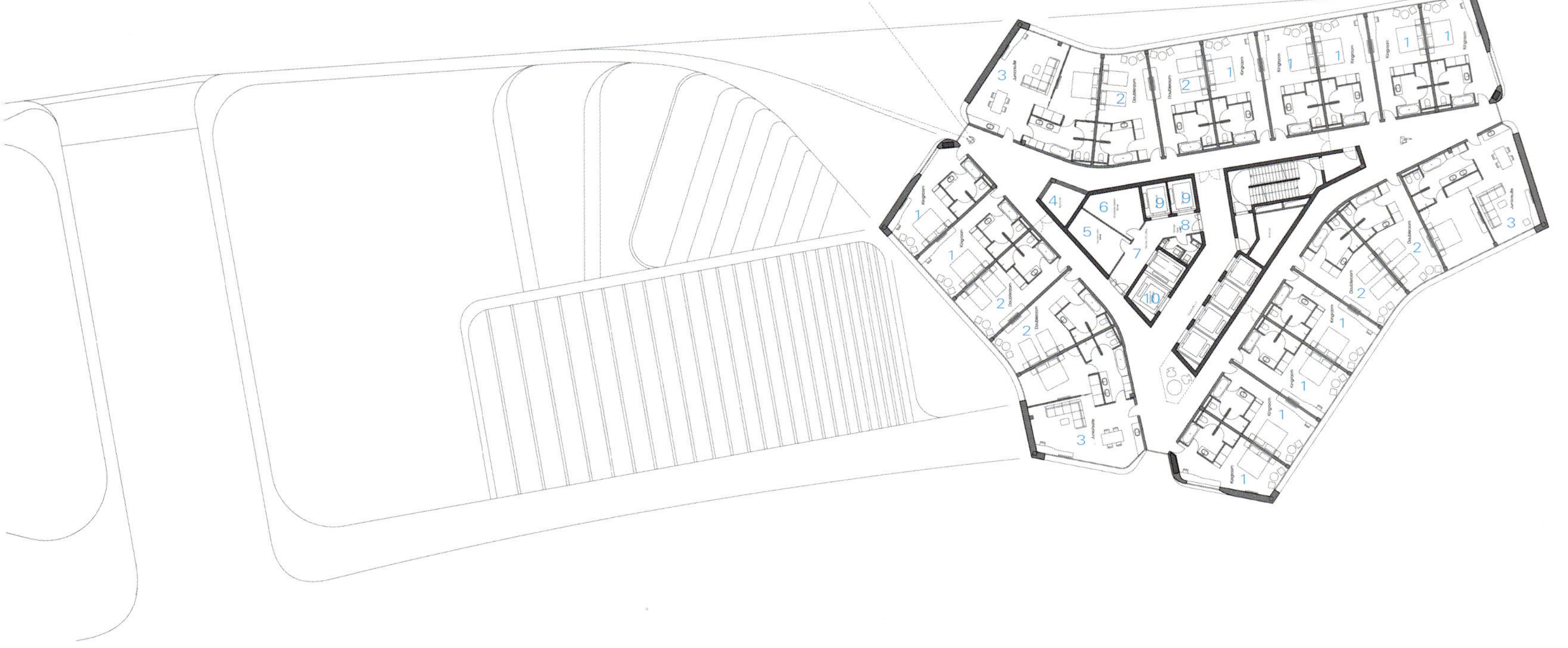

标准楼层平面图
Typical floor plan

1 总统套房
2 套房
3 小套房
4 技术设备室
5 12.6m²服务室
6 12.7m²员工休息室
7 电梯厅
8 5.1m²卫生间
9 电梯
10 13人快速电梯
11 客用电梯

剖面图
Section

1 spa室
2 游泳池
3 多功能区域
4 会议室
5 小舞厅
6 全天候餐厅
7 展示厨房
8 主厨房
9 停车场

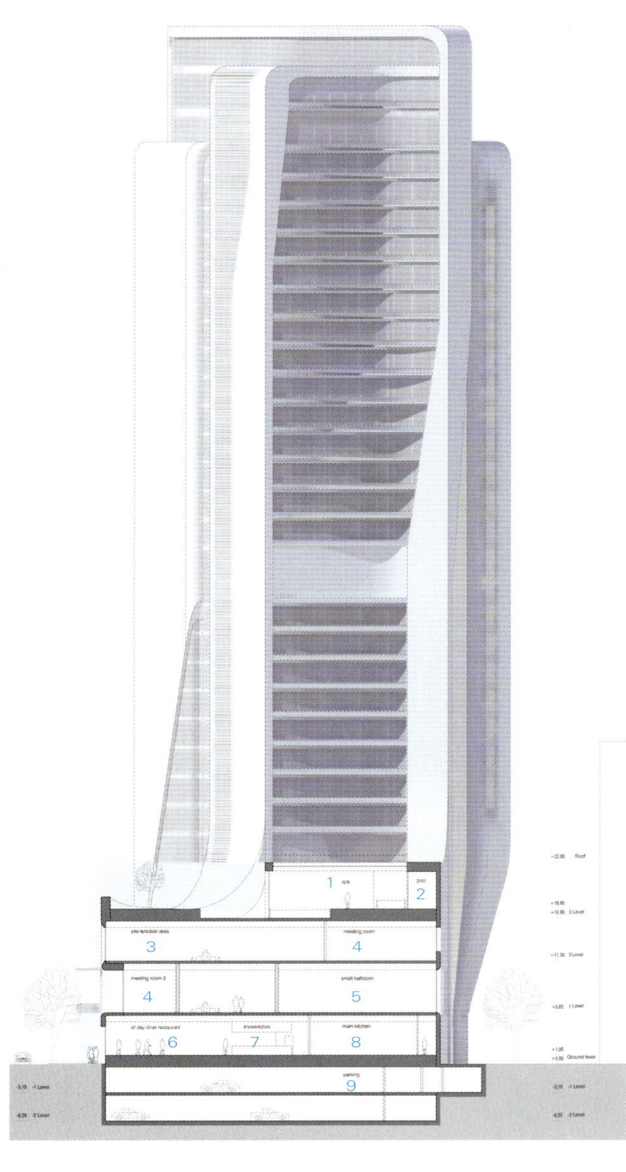

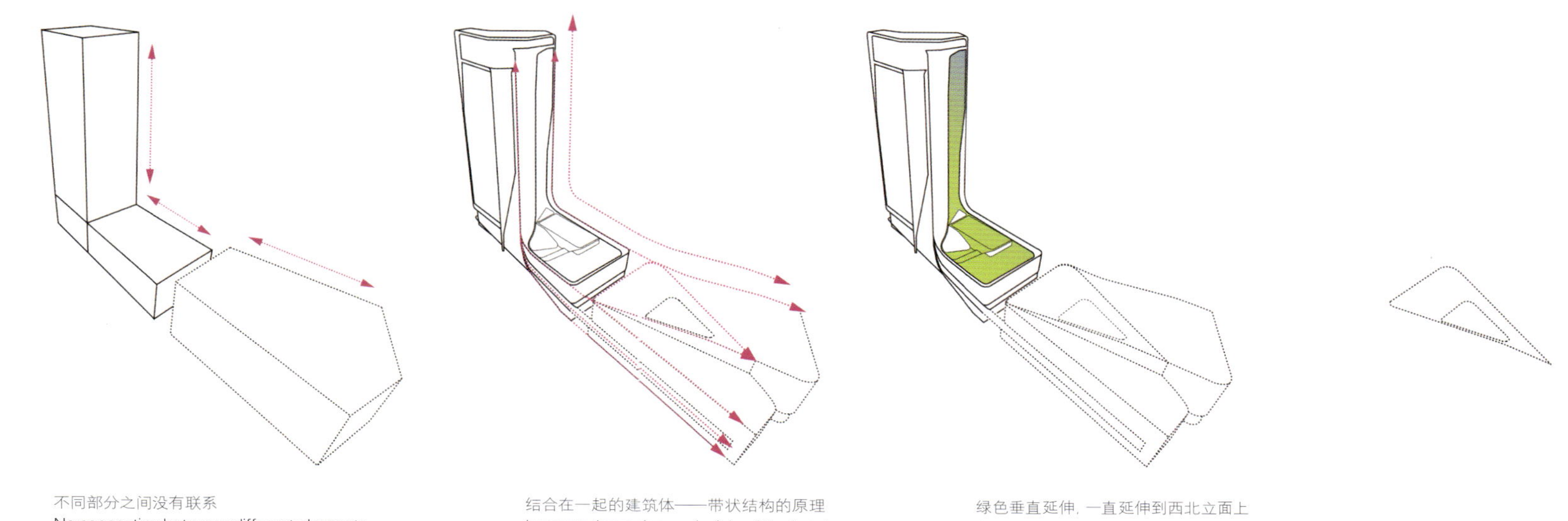

不同部分之间没有联系
No connection between different elements

结合在一起的建筑体——带状结构的原理
Integrated massing – principle of the bands

绿色垂直延伸，一直延伸到西北立面上
Green continues vertically to the north-west façade

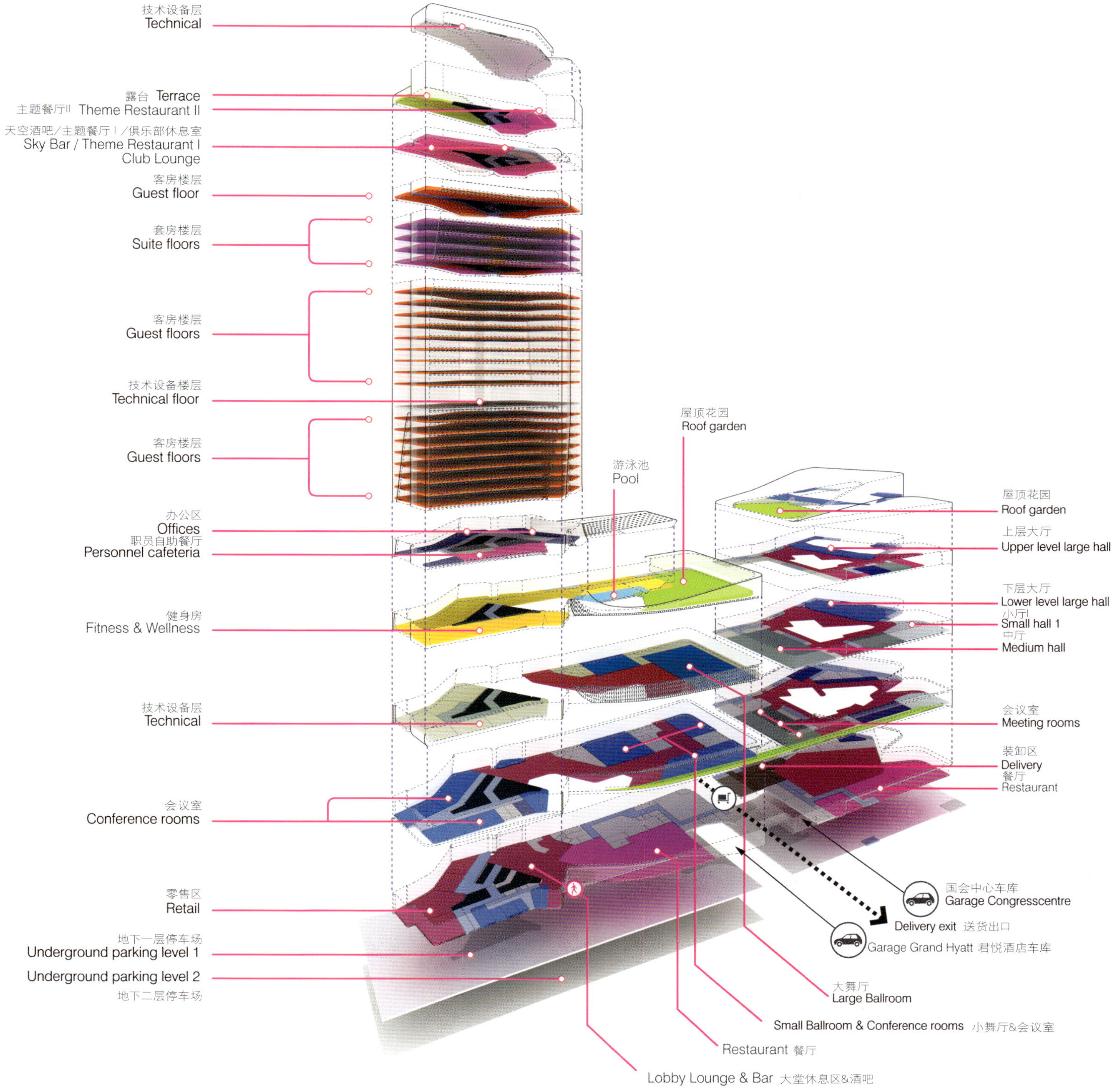

技术设备层
Technical

露台 Terrace
主题餐厅II Theme Restaurant II

天空酒吧/主题餐厅I/俱乐部休息室
Sky Bar / Theme Restaurant I
Club Lounge

客房楼层
Guest floor

套房楼层
Suite floors

客房楼层
Guest floors

技术设备楼层
Technical floor

客房楼层
Guest floors

办公区
Offices
职员自助餐厅
Personnel cafeteria

健身房
Fitness & Wellness

技术设备层
Technical

会议室
Conference rooms

零售区
Retail

地下一层停车场
Underground parking level 1

Underground parking level 2
地下二层停车场

屋顶花园
Roof garden

游泳池
Pool

屋顶花园
Roof garden

上层大厅
Upper level large hall

下层大厅
Lower level large hall

小厅I
Small hall 1

中厅
Medium hall

会议室
Meeting rooms

装卸区
Delivery

餐厅
Restaurant

国会中心车库
Garage Congresscentre

Delivery exit 送货出口

Garage Grand Hyatt 君悦酒店车库

大舞厅
Large Ballroom

Small Ballroom & Conference rooms 小舞厅&会议室

Restaurant 餐厅

Lobby Lounge & Bar 大堂休息区&酒吧

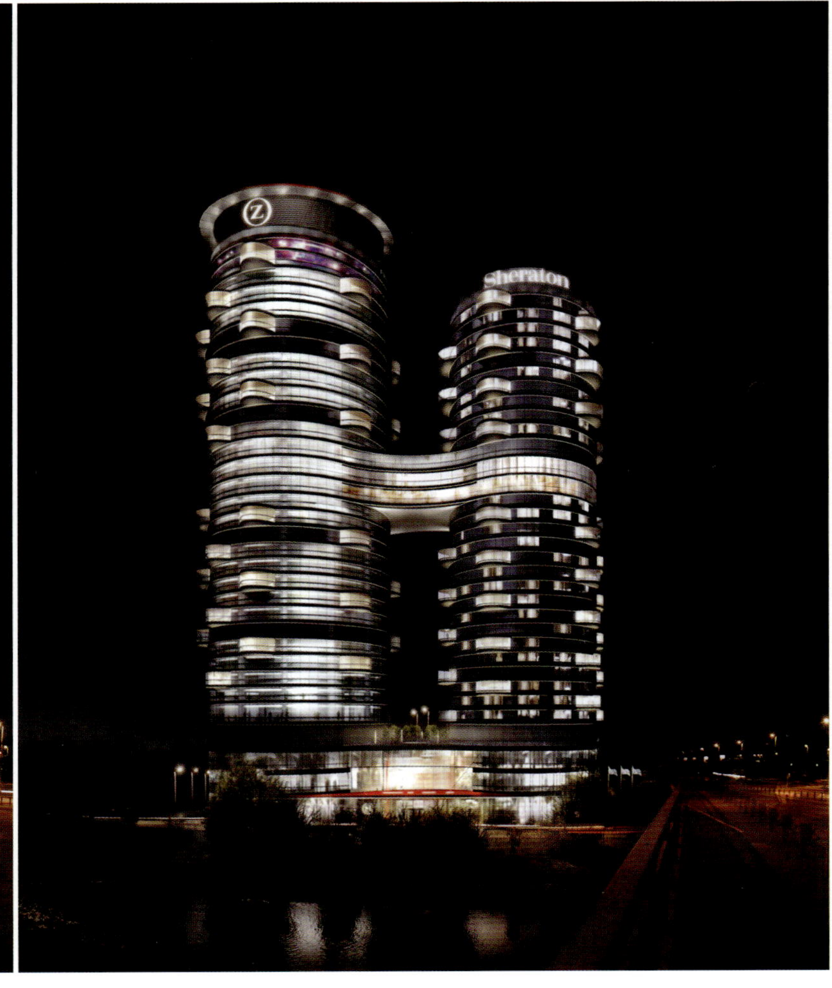

Z双塔
Z TOWERS

建筑师：NRJA (f.l.tadao & Luksevics, Ltd.)
结构：Engineer group „Kurbads", Ltd.
项目地点：Riga, Latvia
用途：办公、酒店
总楼面面积：100 559.68 m²
用地面积：12 766 m²
绿色屋顶露台面积：1790 m²
建筑密度：559.7 %
最大建筑高度：120 m
获奖情况：竞赛一等奖

这两座相连的创新性大楼成为波罗的海地区杰出的建筑地标。Z双塔将提供VIP级和A级办公空间。两座相连的摩天大楼具有现代的外观、优雅的室内，能欣赏到道加瓦河与里加老城区壮丽的景色。完美的服务加上雅致的氛围使这里成为波罗的海地区从事商务活动的绝佳地点。

Z双塔有多个会议大厅、一个商务中心、一流的酒店、商店、餐厅、咖啡厅、美容店，所有商务所需的服务和令客人感觉安逸的设施在这里都能找到。最高的大厦达到了120m。

The innovative construction of the two connected buildings is bound to become one of the outstanding architectural landmarks in the Baltic Region. The "Z Towers" development will offer VIP and A-class office space. Its two interconnected skyscrapers have been designed with a modern exterior, elegant interior and a breathtaking view of the River Daugava and Riga Old Town. Impeccable service combined with a refined atmosphere makes it the most prestigious place to conduct business in the Baltic Region.

"Z Towers" will offer several conference halls, a business centre, a first class hotel, shops, restaurants, cafes, beauty salons – everything required for your business and the comfort of your visitors. The highest tower will reach 120 metres.

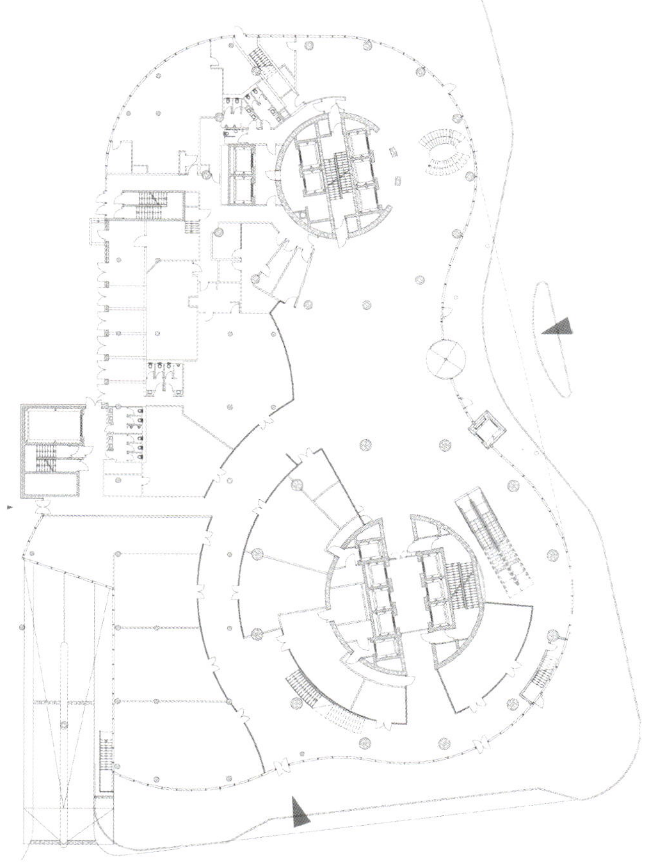

一层平面图
Ground floor plan

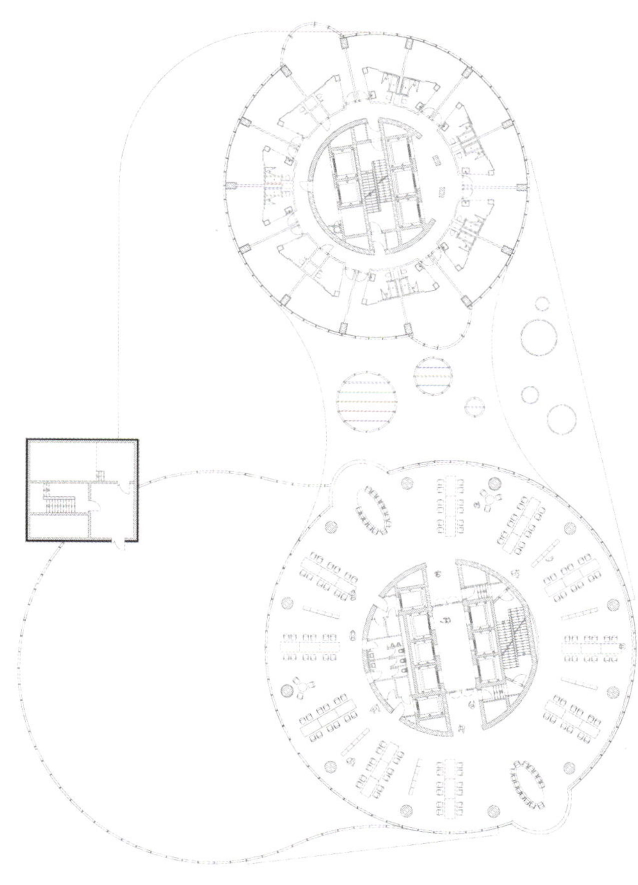

标准层平面图
Typical floor plan

瑞典马尔默的音乐会、会议和酒店建筑
CONCERT, CONGRESS AND HOTEL, MALMÖ, SWEDEN

建筑师：schmidt hammmer lassen architects

景观建筑师：SLA

承包商：Skanska Sverige AB

参与设计建筑师：Akustikon

甲方：Skanska Sverige AB

新的建筑综合体位于马尔默的Universitetsholmen。建筑由多个立方体体量构成，体量之间相互错开，大小不一，以满足朝向的需求并呼应周围城市建筑的高度。立面具有统一的设计，使建筑体呈现出雕塑般的外观。

主入口设置在建筑的北部。入口设置了经典的凉廊，朝向前方的广场。你可以在南面从围绕着运河的步行道直接进入建筑中。建筑中的不同功能以单独的构成形式呈现——就像一座迷你城市。贯穿整个建筑一层的走廊就像城市中的街道，把每个功能区联系起来。就像有着围绕着室内广场和露天广场的弯曲、狭窄街道的中世纪城市一样，建筑的走廊围合出小型的集会场所和休息处，人们可以在此驻足、稍坐和欣赏运河和公园的景色。

在建筑的室内，分别包含大交响乐厅、多功能厅和会议大厅的三个体量自成一体。建筑的形状就像具有温暖红色的三维层压木结构。

新的文化中心成为了具有表现力的动感开放的建筑，其内部功能和建筑本身都具有多样性的特点。建筑设计的出发点是带有清晰的功能构成、可及和开放的一层平面的现代斯堪地维亚建筑传统。建筑成为了马尔默的视觉焦点和地标建筑——一处展现城市精神的场所，以建筑的形式表达了城市的多样性和热情好客。

The new building complex is situated on Universitetsholmen in Malmö. The building consists of a composition of cubic volumes that are mutually twisted and given different sizes to meet the directions and building heights of the surrounding city. The facades are designed with a homogeneous expression to make the composition appear as one architectonic sculpture.

The main entrance is found at the northern part of the building, with a classic loggia-motif facing the plaza in front. From the south you enter the building directly from the promenade that runs along the canal. The different functions in the building are organised like separate elements - as a little city. Here, the lobby becomes the street that runs through the whole ground floor plan and ties everything together. Like the medieval cities, with curved and narrow streets, organised around plazas and squares, the lobby is designed to form small gathering places and recesses where it is possible to stop, sit and enjoy the view to the canal and the park.

From the inside, the three volumes that hold the large symphony hall, the flexible hall and the conference hall, will stand as clearly defined elements. The building shapes appear as a three dimensional composition in laminated wood with warm, red colours.

The new cultural centre becomes an open, expressive and dynamic building that is manifold in both its activities and its architecture. The point of departure for the building design is the modern Scandinavian architectural tradition with the clear functional organisation and the accessible and open ground floor lay-out. The building becomes the focal point and a landmark for Malmö - a place where the spirit of the city, the diversity and the intimacy is given an architectonic expression.

慕尼黑某酒店建筑
A HOTEL COMPLEX IN MUNICH

建筑师：schmidt hammer lassen architects
甲方：Jost Hurler Beteiligungs- und Verwaltungsgesellschaft GmbH & Co. KG

草图
SKETCH

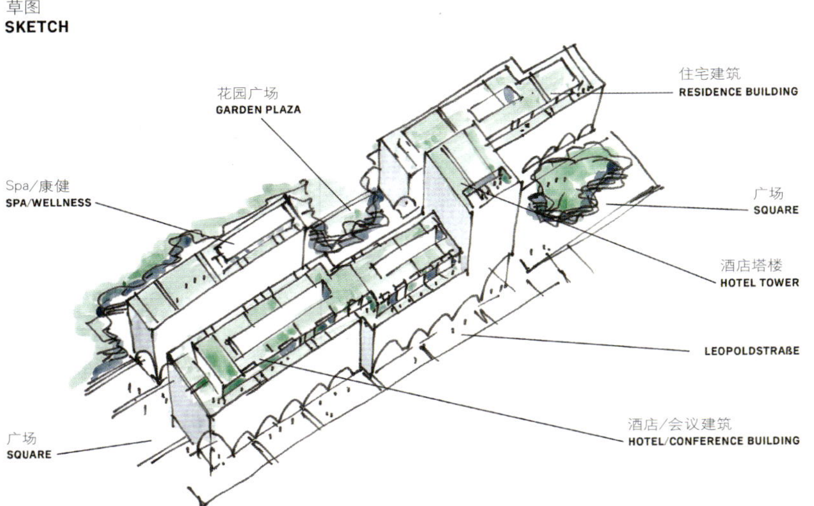

场地平面图
SITE PLAN

酒店建筑位于新开发的Schwabinger Tor区，沿着Leopoldstraβe的北部设置。竞标的难点就是将慕尼黑的建筑特色引入到城市这一全新与现代的区域中来。schmidt hammer lassen architects的设计理念有三个主旨：对这座带有拱桥、拱顶和拱形游廊的历史城市进行分析；总体规划与城市的林荫道、广场和狭窄街道的紧密联系；对市民与酒店客人的总体感受的关注。

酒店每间房间的设计都是独一无二的。通过将宽敞奢华的浴室（私人水疗房间）与日光和看向室外绿色景观的视线结合在一起，酒店的氛围变得非常轻松，并让人印象深刻。根据Kim Holst Jensen的说法，日光的利用是整个建筑的主题。日光为住客提供了独特的体验，并为一般意义的高端酒店设定了新的标准。

建筑立面垂直的薄板具有诗一样的韵律，加之开放的一层平面，这些都为城市空间增添了某种特质。在参考了斯堪地维亚建筑传统中对日光、透明度和简约性的利用方法，schmidt hammer lassen architects所创作出这座出色并富于表现力的建筑。传统的价值观与现代的设计融为一体，其结晶就是有着自身独特风格的酒店建筑。

The hotel complex is situated in the newly developed Schwabinger Tor area, along the northern part of the Leopoldstraβe. The challenge in the competition was to bring the architectural characteristics of Munich into this new and modern part of the city. schmidt hammer lassen architects' design concept has its origin in three themes: the analysis of the historical city with its arches, vaults and arcades; a close relation to the masterplan with its boulevards, plazas and narrow streets; and a focus on human beings as well as the overall experience offered to the hotel guests.

The hotel rooms have been designed to become one-of-a-kind. By integrating large and luxurious bathrooms – as a kind of private spa-rooms – with daylight and visual connection to the outside greenery, the atmosphere in the hotel rooms becomes relaxing and impressive. According to Kim Holst Jensen, the use of daylight is a general theme throughout the building complex. It offers the hotel guests a unique experience and sets a whole new standard for high-end hotels in general.

The building facades with their poetic rhythm of vertical lamella and the open ground floor level add something to the urban spaces of the city. With reference to the Scandinavian architectural tradition of working with daylight, transparency and simplicity schmidt hammer lassen architects has created an outstanding and expressive building complex. The classic values are combined with a modern design and the result is a unique and exclusive hotel with its very own style.

意大利里米尼海滨多功能开发项目
RIMINI SEAFRONT AND MIXED USE, ITALY

建筑师：Foster+Partners

甲方：GECOS Generale Costruzioni Spa

里米尼的新型海滨发展规划目的是加强城镇中心与海滨的联系，以及创造一个新的旅游景点，全年都能吸引来自世界各地的游客。这个规划包括新建一条海滨大道、酒店大厦等相关设施及公共区域，这将延续里米尼的海滩历史文化以及现有的城市组织。海边在特定的时间将成为步行区，并直接连通到带状的公园或绿化带中，在酷暑盛夏之中给人们带来难得的清凉。费里尼电影博物馆位于酒店大楼的基座中。酒店的曲线型外形引人注目，屹立在海滨项目基地上。该项目还采用了一些雨水采集、太阳能发电等新技术，为小镇建立一个长期可持续发展的商业与环境战略。

The new waterfront development for Rimini is designed to strengthen the relationship between the town centre and the seafront and to create a year-round attraction for an international tourist industry. The scheme comprises a new seafront promenade with a mix of related activities and public spaces including a hotel tower, which will extend Rimini's historic beach culture and continue the existing urban grain. The waterfront will be pedestrianized at certain times and will link directly to a linear public park – or green spine – which will provide much needed shade during the hotter months. A hotel tower includes space for a Fellini film museum at its base. Its curving form is a striking marker which anchors the wider project. The scheme will use new technologies, such as rainwater collection and photovoltaics, to establish a long-term, sustainable commercial and environmental strategy for the town.

埃及开罗的“银色云团”
“SILVER CLOUD” IN CAIRO, EGYPT

业主：GB AUTO SAE (Ghabbour), Cairo, Egypt
本地建筑师：MZECH – Medhat Abozied Egyptian Consulting House
结构工程师：B+G Ingenieure, Bollinger und Grohmann GmbH, Frankfurt, Germany
能源与环境设计：Arup GmbH, Prof. Brian Cody, Berlin, Germany
规划设计：COOP HIMMELB(L)AU
Wolf D. Prix / W. Dreibholz & Partner ZT GmbH
主设计师：Wolf D. Prix
项目负责人：Paul Kath
参与项目建筑师：Volker Kilian
设计建筑师：Sophie Grell
项目团队：Robin Heather, Martin Jelinek, Anja Sorger, Vanessa Castro Vélez, Iris Michailou, Martin Neumann, Suleimann Alhadidi, Jenny Chow, Thomas Hindelang
建模：Paul Hoszowski
效果图制作：Jens Mehlan/ Jörg Hugo, COOP HIMMELB(L)AU
线图制作：COOP HIMMELB(L)AU

建筑的外表皮由三维钢架组成，骨架外侧被不锈钢或铝板所覆盖，目的是想得到让人眼前一亮的建筑，哪怕在行驶的车中惊鸿一瞥。汽车展示广场悬挑的体量创造的阴影本身就在附近的广场提供了一处舒适的乘凉之所。该项目为一个空间连续体，其丰富的功能与宽阔的空间相辅相成，由下列功能组成：

· 汽车展示空间，舞台和观众席
· 商业中心和电影院
· 美食街和咖啡馆
· 底层的停车场、配送站和处理站

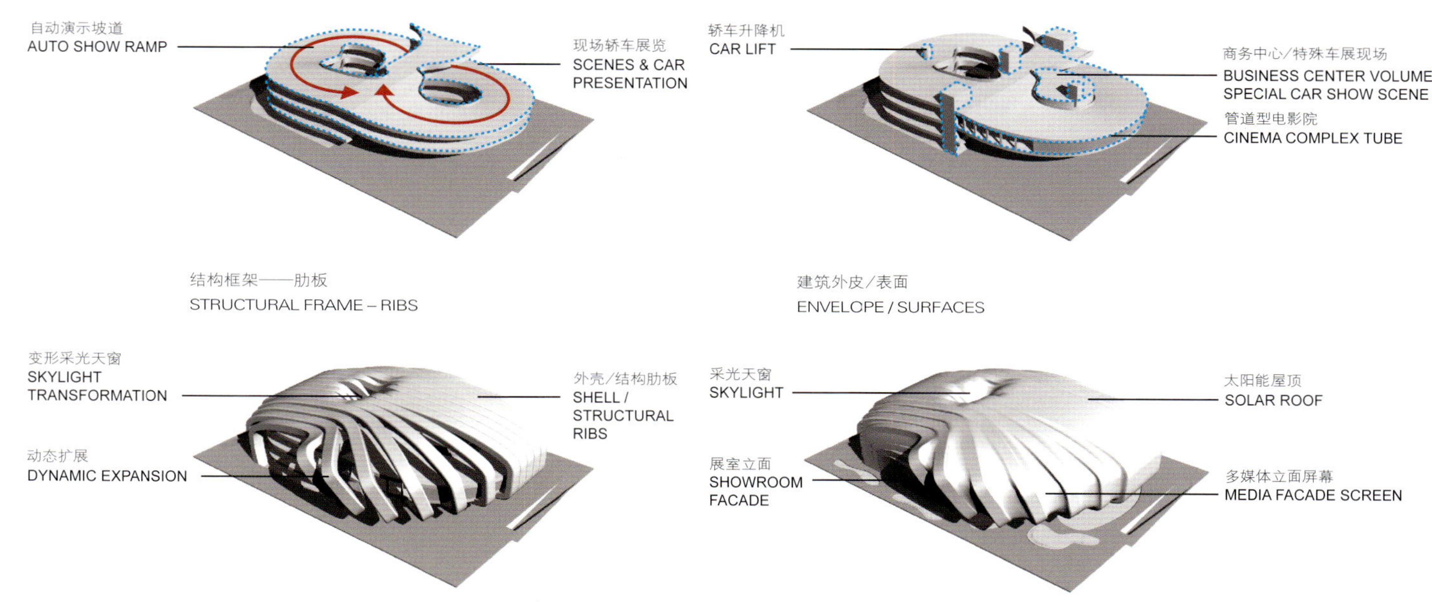

The outer skin of the building consists of three-dimensionally deformed steel frames covered with stainless steel or aluminum panels, in order to create a building envelope which will be instantly recognizable even if just seen for seconds out of a moving car. The shadow created by the cantilever of the building volume of the Automotive Showroom Mall itself provides a pleasant climate in the nearby plaza. The project combines the following functions inside one single spatial continuum, with synergetic effects of functionality and spatial richness:

· car exhibition spaces, show stage and auditorium
· business center and cinema complex
· food court and cafés
· base for parking, delivery and disposal

银色云团/建筑组成
SILVER CLOUD/BUILDING PARTS

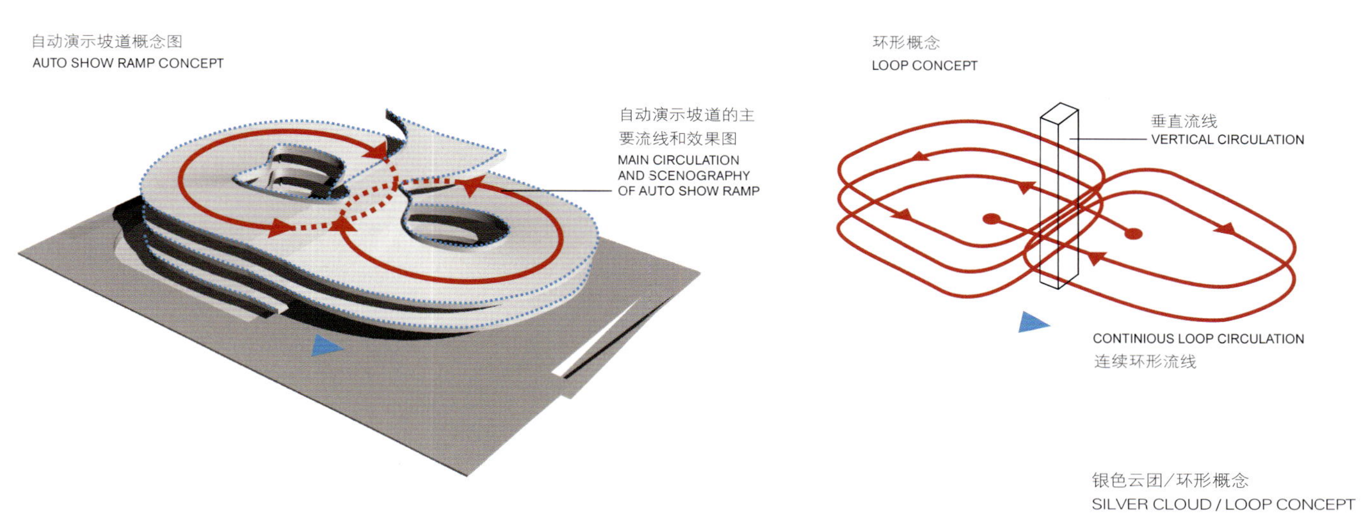

银色云团/环形概念
SILVER CLOUD / LOOP CONCEPT

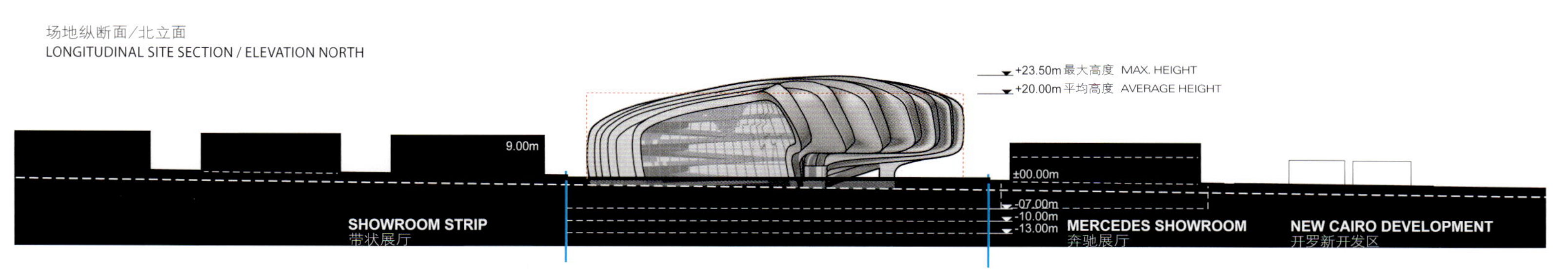

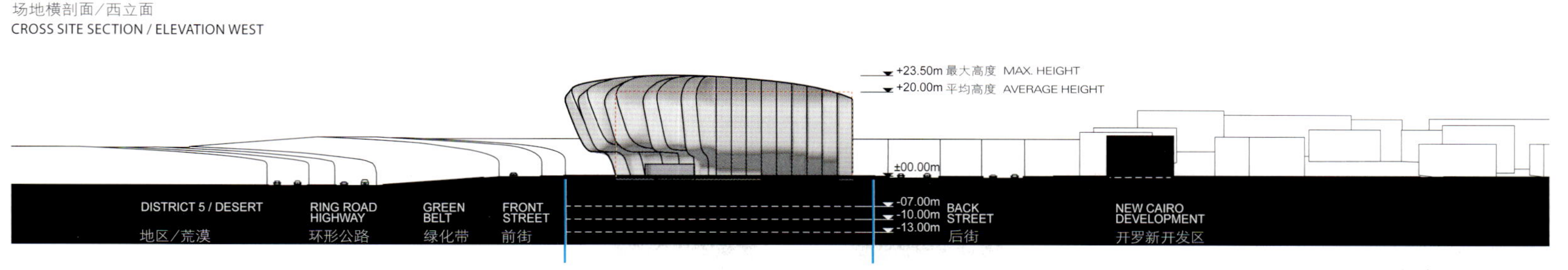

场地剖面图
URBAN SITE SECTIONS

埃及开罗世博会城
CAIRO EXPO CITY, EGYPT

甲方：GOIEF（国际展览及博览会组委会），Egypt
建筑师：Zaha Hadid Architects
设计：Zaha Hadid & Patrik Schumacher
项目指导：Charles Walker
项目建筑师：Viviana Muscettola, Michele Pasca di Magliano
高级工程师：Tariq Khayyat, Kut Nadiadi
结构与机电工程：Buro Happold, London
施工技术：Gleeds, London
交通与物流：Arup, London

扎哈·哈迪德建筑师事务所的开罗世博城提案实现了该城市独特的市政设备的图标式建筑愿景，在开罗市区和机场之间设置了一座展览与会议城。项目为一个主要包括商务酒店在内的国际展览和会议中心。除此之外还增加了两座办公塔楼和一个大型购物中心，形成了多种功能空间的集合，为众多游客提供服务，在各个时期都充满活力。

开罗世博城的规划以创造适应场地周边环境的均质城市聚合体量为目标。对设计纲要进行分析后，ZHA从连接点和功能分布的角度来理解项目的尺度。他们将雕塑般的城市体量规划成小型的建筑群落，其中的建筑可以单独发挥作用并具有自身的体量特征，但是它们又是整体设计的一部分。

Zaha Hadid Architects' proposal for Cairo Expo City delivers an iconic architectural vision for a unique facility for Cairo, a city for Exhibitions and Conference between downtown Cairo and the airport. The project comprises a major international exhibition and conference centre with business hotel. In addition to the above two office towers and a shopping mall are proposed, creating a rich ensemble of diverse functions which caters for multiple audiences and activates the site across different times and days of the week.

The urban strategy of the Cairo Expo pursues the idea of creating a homogeneous urban cluster mass that adapts to the site boundaries. Analysing the brief, ZHA have understood the scale of the project in terms of the connectivity points and the program distribution. They have proposed a carving of the urban mass into smaller clusters that can work as individual buildings and have their own massing features, however relating to part of the overall design.

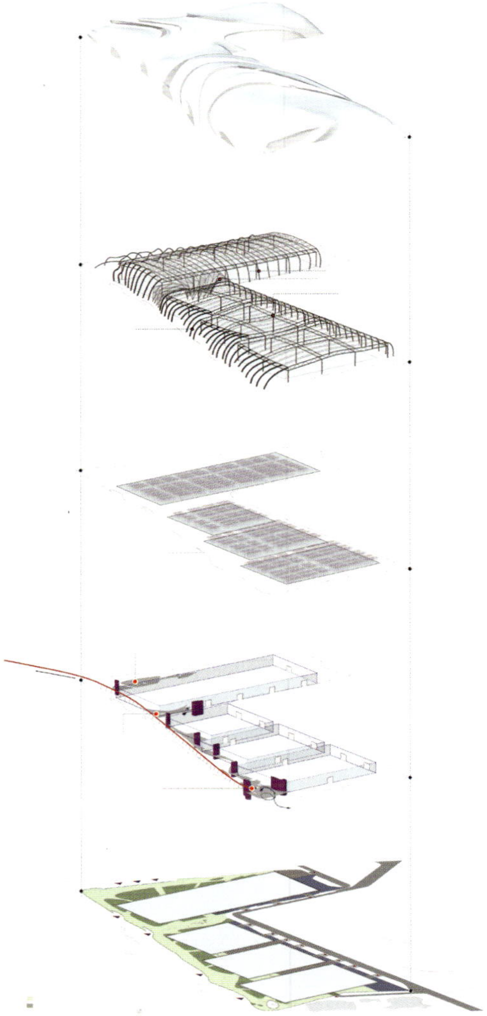

展览中心元素
Exhibition elements

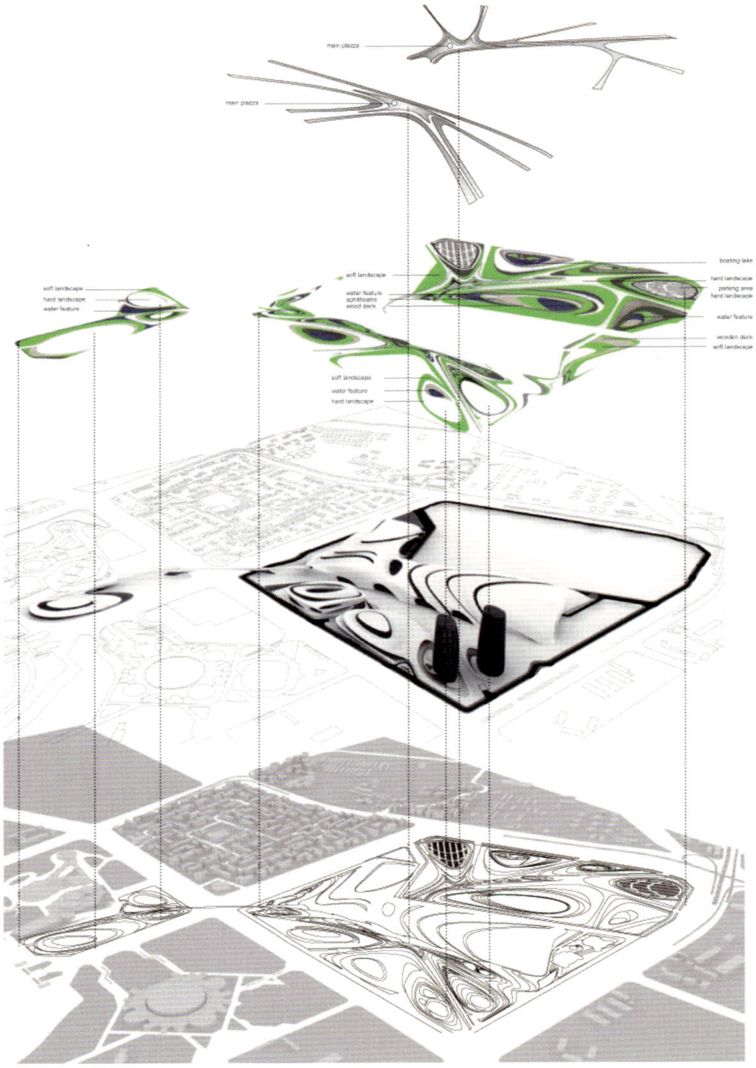

景观元素
Landscape elements

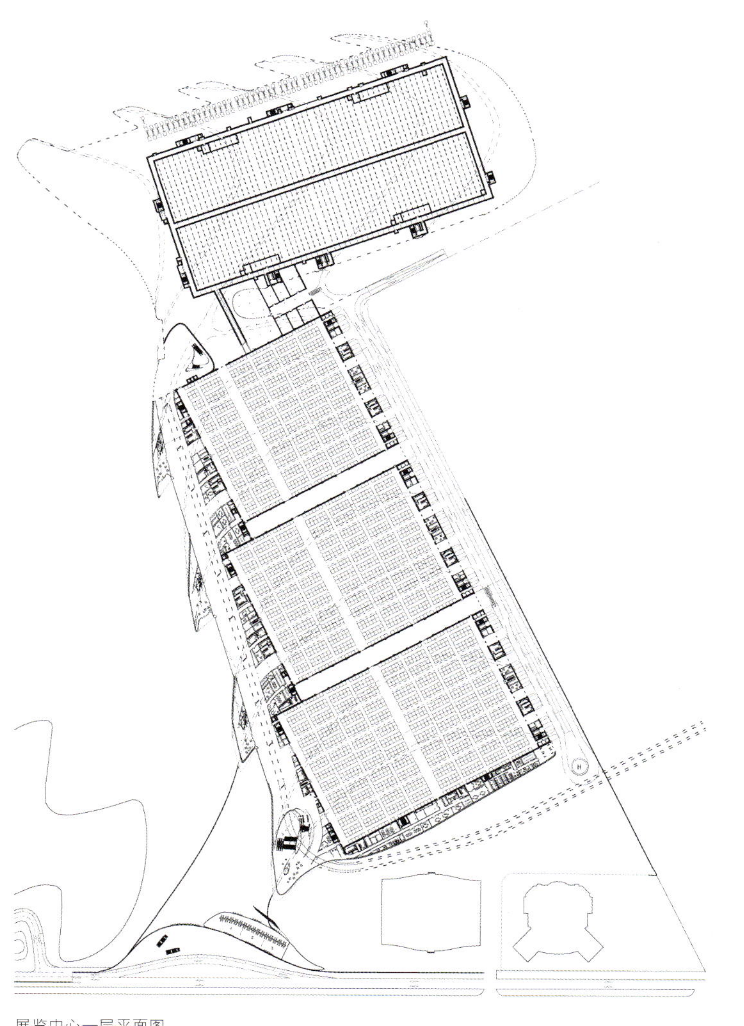

展览中心一层平面图

Exhibition centre ground floor plan

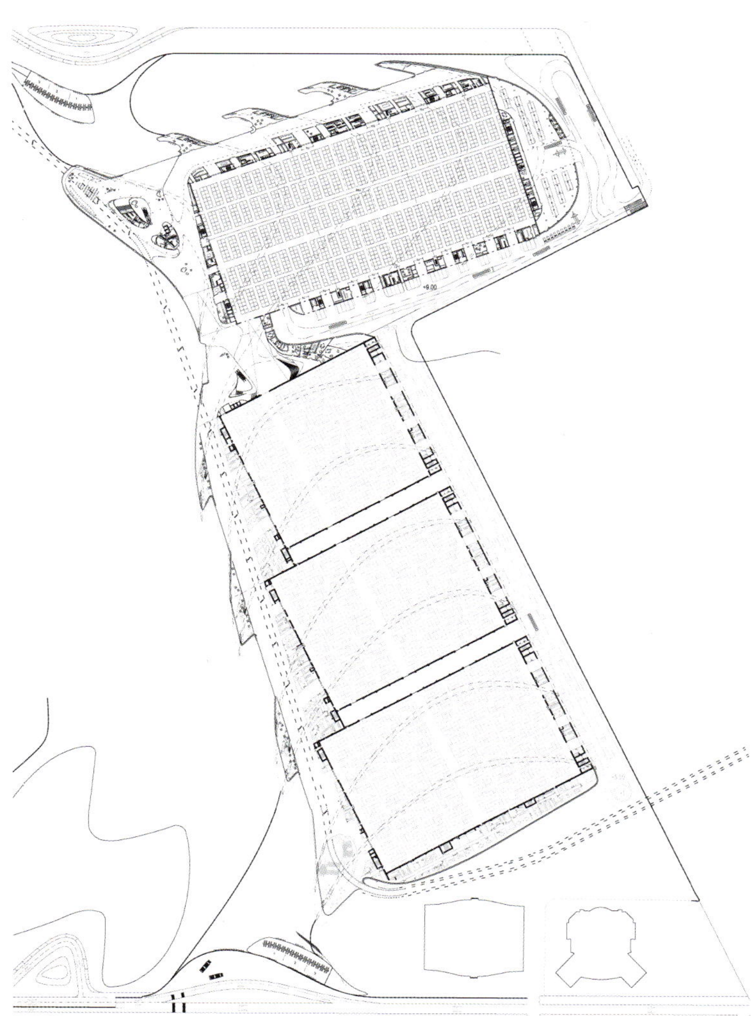

展览中心二层平面图

Exhibition centre first floor plan

会议中心一层平面图

Convention centre ground floor plan

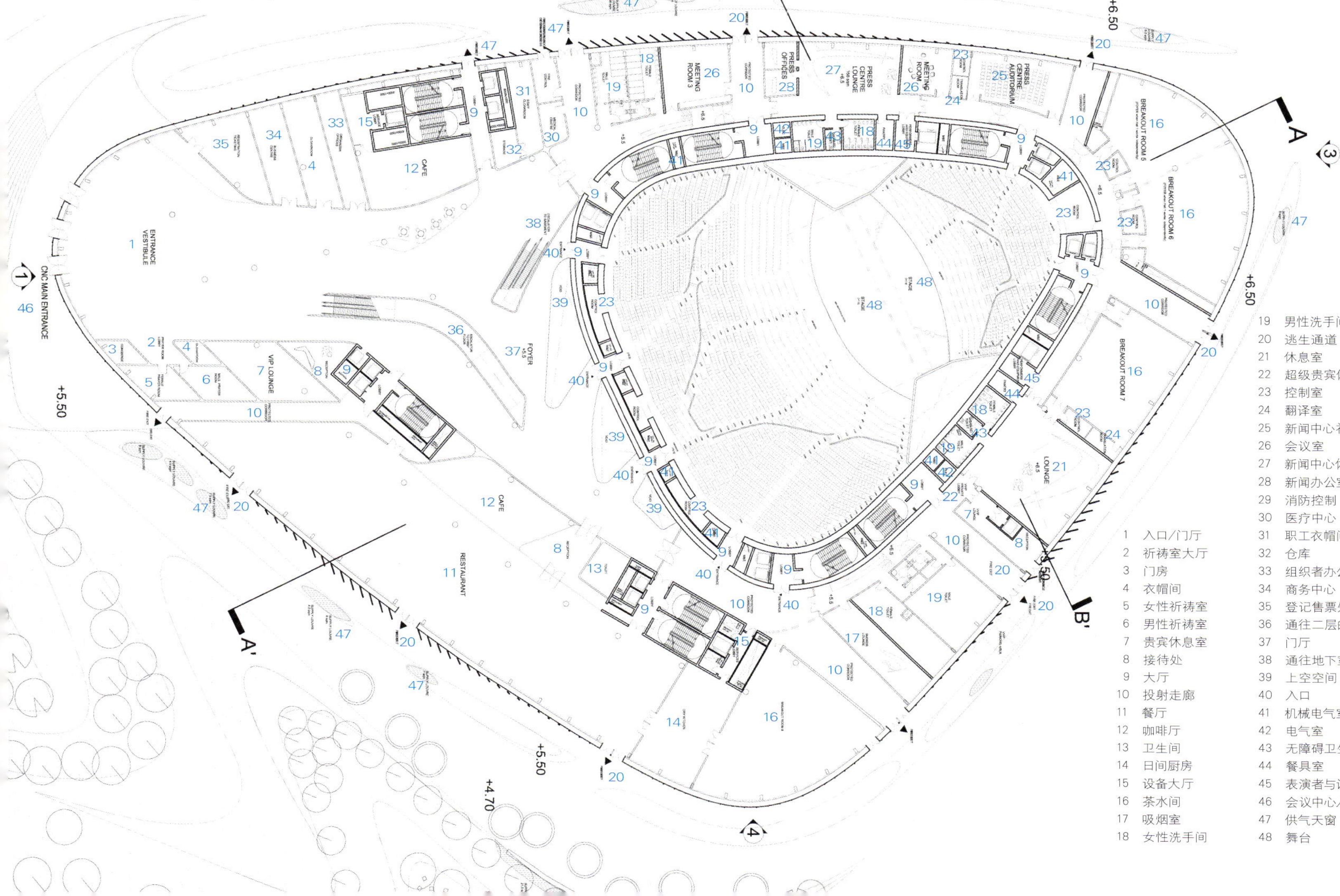

1 入口/门厅
2 祈祷室大厅
3 门房
4 衣帽间
5 女性祈祷室
6 男性祈祷室
7 贵宾休息室
8 接待处
9 大厅
10 投射走廊
11 餐厅
12 咖啡厅
13 卫生间
14 日间厨房
15 设备大厅
16 茶水间
17 吸烟室
18 女性洗手间
19 男性洗手间
20 逃生通道
21 休息室
22 超级贵宾休息室
23 控制室
24 翻译室
25 新闻中心礼堂
26 会议室
27 新闻中心休息室
28 新闻办公室
29 消防控制
30 医疗中心
31 职工衣帽间
32 仓库
33 组织者办公室
34 商务中心
35 登记售票处
36 通往二层的滚梯
37 门厅
38 通往地下室的滚梯
39 上空空间
40 入口
41 机械电气室
42 电气室
43 无障碍卫生间
44 餐具室
45 表演者与设备大厅
46 会议中心入口
47 供气天窗
48 舞台

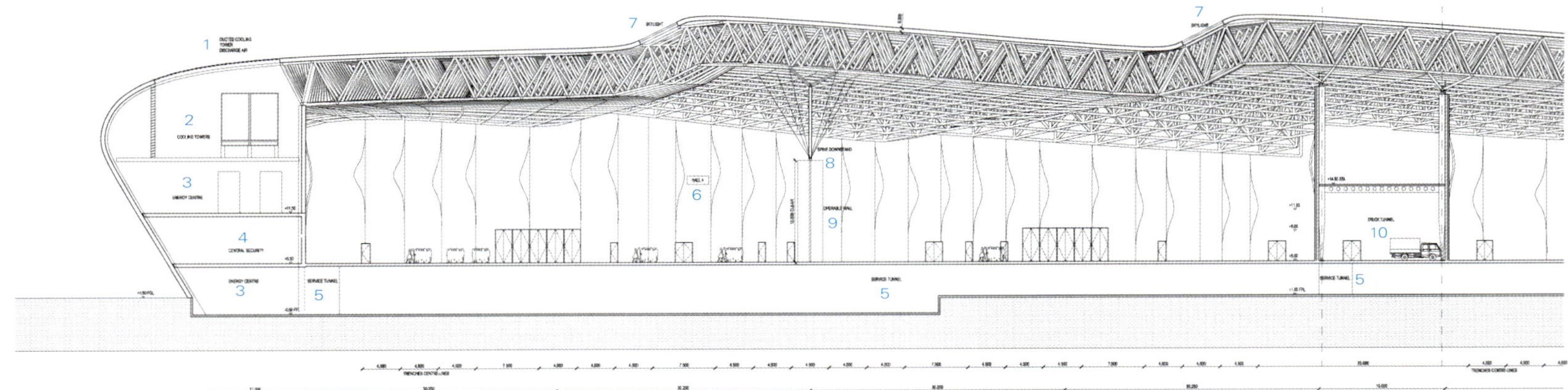

展览中心剖面1
Exhibition centre section 1

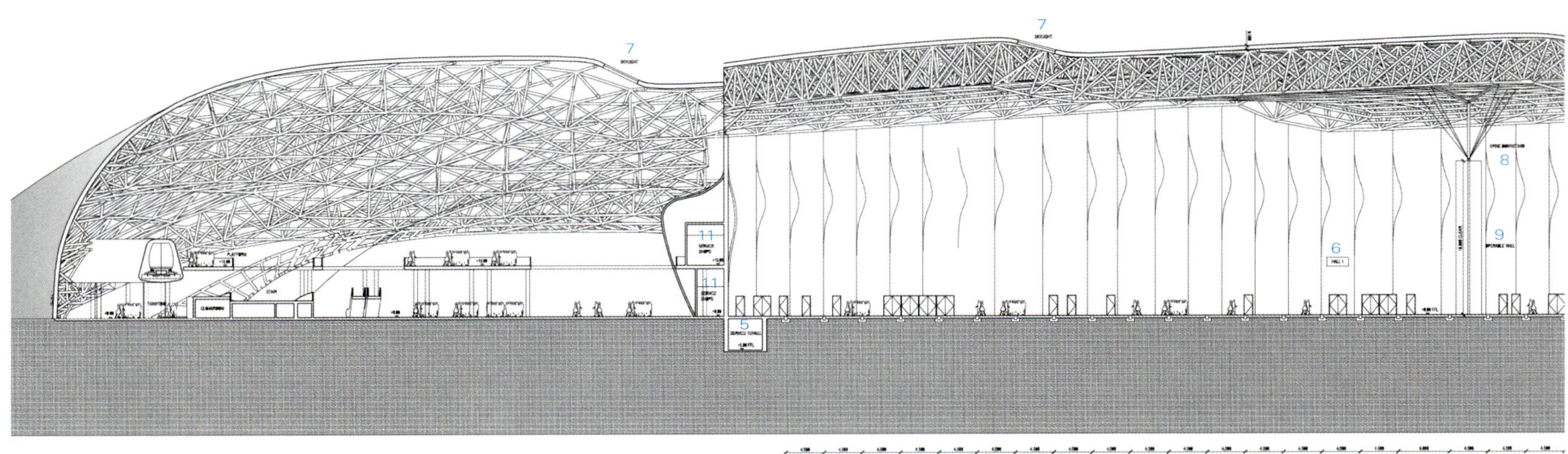

展览中心剖面2
Exhibition centre section 2

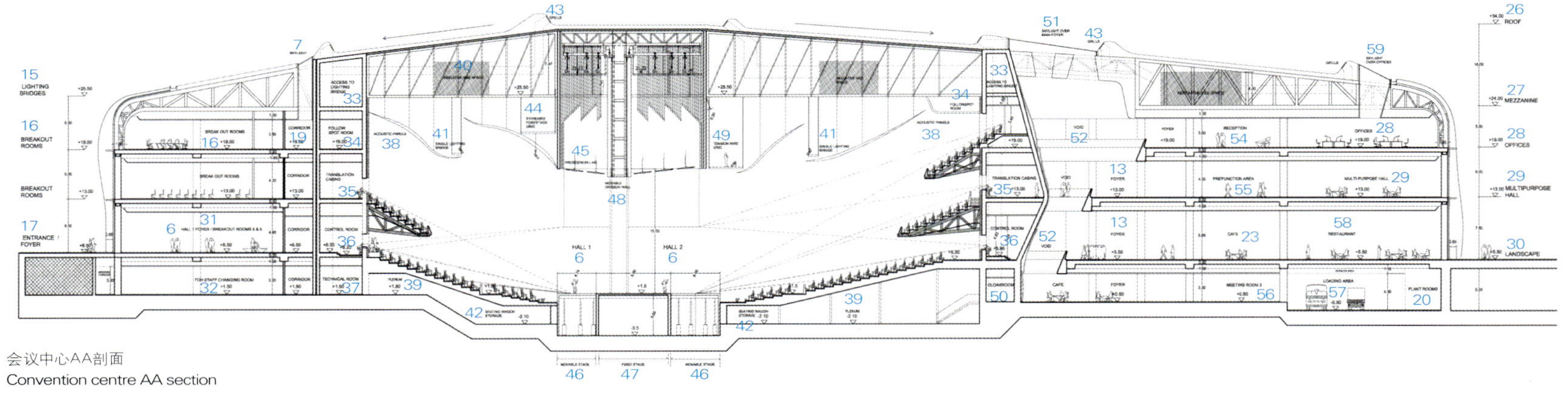

会议中心AA剖面
Convention centre AA section

1 导管冷却塔/排气
2 冷却塔
3 能源中心
4 中心安全室
5 设备通道
6 大厅
7 天窗
8 肋形承重楼板
9 开闭式间壁
10 卡车通道
11 维修车间
12 卫生间
13 门厅
14 海关区
15 舞台安灯天桥
16 分会议室
17 入口/门厅
18 员工衣帽间
19 走廊
20 设备间
21 滚梯
22 礼堂入口
23 咖啡厅
24 投射走廊
25 通向展览中心
26 屋顶
27 夹层
28 办公室
29 多功能大厅
30 景观
31 门厅/分会议室
32 员工更衣室
33 通向舞台安灯天桥
34 追光灯室
35 翻译室
36 控制室
37 技术室
38 吸声板
39 集气室
40 指示性机械与电气室
41 单个舞台安灯天桥
42 坐位手推车仓库
43 金属板
44 标准舞台幕布格栅
45 镜框线
46 移动舞台
47 固定舞台
48 移动分隔墙
49 张力线格栅
50 衣帽间
51 主门厅上的天窗
52 上空空间
53 会议室1
54 接待处
55 前厅
56 会议室2
57 卸货区
58 餐厅
59 办公室上方的天窗
60 用地范围
61 区内道路
62 大型客车停车位
63 消防车道
64 第二装货间
65 新闻中心休息室
66 休息室
67 走廊上方的天窗
68 女性洗手间
69 舞台仓库
70 移动的侧舞台
71 侧舞台
72 仓库
73 礼堂
74 第一装货间
75 入口坡道
76 1号大厅的门厅

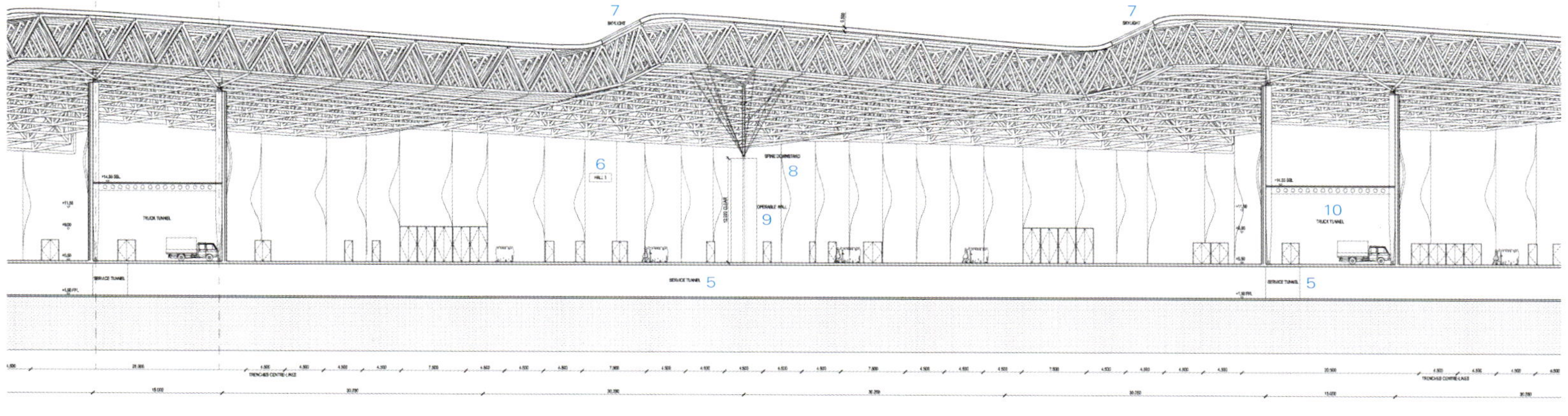

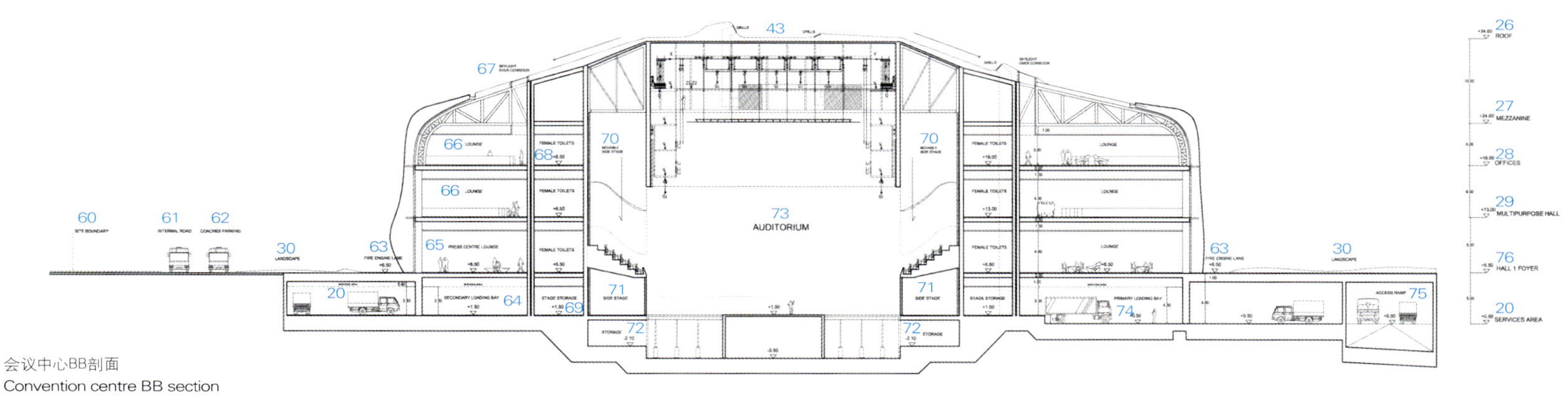

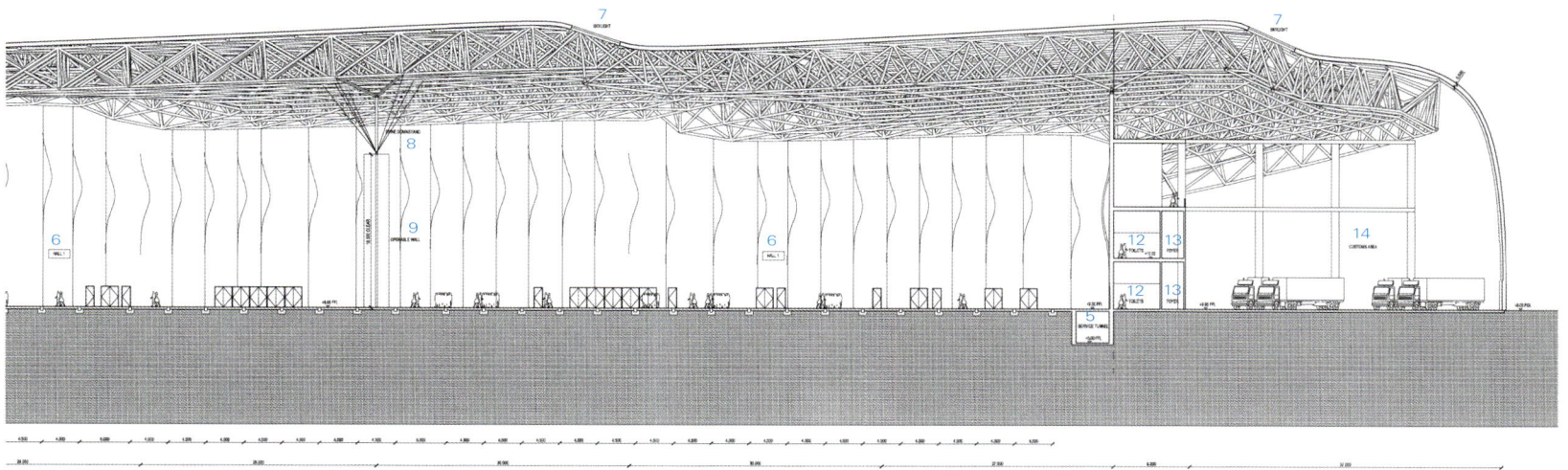

会议中心BB剖面
Convention centre BB section

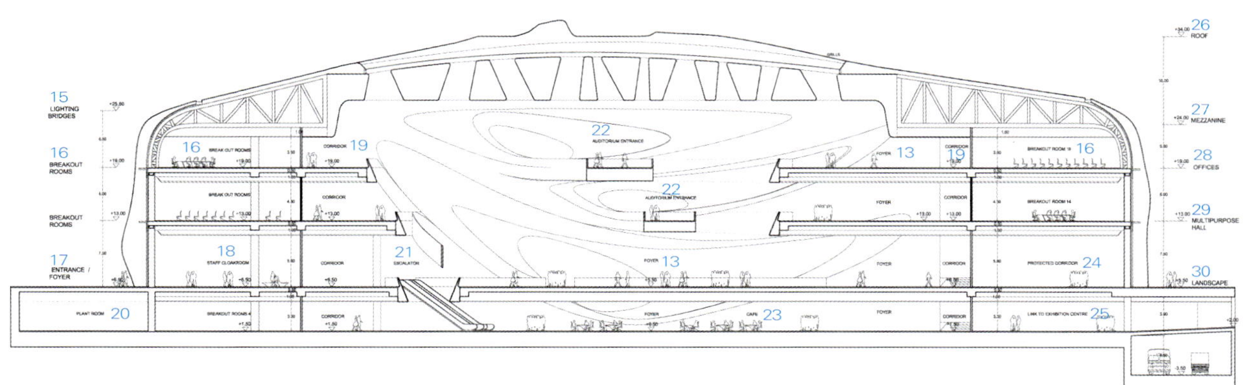

会议中心CC剖面
Convention centre CC section

展览中心立面图
Exhibition centre elevation 2

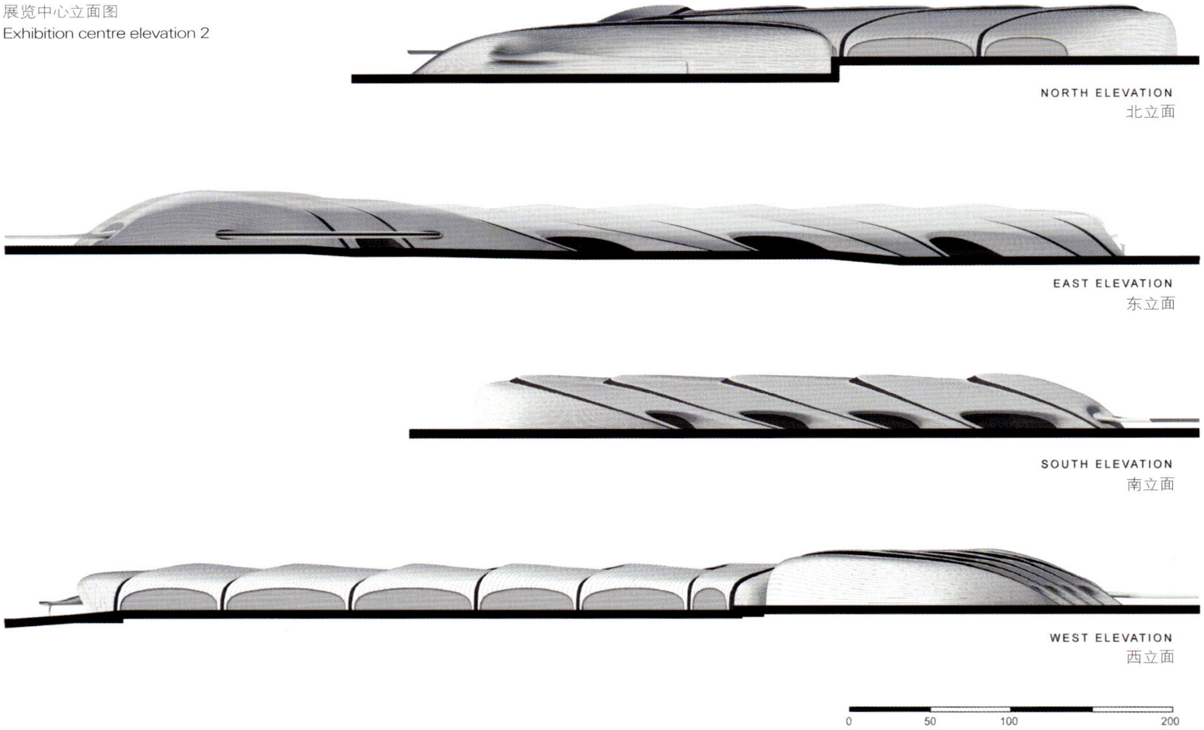

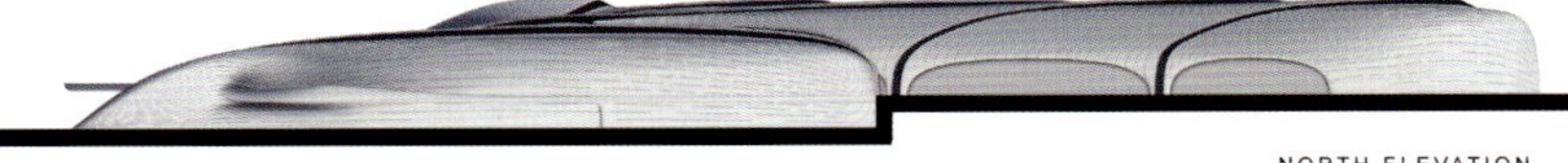

NORTH ELEVATION
北立面

EAST ELEVATION
东立面

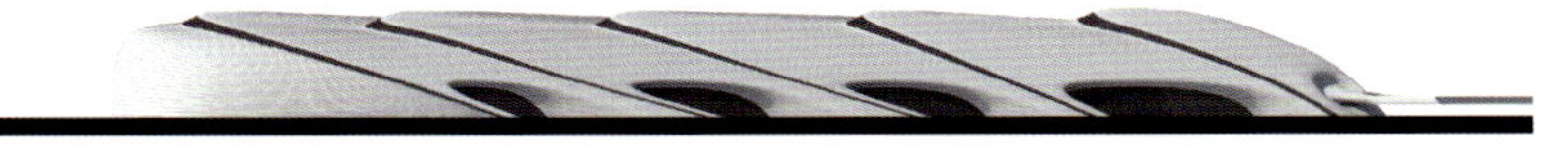

SOUTH ELEVATION
南立面

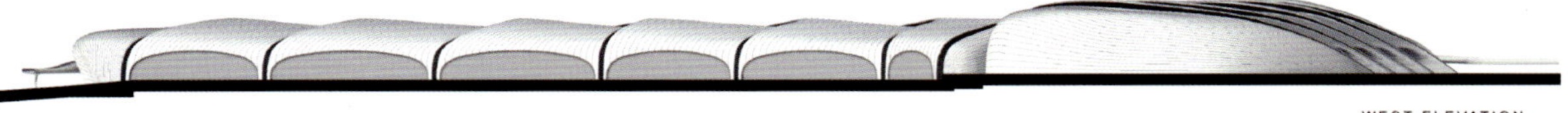

WEST ELEVATION
西立面

0 50 100 200

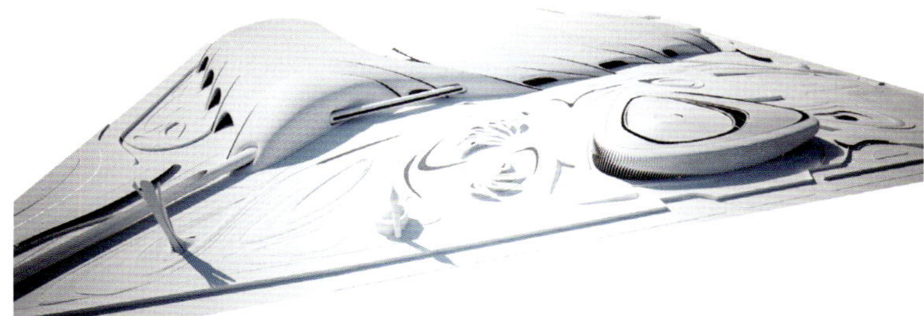

梅赛德斯·奔驰大酒店
MERCEDES BENZ HOTEL TOWER

建筑师：OFIS
项目负责人：Rok Oman, Spela Videcnik
项目团队：Robert Janez, Janez Martincic, Janja Del Linz, Katja Aljaz, Andrej Gregoric
项目地点：Yerevan, Armenia
结构：钢筋混凝土与金属结合
外立面：金属网
窗户：金属窗框双层玻璃窗
室内装修：混凝土、天然石材、木材

梅赛德斯·奔驰大酒店以圣经中的亚拉腊山为背景，地理位置得天独厚，因此可以重新布局，设计出与众不同的特点，有望成为耶烈万的象征和当代地标性建筑。各功能空间混合在一起，需要复杂的组织布局，无论内部还是外部。

较高的塔楼是酒店和商务中心（购物、零售、会议……），较低的塔楼为公寓（住宅一商务公寓，私人公寓……）。公共功能空间连接着两座塔楼，在此可以购物、举办展览、就餐，这里还连通着酒店大堂和商务入口。泊车主要集中在地下车库，部分临时停车以及出租车/巴士接送宾客可以停在室外。

The prominent location and dominating position with biblical Mt. Ararat as background "wallpaper" represents a chance for unique rearrangement with its own identity and could become a symbol and landmark of contemporary architecture in the city of Yerevan. The mixture of programs and relations inside the Mercedes Benz Hotel Tower calls for complex organisation – both inside and outside.

Higher tower is hotel and business centre (shopping, retail, convention...), lower is occupied with apartment program (apart-business condos, private apartments...). Public program is connecting the towers, here shopping, exhibition and restaurants are combined with hotel lobby and business entrance. Parking is mainly in garage under the ground floor, part of the fast parking and taxi/bus are outside at the Plato.

耶烈万圆环示意图
circles inside Yerevan

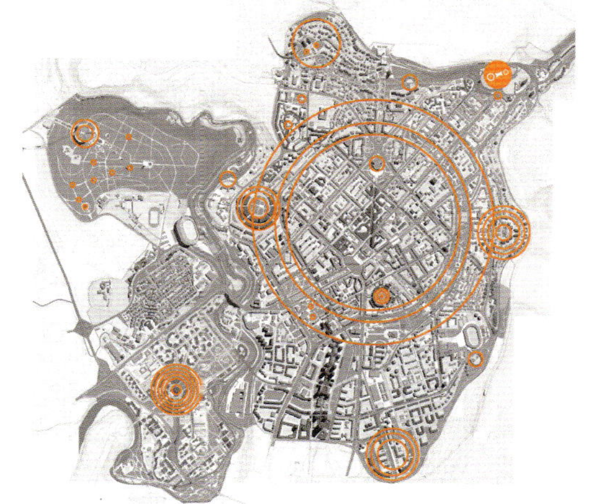

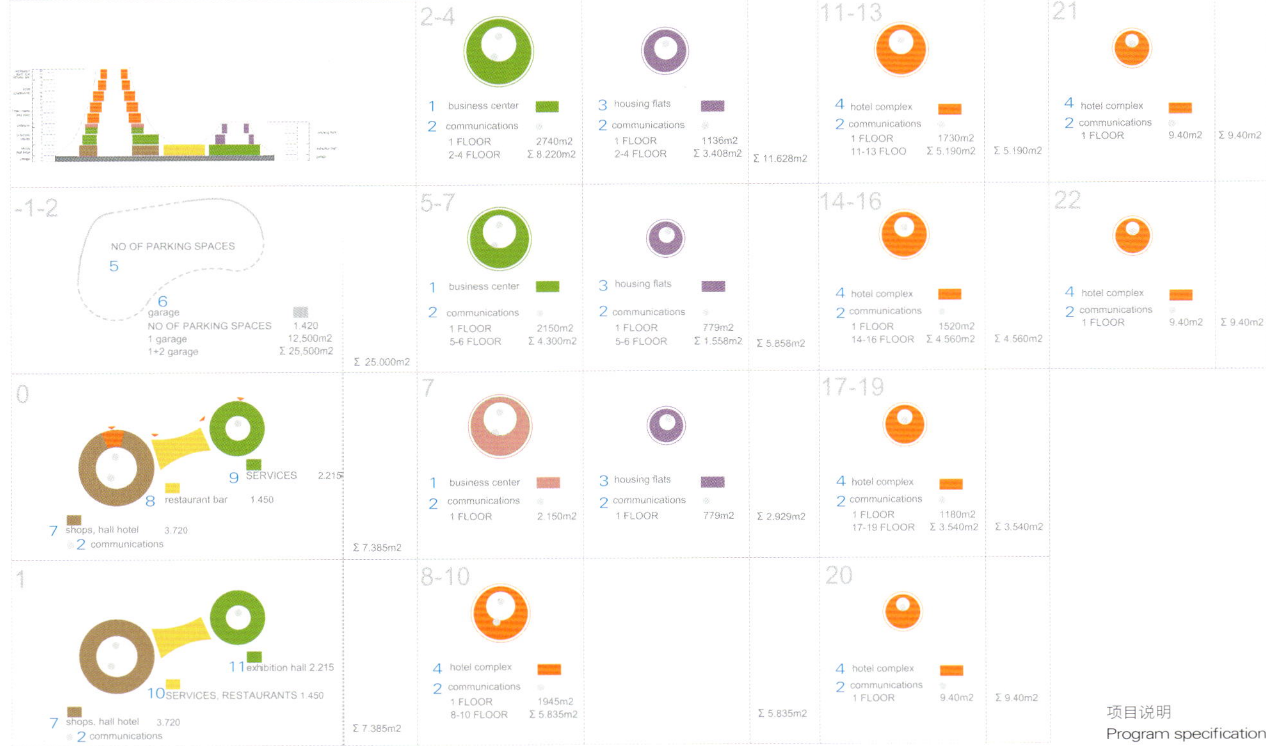

1 商务中心
2 交流区域
3 住宅公寓
4 酒店
5 停车位
6 车库
7 商店、酒店大堂
8 餐厅、酒吧
9 服务区
10 服务区、餐厅
11 展览大厅

项目说明
Program specification

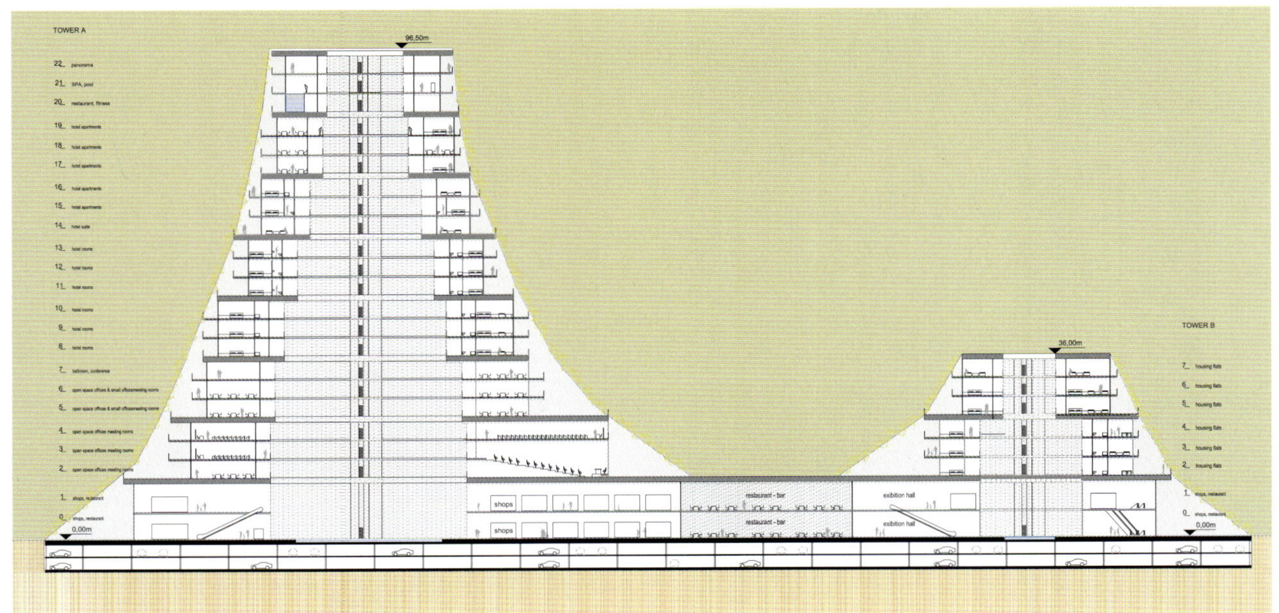

A座 Tower A
0 商店、餐厅
1 商店、餐厅
2 开放式办公室、会议室
5~6 开放式办公室&小型办公室、会议室
7 舞厅、会议大厅
8~13 酒店客房
14 酒店套房
15~19 酒店公寓
20 餐厅、健身中心
21 SPA中心、游泳池
22 全景观光层

B座 Tower B
0 商店、餐厅
1 商店、餐厅
2~7 住宅公寓

剖面图
Section

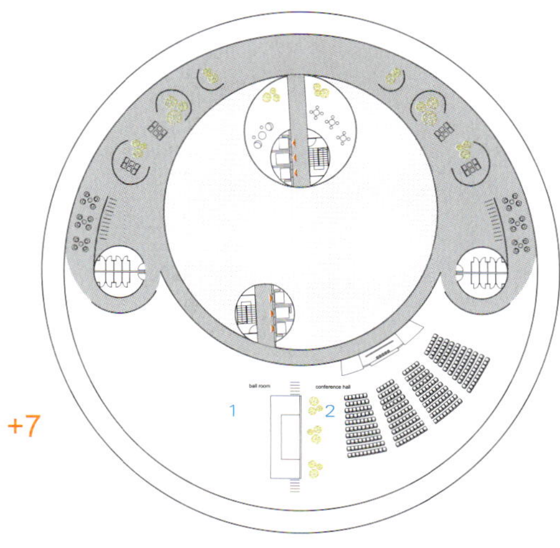

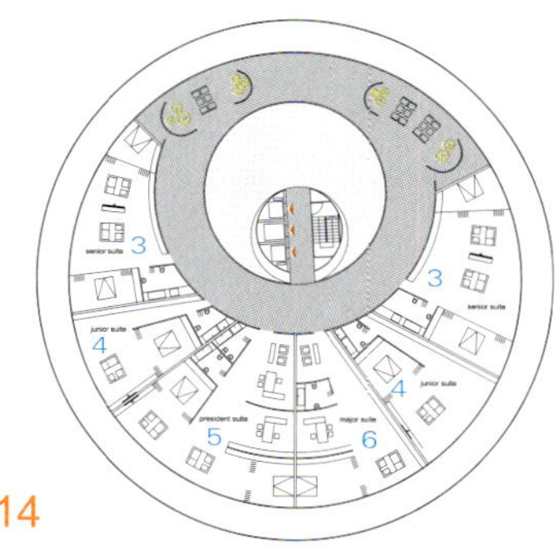

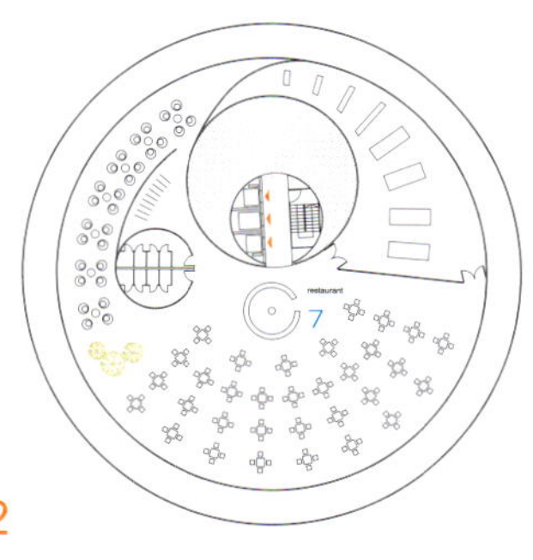

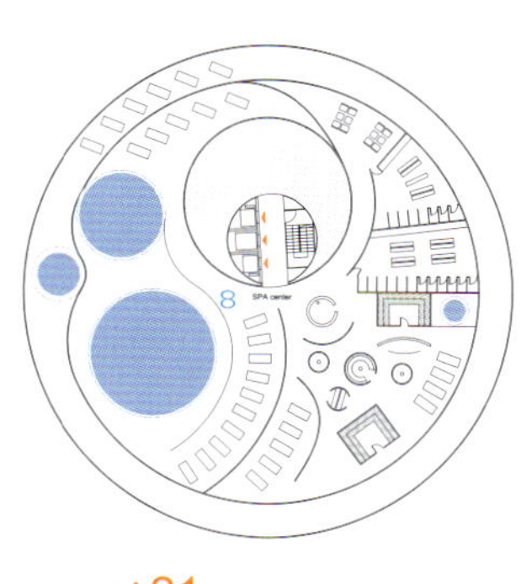

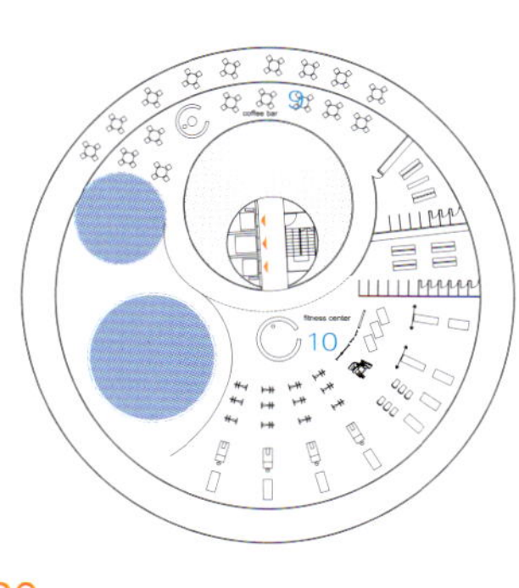

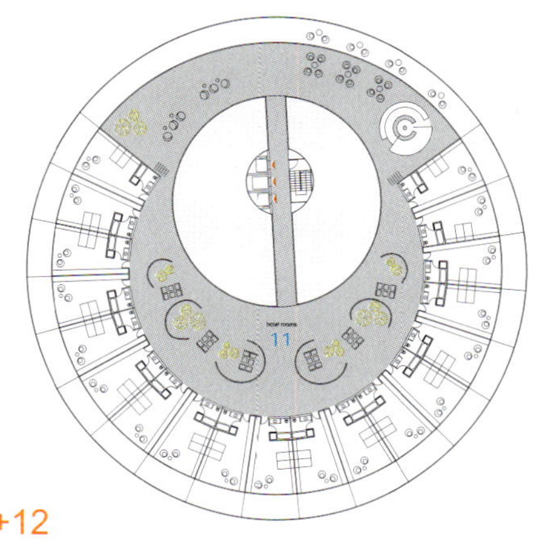

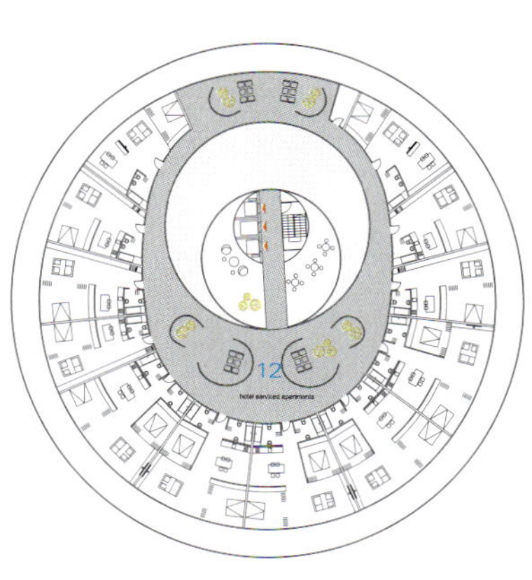

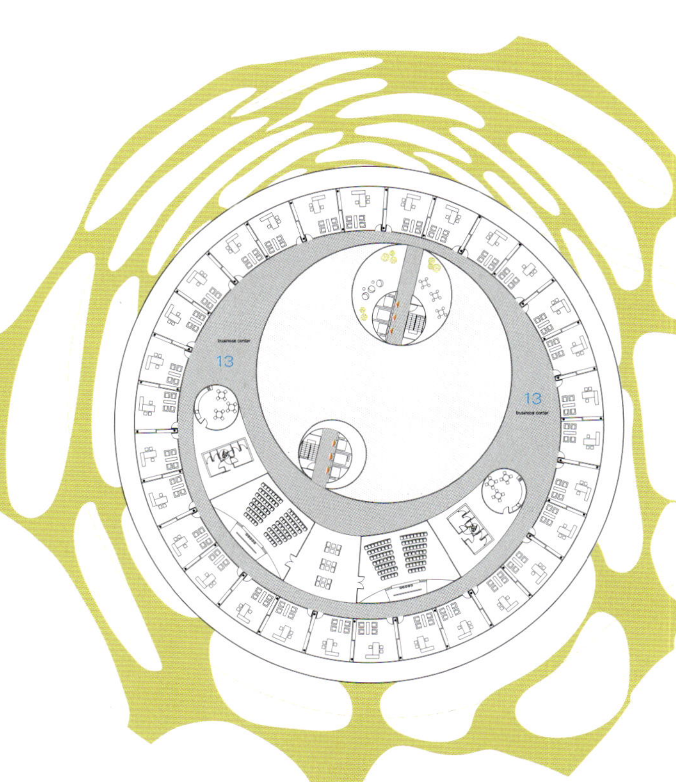

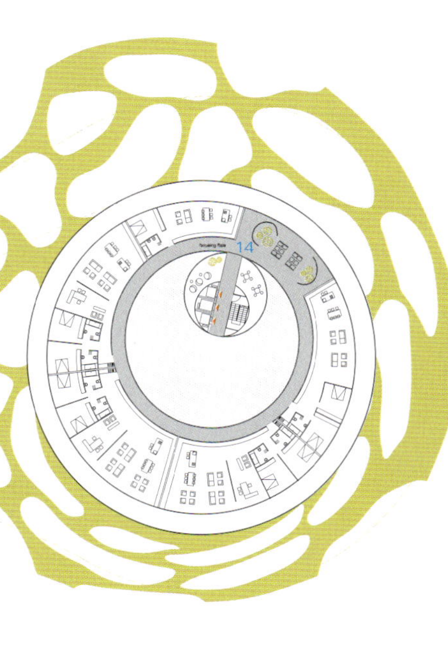

+4

1 舞厅
2 会议大厅
3 大套房
4 小套房
5 总统套房
6 套房
7 餐厅
8 SPA中心
9 咖啡吧
10 健身中心
11 酒店客房
12 酒店服务公寓
13 商务中心
14 住宅公寓

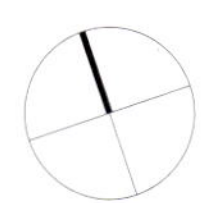

0 10 20 m

楼层平面图
Floor plans

南立面
South Elevation

北立面
North Facade

西立面
West Elevation

东立面
East Elevation

荷兰Zeeburgereiland的粮仓改造项目
SILOS ZEEBURGEREILAND IN THE NETHERLANDS

甲方：AmvestBlauwhoed
建筑师：NL Architects: Pieter Bannenberg, Walter van Dijk, Kamiel Klaasse
项目经理：Arne van Wees, Gen Yamamoto
设计团队：Tomas Amtmann, Joost Luub, Yuichi Tanaka, Yannick van Haelen, Ivar van der Zwan
承包商：De Nijs en Zonen
设备：VIAC Installatie Adviseurs

Zeeburgereiland的粮仓正在寻找新的落脚处。污水处理设备的重新布置引发了场地区域的新发展。"岛屿"现在可以用来居住。它位于阿姆斯特丹中心和名为艾瑟尔堡的最新扩建部分之间的重要位置上。岛屿的交通非常便利，它直接与A10公路相连接。

三座粮仓将被保留下来。其中一个将被改建成办公楼，但这并不是此次竞标的一部分。这座办公楼将与其他两座粮仓保持密切的联系，这一点非常重要。它们将构成一个体系。在本次规划中，粮仓将被改造为攀援、体育和文化设施。原有结构的高度将扩建到最大，以充分利用景观，并在规划的城市条件中创造一种强烈的存在感。

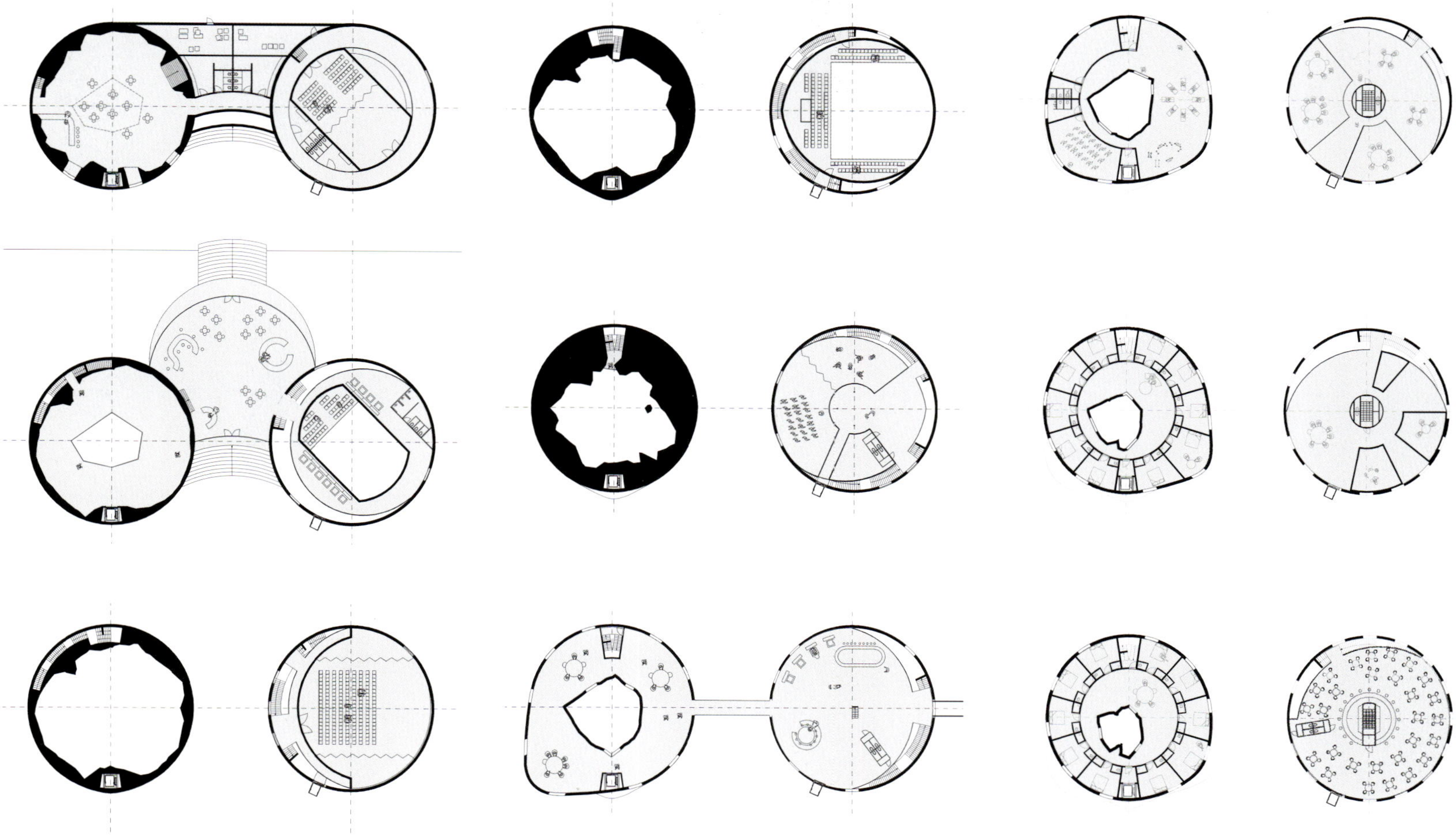

楼层平面图
Floor plans

The silos on Zeeburgereiland are looking for a new destination. The relocation of the sewage treatment plant sparked new developments in the area: the Island now becomes inhabitable. It is strategically positioned in between the center of Amsterdam and its latest expansion called IJburg. The Island is very accessible; it directly links to the ring road A10.
Three silos will remain. One will be developed into an office building and is not part of this competition. But it is essential that this tower will be conceived in close cooperation with the two other silos; they should form a set. In this proposal the silos will be dedicated to Climbing, Sports and Culture.The existing structures will be extended to the maximum height to benefit from the views and to create monumental substance in the projected urban condition.

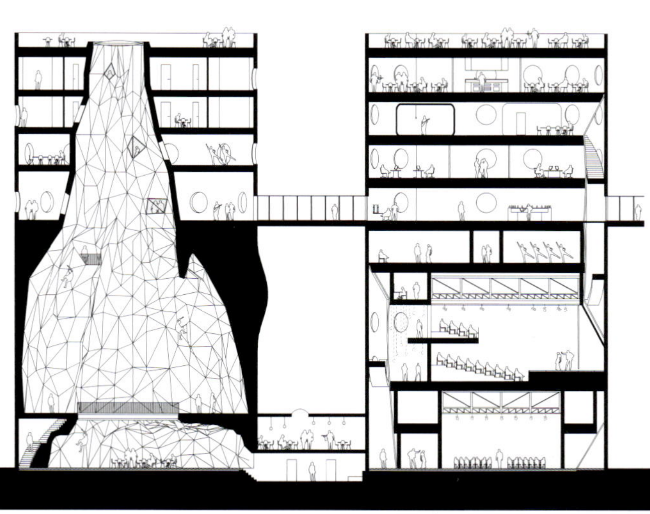

剖面图
Sections

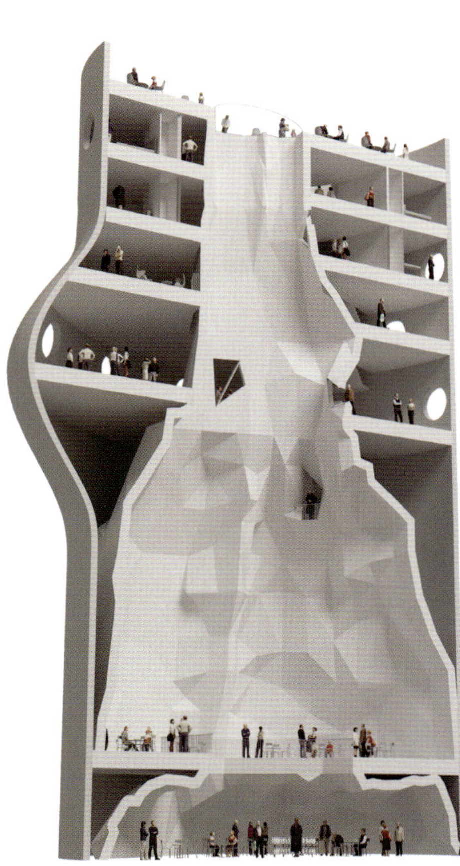

北京银河SOHO
THE GALAXY SOHO IN BEIJING

建筑师：ZAHA HADID ARCHITECTS
甲方：SOHO China Ltd.
设计：Zaha Hadid with Patrik Schumacher
项目指导：Satoshi Ohashi and Cristiano Ceccato
项目建筑师：Yoshi Uchiyama
项目经理：Raymond Lau

银河SOHO项目位于北京城的中心地区，是一处面积为3.3万平方米，集办公商业和娱乐为一体的综合建筑，它将成为北京这座充满活力的城市的一部分。设计受北京城市规模之大的启发。建筑由五个连续、流动的体量构成，它们彼此之间分离、融合或由拉伸的桥梁连接起来。五个体量在各个方向上彼此呼应，形成了全景建筑，没有边角或生硬的过渡来打破形式构成的流动性。

建筑巨大的内庭院是对由庭院形成连续开放的内部空间的传统中国建筑的呼应。在银河SOHO项目中，建筑不再是由刚性的建筑体构成，而是由互相结合的体量构成，形成了连续的相互适应和各个建筑间的顺畅流动。设计中移动的平台相互影响，产生了深深的"沉浸"与"包围"感。用户进入到建筑中，就会发现全部遵循连续曲线-线性形式逻辑的私密空间。

银河SOHO最下面的三层容纳商业和娱乐等公共设施。其上的楼层为创新型企业的办公聚集地。建筑的顶部为酒吧、餐厅和咖啡馆这些可观赏城市壮美街景的场所。不同的功能区通过私密的内部空间相连。而内部空间也与城市相连接，这有助于将银河SOHO打造成北京主要的城市地标。

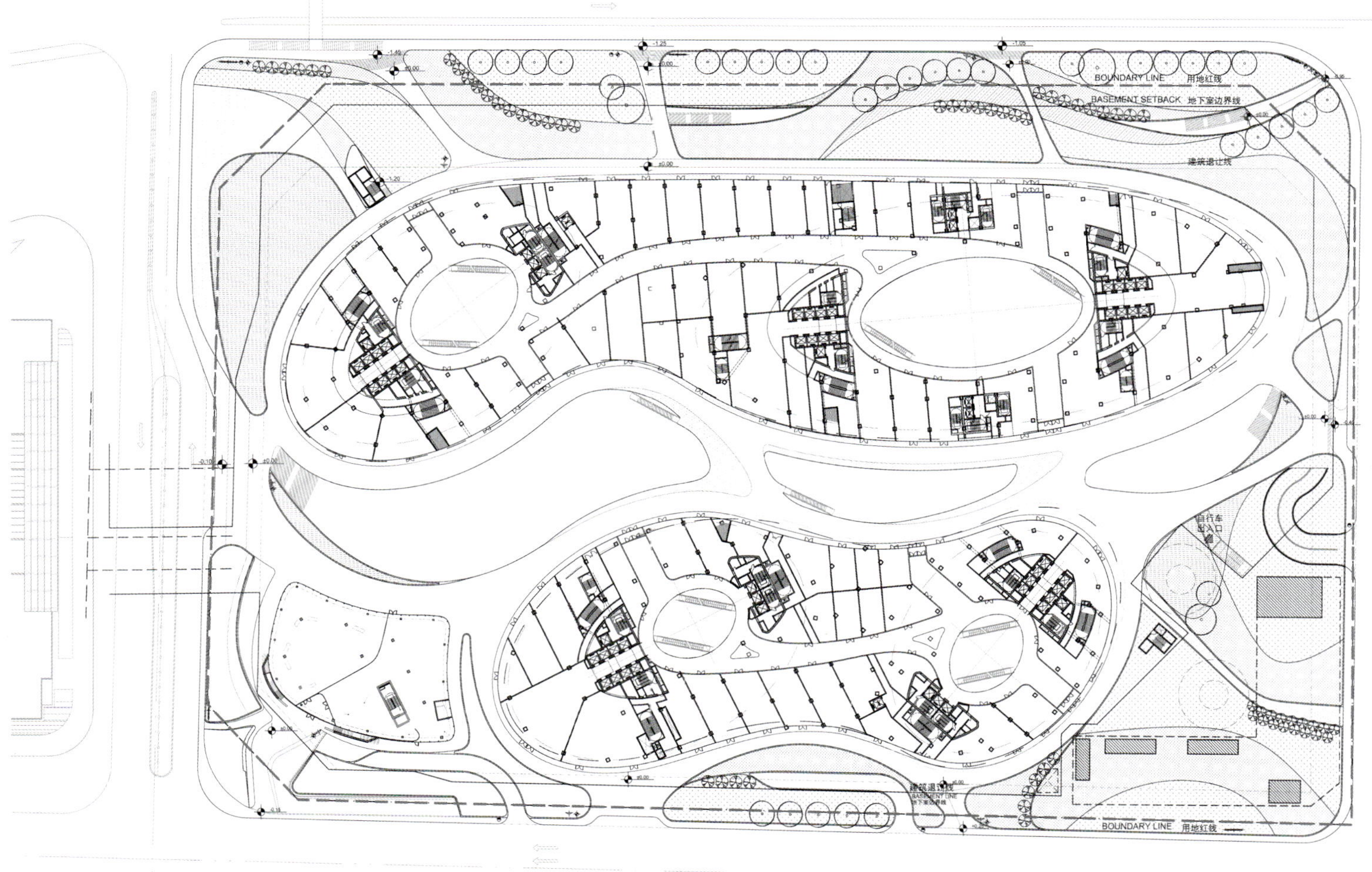

一层平面图
Ground floor plan

The Galaxy SOHO project in central Beijing for SOHO China is a 330,000 m² office, retail and entertainment complex that will become an integral part of the living city, inspired by the grand scale Beijing. Its architecture is a composition of five continuous, flowing volumes that are set apart, fused or linked by stretched bridges. These volumes adapt to each other in all directions, generating a panoramic architecture without corners or abrupt transitions that break the fluidity of its formal composition.
The great interior courts of the project are a reflection of traditional Chinese architecture where courtyards create an internal world of continuous open spaces. Here, the architecture is no longer composed of rigid blocks, but instead comprised of volumes which coalesce to create a world of continuous mutual adaptation and fluid movement between each building. Shifting plateaus within the design impact upon each other to generate a deep sense of immersion and envelopment. As users enter deeper into the building, they discover intimate spaces that follow the same coherent formal logic of continuous curve-linearity.
The lower three levels of Galaxy SOHO house public facilities for retail and entertainment. The levels immediately above provide work spaces for clusters of innovative businesses. The top of the building is dedicated to bars, restaurants and cafés that offer views along one of the greatest avenue of the city. These different functions are interconnected through intimate interiors that are always linked with the city, helping to establish Galaxy SOHO as a major urban landmark for Beijing.

塔林市政厅
TALLINN TOWN HALL

建筑师：BIG
合伙负责人：Bjarke Ingels
项目负责人：Jakob Lange
类型：Competition
甲方：Union of Estonian Architects
合作者：Adams Kara Taylor, Grontmij Carl Bro, Ramboll, Allianss Arhitektid Oü
面积：28 000 m^2
项目地点：Tallinn, Estonia

健全的政府管理和众人参与的民主制度依赖于二者的双向透明度。这需要政治上对民众问题、需求、渴望的透彻观察，也需要民众对政治过程的观察。

新塔林市政厅以一种毫不夸张的方式体现了双向透明度。不同的公共部门在公共服务商业中心之上形成了一个多孔的雨篷，让日光和人们的视野能够从中穿透。

办公人员并不是离群众很远只会在厚厚的墙里作决定的行政官员，而是把自己暴露在采光井下、庭院中央，从商业中心的任何角度都能看到他们的日常办公。透过全景窗，外面的市民们可以看到城市行政工作在运转中。

办公人员也将能够看到商业中心，确保城市和城市里的市民没有从眼前和头脑中消失。

Good governance and participatory democracy are dependent on transparency in both directions. It requires adequate political overview of the problems, demands and desires of the public, as well as public insight into the political processes.

The new town hall of Tallinn will provide this two way transparency in a very literal way. The various public departments form a porous canopy above the public service market place allowing both daylight and view to permeate the structure.

The public servants won't be some remote administrators taking decisions behind thick walls, but will be visible in their daily work from all over the market place via the light wells and courtyards. From outside the panoramic windows allow the citizens to see their city at work.

In reverse the public servants will be able to look out and into the market place's making sure that the city and its citizens are never out of sight nor mind.

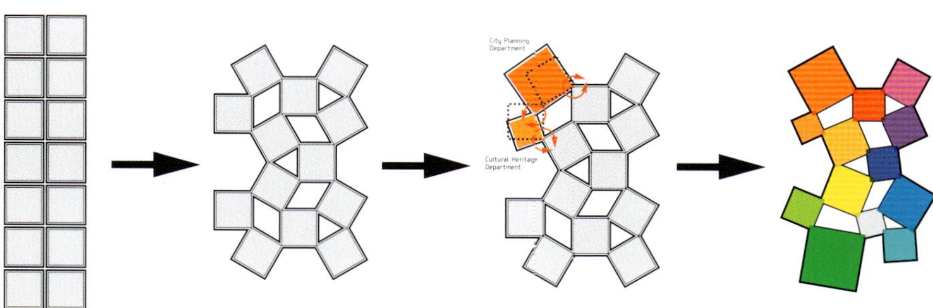

紧凑的组织
COMPACT ORGANIZATION

多孔的组织
POROUS ORGANIZATION

不同的项目
DIFFEFENTIATED PROGRAM

公共村落
THE PUBLIC VILLAGE

- 教育部 Education Department
- 环境部 Environment Department
- 城市事业部 City Enterprise Department
- 市政公安部 Municipal Police Department
- 住房经济部 Housing Economy Department
- 城市规划部 City Planning Department
- 文化遗产部 Cultural Heritage Department
- 土地问题部 Land Issues Department
- 市政工程部 Municipal Engineering Department
- 社会福利保健部 Social Welfare and Healthcare Department
- 交通部 Transport Department
- 议会部门 Council
- 主要交通流线 Main Circulation

连接方式
CONNECTIONS

每个部门都进行了旋转，以在各部门之间创造最大的连接，同时为内部庭院留出空间。
Each department is rotated in order to create maximum connections between the units and at the same time leave place for internal courtyards

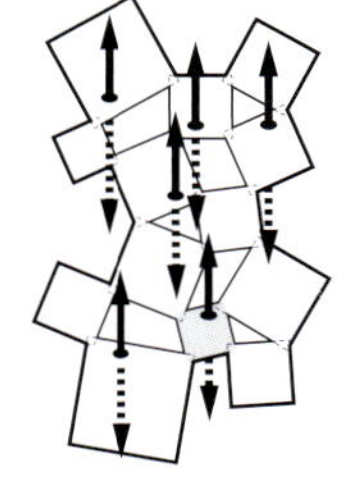

电梯核心筒
ELEVATOR CORES

六个电梯核心筒将一层和上面各个楼层连接起来。市民可以乘坐电梯到达各部门的柜台及所要办理的公共事务的办公室。
6 elevator cores are linking the ground floor to the upper floors. A citizen would go to the counter of the respective department and get accompanied to the office of their public official.

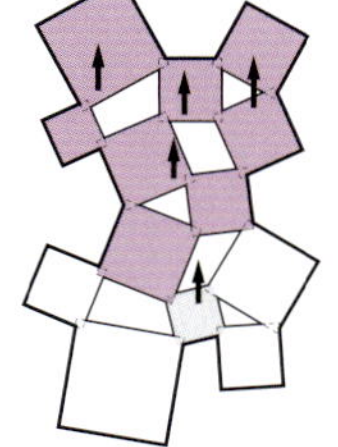

市政办公室
CITY OFFICE

各部门下面的一层主要由市政办公室占据。这些办公室能灵活地满足各种变化的需求，随时联系不同部门。
The floor below the departments is mainly occupied by the city offices. The city offices have maximum flexibility to address the changing demands and connect to the different departments at different times.

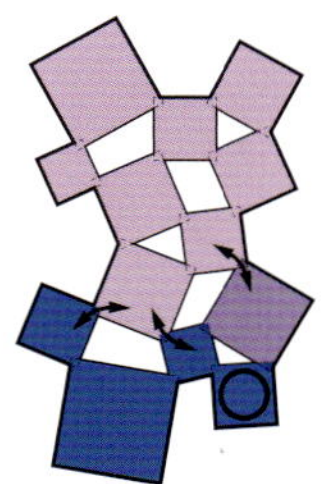

二层连接方式
FIRST FLOOR CONNECTIONS

市政府和议会部门位于二层，就在议会大厅的正下方。
The City Government and the Council are located on the first floor right under the Council Hall.

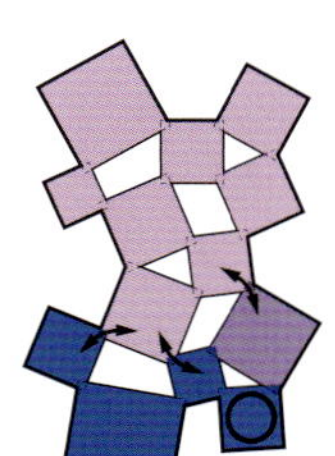

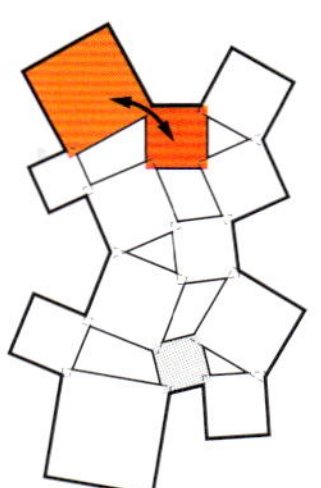

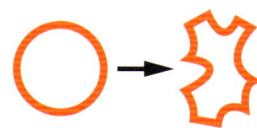

部门连接方式
DEPARTMENT CONNECTION

城市规划部和文化遗产部位于三层和四层，位置相邻。
The City Planning Department and the Cultural Heritage Department are closely connected on the second and third floor.

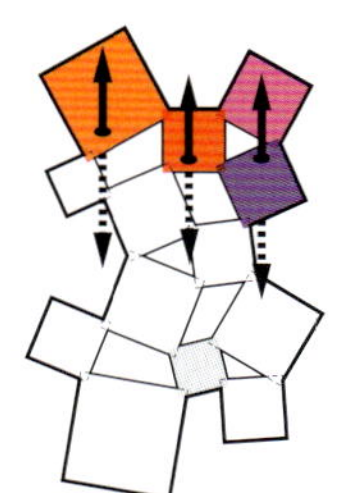

存档部门电梯
ARCHIVE ELEVATORS

指定的存档部门电梯将城市规划部、文化遗产部、土地问题部以及市政工程部直接与地下室的存档部门连接起来。
Designated archive elevators connect from the City Planning, the Cultural Heritage, the Land Issues and the Municipal Engineering Department directly to the heart of the archive in the basement.

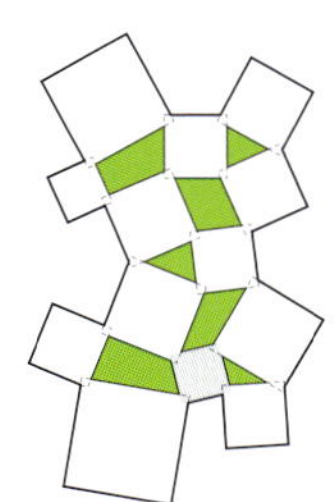

庭院
COURTYARDS

七个大采光井使日光能够照射到一层的市场。
The seven grand light wells allow daylight to enter the market place at the ground floor.

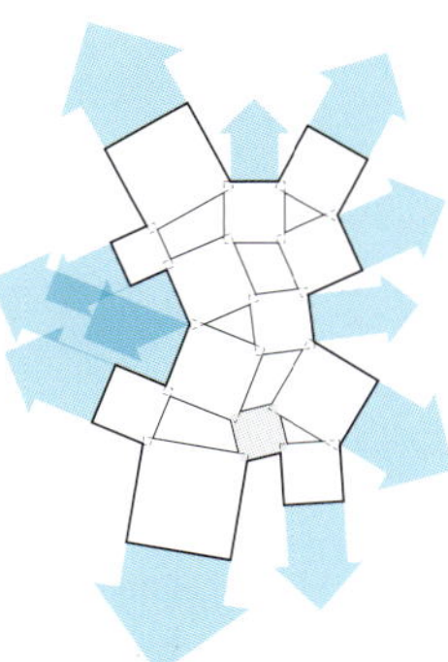

全景视野
PANORAMIC VIEWS

大型的洞口使市政厅办公室内的工作人员可以看到整个城市的景色。
The large openings offers views of the entire city from within the offices in the Town Hall.

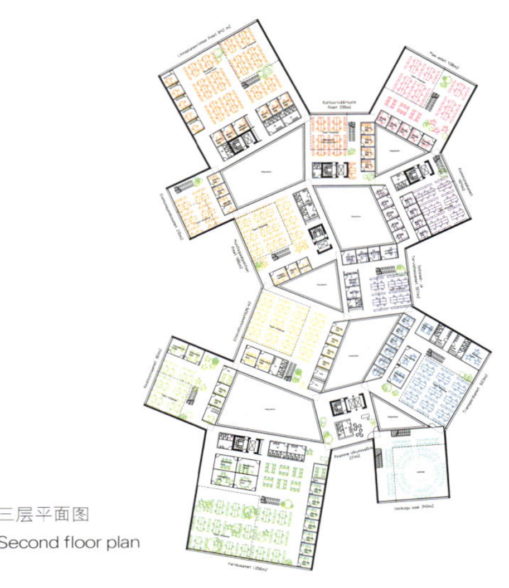

三层平面图
Second floor plan

四层平面图
Third floor plan

五层平面图
Fourth floor plan

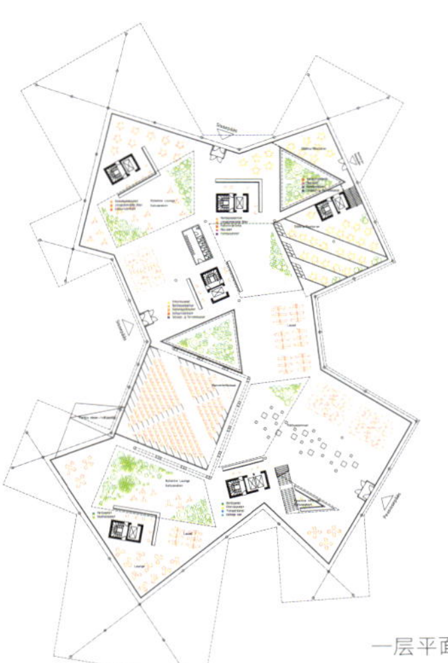

一层平面图
Ground floor plan

二层平面图
First floor plan

总平面图
Site plan

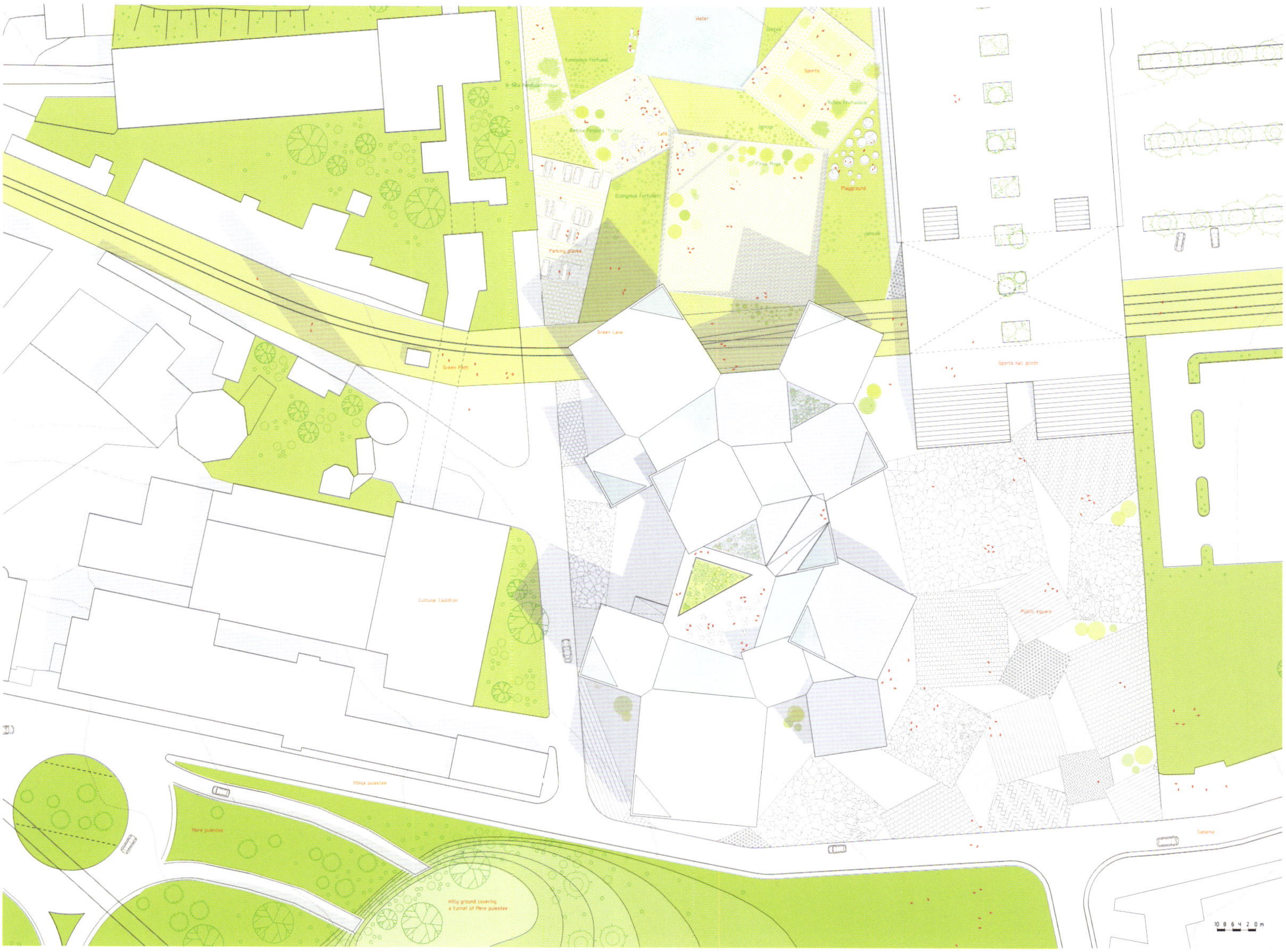

剖面图

Section

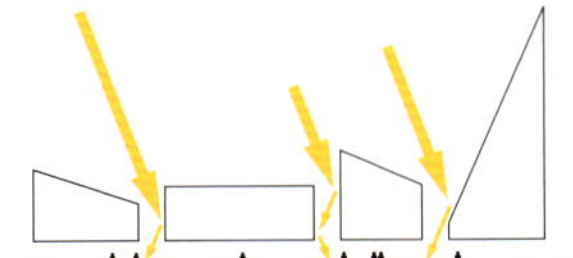

庭院

COURTYARDS

各部门之间距离较大，使自然光可以进入下面的市场。

The distance between the departments allows natural daylight to enter the marketplace below.

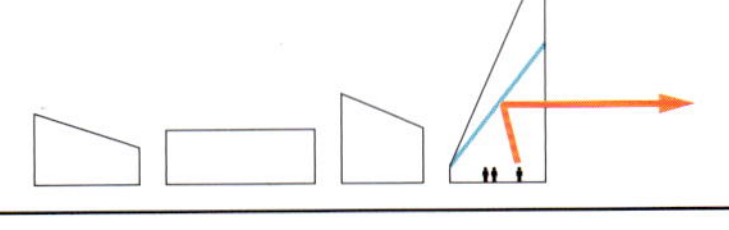

向外的视野

OVERVIEW

议会大厅内的一面大大的镜子可以使下面的公务员看到远处城市的景色。

A large mirror in the Council Hall above the public servants gives them a periscope view of the city.

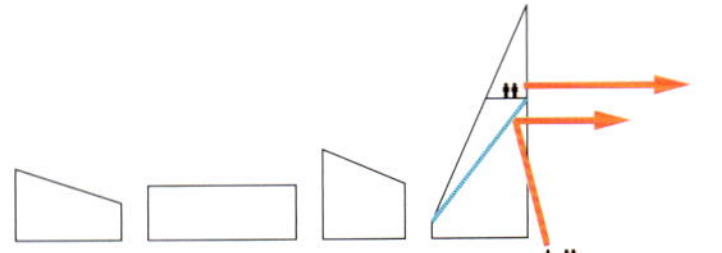

民主视野

DEMOCRATIC VIEW

在市政厅建筑的上方，市民们可以观赏到塔林市壮丽的景色。通过议会大厅内的镜子也能看到相同的景色。

From the top of the Town Hall tower the citizens can enjoy the great view of Tallinn. The same view can be experienced via the Council Hall mirror.

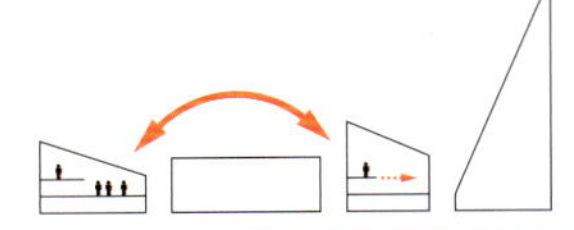

空间灵活性

SPATIAL FLEXIBILITY

每个部门都可以通过扩大中间楼层或与设置另一个办公室来增加20%的面积。

Each department can grow up to 20% by enlarging their mezzanine or changing building with another office.

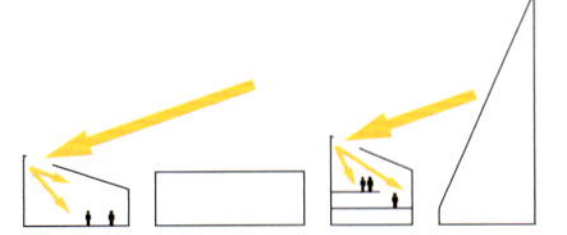

天窗

SKYLIGHTS

每个部门上方的窗户都能捕获位置较低的太阳照射的阳光，将漫射光线反射到办公室内。

Windows at the top of each department catches the low sun and reflects diffuse light into the offices.

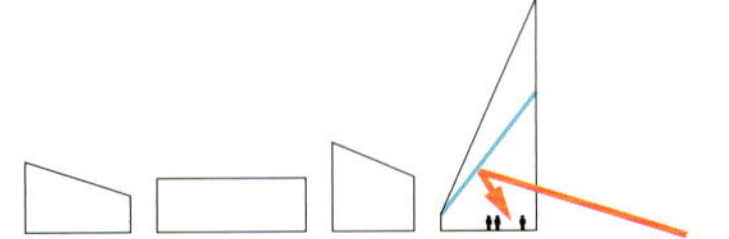

向内的视野

INSIGHT

市政厅外面的市民也能看到建筑内部的行政工作。

From outside the Town Hall the citizens can get an insight into the political work.

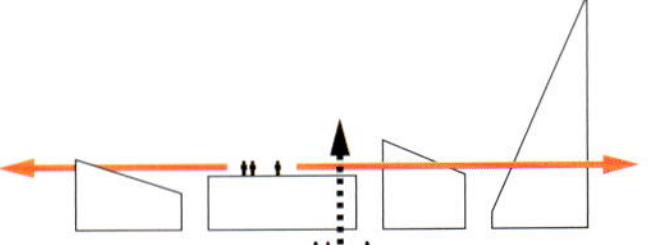

公共屋顶露台

PUBLIC ROOF TERRACE

一层的餐厅有一个公共电梯，可以将市民带到屋顶的咖啡厅，从这里可以360°观看塔林市的风光。

From the ground floor restaurant a public elevator brings the citizens to the roof café. From here they can enjoy the 360 degree panorama of Tallinn.

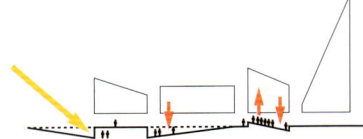

地面

THE CARPET

一层有通向厨房的入口，并为厨房提供日光。倾斜的广场是主会议大厅。

The ground floor is manipulated to create access and light for the kitchen. A tilted square becomes the main conference hall.

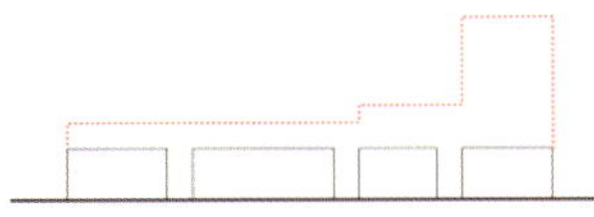

部门布局

DEPARTMENT LAYOUT

各部门根据各自需要布局。

The Departments are arranged according to their individual needs.

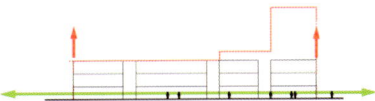

市场

MARKET PLACE

抬高体量，在市政厅下创建一个新的公共场所。

The volumes are elevated creating a new public under the Town Hall.

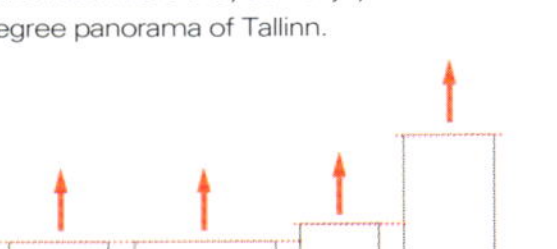

高度

HEIGHTS

建筑主要体量根据建筑规范抬高。

The main volumes of the building are raised according to the building regulations.

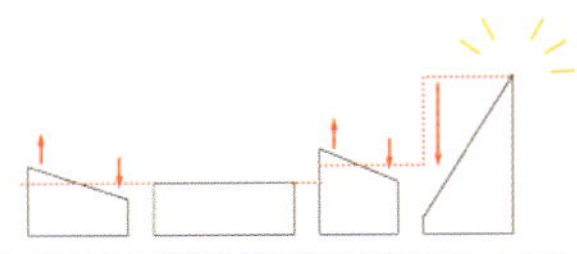

民主的大楼

DEMOCRATIC TOWER

抬高和降低屋顶，只有尖顶超过了建筑限高。

The roof of the volumes are raised and lowered, so only spires break the building height limits.

芬兰海湾
GULF OF FINLAND

绿化带
GREEN BELT

塔林市政厅的新基址位于中世纪城市外围的绿化带的边缘。
The new site for the Tallinn Town Hall is located on the edge of the green belt surrounding the medieval city.

绿色链
THE GREEN LINK

新基址将绿化带与滨水区连接起来。
The site offers the possibility of linking the green belt to the waterfront.

绿色连接
GREEN CONNECTIONS

基址将绿化带与滨水区连接起来，还将铁路转变成从西到东的主连接路线。
The site offers the possibility to connect the green belt to the waterfront and turning the railroad path into the main connection from east to west.

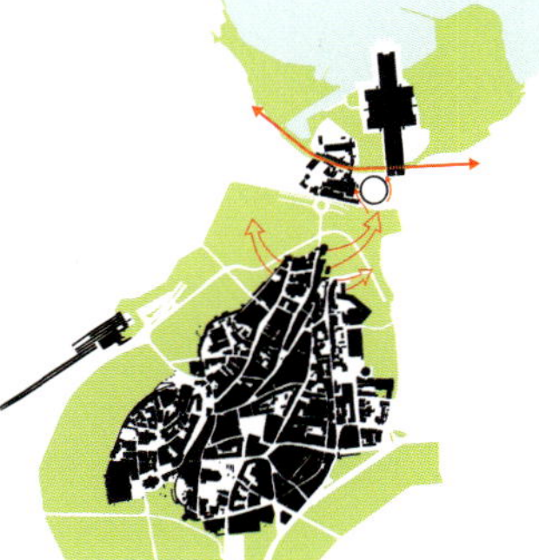

公共连接
PUBLIC CONNECTIONS

重建交通后，绿色环将连接中世纪城市与新塔林市政厅。我们将在老铁路南端延伸铁路，确保从西到东的交通顺畅。
After completing the restructuring of the traffic, the green ring will connect the medieval city with new Tallinn Town Hall. We propose to build south of the old railway to ensure the strong connection from east to west.

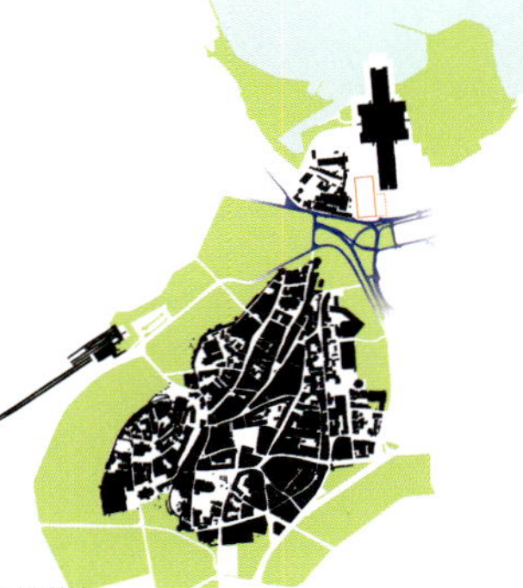

关键节点
GORDIAN KNOT

目前的交通是一个障碍，将中世纪城市与滨水区域分离开。
The traffic currently acts as a barrier, separating the medieval city from the waterfront.

交通解决方案
TRAFFIC SOLUTION

将环路掩埋后，繁忙的交通将会消失，骑自行车和步行来这里的人们将更加方便。
By submerging the ring road the heavy traffic will disappear and bikes and pedestrians will get easy access to the area.

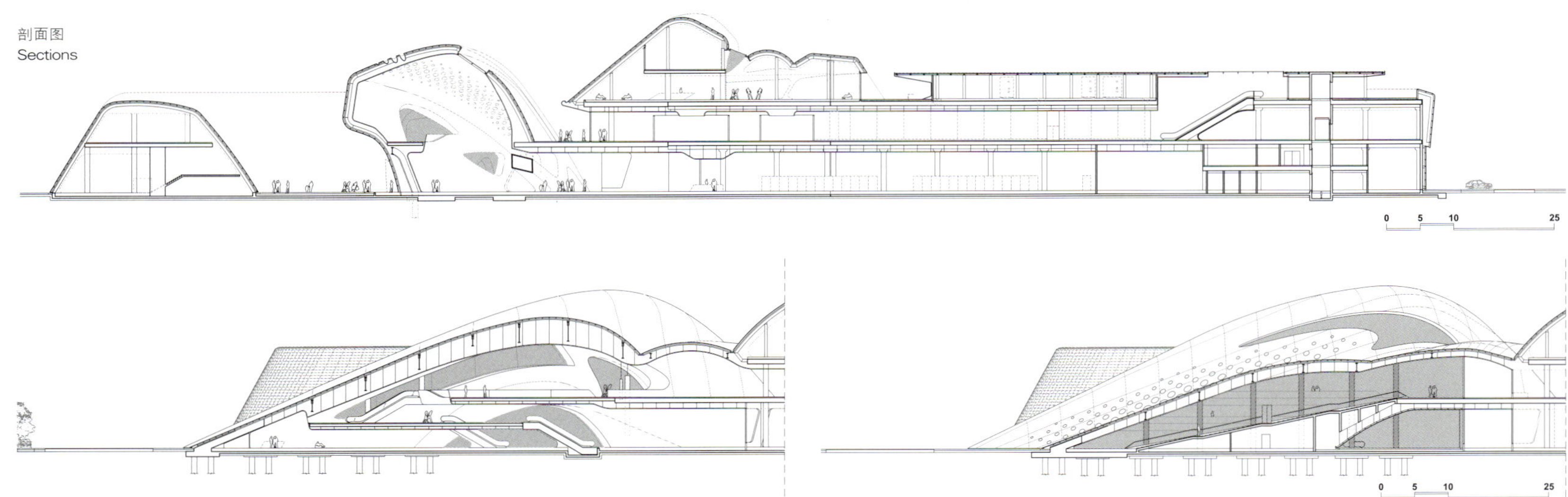

剖面图
Sections

意大利JESOLO MAGICA零售与商务中心
JESOLO MAGICA - RETAIL & BUSINESS CENTRE, ITALY

甲方：Home Group
建筑师：Zaha Hadid Architects
设计：Zaha Hadid with Patrik Schumacher
项目主管：Gianluca Racana
项目建筑师：Paolo Matteuzzi

Jesolo是意大利最著名的海滨疗养胜地之一。Jesolo Magica零售与商务中心的设计充分利用了场地邻近威尼斯的优势。项目致力于成为城市革新与再生的催化剂——为Jesolo城进一步发展为会议与度假胜地提供良好的契机。设计由连续流动的空间组成。这些空间使人们对城市全新的发展充满信心。

Jesolo Magica建筑群各不相同的部分共同组成了连续的建筑场地，彼此之间分离，但又以不断变化的效果在空间逻辑上相互连接。围绕着零售中心的体量在中心空间周围“开放”，就像花瓣一样。酒店建筑是最后一片花瓣，框起了附近泻湖的景观。

除了办公室、零售空间和餐馆外，Jesolo Magica项目还包括一家带会议中心、水疗馆、夜总会和室外活动空间的酒店。

Jesolo is one of Italy's most established seaside resorts and the design of Jesolo Magica makes full advantage of its location near the Venice Lagoon. The project aims to be the catalyst for reinvention and regeneration – giving the of the town of Jesolo an excellent opportunity to further develop as a conference and holiday destination. The design creates a continuum of fluid space that instigates a renewed sense of possibility.

The disparate elements of the Jesolo Magica complex fit together to form a coherent field of buildings, each one separate - but logically connected to the next in a continually changing ensemble. The volumes encompassing the retail centre "open-up" around a central space, like the petals of a flower. The hotel building forms the final "petal", framing the views over the adjacent lagoon.

In addition to offices, retail spaces and restaurants, the Jesolo Magica project features a hotel with conference center, spa, nightclub and outdoor spaces for events.

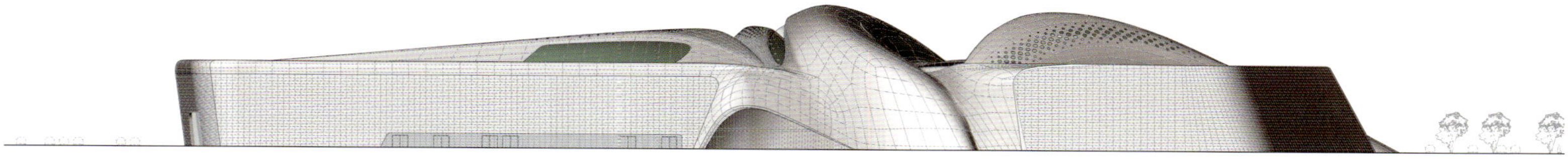

0 5 10 25

东立图
East elevation

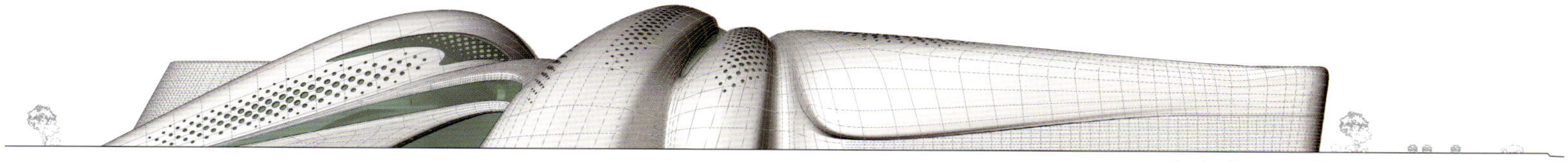

0 5 10 25

西立图
West elevation

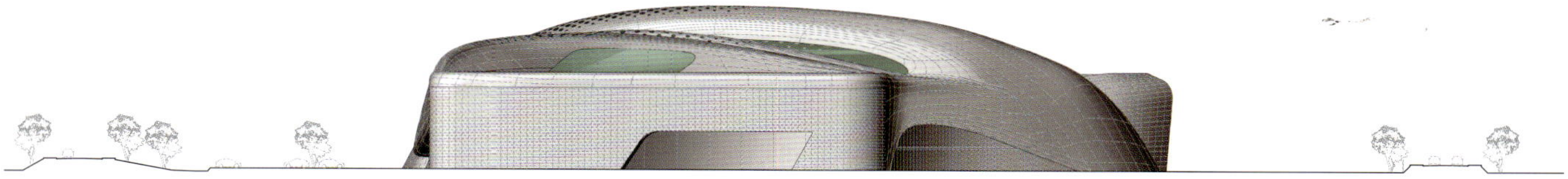

南立图
South elevation

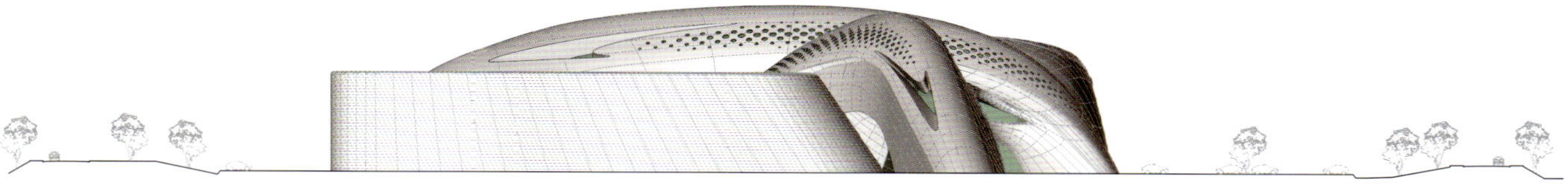

北立图
North elevation

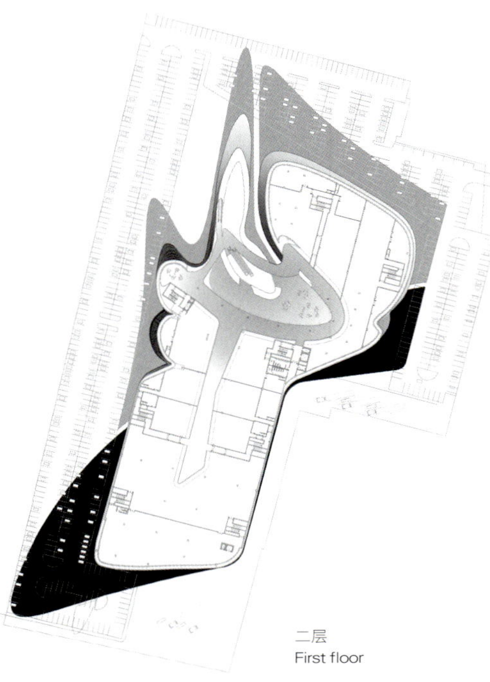

二层
First floor

三层
Second floor

屋顶
Roof plan

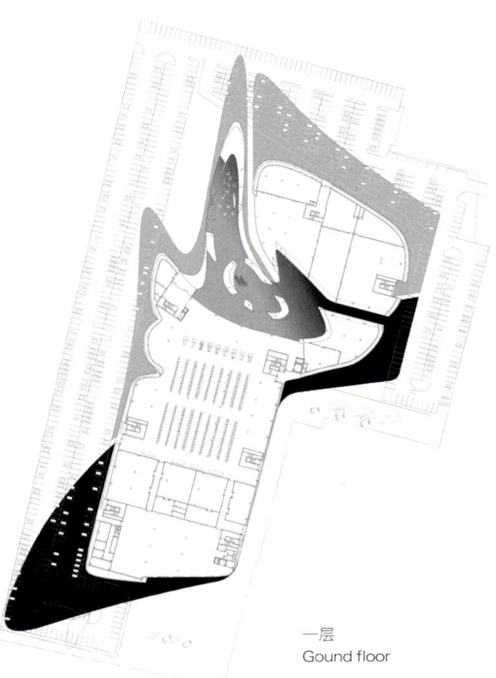

一层
Gound floor

苏黎世机场的弧形建筑
THE CIRCLE AT ZURICH AIRPORT

建筑师：Zaha Hadid

苏黎世国际机场位于瑞士的克洛滕，建筑师参加竞标设计的这座具有休闲和商务功能的弧形建筑是对机场的扩展。建筑位于机场主建筑和一座满是绿树的小山之间的一块弧形场地上。

建筑采用了"内部城市化"的设计，这种设计既促进了内部各个不同功能区域的相互交流，也使各类设施向公众开放。与此同时，通过对所有的功能区域进行竖向分层，使其能够提供合理的服务并保障其自身的顺畅运行。

问题的关键是在这个复杂的建筑体系中，如何清晰地规划通行路线及其与进入建筑内纵横双向交通流线之间的联系。这些通行路线将与贯穿建筑的三个主要上空空间或"峡谷"相关，并与开放式屋顶融为一体。

这些"峡谷"具有三项重要的功能：首先是标明了综合建筑的主要入口，其次是加强了（内部纵横双向）主要交通流线，最后是通过创造高度各异的楼层以提供各种不同的使用功能。

The Circle at Zurich Airport was a design competition for an extension to Zurich International Airport in Kloten, Switzerland. The building occupies a semi-circular site between the airport's main buildings and a wooded green hill.

The building's "interior urbanism" encourages interaction amongst different programmatic modules and sharing of facilities open to the public, whilst maintaining a practical vertical stacking for each module, aimed at rationalizing services and circulation.

Critical importance has been assigned to clarity of access and to the relation of entry to both horizontal and vertical circulation within the complex. These will be related to three major voids or "canyons" cutting through the building section and merging in an open plan top floor.

These "canyons" cover three critically important functions: to indicate major points of entry into the complex, to reinforce the major circulation routes (both horizontal and vertical) and to create a variety of depths of floor plate appropriate to the mix of uses to be accommodated.

ZURICH AIRPORT
ORACLE
LA PRAIRIE

意大利Pregasina的山地自行车酒店
HIDING IN TRIANGLES, PREGASINA, ITALY

建筑师：Dipl. Ing. Philip Modest Schambelan, www.scham.be, Dipl. Ing. Anton Fromm, www.copypasters.com

顾问：Dipl. Ing. Jörg Höfer, TU Dresden

这座新建的山地自行车酒店为运动爱好者提供了一处独一无二的住宿场所，也提供了一幅令人难忘的美景，以及快速到达多种自行车路线的方式。Pregasina坐落在加尔达湖最北端上方500m的一个高原上，一座小山成为观赏阿尔卑斯山全景的非常碍眼的障碍物。为了能欣赏到全景，这座新建酒店就坐落在小镇东南端的陡峭山脊上。酒店给人带来了一次别具魅力的居住体验，游客无不被其新颖的设计、通道系统以及美学思想所折服，对阿尔卑斯地区的极限运动爱好者来说，这里将成为一处迷人的新落脚点。

The new mountain bike hotel offers sports enthusiasts a unique place to stay, an unforgettable view and a fast connection to an impressive variety of routes, trails and single tracks. Pregasina is located on a plateau 500 meters above the northern tip of Lake Garda. A hill is a line-of-sight obstruction to an extensive view of the Alps. To totally exploit the panoramic view the new hotel is situated on a steep mountain ridge on the southeastern edge of the small town. The mountain bike hotel is an attractive residential experience, which captivates guests with its impressive design, access system and aesthetics, and will become the newest magnet for extreme sports enthusiasts in the Alps.

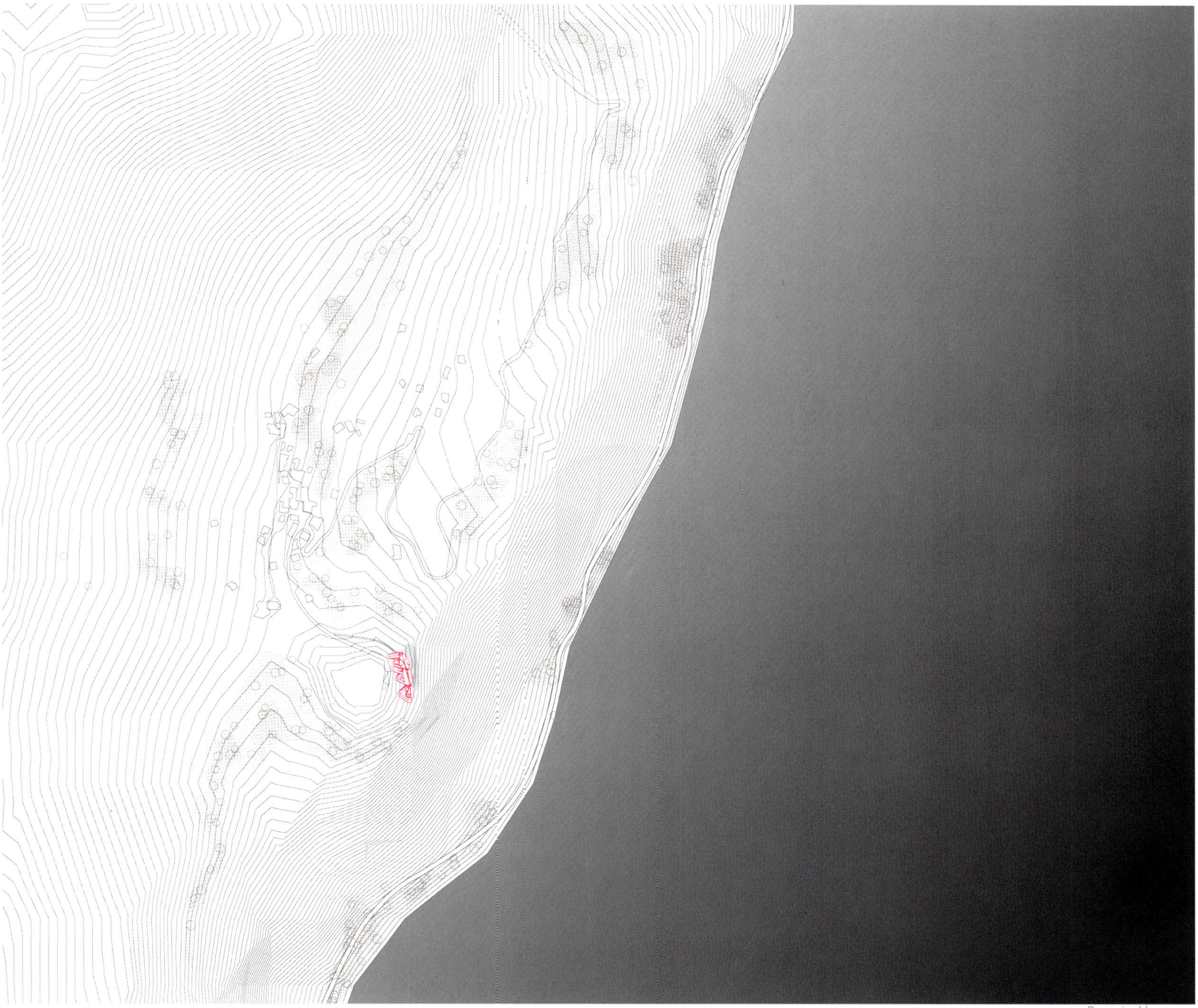

Pregasina村

Village of Pregasina

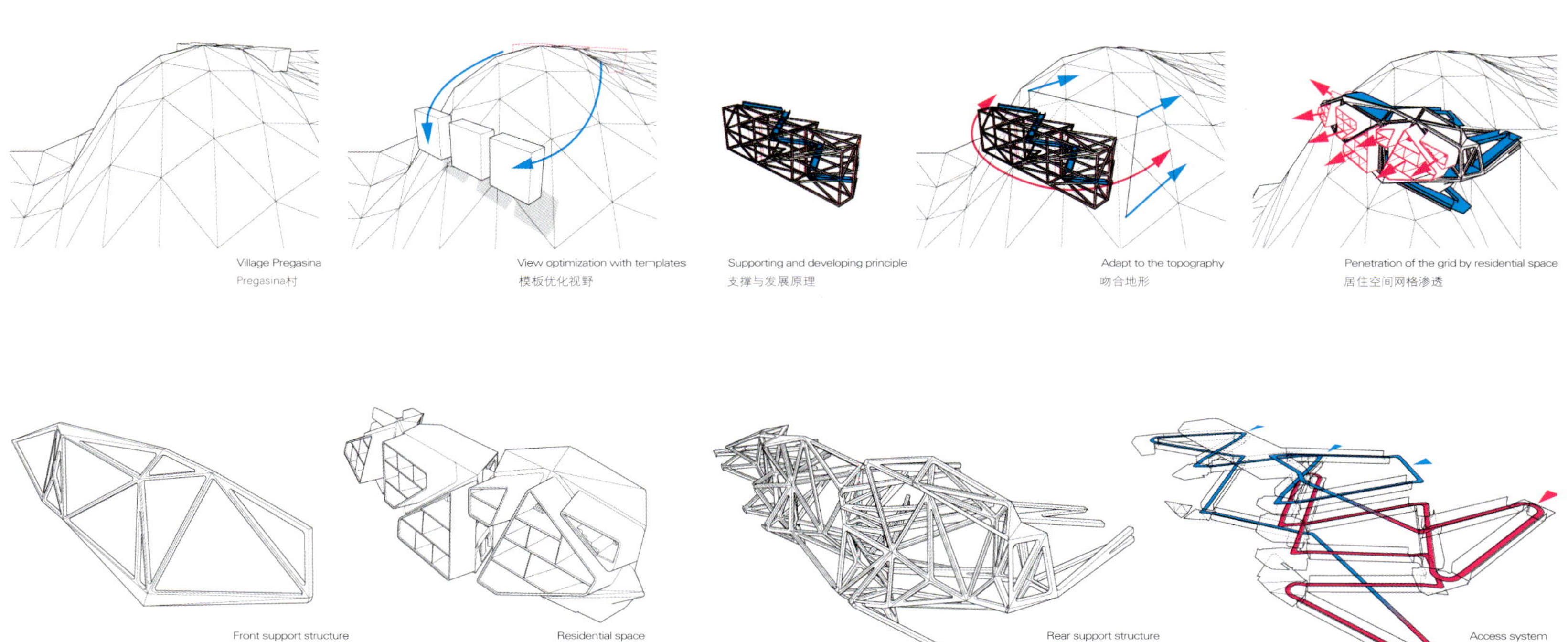

Village Pregasina
Pregasina村

View optimization with templates
模板优化视野

Supporting and developing principle
支撑与发展原理

Adapt to the topography
吻合地形

Penetration of the grid by residential space
居住空间网格渗透

Front support structure
前部支撑结构

Residential space
居住空间

Rear support structure
后部支撑结构

Access system
通道系统

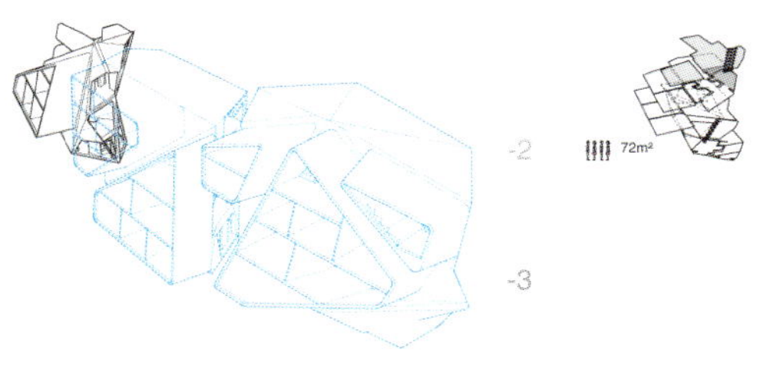

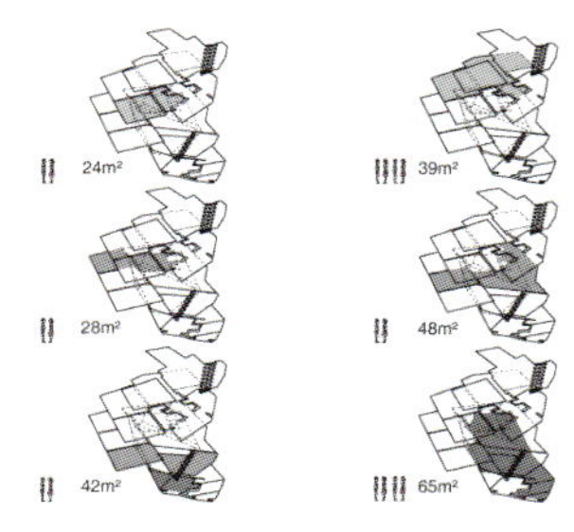

1号楼位置与规模
Location and size of House 1

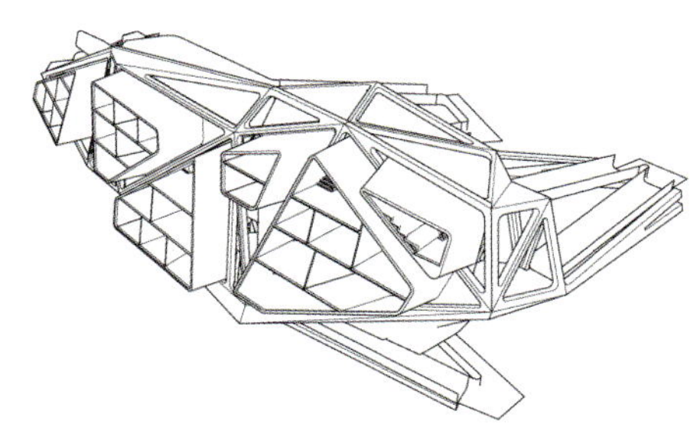

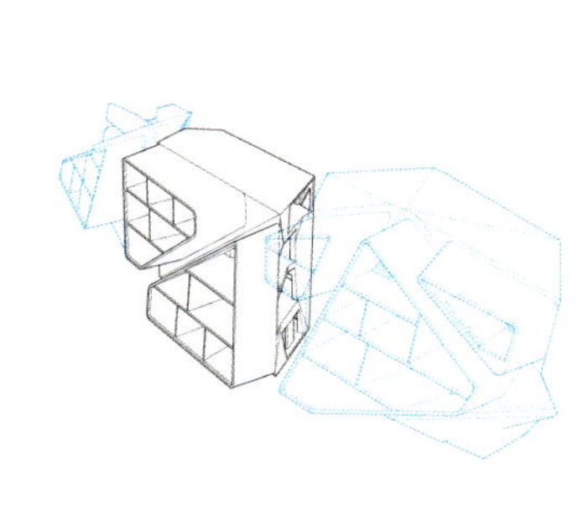

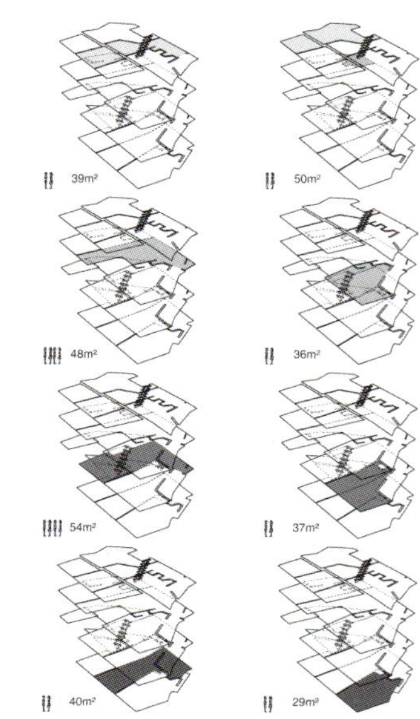

2号楼位置与规模
Location and size of House 2

520 m NN

地形剖面
Terrain section

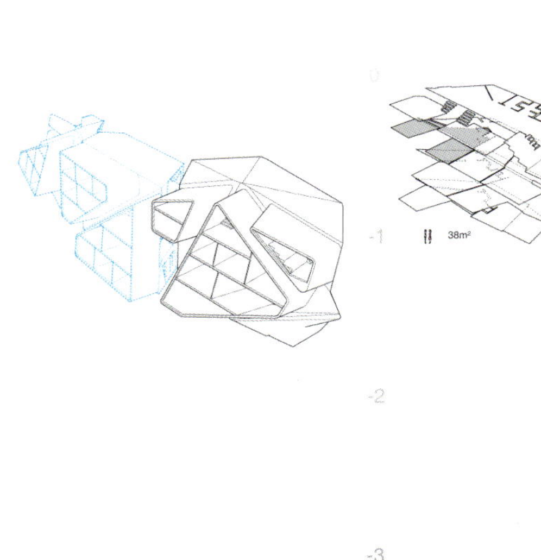

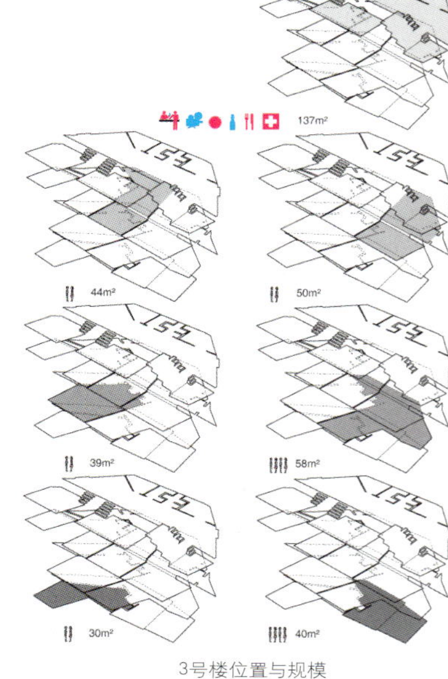

3号楼位置与规模
Location and size of House 3

60m NN

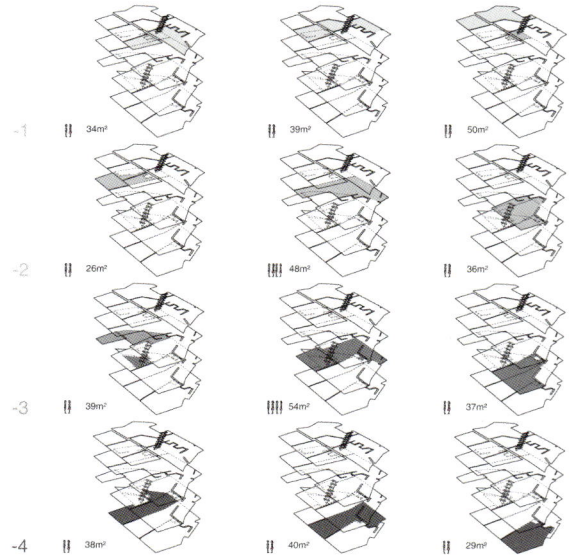

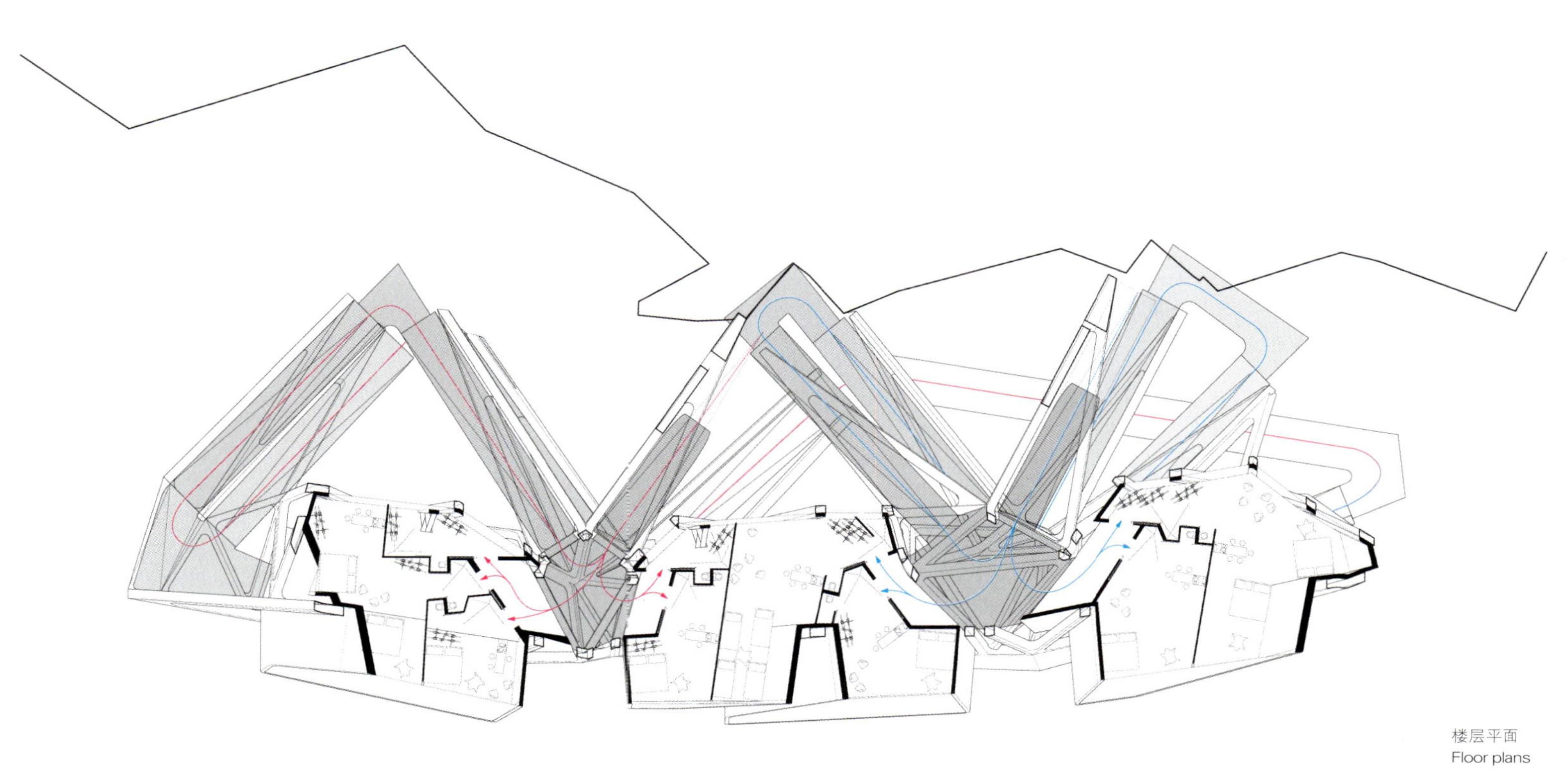

楼层平面
Floor plans

南向视图
View towards south

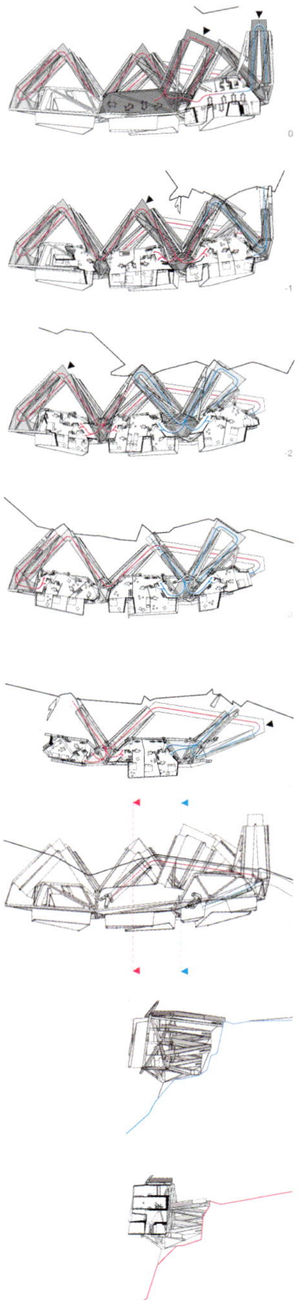

剖面
Sections

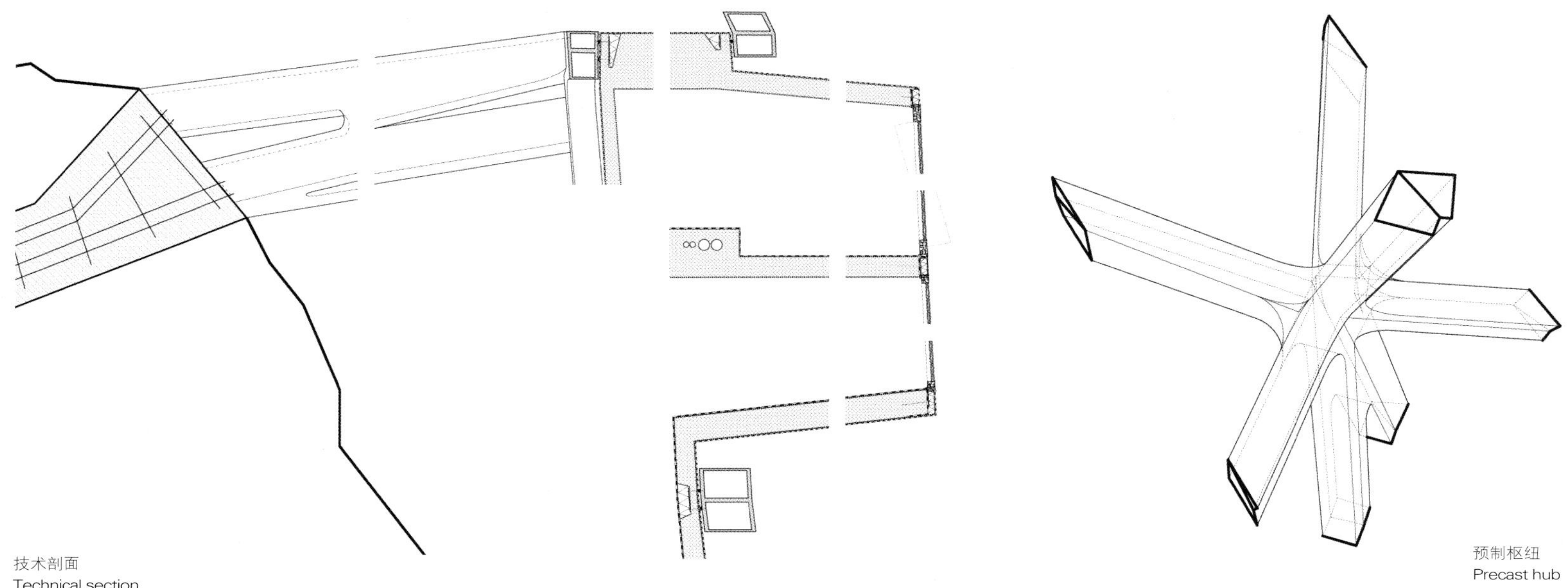

技术剖面
Technical section

预制枢纽
Precast hub

防雨设施细部
Detail of Fall Protection

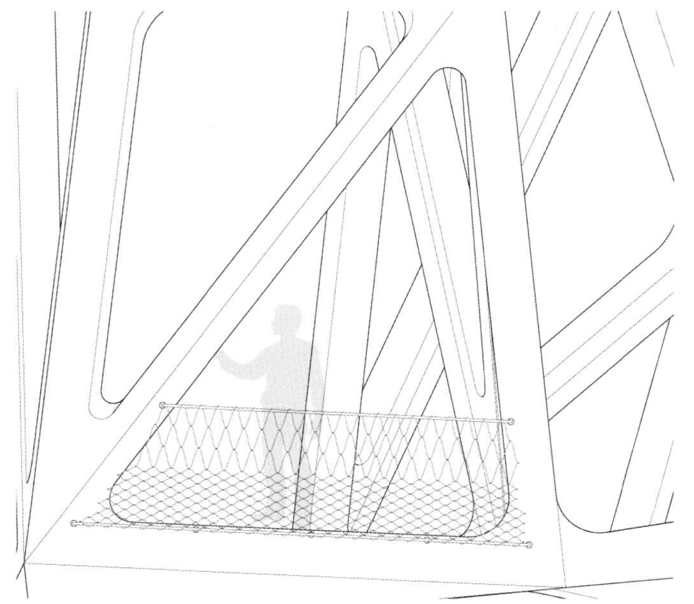

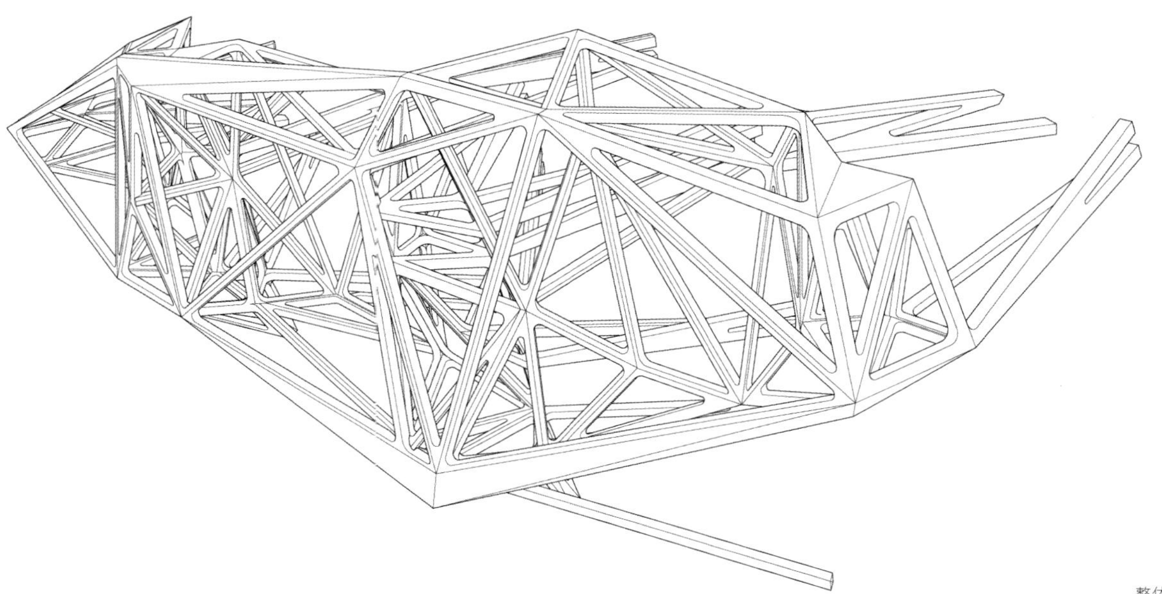

整体三维结构
The whole 3D Structure

卢布尔雅那市政中心
CITY MUNICIPALITY, LJUBLJANA

建筑师：OFIS
项目负责人：Rok Oman, Spela Videcnik
项目团队：Andrej Gregoric, Janez Martincic, Magdalena Lacka, Katja Aljaz
甲方：City Municipality Ljubljana
项目地点：Ljubljana, Slovenia
项目类型：办公及公共项目
地上建筑：
新建筑：42 288m²
原有建筑：16 868m²
地下建筑（新建筑）：20 800m²
投资：67 430€
获奖情况：竞赛三等奖

街道—交通区域组织
Street-traffic areas organisation

街道—交通区域
车库出入口
人行道、自行车道和干涉区域

LEGEND:
street - traffic areas
enter/exit garage
pedestrians, cyclists and intervention zone

项目分布
Program distribution

A 公共部门
B 公共部门
C 公共部门
D 检查
E 部门
F 公共部门
G 市政中心主厅/政府部门
H 卢布尔雅那市政中心

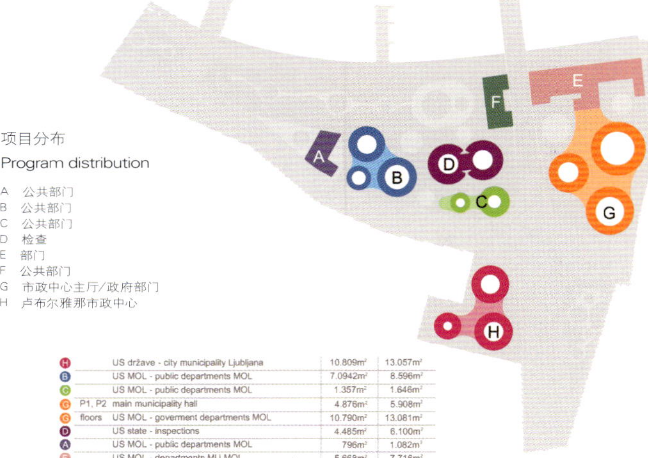

H		US države - city municipality Ljubljana	10.809m²	13.057m²
B		US MOL - public departments MOL	7.0942m²	8.596m²
C		US MOL - public departments MOL	1.357m²	1.646m²
G	P1, P2	main municipality hall	4.876m²	5.908m²
G	floors	US MOL - goverment departments MOL	10.790m²	13.081m²
D		US state - inspections	4.485m²	6.100m²
A		US MOL - public departments MOL	796m²	1.082m²
E		US MOL - departments MU MOL	5.668m²	7.716m²

公共内广场
Public inner squares

新建筑足迹
原有建筑足迹
建筑布局
内庭院
公共与半公共项目大厅
室外布局
广场布局
道路区域
绿色植物与停车位布局

交通流线
Fluidity

新建筑足迹
原有建筑足迹
街道—交通区域
人行道、自行车道和干涉区域
人行流线区域
公交车站
河流港口

总平面图
Site plan

主厅组织示意图

main Hall organisation

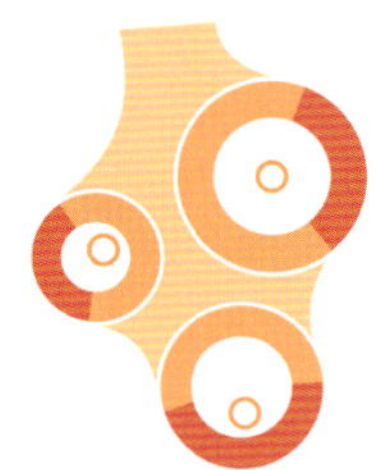

受控区域与公共区域分级
gradation of controlled / public areas

- public program - self service terminal
  公共项目——自助终端
- semi public programs - counters
  半公共项目——柜台服务
- controlled areas - offices
  受控区域——办公室
- reception to controlled offices
  办公室接待处

部门组织示意图
diagrams of department organisation

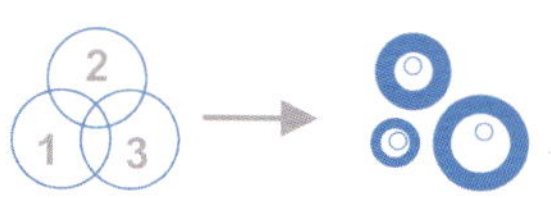

分离的办公室
separate offices

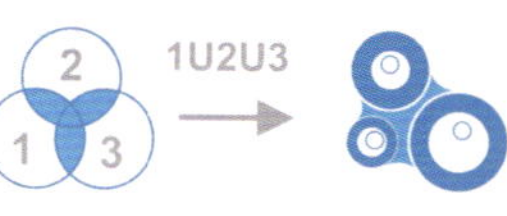

带有共用设备（会议室、厨房、档案室、演示室）的办公室
offices with shared facilities (meeting space, kitchenette, archives, presentation rooms)

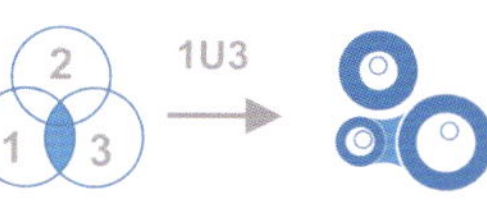

带有共用设备（会议室、厨房、档案室、演示室）的办公室
offices with shared facilities (meeting space, kitchenette, archives, presentation rooms)

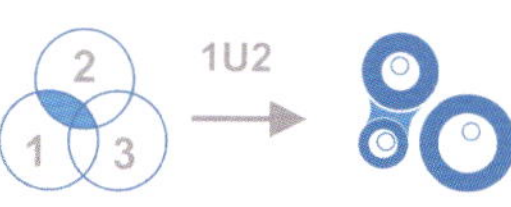

带有共用设备（会议室、厨房、档案室、演示室）的办公室
offices with shared facilities (meeting space, kitchenette, archives, presentation rooms)

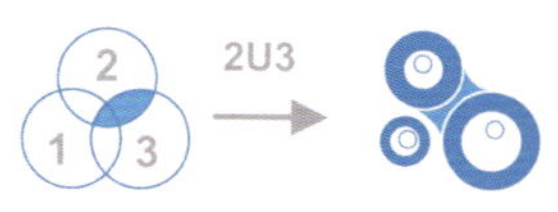

带有共用设备（会议室、厨房、档案室、演示室）的办公室
offices with shared facilities (meeting space, kitchenette, archives, presentation rooms)

本案是新卢布尔雅那市60 000m^{2}的市政中心的竞标项目。设计任务书建议若干个部门定址同一基地，但各自拥有自己的办公楼。市政中心的中央是一个主厅，市民可以在此办理相关事务。

基地毗邻新卢布尔雅那市中心，位于河边，在此区域已有部分受保护的建筑。

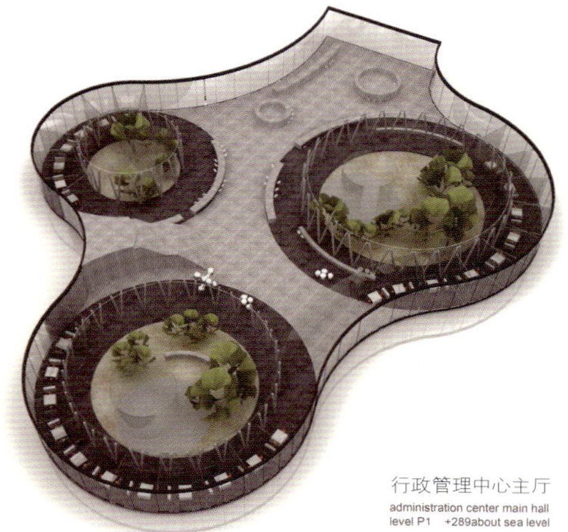

行政管理中心主厅
administration center main hall
level P1 +289about sea level

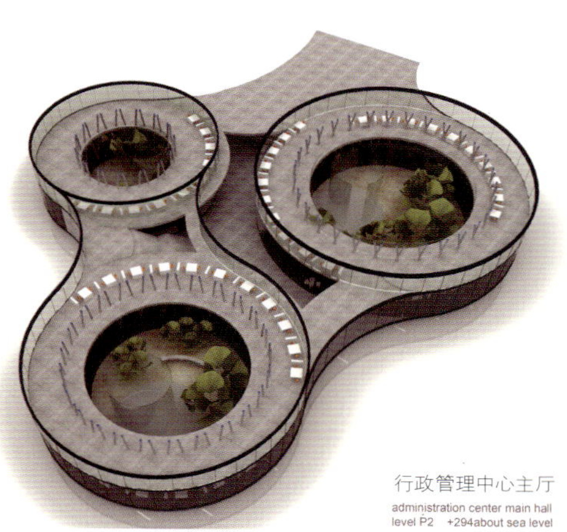

行政管理中心主厅
administration center main hall
level P2 +294about sea level

公共区域到受控区域的分级
gradation from the public access to controlled

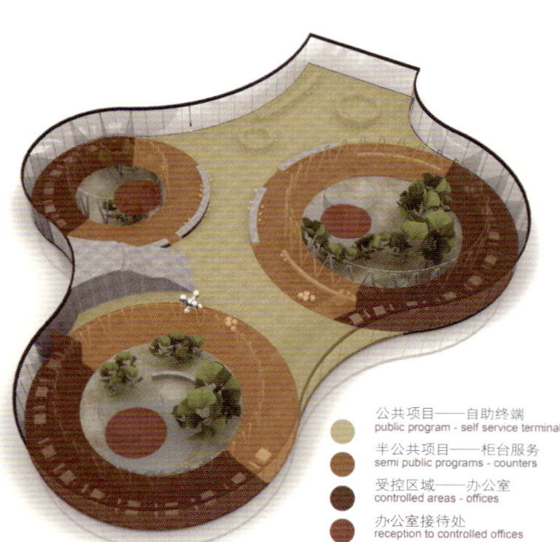

- 公共项目——自助终端
  public program - self service terminal
- 半公共项目——柜台服务
  semi public programs - counters
- 受控区域——办公室
  controlled areas - offices
- 办公室接待处
  reception to controlled offices

The competition was for new Ljubljana administration centre of 60 000m^2. The brief proposed several departments to move to the same site but occupy different buildings. The heart of the centre is the main hall where citizens could also arrange all the documents.

The site is just on the edge of the Ljubljana city centre, by the river and already occupied by some existing protected buildings.

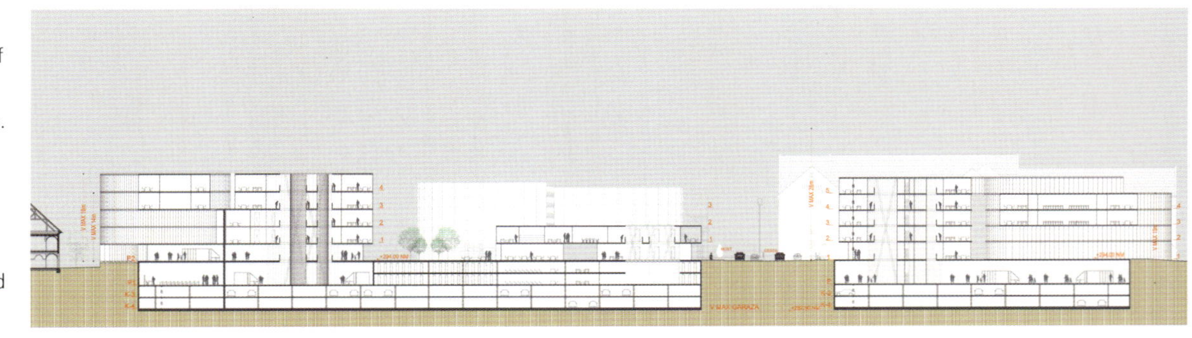

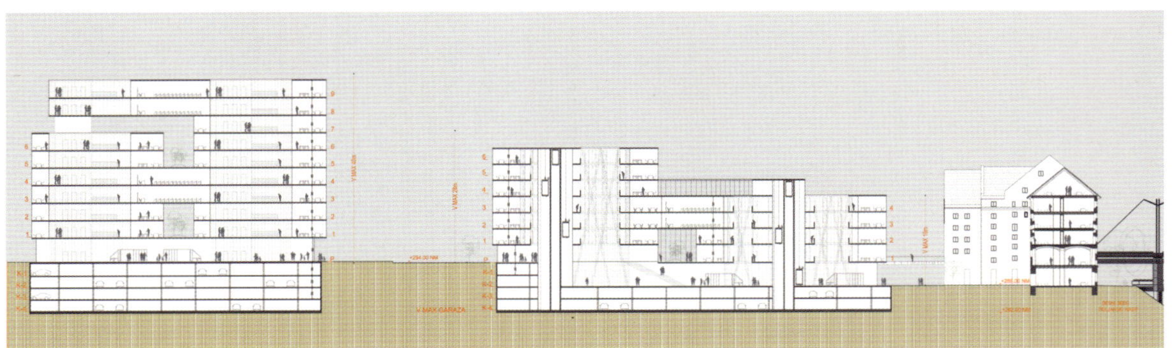

剖面图
Sections

C1
B1

A1
B1

A1

A1

中国北京国际投资广场
INTERNATIONAL INVESTMENT SQUARE, BEIJING, CHINA

建筑师：UNStudio

Ben van Berkel, Caroline Bos, Astrid Piber with Hannes Pfau, Luis Etchegorry, Hans Peter Nuenning and Kristina Madsen, Malaica Cimenti, Junjie Yan, Shi Yang, Veronica Baraldi, Nannang Santoso, Albert Gnodde, Ramon van der Heijden, Jeorg Lonkwitz

结构工程顾问：ARUP Shanghai, China; Arup International Consultants (Shanghai) Co., Ltd.

成本评估：WT Partner, Beijing China

甲方：Beijing Guorui Real Estate Development Co., Ltd.

项目地点：Chaoyang Road, Chaoyang District, Beijing

建筑面积：420 000 m^2

建筑体量：140000 m^3

基地面积：118 975 m^2

用途：综合用途Soho (170 000 m^2)、办公 (170 000 m^2)、商业酒店 (30 000 m^2)、零售 (20 000 m^2)、公寓 (30 000 m^2)、音乐厅 (5000 m^2)

国际投资广场是北京的新兴区域，孕育着现代都市发展的蓝图，提供给人们集商业与居住为一体高效持久的新天地。新区域地处五环六环之间，位于经济技术开发区的核心地带。

The International Investment Square is a new area in Beijing, envisioned as a modern urban development providing an efficient and sustainable location for combining business and living. The new neighbourhood is situated between the Fifth and Sixth Ring of Beijing, in the core of the Economic Technological Development Area.

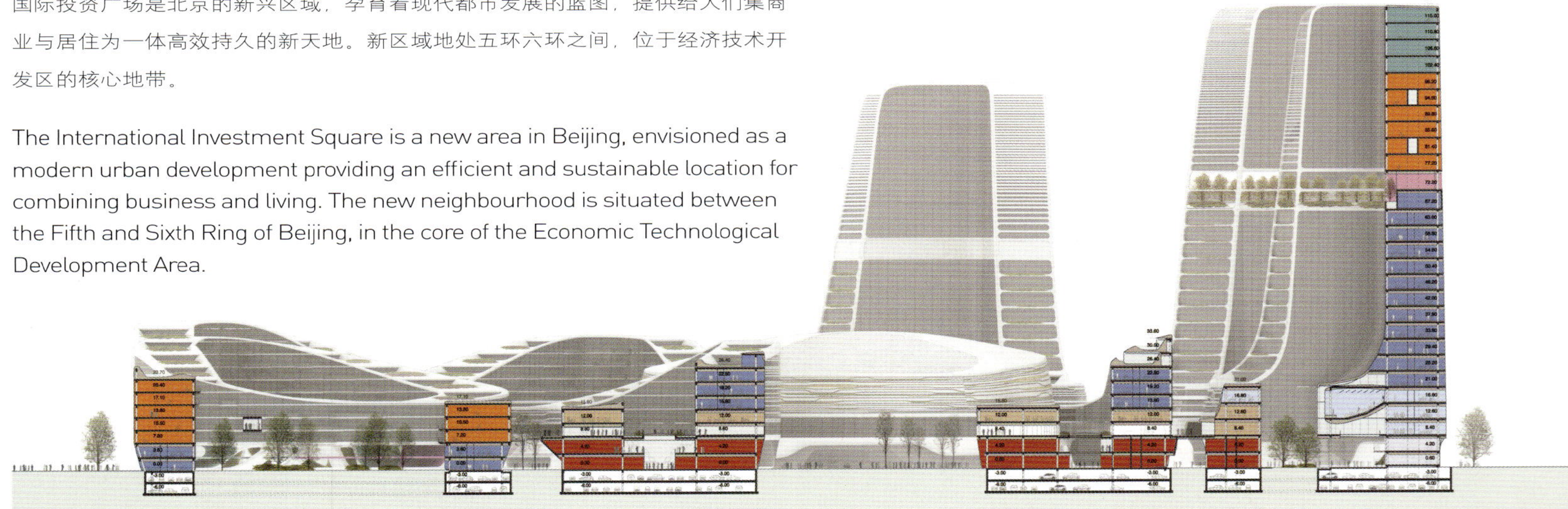

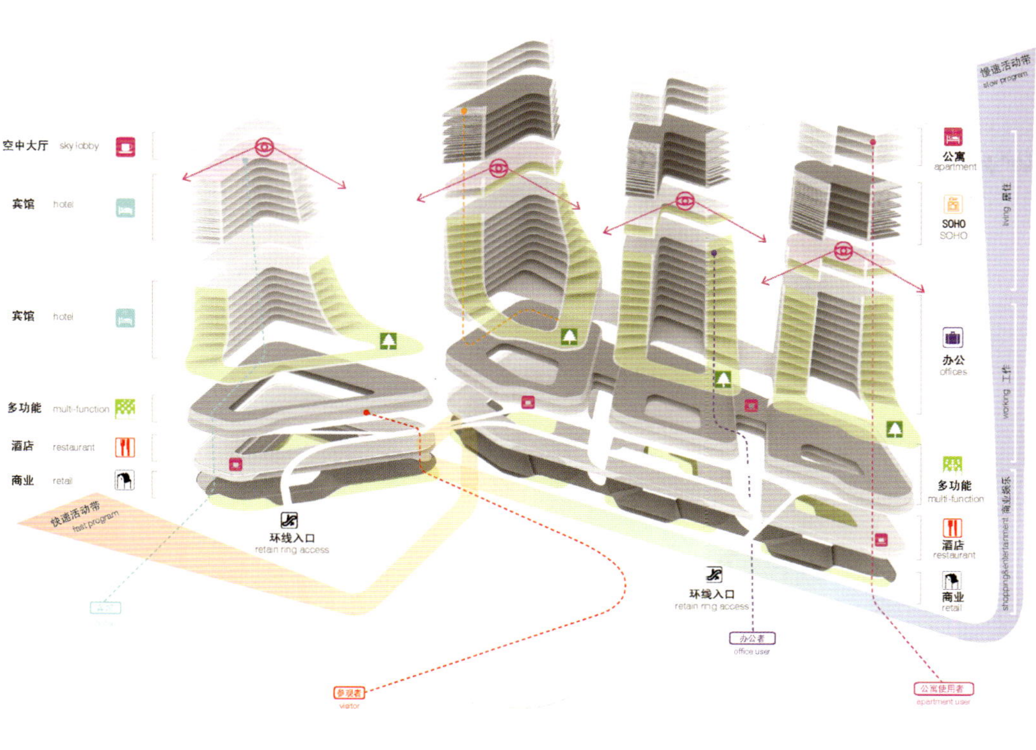

分形 层次一 Subdivision Level 01

分形 层次二 Subdivision Level 02

分形 层次三 Subdivision Level 03

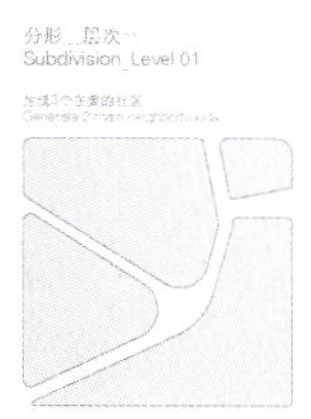

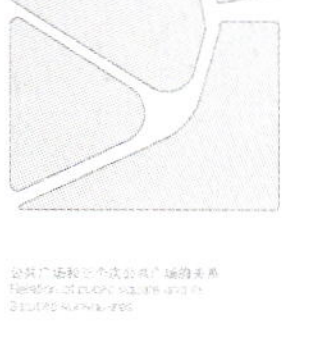

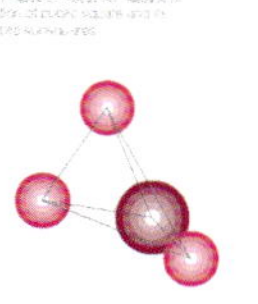

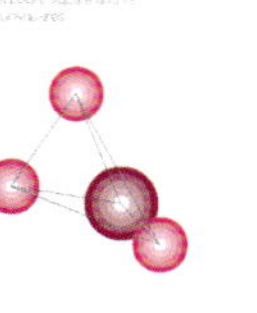

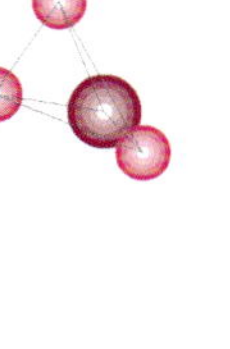

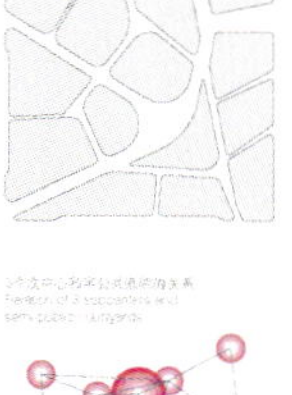

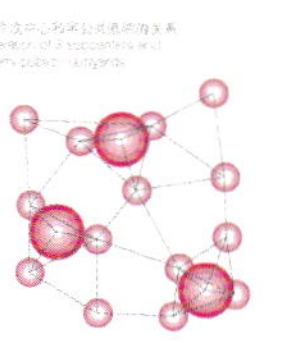

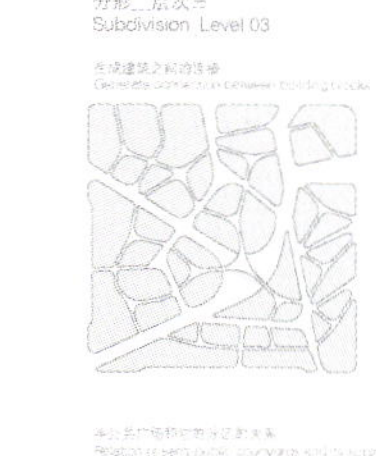

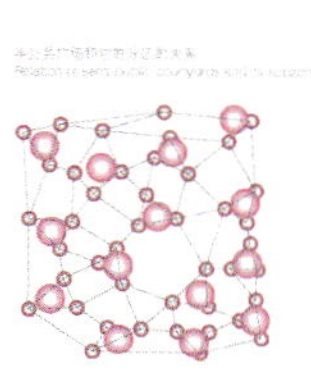

全部连接与关系的网络 Network of all Relations and Connections

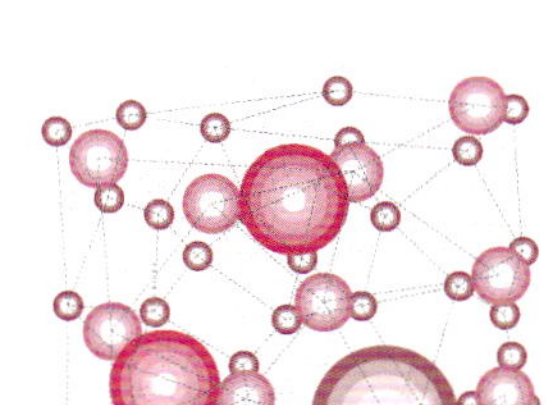

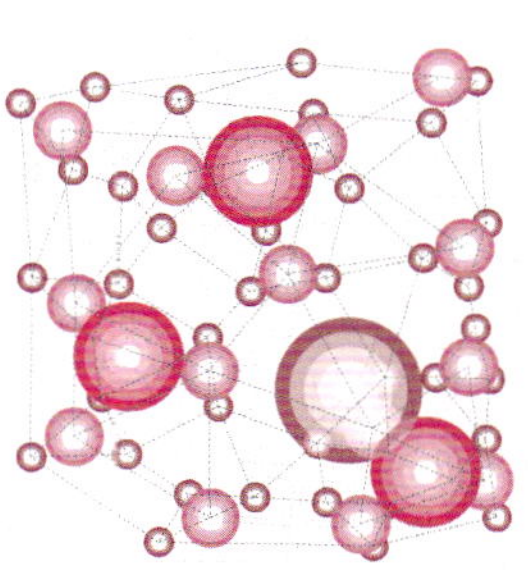

动态网络 Network of movement

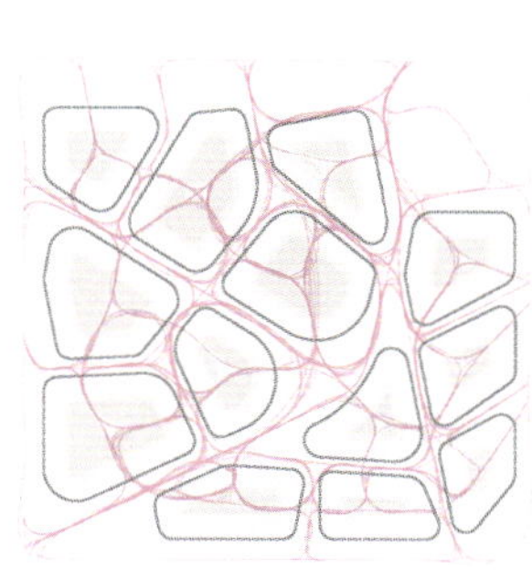

建筑形体的区别 Differentiation of Building blocks

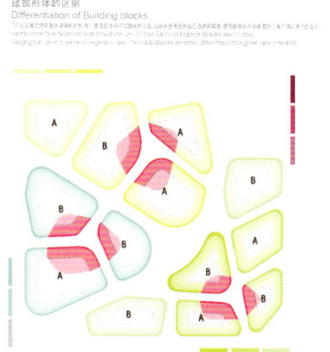

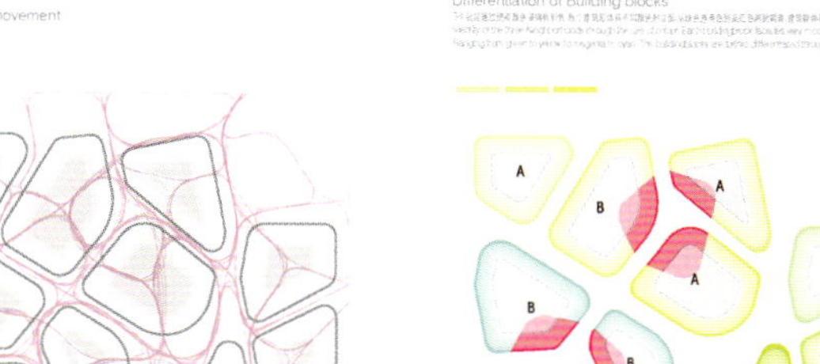

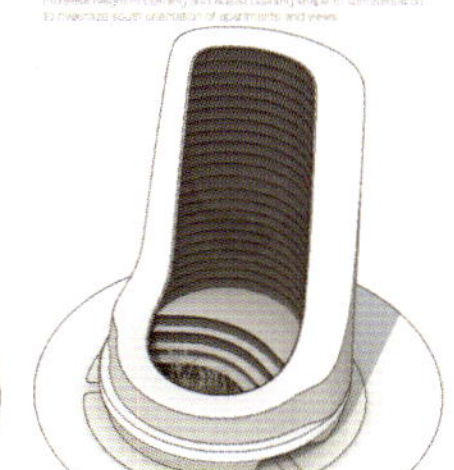

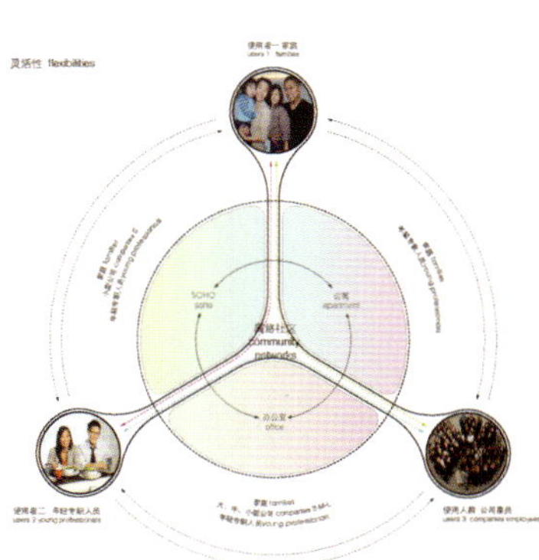

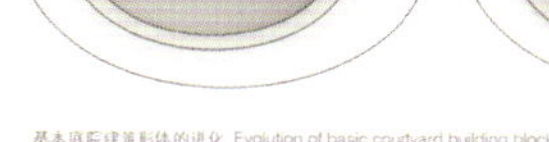

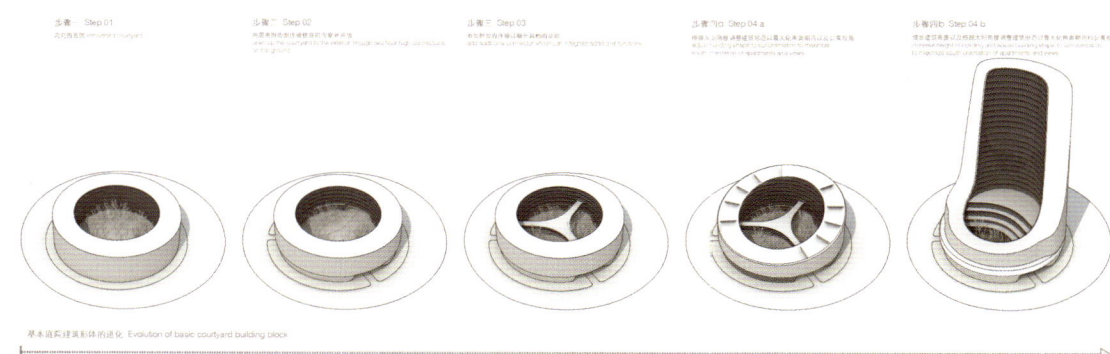

基本庭院建筑形体的进化 Evolution of basic courtyard building block

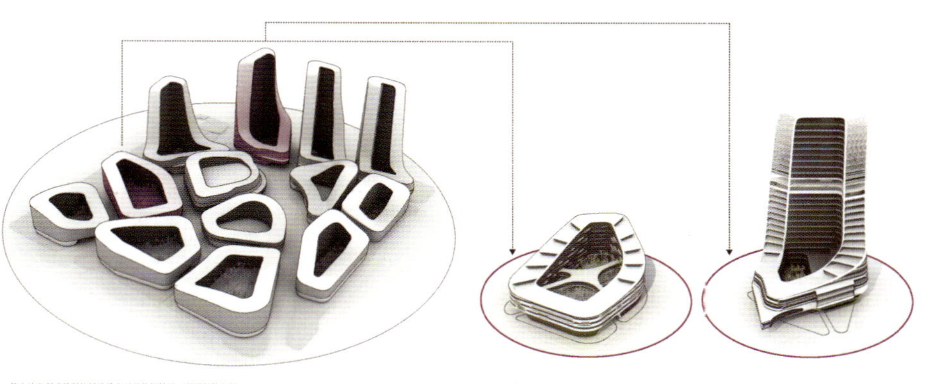

基本的环形建筑形体被填充以及根据社区分形逻辑的变形
Basic ring building block populated onto site and deformed according to neighborhood subdivision logic

低层庭院建筑形体 Lowrise courtyard building block

高层建筑形体 Highrise building block

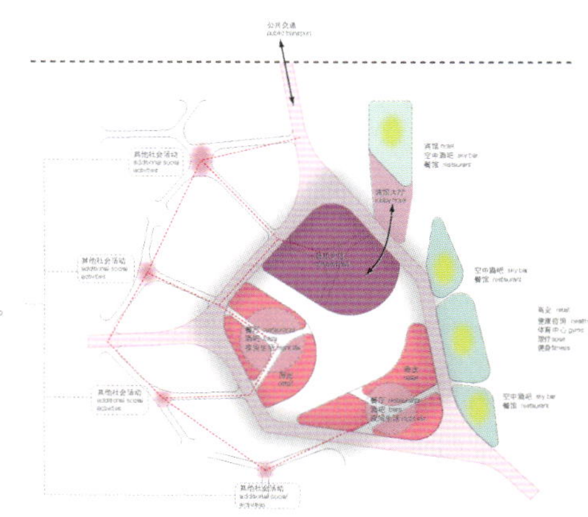

赫尔辛基的水晶地标建筑
A CRYSTALLINE LANDMARK FOR HELSINGBORG

甲方：Midroc Property Development（瑞典）
建筑师：schmidt hammmer lassen architects
项目合作人：Kim Holst Jensen, schmidt hammer lassen architects
项目建筑师：Kristian Lars Ahlmark, schmidt hammer lassen architects
承包商：Midroc Property Development
景观建筑师：schmidt hammer lassen architects in collaboration with Masu Planning Landscape（丹麦）
执行顾问：Sweco（瑞典）

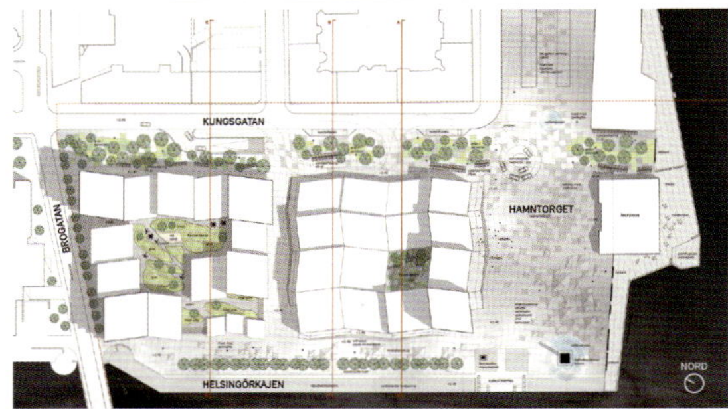

建筑包括位于赫尔辛堡最具魅力的区域——市中心的前Ångfärjan渡轮码头，包括16 900m^{2}的会议和酒店设施、17 100m^{2}的住宅。新的设施将在贯穿城市滨水区散步道的当前发展中起到重要的作用。

建筑以将网格变形为水晶的表达方式为特点，因此得到“盐晶体”的昵称。建筑东南端12层的酒店体量将成为城市的新地标。

会议中心从另一端的3层开始，逐渐变为12层的酒店。酒店包含230间客房。会议和酒店中心沿着海边散步道设置，以此来与公寓大楼相呼应。公寓大楼与会议中心之间由一条小的人形步道分隔开来。公寓大楼也有同样的“盐晶体”网格，于此同时立面交错的长方形图案加强了其结构开放、轻盈的感觉。

The project consists of a 16,900 m^2 congress and hotel facility and 17,100 m^2 housing on the most attractive area of Helsingborg – the former central ferry dock Ångfärjan in the city centre. The new facilities will play an important part in the ongoing development of the promenade running the length of the city waterfront.

The building is characterised by a deformation of the grid into a crystalline expression that has coined the nick name "The Salt Crystals". The twelve-storey hotel volume in the south east corner will become the new landmark of the city.

The congress centre starts in three stories at the opposite end and grows gradually to become the twelve-storey hotel. The hotel has 230 rooms. The Congress and Hotel Centre run along the sea promenade to meet the apartment blocks that are separated from the congress centre by a small pedestrian street. The apartments have the same "salt crystal" grid, while the facades have a shifting rectangular pattern to reinforce their open and light structure.

绿色综合建筑
GREENSIDE OUT

建筑师：JA Joubert Architecture
总建筑师：Marc Joubert
项目团队：Jeroen de Loor, Marian Dusinsky, Alessandro Guida, Kim Byungchan
甲方：Eurocol
项目地点：Tirana, Albania
面积：9600 m² + 3000 m²停车场
用途：多功能建筑
获奖情况：邀请赛二等奖

本案项目空间可灵活分配，主要分为一层的商业空间以及高层的公寓空间。有两层为办公空间，可随意缩小与扩大。办公空间通过建筑内的两个中央广场进入，也可使用楼梯或独立的电梯直接从停车场进入。建筑围绕着两个广场而设，为公寓提供了室外空间，为办公空间提供了入口和服务区域，为商业中心提供了室外空间。这两个广场可根据天气情况关闭玻璃屋顶。

The program can be distributed flexibly, with the main division between commercial spaces on the ground floor and the cores reserved for the apartments on the higher levels. Two floors are reserved for office space, these could easily be reduced or enlarged. The office spaces are accessed through the two central plazas in the building, allowing for direct access with stairs as well as a separate elevator core running straight from the parking garage. The buildings are focused around these two openings, providing outdoor spaces for the apartments, entrance and service areas for the offices and outdoor spaces for the commercial centre. These plazas can be closed off with glass separations depending on weather conditions.

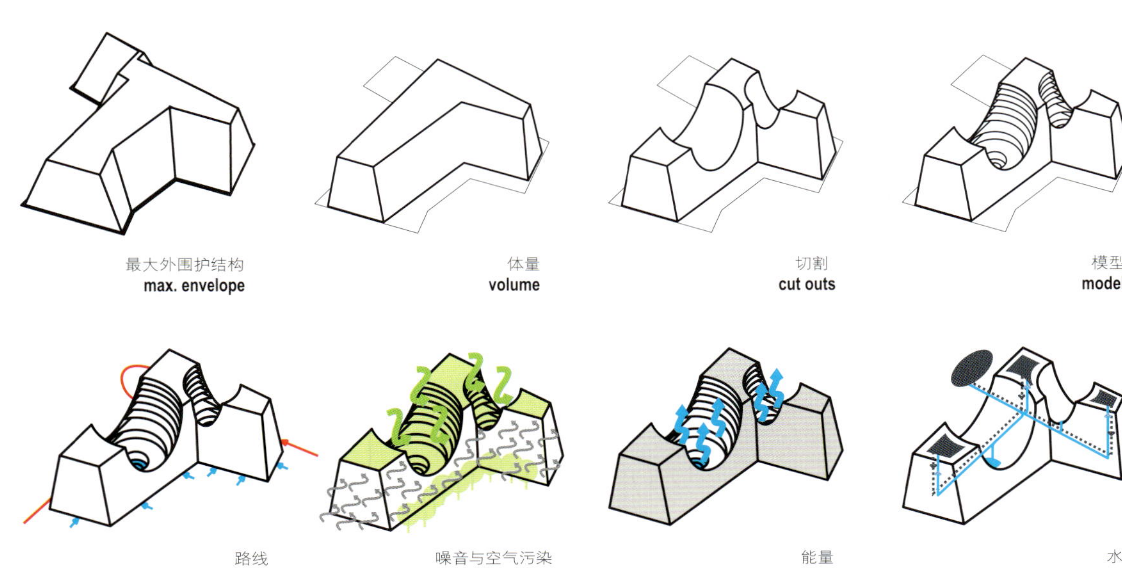

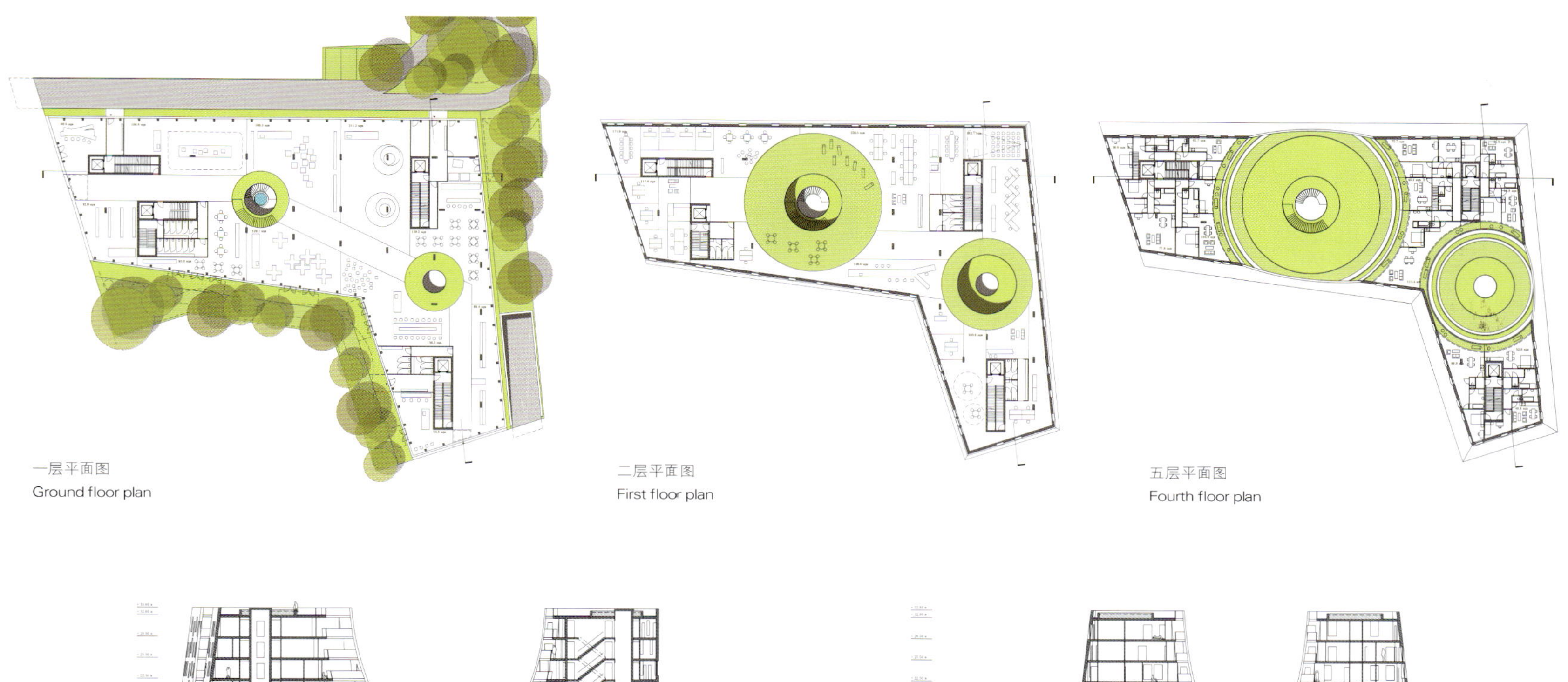

一层平面图
Ground floor plan

二层平面图
First floor plan

五层平面图
Fourth floor plan

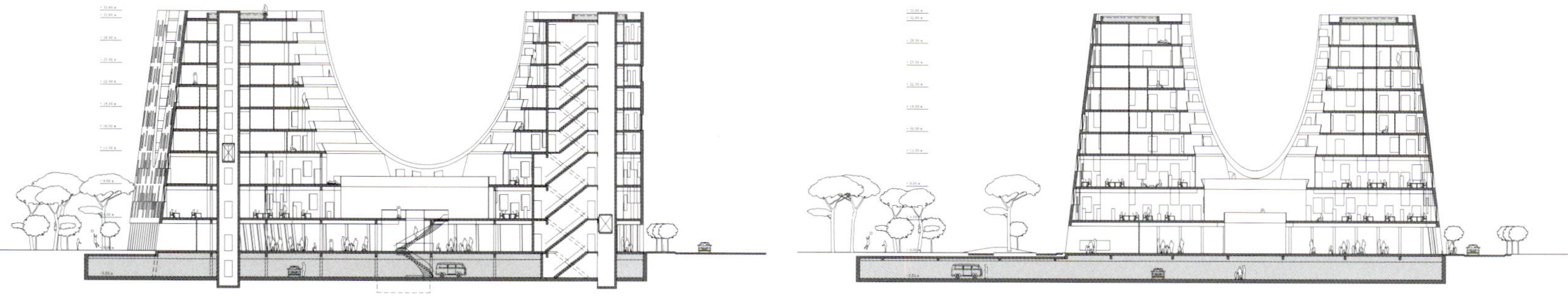

剖面图
Sections

里瓦酒店
RIVA HOTEL

建筑师：Foster+Partners

项目团队：Norman Foster, Grant Brooker, Hugh Stewart, Christopher Hammerschmidt, Diogo Bleck, Tim Kemp, Li-Jun Lin, Julian Sattler, Yusuke Tsutsui

甲方：Riva Bowl Limited

结构工程师：Adams Kara Taylor

机械工程师：Adams Kara Taylor

景观设计：Whitelaw & Turkington

里瓦酒店作为该地区唯一一家五星级酒店，将为包括当地社会团体及商务人士在内的人群提供会议设施，其设施比伦敦任意一家酒店都全面，也为使用希思罗机场的旅客提供服务。这家酒店的特点是拥有独特的分层玻璃外壳设计，这个设计使日光完全涌入到公共空间，非常有助于节能。

The only five-star hotel in the area, it will offer a range of services, including the most extensive conference facilities of any London hotel, to serve the local community and businesses, as well as passengers using Heathrow. The building is characterised by a distinctive layered glass shell, which floods the public spaces with daylight and contributes to a highly efficient energy strategy.

代夫特市政厅与火车站
DELFT CITY HALL AND TRAIN STATION

建筑师：Mecanoo architecten, Delft, The Netherlands
结构工程师：ABT bv, Delft
机械工程师：Deerns Raadgevende Ingenieurs B.V., Rijswijk
成本顾问：Basalt bouwadvies bv, Nieuwegein
风险分析顾问：Aboma Keboma, Ede
立面系统顾问：Permasteelisa Central Europe, Heerlen
甲方：Ontwikkelingsbedrijf Spoorzone, Prorail, Municipality of Delft
用途：30 000 m^2 市政办公空间及公共大厅，火车站，火车站大厅内有商店、售票处、餐厅，地下站台有4500 m^2

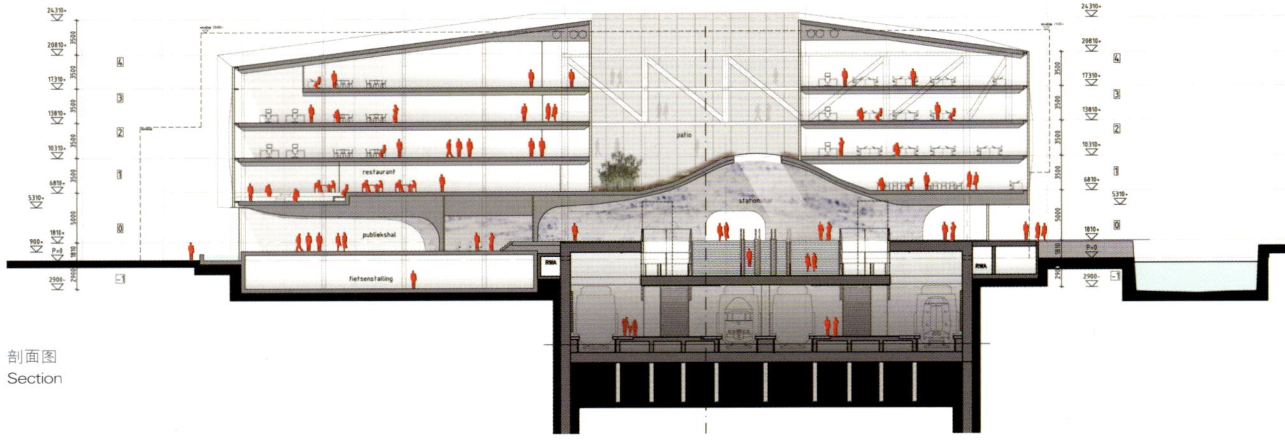

剖面图
Section

市政厅的玻璃表皮反射着荷兰的天空，也使建筑变得透明。立面上的斜纹图案创造了钻石一样的外观。玻璃体量上的切口形成小巷的图案，设计灵感来自老代夫特错综复杂的街道格局。建筑的水平分割、透明玻璃的基座和架高的市政厅办公区清晰地将公共与私密区分隔开。二层上设有多功能空间、餐厅、会议室以及市长与市议员使用的房间。连接市政厅和车站大厅的是一个拱形的天花板，使用了代夫特蓝色陶瓷，二者通过一面玻璃墙和两个体量巧妙地分隔开，这两个体量容纳了公共柜台、办公室、会客室、通往上层办公室的电梯与楼梯、商业设施以及卫生间。

The glass skin of the city hall reflects the Dutch skies of Vermeer and also makes the building transparent. A pattern of diagonals within the facade creates a diamond like appearance. Incisions in the glass volume form a pattern of alleyways inspired by the intricate structure of streets in Old Delft. The horizontal division of the building - with a plinth of clear glass and raised office landscape of the municipal offices - provides a clear distinction between public and private. On the first floor lay multifunctional spaces, a restaurant, conference rooms and rooms for the Mayor and the Alderman. A vaulted ceiling, featuring scenes in Delft Blue ceramics, connects the city hall and the station hall. The station hall and city hall are subtly separated by a glass wall and two volumes with public counters, front office, interview rooms, stairways and elevators to the upper offices, commercial facilities and sanitary facilities.

纯粹绿色办公建筑 "JUST GREEN"

建筑师：NRJA
项目地点：Matrozu str, Riga, Latvia
项目团队：Uldis Luksevics, Linda Leitane
基地面积：4170 m^2
总建筑面积：6250 m^2
效果图制作：NRJA

既然不能与城市网格相连，那么就让这座建筑成为使用自然能源的建筑。这座位于里加的办公建筑被设计成绿色建筑，具有多种可替代能源，对使用者来说，无论在建筑上还是在环境上都属于友好型的。

建筑的主要区域围绕在中庭周围，中庭有一个螺旋形坡道，因此能保证建筑内的垂直交通。作为中心区域的中庭也强调了“绿色”的能源流动。

电力由安装在屋顶的竖轴风力发电机和光电板系统提供。雨水收集在屋顶上，用于植物灌溉。建筑通过地下水集热器、开放式集热水库和集热器来收集热量。建筑正立面可以进行自然通风。

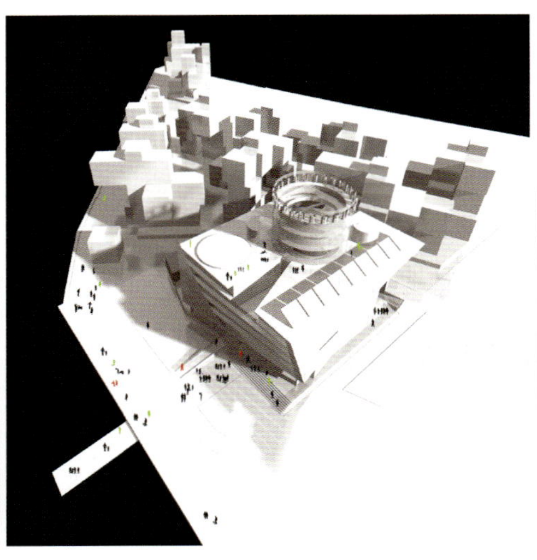

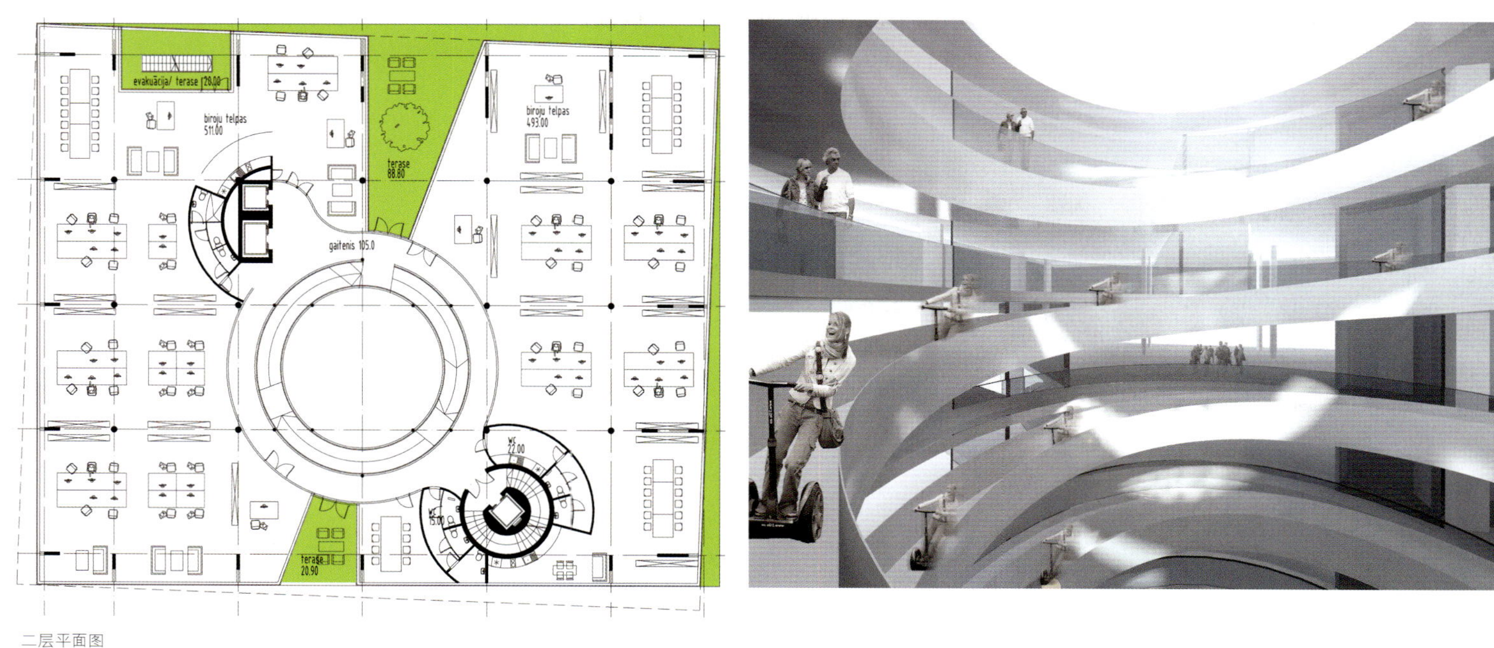

二层平面图
First floor plan

When connection to the city network almost impossible, it is time to switch to the natural energy sources. This office building in Riga, is designed to be green - it is full set of alternative energy sources and architecturally and environmentally friendly to its users.
The main areas are located around an atrium surrounded by a spiral ramp; therefore it is easy to ensure the vertical movement through building. This atrium - central area also emphasizes "green" energy movement.
The electricity is provided by the roof mounted vertical axis wind generators and photovoltaic panel system. Rainwater is collected on the roof and used for plant irrigation. The building is using groundwater heat collector, open water heat reservoirs and collector. The front side of the building provides natural ventilation.

1 光电系统
2 普通电力转换与电池系统
3 建筑走廊类型的正立面
4 新风进入
5 地下水位
6 水供应系统，与承重结构桩建在一起
7 为了引入日光和合理的交通流线，中庭内设置螺旋楼梯
8 竖轴风力发电机
9 植物能减少中庭内的二氧化碳，增加氧气
10 机械送风装置在热交换器内加热进入的新风
11 机械送风装置在中庭上方收集空气并将其输送给热交换器，在冬季为新风预热
12 雨水收集装置
13 为中庭供暖的热空气
14 露台
15 照射到中庭的日光
16 地下停车场自然通风
17 恢复装置

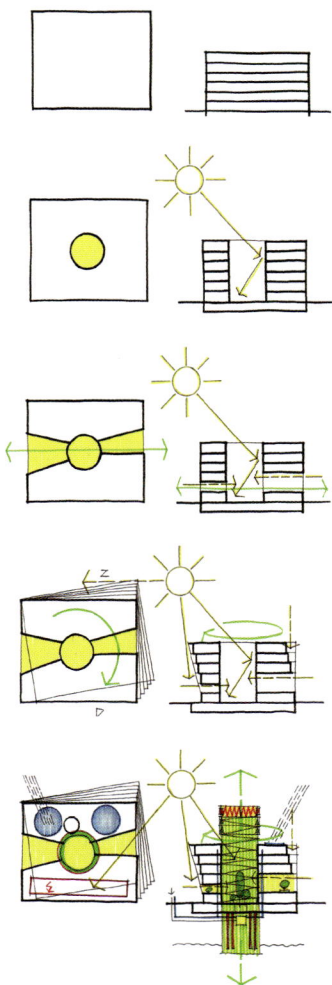

光照、通风示意图
Sunlight & ventilation diagram

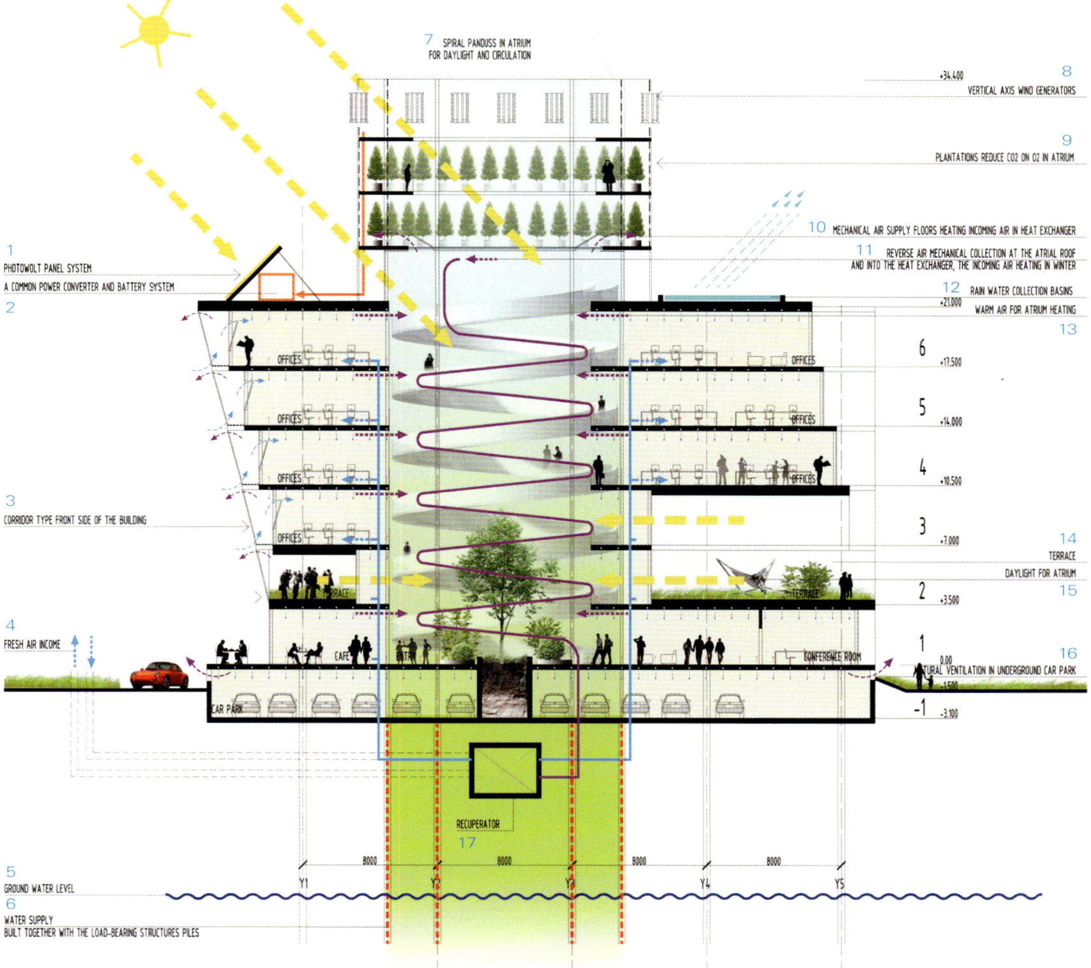

剖面图
Section

意大利米兰Porta Volta城市改造项目
PORTA VOLTA FONDAZIONE FELTRINELLI IN MILAN, ITALY

业主：Feltrinelli, Milan, Italy
合伙人：Jacques Herzog, Pierre de Meuron, Stefan Marbach
参与项目建筑师：Andreas Fries (Associate), Mateo Mori Meana
项目团队：Liliana Amorim Rocha, Nils Büchel, Claudius Frühauf, Monica Leung, Adriana Müller, Carlos Viladoms
合作建筑师：SD Partners, Milan, Italy
结构：Zaring, Milan, Italy
机电工程：Polistudio, Riccione, Italy

作为Porta Volta地区城市改造的一部分，Fondazione Giangiacomo Feltrinelli打算将其所在地搬到米兰的北部中心，费尔特里内利集团把这座建筑当成了可以举办多种活动的理想环境。Porta Volta地区的总体计划由费尔特里内利集团的两座新办公楼和一片丰富的绿地组成，在对周边地区产生积极影响的同时，也具有重要的战略意义。Fondazione的一层容纳了主入口、自助餐厅和书店，二层有双层高的多功能空间，三层是办公区域。Fondazione顶部的阅览室为研究人员和有兴趣的公众提供了去安全可靠的地下档案库中研究历史收藏文献的机会。

As part of the redefinition of the area Porta Volta, Fondazione Giangiacomo Feltrinelli intends to relocate its seat to the northern centre of Milan, which the Feltrinelli Group considers as an ideal environment for the foundation's multiple activities. The overall master plan for Porta Volta, consisting of the Fondazione, two new office buildings, and a generous green area, holds an important strategic potential for creating a positive impact on the surrounding area. The ground floor of the Fondazione accommodates the main entrance, cafeteria and book store, followed by the double height multi-functional space on the first floor, and an office area on the second floor. The reading room on top of the Fondazione offers researchers and interested public the opportunity to study documents from the historical collection stored in the secure underground archive.

美国华盛顿特区的捷克大使馆
EMBASSY OF THE CZECH REPUBLIC IN WASHINGTON DC, USA

建筑师：Marek Chalupa, Stepan Chalupa, Tomas Havlicek, Michal Rosicky, Tomas Horalik, Jakub Chuchlik / Chalupa architekti
参与合作建筑师：Adam Gebrian / AG-ENT, Zdenek Sendler / Atelier zahradni
a krajinarske architektury, Petr Babak / Laborator, Jiri Cajthaml / PBA, Ivan Nemec / Nemec Polak spol. s r.o., Vladimir Jenik / Architektonicke modely a design
效果图：Vit Musil + Radim Petruska / miss3
业主：捷克共和国外交部

大使馆的设计来自于其独特的自然环境，建筑不是主角，而是环境的陪衬。建筑把基地分为三个部分。首先进入一条圆形车道，其简朴而高雅的特征更加衬托出磨砂玻璃正立面如窗帘一般的外观。其次，建筑划分出私人花园，以此联系公寓、办公室和会议室。最后，还有一个具有代表性的园林空间，构成了项目概念的核心。花园很大，与使馆的主要休息室有机地连接在一起，它位于新建筑和原有大使官邸之间，加强两者的联系。设计应该试图提供我们喜欢的建筑语言，也就是我们希望看到的世界：开放、自信、友善、乐于助人、尊重，注意爱护自然环境，牢牢扎根于深厚的文化传统，并尊重民主原则。

The design of the Embassy of the Czech Republic pays tribute to its unique natural setting, and the building itself is only an adjunct, not the main actor. Its form divides the site into three separate parts. Firstly what emerges is a circular driveway space, the austere elegance of which underlines the drapery-like front facade made of frosted glass. Secondly, it creates a private garden space linked to the apartments and offices used for more working-like meetings. Finally, there is the representational garden space which forms the conceptual core of the project. The garden is generously dimensioned, organically connected with the main lounges of the Embassy, and mediates a strong contact between the new building and the existing residence of the ambassador. The design should attempt to show what we would like to be like, or how we wish to be seen to the world: open, confident, friendly, helpful, and respectful, considerate to Nature and the environment in general, firmly rooted in rich cultural traditions and with respect to democratic principles.

representational garden
driveway

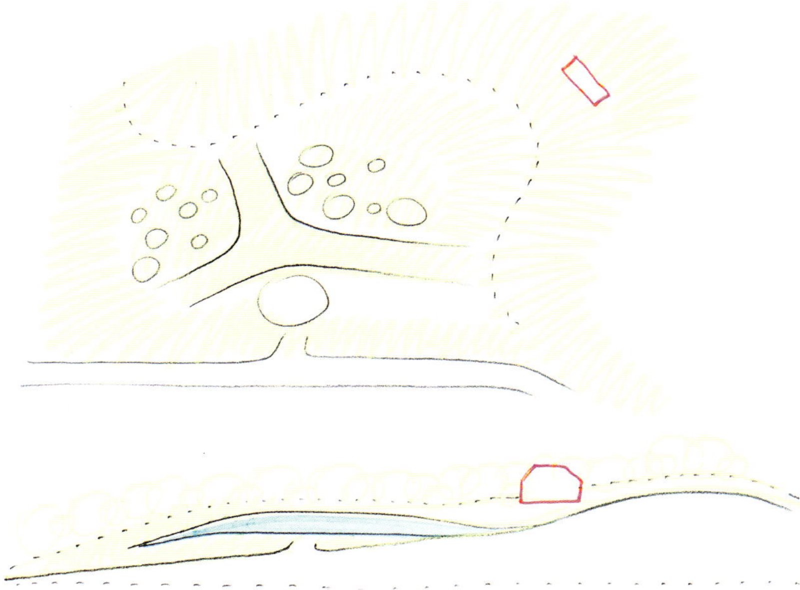

viewpoint

viewpoint
entrance hall
representational dining room
representational hall
private garden
residence
of the ambassador

viewpoint
representational garden

representational garden
private garden
driveway

开普敦的OYES椅子
OYES CHAIR, CAPE TOWN

建筑师：Hofman Dujardin Architects, Amsterdam
地毯设计：EGE, Copenhagen
Onstein Textiel Agenturen, Blaricum
效果图：A2Studio, Rotterdam
摄影：Matthijs van Roon, Amsterdam
慈善项目：Dietiker AG, Switzerland
AIT magazine, Germany
SV interior group, The Netherlands
特别鸣谢：Rene Magritte, Schuiten & Peeters and Google Maps

德国的AIT杂志从欧洲挑选了100位建筑师及室内设计师，邀请他们重新设计由Dietiker公司出品的“ONO”椅子。这把椅子成为760m高的建筑物模型，该建筑也是开普敦的城市象征。“ONO”椅子最多可以坐两个人，而OYES椅子的表面总计达80万平方米，可以容纳约50 000人。四只纤细的椅脚（25m × 30m）触及开普敦的商业区和港口。每个椅脚都影响着城市的现有结构，把不同地区的人们联系起来。OYES椅子也是慈善的象征，因为这座世界第二高的建筑为城市中无家可归的人们提供了庇护场所。

The German magazine AIT invited 100 selected architecture and interior design offices across Europe to redesign the "ONO" chair produced by the Dietiker company. The chair becomes a model of a building with a height of 760m, an urban icon for Cape Town. While the ONO chair can be used by a maximum of two people, the OYES chair with a surface of 800 000m² can host about 50 000 people. Four slim feet of 25m×30m will touch Cape Town in the business district and the harbor. Each foot will intervene in the existing urban fabric, connecting people from different areas. The OYES chair becomes the iconic symbol of charity, the second tallest building of the world is the shelter of the homeless people of the townships.

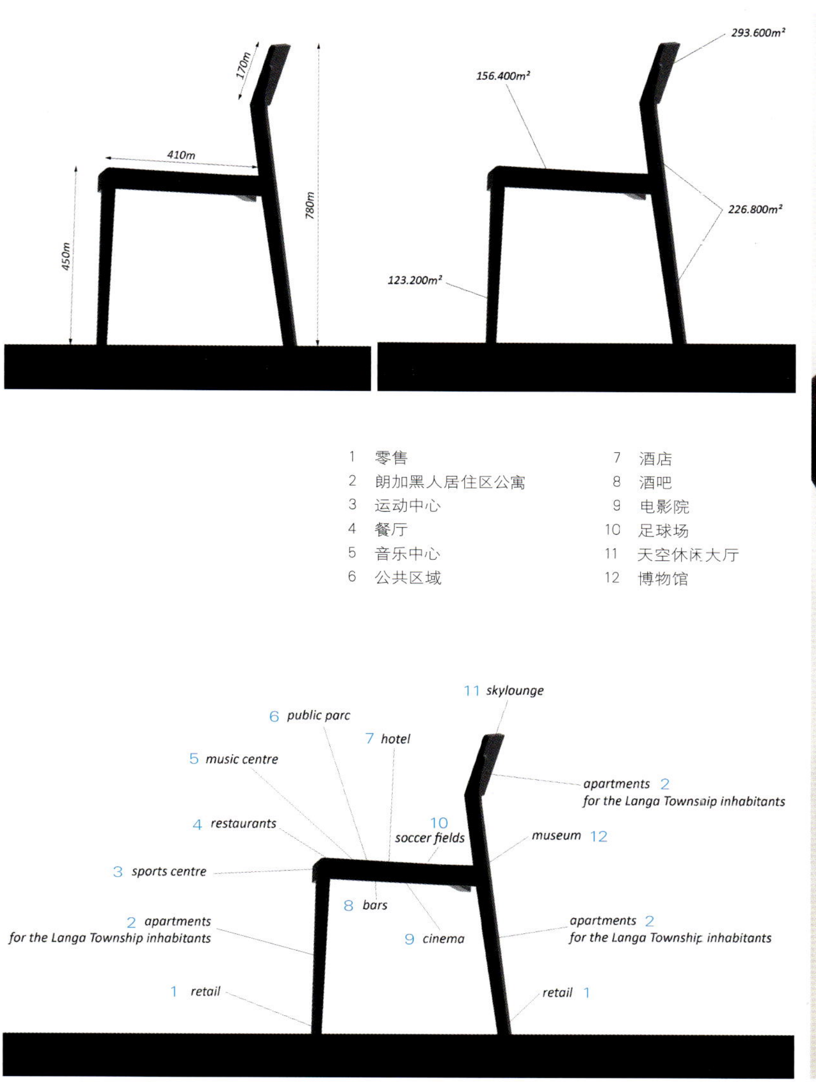
170m
410m
450m
780m
293.600m²
156.400m²
226.800m²
123.200m²
1 零售
2 朗加黑人居住区公寓
3 运动中心
4 餐厅
5 音乐中心
6 公共区域
7 酒店
8 酒吧
9 电影院
10 足球场
11 天空休闲大厅
12 博物馆
11 skylounge
6 public parc
7 hotel
5 music centre
apartments 2
for the Langa Townsaip inhabitants
4 restaurants
10
soccer fields
museum 12
3 sports centre
8 bars
2 apartments
for the Langa Township inhabitants
9 cinema
apartments 2
for the Langa Township inhabitants
1 retail
retail 1

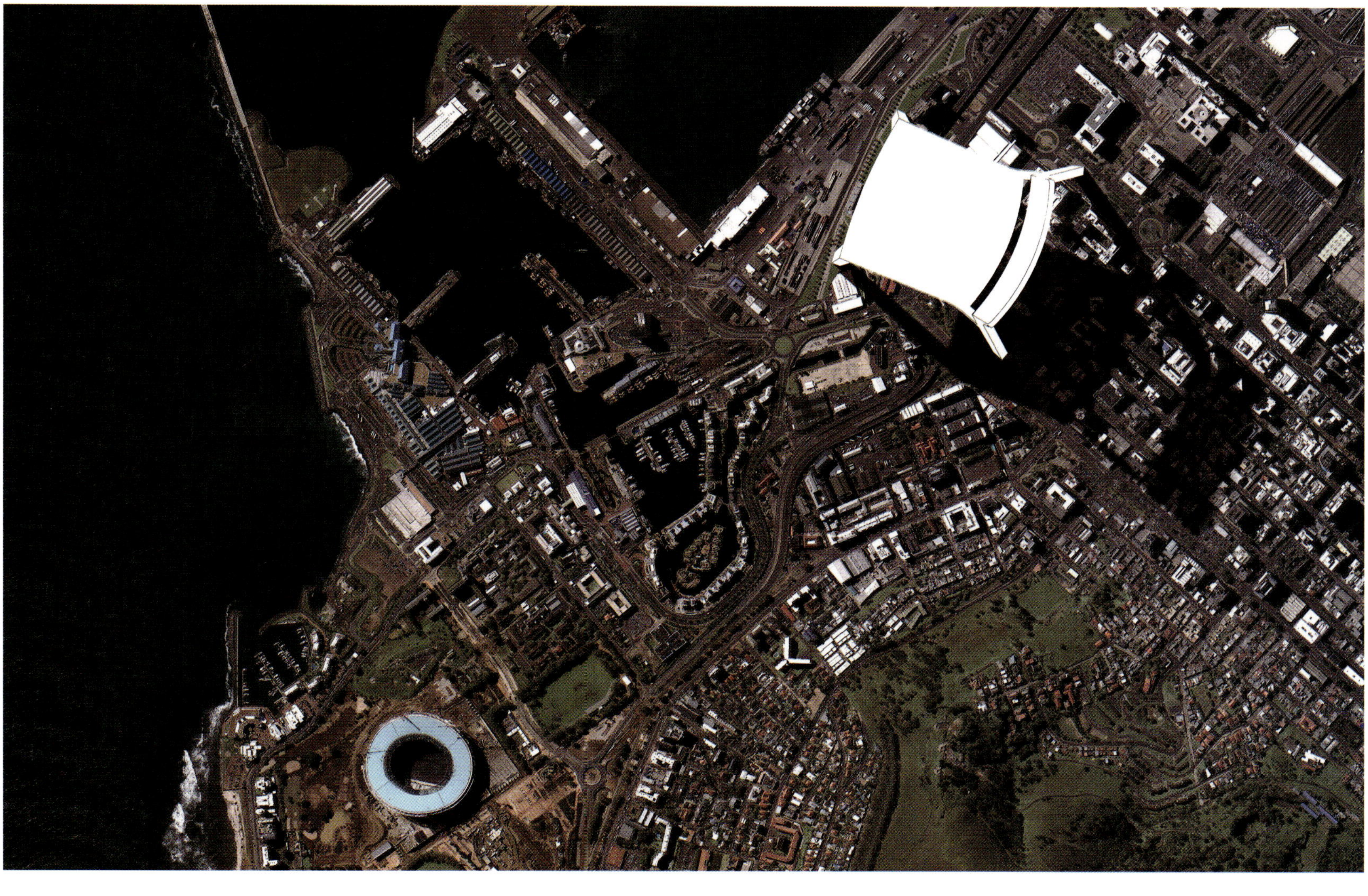

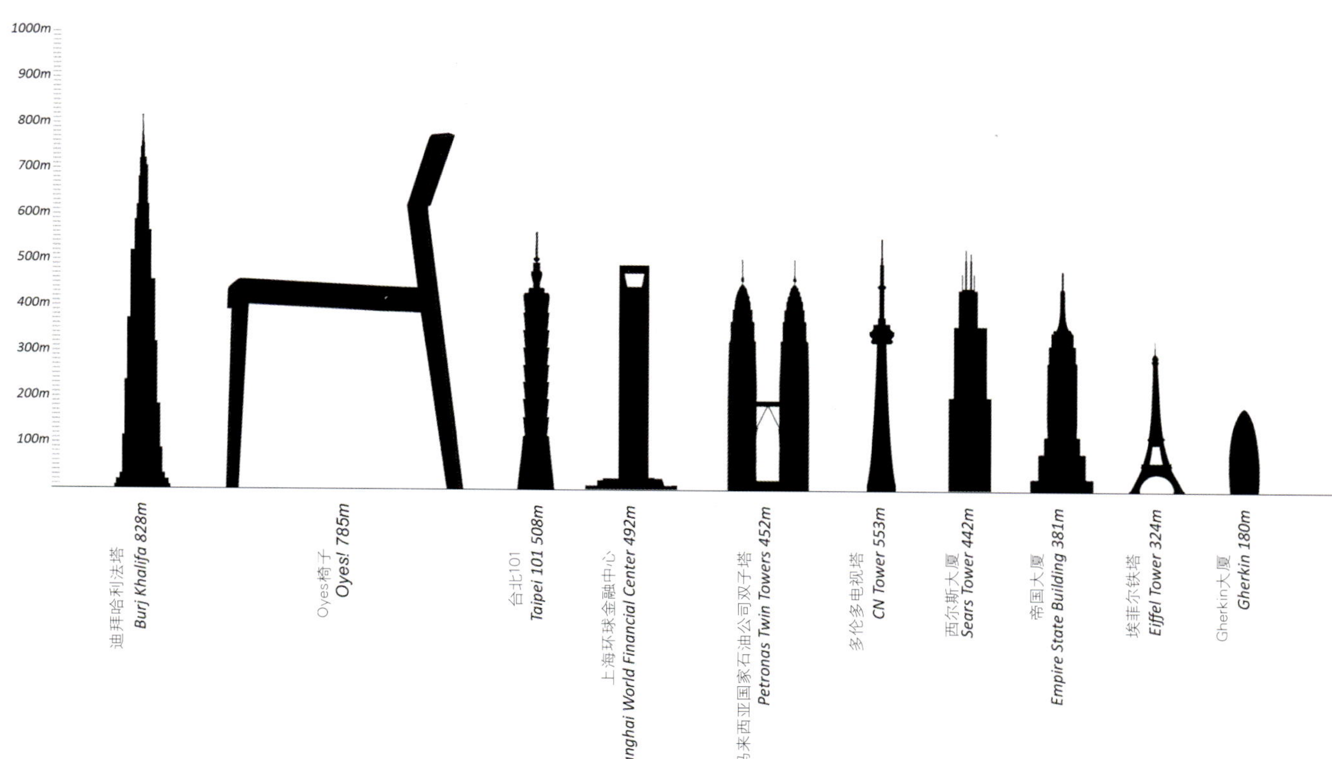
1000m
900m
800m
700m
600m
500m
400m
300m
200m
100m
迪拜哈利法塔
Burj Khalifa 828m
Oyes椅子
Oyes! 785m
台北101
Taipei 101 508m
上海环球金融中心
Shanghai World Financial Center 492m
马来西亚国家石油公司双子塔
Petronas Twin Towers 452m
多伦多电视塔
CN Tower 553m
西尔斯大厦
Sears Tower 442m
帝国大厦
Empire State Building 381m
埃菲尔铁塔
Eiffel Tower 324m
Gherkin大厦
Gherkin 180m

100

100 Global Architectural Schemes

国 际 最 IN 建 筑 设 计

100

住宅建筑

RESIDENTIAL
BUILDINGS 3

奥尔堡的能源产出住宅
HOUSING+ IN AALBORG

建筑师：C. F. Møller Architects
景观建筑师：Vogt Landschaftsarchitekten
工程师：Moe & Brødsgaard
参与设计建筑师：Cenergia
图片：C. F. Møller Architects
地点：Nørre Sundby, Aalborg, Denmark
甲方：Enggaard A/S

能源产出住宅项目以零能耗住宅方案作为其宏伟目标，这其中也包括居住者日常所需的初级能源消耗。住宅群届时将全部依靠可再生能源。包括60个单元的住宅采用了倾斜式的体量，高度从12层过渡到4层，形成了巨大的南向屋顶平面，这对太阳能的利用非常有利。而且规格对住宅单元来说也非常适宜。最优化的建筑形式使其具有了成为地标的外形，在奥尔堡的桥梁之间彰显自己的风采。

屋顶平面一路延伸到Limfjord峡湾中，并在交汇处设置一处公共露台和一间咖啡厅。屋顶的延伸强调出自身引人注目的外形，而且整个屋顶都安置了太阳能电池、太阳能供暖设备和这二者的组合设备，成为建筑的发电厂。

住宅按照被动住宅的标准建造，减少了供暖和热水需求的能量，转而以太阳能和由峡湾中水温来操控的热泵来补充这部分能源需求。3m宽、12m高的超保温水罐用来储存白天生成的能量。1200m²的太阳能电池可以满足每户年均1740Kwh、共104 400Kwh的电力需求。建筑无需与外界的热电联产设备联通。4个垂直的低噪音风力涡轮机利用场地的风能生成额外的电能，用于给电力汽车充电。

The Housing+ concept sets the ambitious target of a zero-energy housing scheme, which also includes the tenant's primary household energy consumption. The complex will thus be 100% relying on renewables.The 60 units take the form of a sloped volume, from 12 to 4 storeys, creating a large south-facing roof-plane, ideal for solar energy, and just the right size to supply the housing units. This optimized shape also creates a landmark silhouette, prominently positioned between Aalborg's bridges.

The roof-plane stretches all the way into the Limfjord, where it shelters a public gazebo and café. The extension of the roof underlines the dramatic shape of the building, and the entire surface of the roof becomes the building's power plant using solar cells, solar heating and a combination of the two.The housing is built to passive-house standards, ensuring reduced energy consumption for heating and hot water supply, which can thus be covered by the solar array and heat pumps operating on fjord water temperatures. A 3 meter wide by 12 meter tall highly insulated water-tank is integrated to store the generated energy during daytimes. The 1200 m2 solar array produces sufficient power to cover the annual 1740 Kwh electricity demand of each unit, a total of 104.400 Kwh. The building need not be connected to an external CHP. 4 vertical low-noise wind turbines take advantage of the windy location for additional power generation, and to recharge electric cars.

HOUSING+ AALBORG = BASIC CONCEPT (NEED TO HAVE) + ADDITIONAL ENVIRONMENTAL FEATURES (NICE TO HAVE)

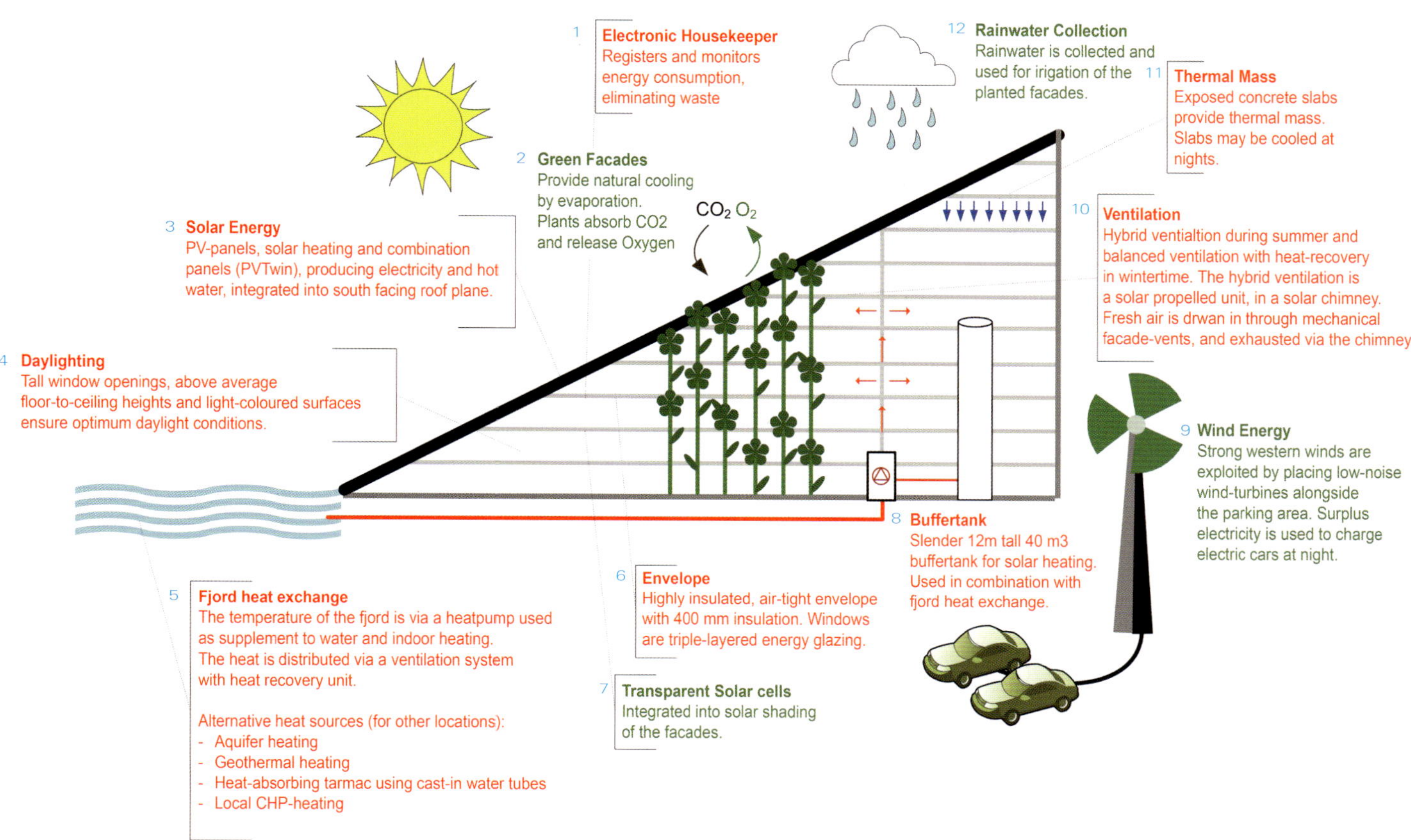

1 Electronic Housekeeper
Registers and monitors energy consumption, eliminating waste
2 Green Facades
Provide natural cooling by evaporation. Plants absorb CO2 and release Oxygen
CO_2 O_2
3 Solar Energy
PV-panels, solar heating and combination panels (PVTwin), producing electricity and hot water, integrated into south facing roof plane.
4 Daylighting
Tall window openings, above average floor-to-ceiling heights and light-coloured surfaces ensure optimum daylight conditions.
5 Fjord heat exchange
The temperature of the fjord is via a heatpump used as supplement to water and indoor heating.
The heat is distributed via a ventilation system with heat recovery unit.
Alternative heat sources (for other locations):
- Aquifer heating
- Geothermal heating
- Heat-absorbing tarmac using cast-in water tubes
- Local CHP-heating
6 Envelope
Highly insulated, air-tight envelope with 400 mm insulation. Windows are triple-layered energy glazing.
7 Transparent Solar cells
Integrated into solar shading of the facades.
8 Buffertank
Slender 12m tall 40 m3 buffertank for solar heating. Used in combination with fjord heat exchange.
9 Wind Energy
Strong western winds are exploited by placing low-noise wind-turbines alongside the parking area. Surplus electricity is used to charge electric cars at night.
10 Ventilation
Hybrid ventialtion during summer and balanced ventilation with heat-recovery in wintertime. The hybrid ventilation is a solar propelled unit, in a solar chimney. Fresh air is drwan in through mechanical facade-vents, and exhausted via the chimney.
11 Thermal Mass
Exposed concrete slabs provide thermal mass. Slabs may be cooled at nights.
12 Rainwater Collection
Rainwater is collected and used for irigation of the planted facades.

奥尔堡的能源产出住宅基础概念（必需）+附加的环境特色（建议添加）

1 电器设备
注册并监督能源消耗，减少废物排放

2 绿色立面
提供蒸发自然制冷
植物吸收CO_2，释放O_2

3 太阳能
电光板、太阳能供暖装置和热电混合装置被整合在南立面中，生成电能和热水

4 日照
窗洞位置较高，超过平均楼层高度
浅色的玻璃表面确保最佳的日照条件

5 峡湾热交换
峡湾的水温所蕴含的能量通过热泵进行利用，作为建筑用水及室内供暖的补充

供选择的热源（其他区域）
含水土层供暖
地热供暖
内铸水管的沥青进行热吸收
当地的热电联产设备

6 外围护结构
高保温的气密性外围护结构，带400mm保温层。窗户为三层节能型玻璃

7 透明的太阳能电池
融入到立面的遮阳设备中

8 缓冲水罐
高12m、容积40m^{3}的细长水罐用于太阳能供暖。与峡湾的热交换一同使用

9 风能
停车场旁的低噪音风力涡轮机利用强劲的西风发电。生成的电力用于在夜晚为电力汽车充电

10 通风
夏季为混合通风设备，冬季为带热交换设施的平衡通风设备。混合通风设备为利用太阳能烟囱效应的太阳能驱动设备。新鲜空气通过机械方式从立面通风口引入。废气通过烟囱效应排到室外

11 蓄热体
裸露的混凝土作为蓄热体。混凝土板可以在夜晚进行冷却

12 雨水收集
雨水被收集起来用于绿化屋顶的灌溉

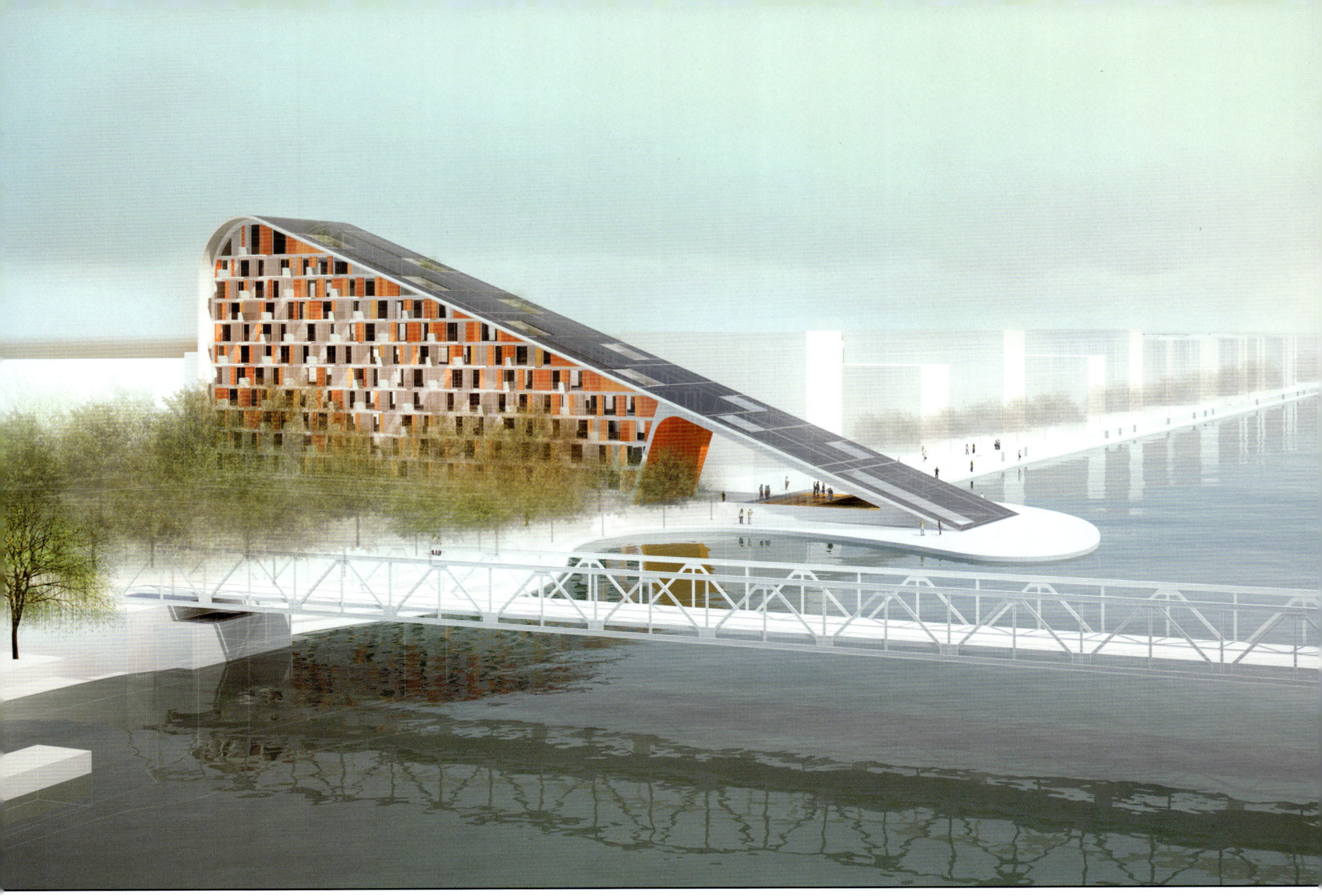

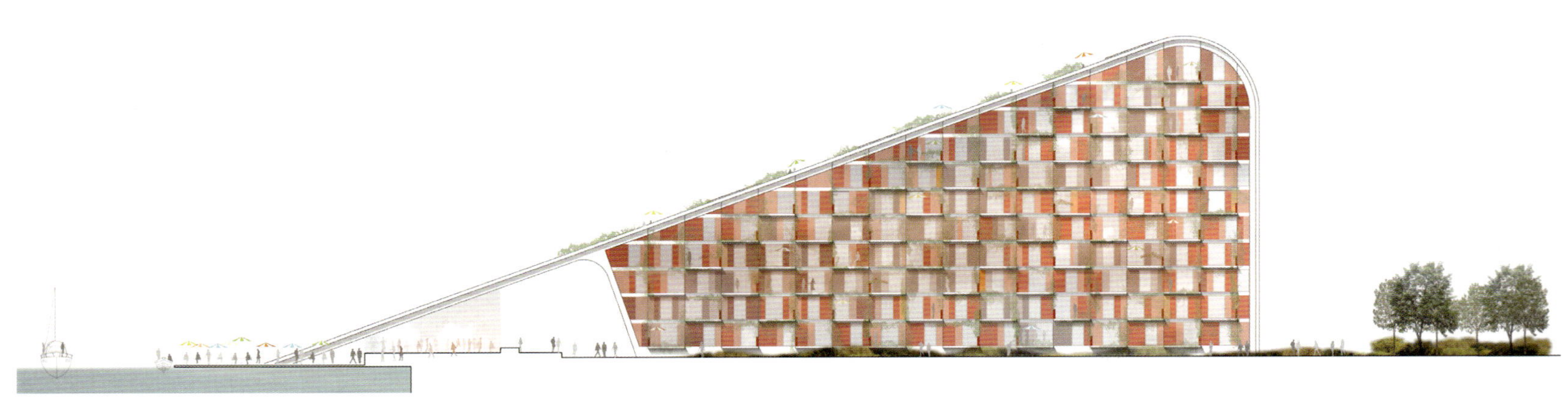

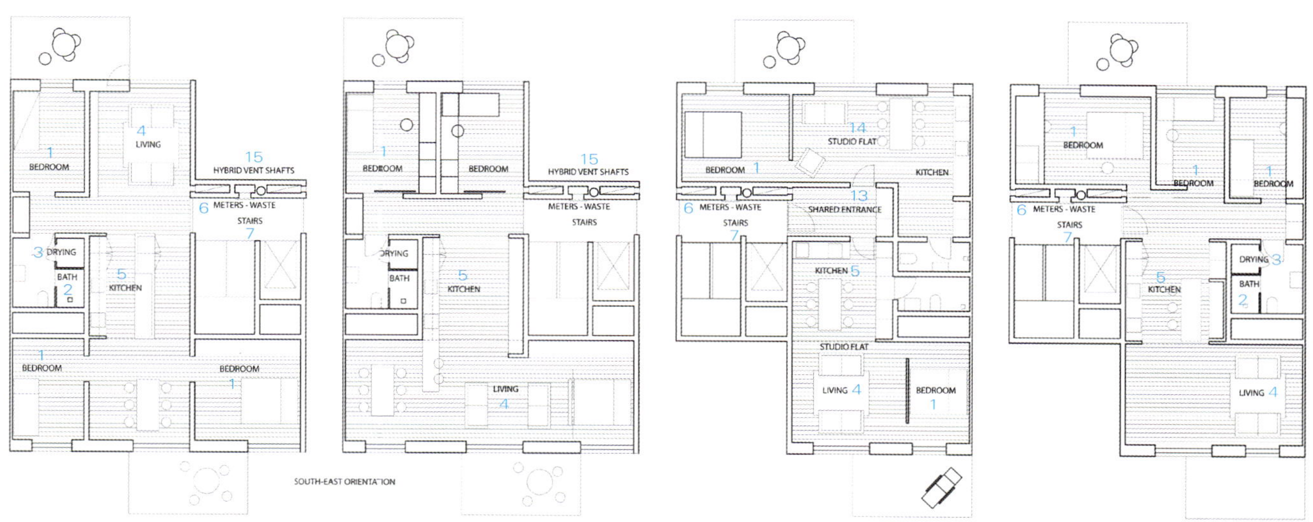

单元图例1
基本的双朝向布局
总面积109.5m²

UNIT EXAMPLE 1

BASIC DUAL ASPECT LAYOUT
GROSS AREA 109,5m²

单元图例2
厨房/起居室面积较大
朝向东南
总面积109.5m²

UNIT EXAMPLE 2

LARGER KITCHEN/LIVING TOWARDS SOUTH-EAST
GROSS AREA 109,5m²

单元图例3
厨房/起居室面积较大
朝向东南
总面积109.5m²

UNIT EXAMPLE 3

TWIN STUDIO-FLAT LAYOUT FOR FLEXIBLE USE AND VARIED TENANCIES
GROSS AREA 109,5m²

单元图例4
带有厨房吧台，并有更多的公共空间
总面积109.5m²

UNIT EXAMPLE 4

KITCHEN COUNTER AND MORE COMMON SPACE
GROSS AREA 109,5m²

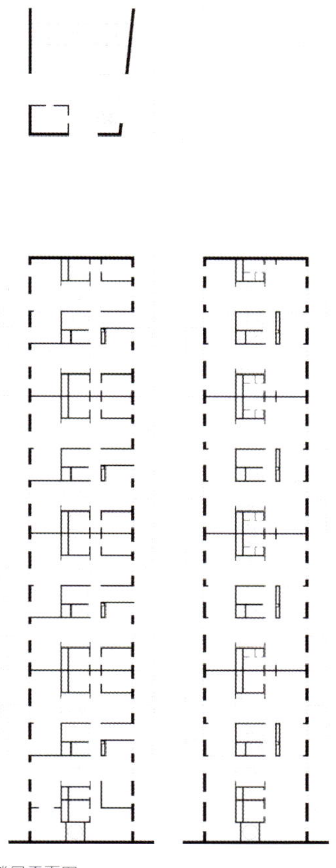

楼层平面图
Floor plans

1 卧室
2 浴室
3 干燥机
4 起居室
5 厨房
6 垃圾处
7 楼梯
8 设备间
9 私人屋顶花园
10 温室（未供暖，覆盖带太阳能电池的玻璃）
11 卧室/起居室
12 中心的缓冲水罐
13 共用入口
14 开放式单元
15 混合通风竖井
16 西南朝向

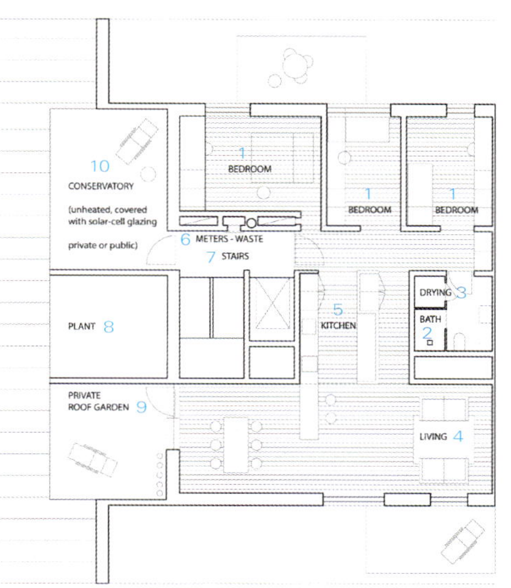

单元图例5
倾斜屋顶下的单元
有通向私人屋顶花园的通道，可观赏峡湾的景观
总面积134m²

UNIT EXAMPLE 5

UNITS UNDER INCLINED ROOF HAVE ACCESS TO PRIVATE ROOF GARDENS WITH FJORD VIEWS
GROSS AREA 134m²

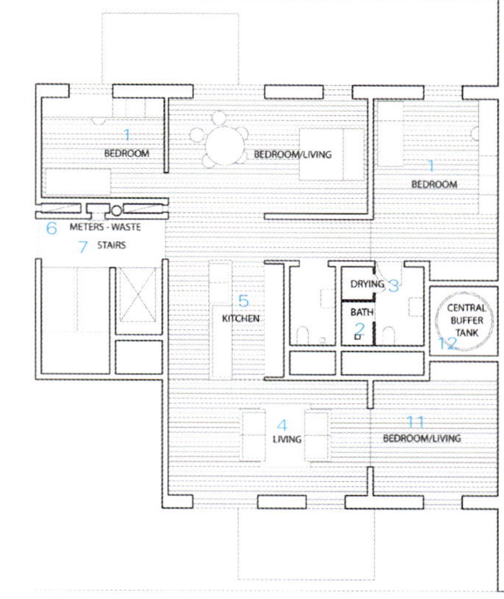

单元图例6
住宅建筑较高一侧的单元
多功能的大型大院，可灵活布局
总面积158.2m²

UNIT EXAMPLE 6

THE TALL END OF THE BUILDING HOUSES LARGER UNITS WITH MULTIPLE OPTIONS FOR FLEXIBLE LAYOUTS
GROSS AREA 158,2m²

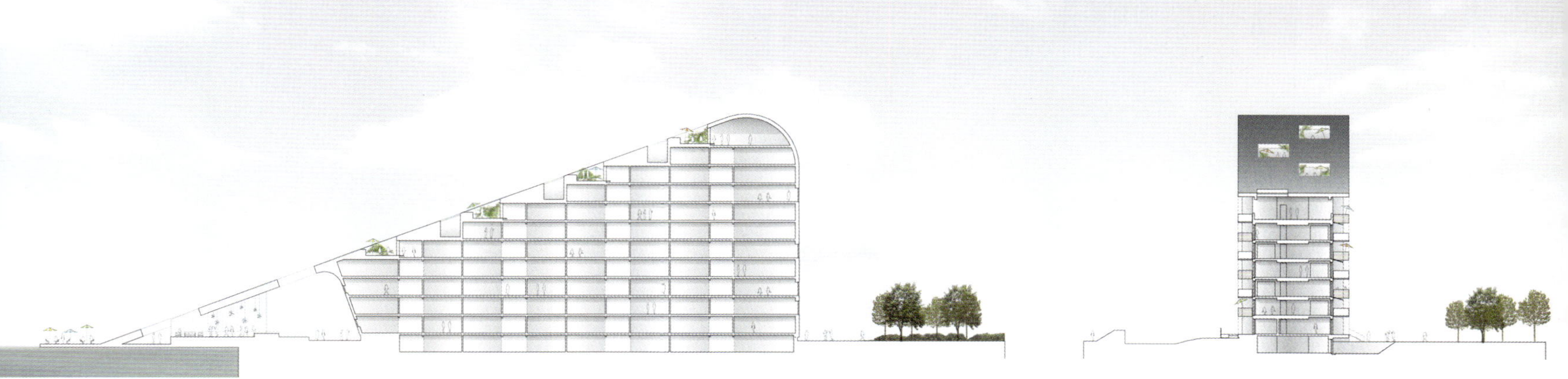

大河之舞——关于新建宿舍的方案
RIVERDANCE – PROPOSAL FOR NEW CTU DORMITORY

建筑师：Philip Modest Schambelan (www.scham.be)

顾问：Prof. Pata, CTU Prague

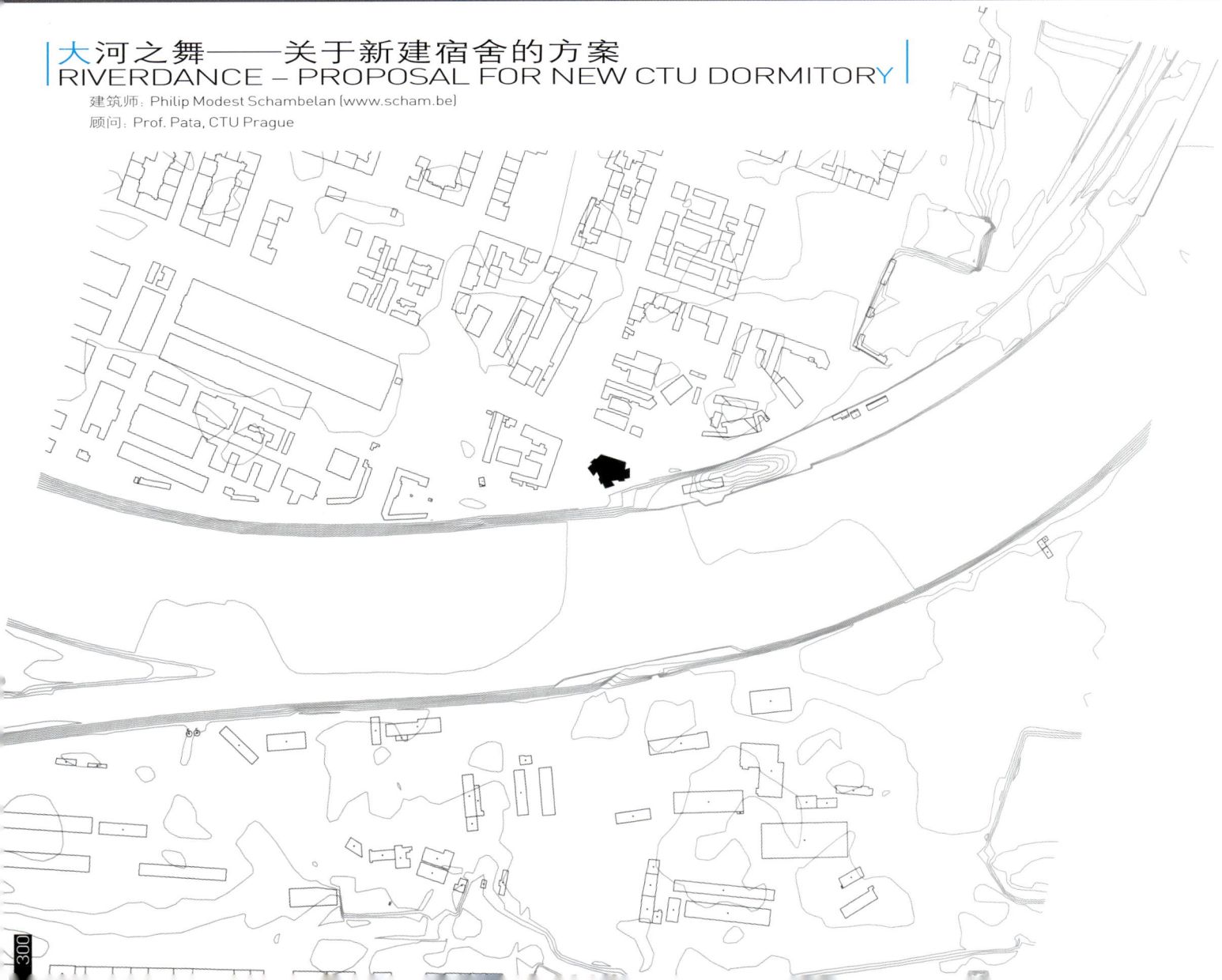

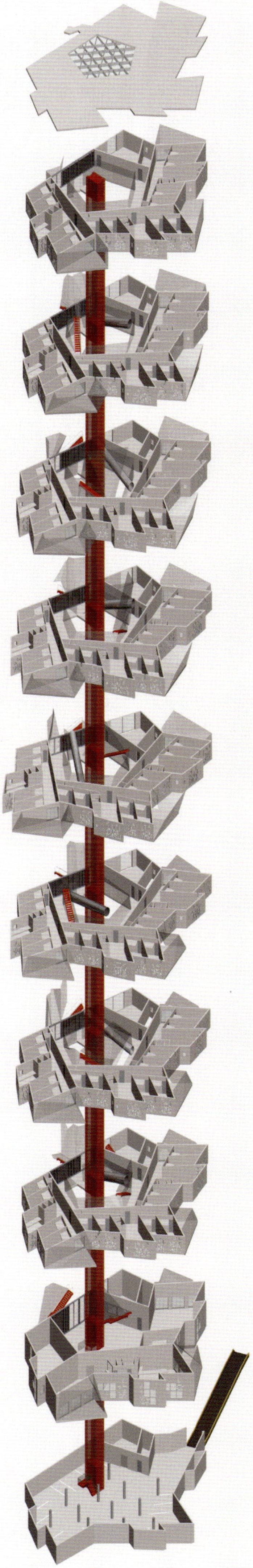

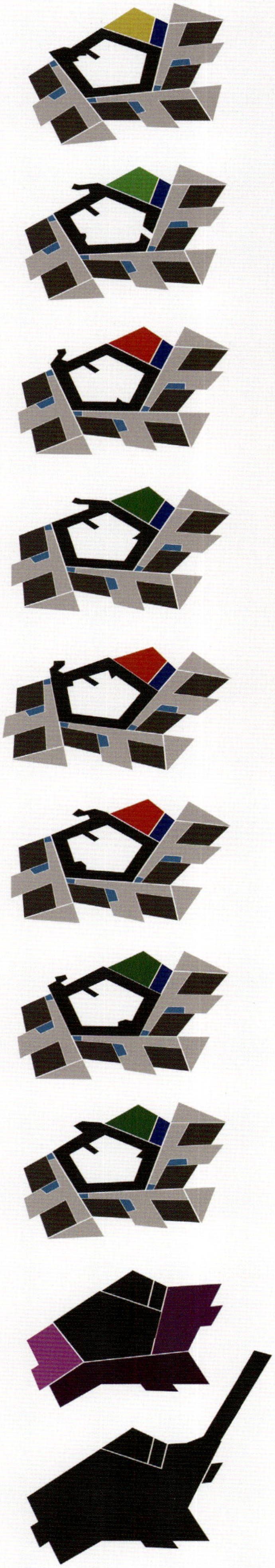

俄罗斯方块拆解
Tetris explosion

新宿舍大楼重新利用了河边闲置的废弃空地，提供了私人与公共空间，运动、学习、聊天场所，艺术以及其他创意空间。大楼北部正对着Holeovice区，虽然它看上去与这个城市的其他建筑一样庄重沉稳，但人们仍可以感受到它的节奏。钢缆支撑着顶棚和阳台，既具有令人印象深刻的外形，也确保了安全性。顶棚中连接每个楼层的滑管加快了楼层之间的沟通。每间公寓的阳台和起居室都面向河边，每个楼层都具有不同的视角。作为设计师的学生们将加强河边的建设发展，它的潜力无限。

Reconquering unused abandoned space near the riverside, the new dormitory offers privacy, community, sports, learning and discussion areas, space for arts and other creativity. The northern part of the building is facing the district Holeovice head-on. While it is kept in a rather calm manner to reflect this part of the city's architecture you can still feel its movement. Steel cable trellis provides the vault and its balconies with both thrill and safeness. Slide tubes connecting the floors inside the vault allow fast communication between floors. Each apartment presents itself with riverside balconies and living rooms. Each angle of view is different, in every apartment on every floor. The students will refresh, enhance and develop the riverside, which is full of forgotten potentials.

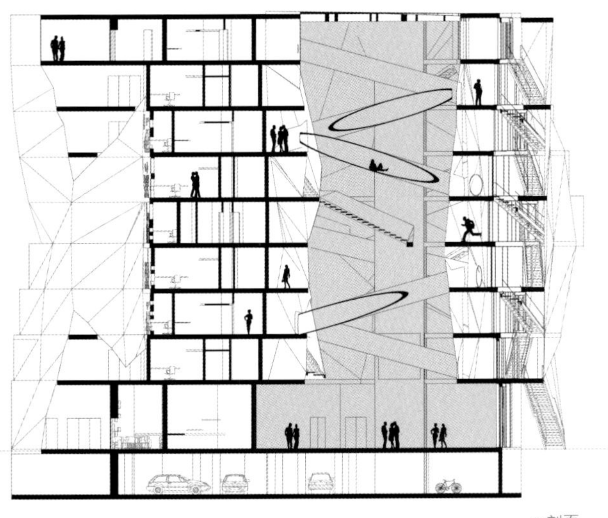

aa剖面
Section aa

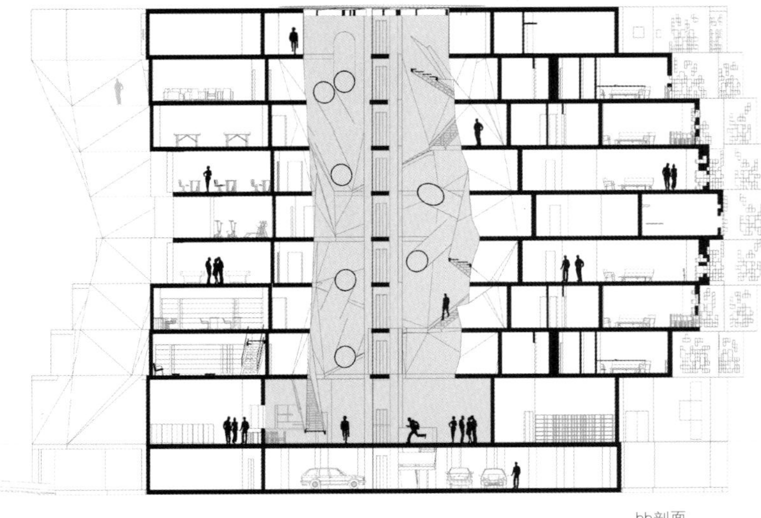

bb剖面
Section bb

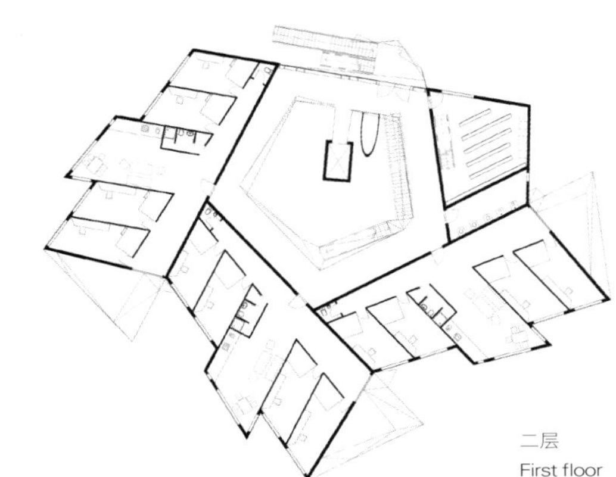

二层
First floor

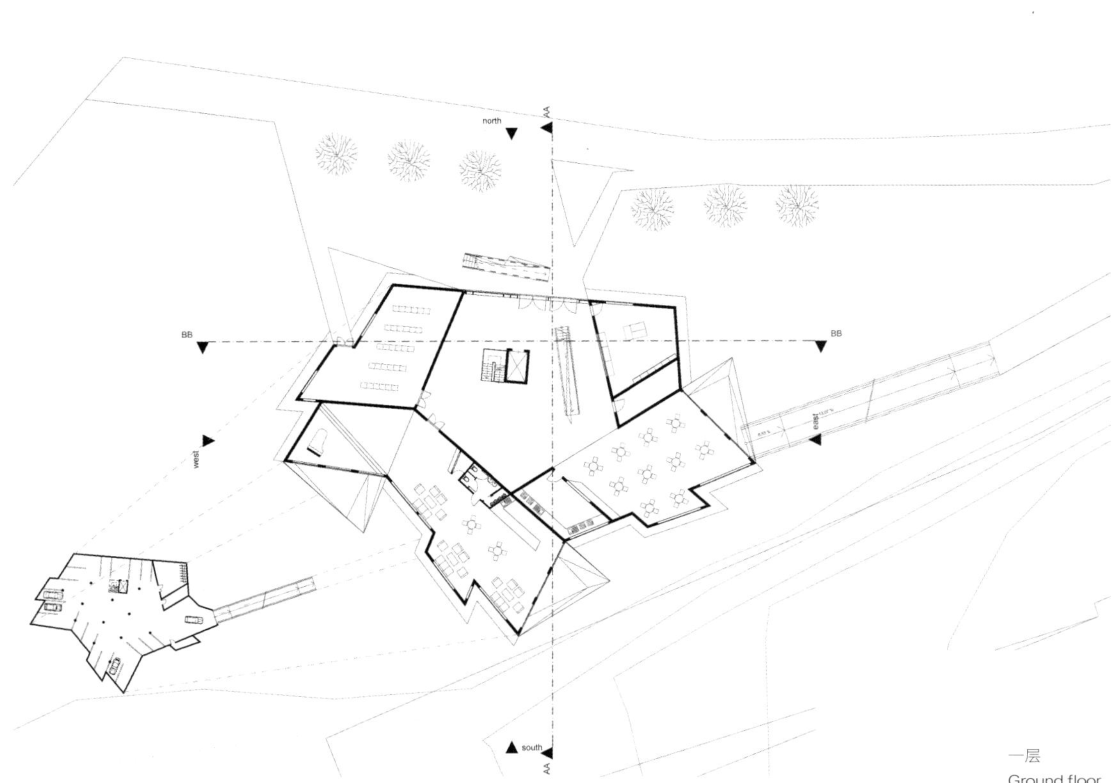

一层
Ground floor

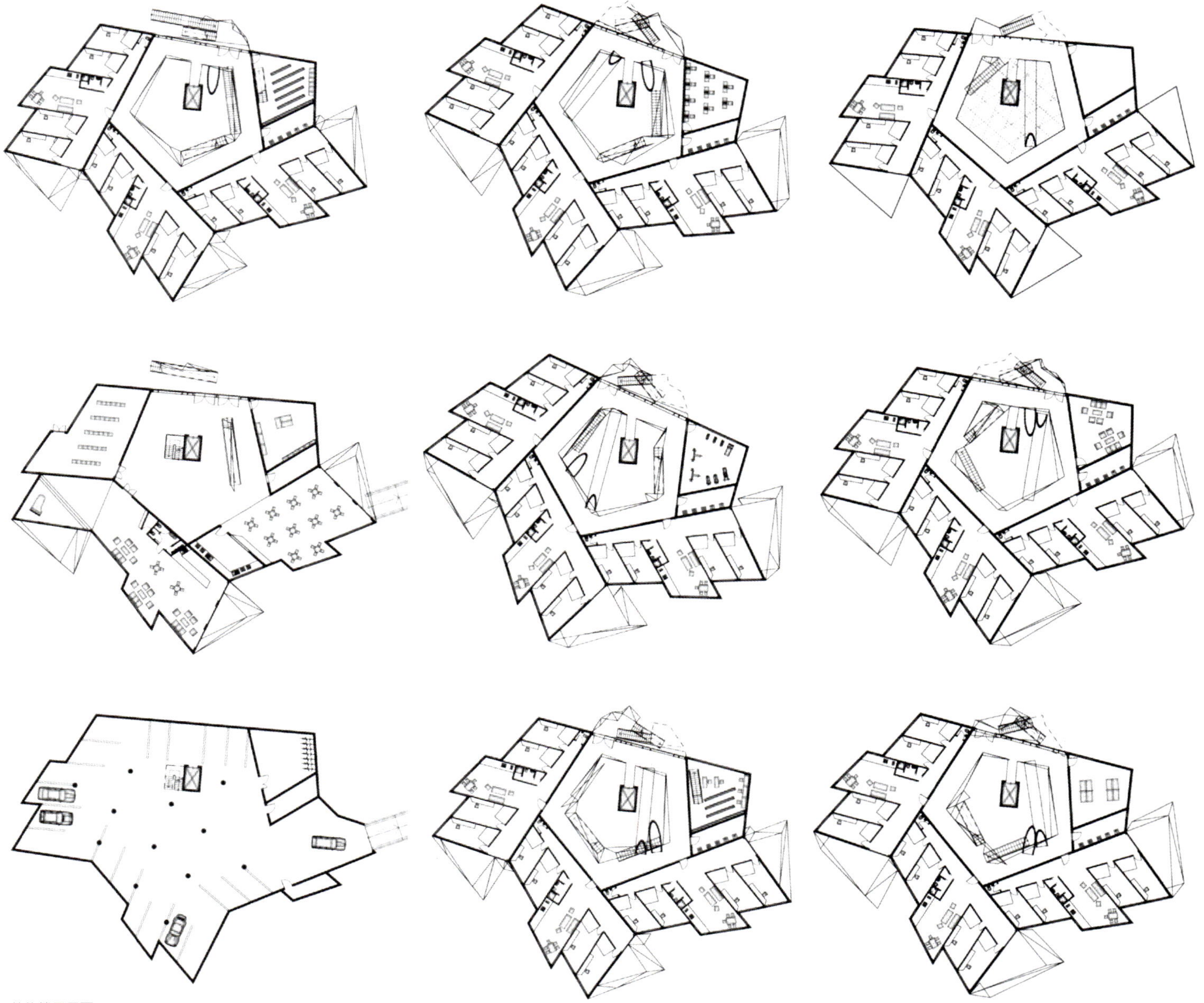

其他楼层平面
Floor plans of other levels

北侧视图
View from North

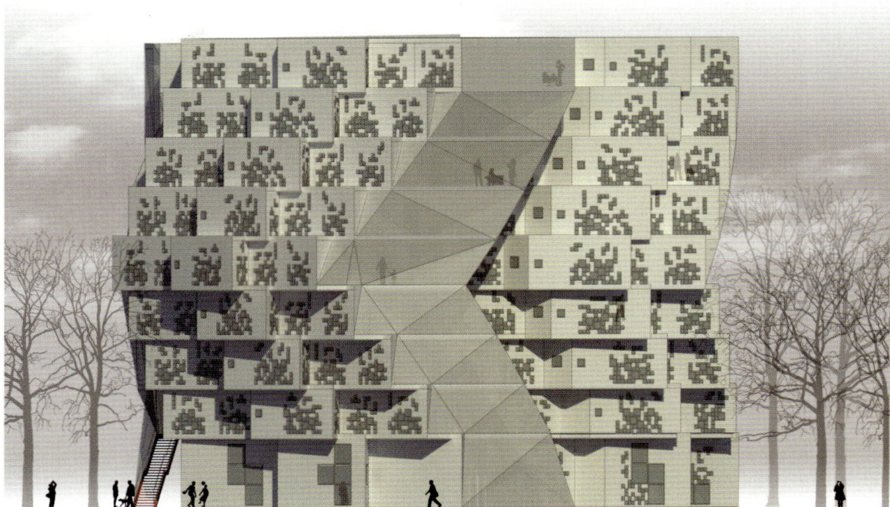

西侧视图
View from West

东侧视图
View from East

南侧视图
View from South

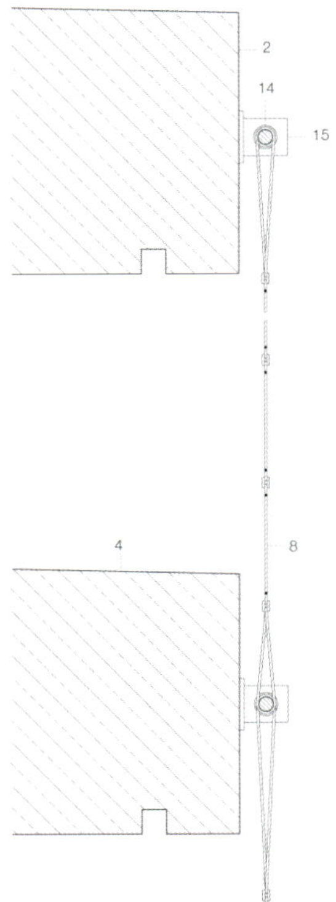

地下室剖面详图
Section details of vault

1 屋顶构造：
2mm单层塑料屋顶密封
200mm刚性泡沫塑料保温层
隔汽层
200mm钢筋混凝土楼板
2 楼板上方的入口阳台：
210~240mm预制混凝土构件
3 保温层加固连接杆
4 防护格子墙：
3mm不锈钢缆
5 双层玻璃窗：
8+12mm浮法玻璃+16mm空腔
6 立柱栏杆立面，安装玻璃窗
7 30/20/4mm不锈钢角钢，嵌入预制混凝土构件
8 210~240mm预制混凝土入口阳台，表面喷砂
9 双层玻璃窗：
6+10mm浮法玻璃+16mm空腔
10 100/75/7mm不锈钢角钢
11 四周密封层
12 120mm刚性泡沫塑料保温层
13 200mm预制混凝土带状入口
370/1450mm板
14 12mm不锈钢边缘钢缆
15 不锈钢锚固件

1 roof construction:
2mm single-layer plastic roof seal
200mm rigid-foam thermal insulation
vapour barrier
200mm reinforced concrete roof slab
2 access balcony cover slab:
210–240mm precast concrete element
3 thermally insulated reinforcement stitching bars
4 trellis as safety barrier:
Ø 3mm stainless-steel cable
5 double glazing:
8 + 12mm float glass + 16mm cavity
6 post-and-rail facade with fixed glazing
7 30/20/4mm galvanized steel angle inserted in precast concrete element
8 210–240mm precast concrete access balcony, sandblasted on top surface
9 double glazing:
6 + 10mm float glass + 16mm cavity
10 100/75/7mm galvanized steel angle
11 perimeter sealing layer
12 120mm rigid-foam thermal insulation
13 200mm precast concrete access-strip slabs 370/1450mm
14 Ø 12mm stainless-steel edge cable
15 stainless-steel anchor piece

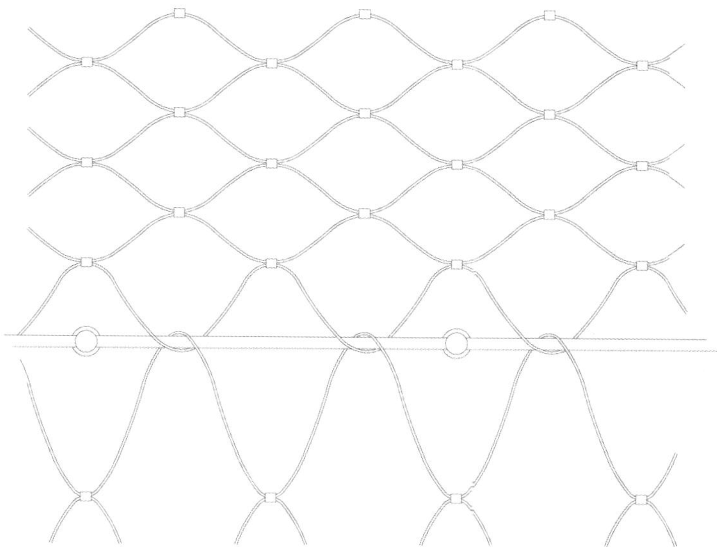

格子墙立面详图
Elevation detail of trellis

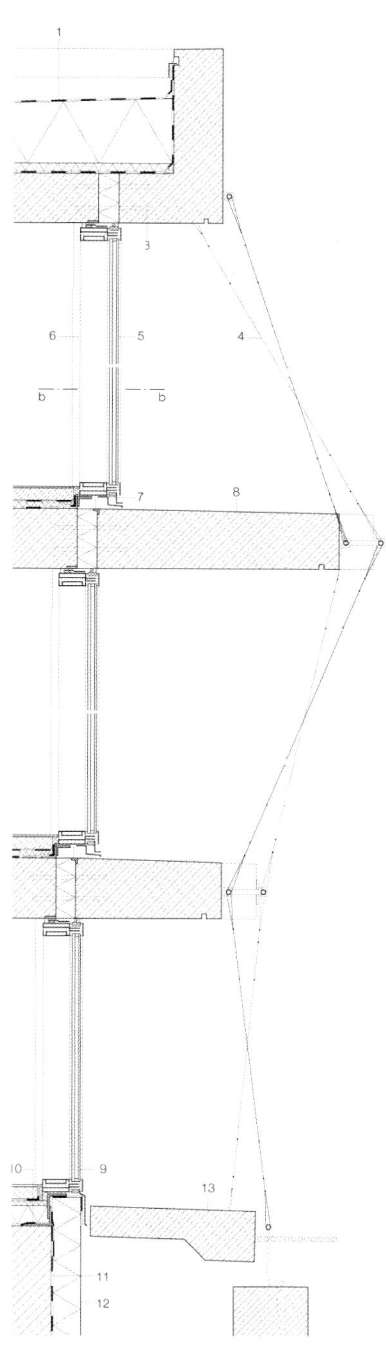

阳台剖面详图
Sectional details of balcony

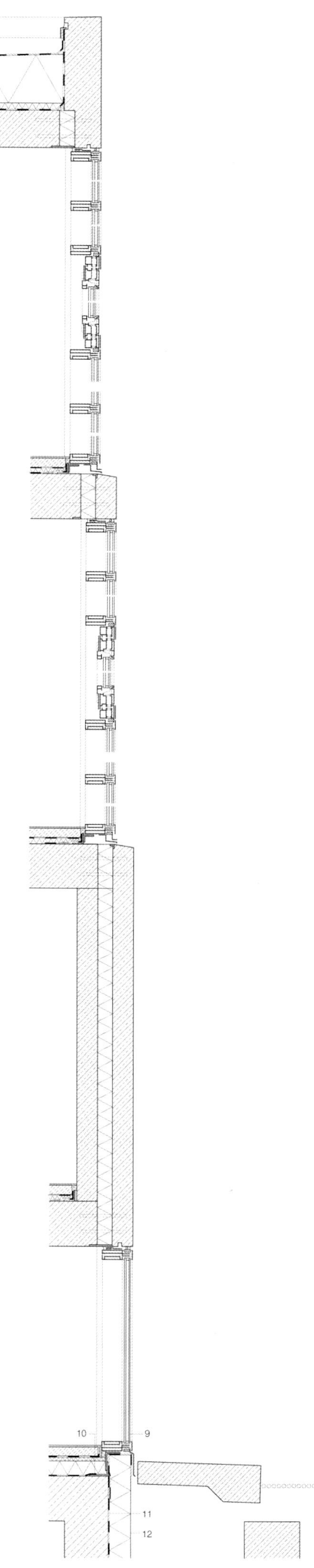

立面剖面详图
Section details of elevation

双巢村庄
TWO NEST VILLAGE

建筑师：OFIS
项目负责人：Rok Oman, Spela Videcnik
项目团队：Janez Martincic, Katja Aljaz, Janja Del Linz, Andrej Gregoric, Katarzyna Bernatek, Robert Janez, Cristian Gheorghe
可持续性工程师：Anthony Martin PE, LEED® AP
Henderson Engineers, Inc., Dallas, Texas
结构工程师：Jaka Zevnik u.d.i.g., ELEA IC d.d., Ljubljana, Slovenia

本案把项目基址分成大致相等的两个区：南区和北区。因为角落位置更适合发挥商业型或混合型开发的作用，北区被预留为二期开发项目。住宅区坐落在比较安静的南区，被规划为巢穴状——公寓群落层层叠起。这种建筑格局形成了完美的屏障，使其免受阳光和风蚀，同时又提供了高质量的居住环境：由于内外部拥有一系列共享区域，该区域具有极佳的采光和通风。

公寓单元或者朝向内部公园，或者朝向外部区域，房间的景观因此具有了私密型（朝向庭院）和开放型（朝向四周——有群山衬托的都市风光）两种选择。

环绕着内部广场的一层公寓设计有中庭和与广场相连的露台。所有其他的公寓设计有凉廊，凉廊上配置有起遮阳作用的小花架。

宏观设计概念：
design concept – macro scale:

1　气流
1. air flow

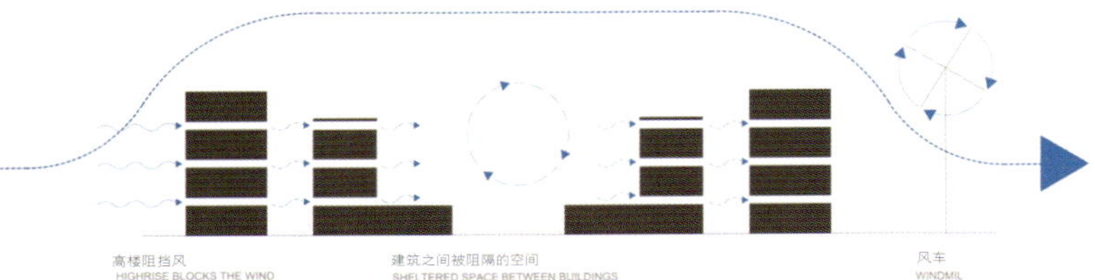

2　遮阳
2. shading

3　自然通风
3. natural ventilation

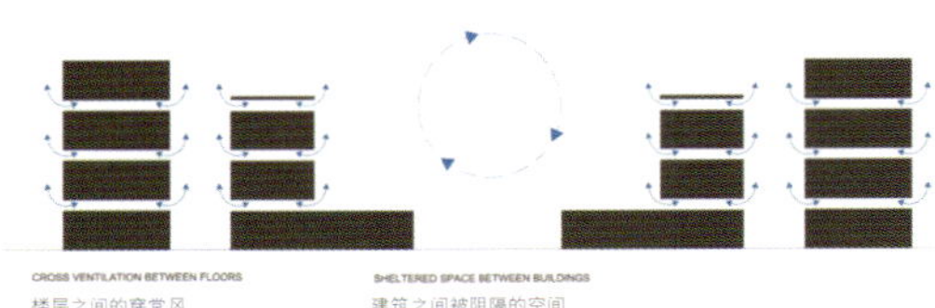

4　水的利用
4. water usage

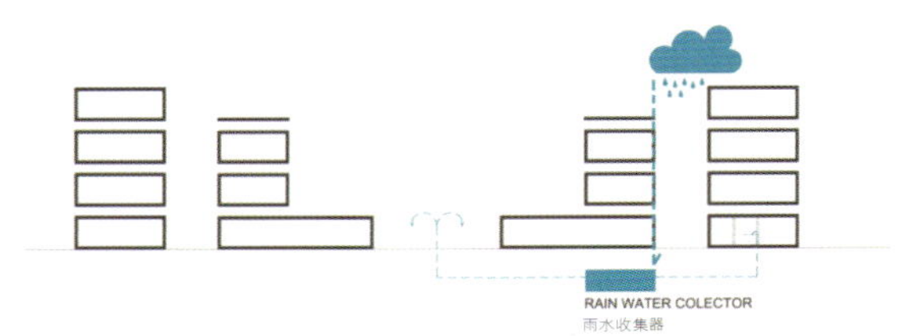

5　水流系统
5. water flow system

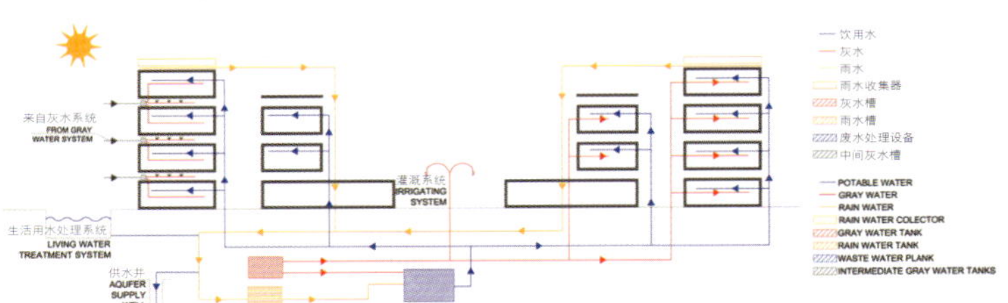

6　排水系统
6. drainage system

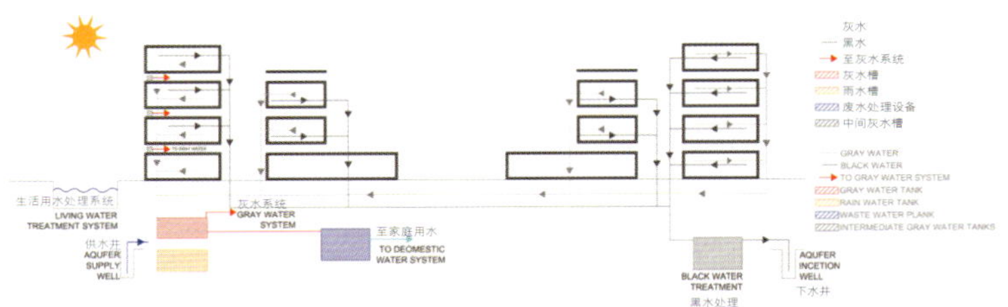

7　光电设备
7. photovoltaic

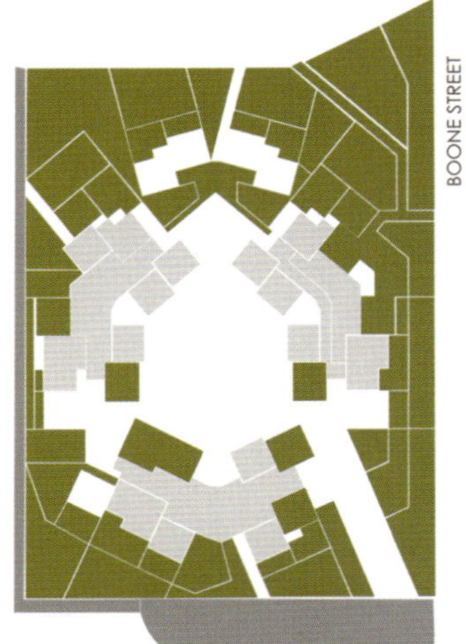

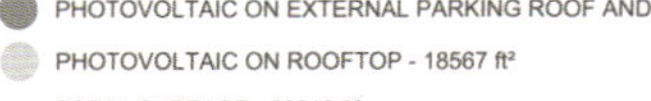

室外停车场屋顶和通道屋顶上的光电板面积1091m² PHOTOVOLTAIC ON EXTERNAL PARKING ROOF AND PASSAGES ROOFS - 11743
屋顶上的光电板面积1725m² PHOTOVOLTAIC ON ROOFTOP - 18567 ft²
总面积2816m² TOTAL SURFACE - 30310 ft²
产生能量600kW ENERGY PRODUCED- 600 kW

The proposal separates the site in two roughly equal halves: North and South. The North is reserved for Phase II of the development, because its corner location is better suited for its anticipated use as a commercial or mixed use development. The housing complex is located in the quieter, southern part of the plot and s organized in the form of a nest – living mound of apartments, stuck one above another. The form makes a perfect shape of a shield, which protects the units from sun and wind, but on the other hand offers a quality environment: well lit and ventilated with a variety of shared areas, both external and internal.

The apartment units are oriented towards the internal park or towards the exterior. The resulting views offer a choice of intimacy (facing the courtyard) and openness (facing the surroundings – urban scenery and mountains in the background).

Apartments on level +1 around the internal park have atriums and terraces connected to the plaza. All other apartments have small loggias which incorporate shading devices with small planters.

公共区域、景观与空间
common areas, landscape and spaces

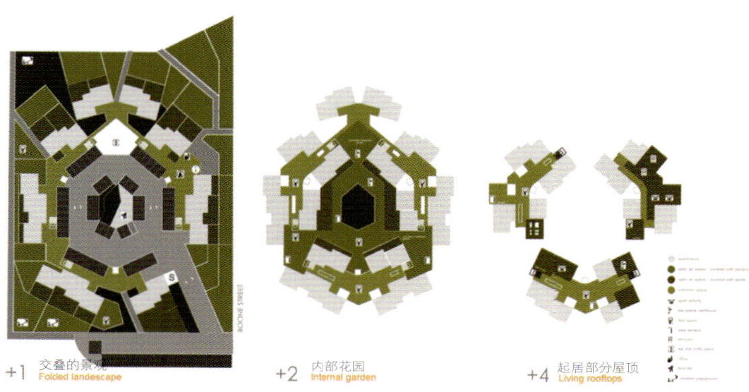

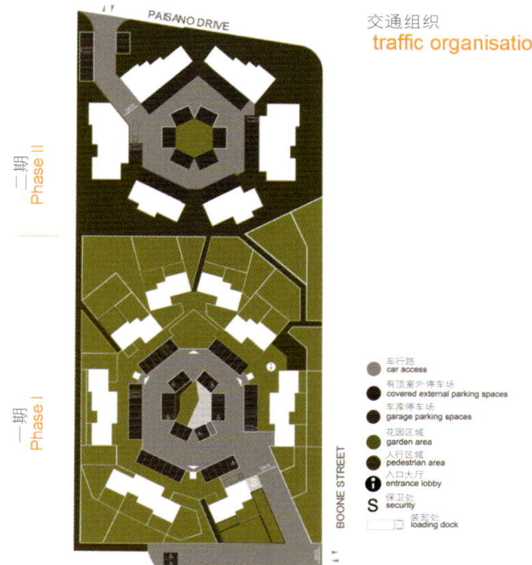

1 大门
2 商店
3 商店/办公室
4 咖啡厅
5 办公室
6 商业区域
7 健身房
8 瑜伽室
9 自行车存放处
10 技术设备室
11 维护室
12 大厅
13 接待处
14 洗衣房
15 花园
16 58个停车位
17 装卸平台
18 装卸处
19 保卫处
20 储存室
21 垃圾存放处
22 公共花园
23 有顶室外停车场
24 22个停车位
25 公共露台
26 内庭院
27 有顶公共空间
28 竖井
29 公共花园
30 公共空间
31 观景露台
32 日光甲板
33 烧烤区
34 露天体育馆
35 眺望台
36 饮茶区

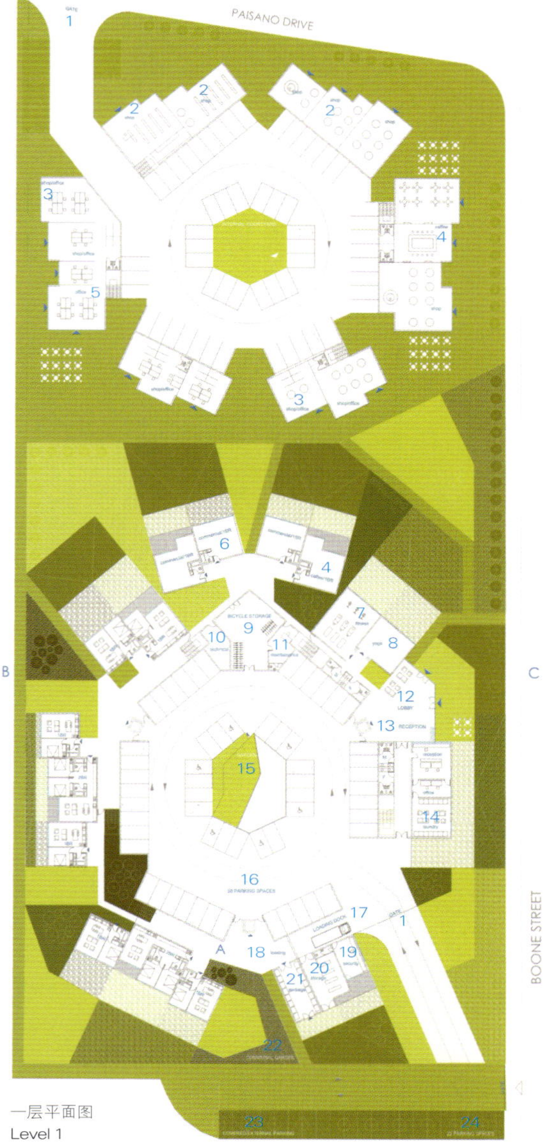

一层平面图
Level 1

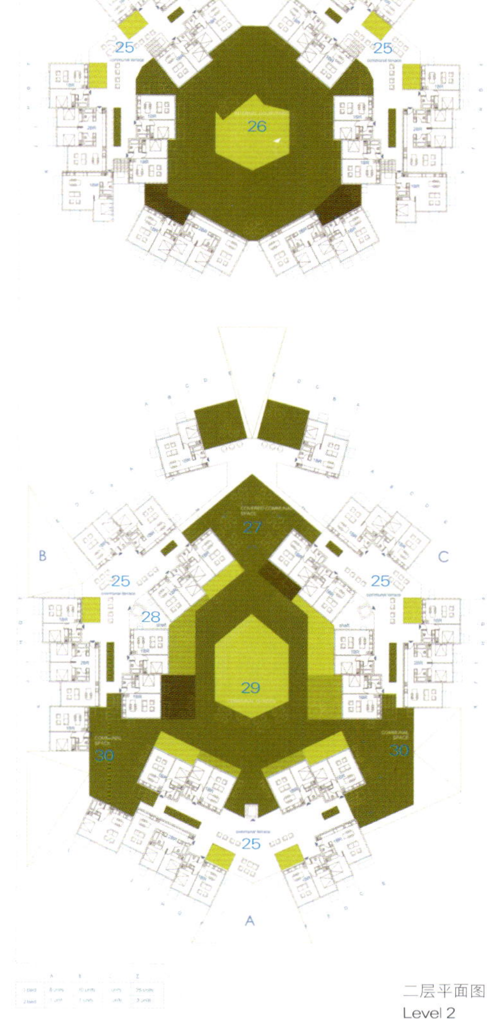

二层平面图
Level 2

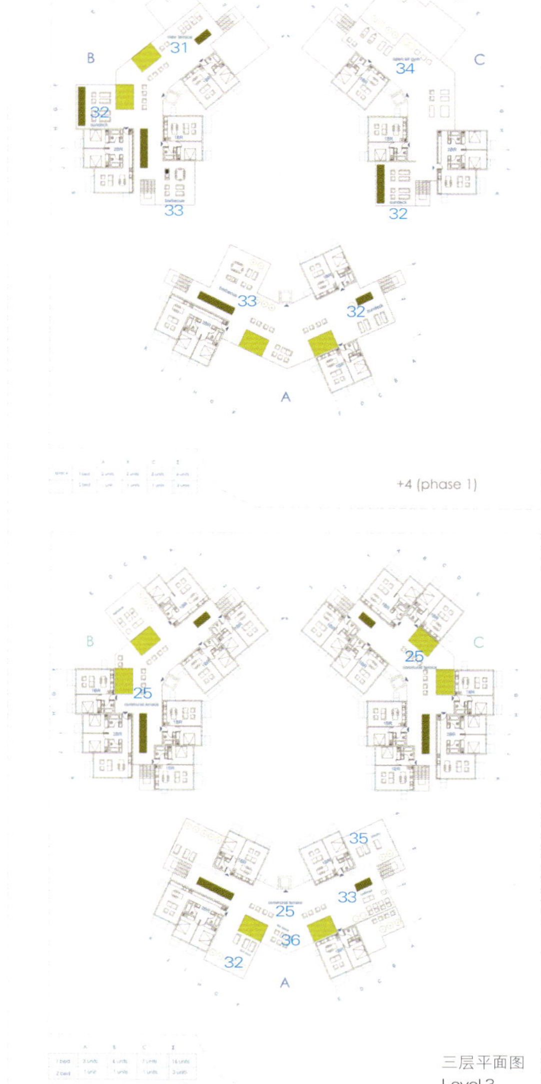

三层平面图
Level 3

无障碍设计

handicapped friendly

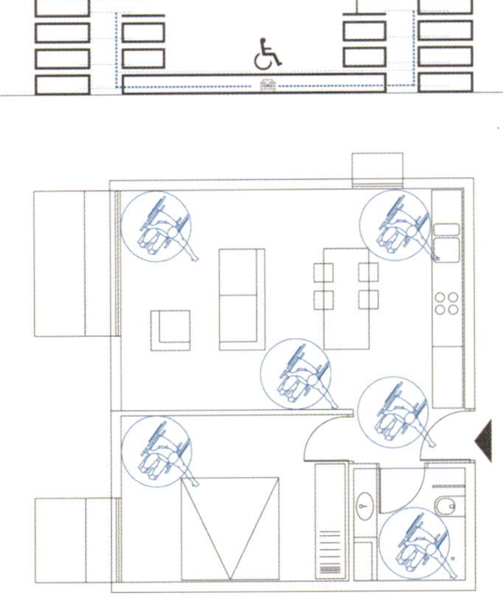

家具元素重复可预制

elements of furniture are...

...repetetive and can be prefabricated

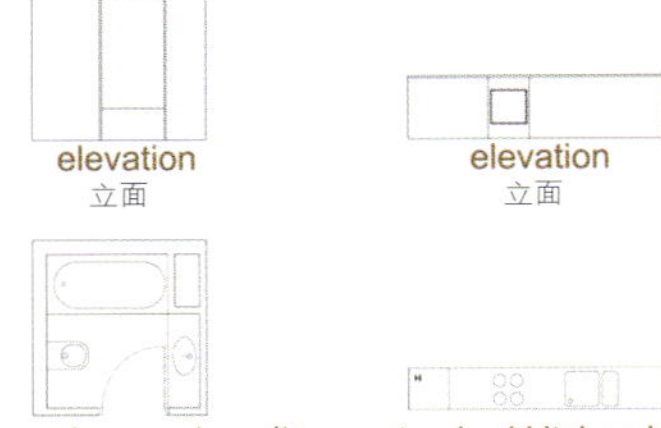

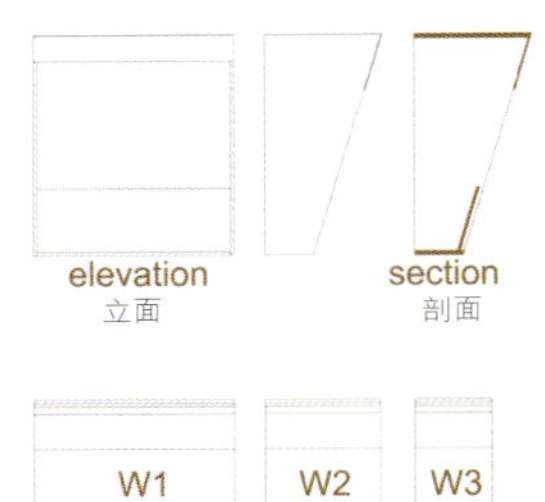

elevation
立面

elevation
立面

elevation
立面

section
剖面

W1 W2 W3

precast concrete units
预制混凝土单元

standard kitchen block
标准厨房构件

standard windows with shading
带遮阳设备的标准窗户

单元重复可预制

units are...

...repetetive and can be prefabricated

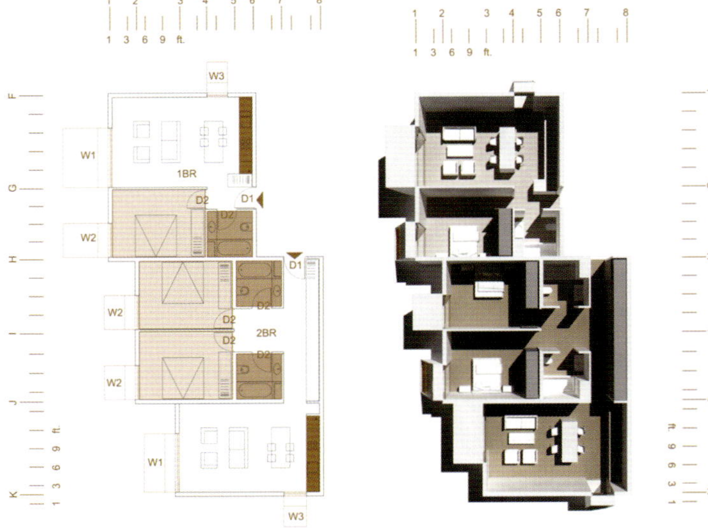

公寓数量与类型

No. and typology of apartments

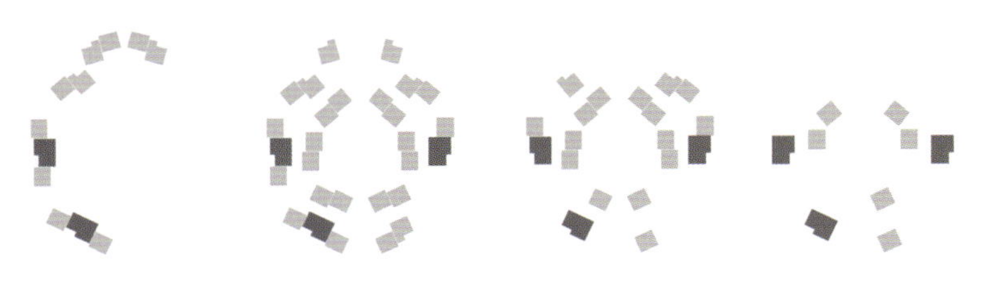

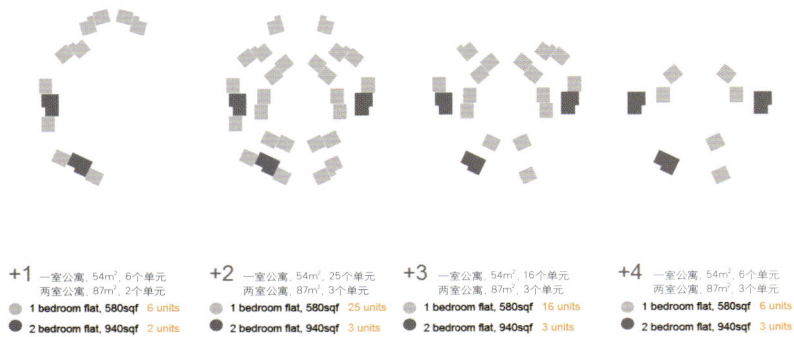

+1 一室公寓，54m²，6个单元
两室公寓，87m²，2个单元
- 1 bedroom flat, 580sqf 6 units
- 2 bedroom flat, 940sqf 2 units

+2 一室公寓，54m²，25个单元
两室公寓，87m²，3个单元
- 1 bedroom flat, 580sqf 25 units
- 2 bedroom flat, 940sqf 3 units

+3 一室公寓，54m²，16个单元
两室公寓，87m²，3个单元
- 1 bedroom flat, 580sqf 16 units
- 2 bedroom flat, 940sqf 3 units

+4 一室公寓，54m²，6个单元
两室公寓，87m²，3个单元
- 1 bedroom flat, 580sqf 6 units
- 2 bedroom flat, 940sqf 3 units

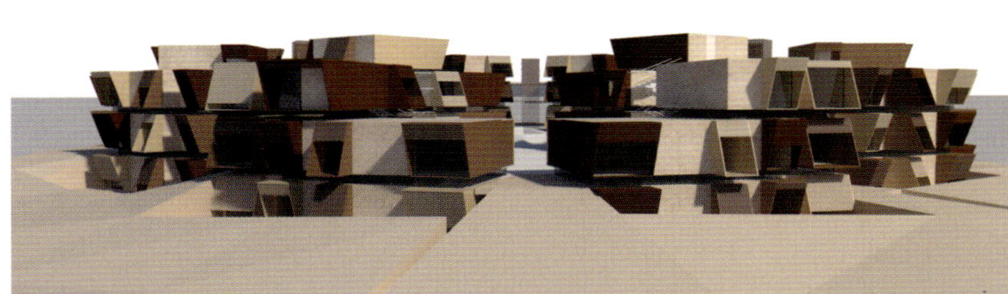

微观设计概念

design concept – micro scale:

1 遮阳构件

1. element of sun protection

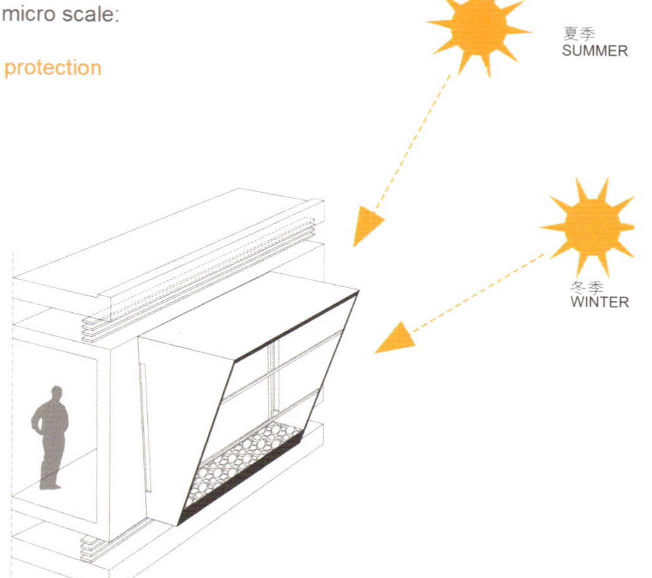

3 自然制冷与通风

3. natural cooling and ventilation

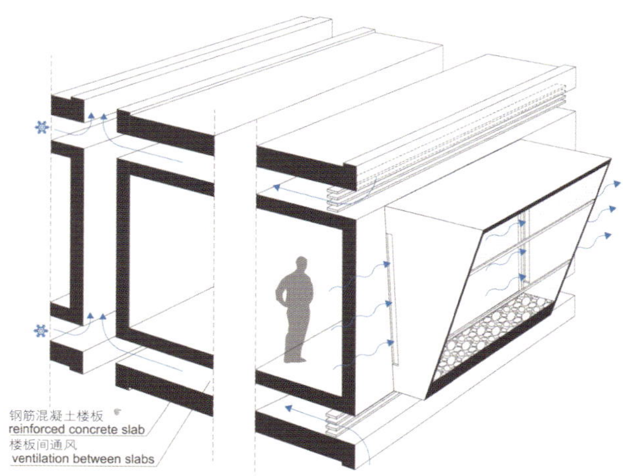

2 材料

2. materials

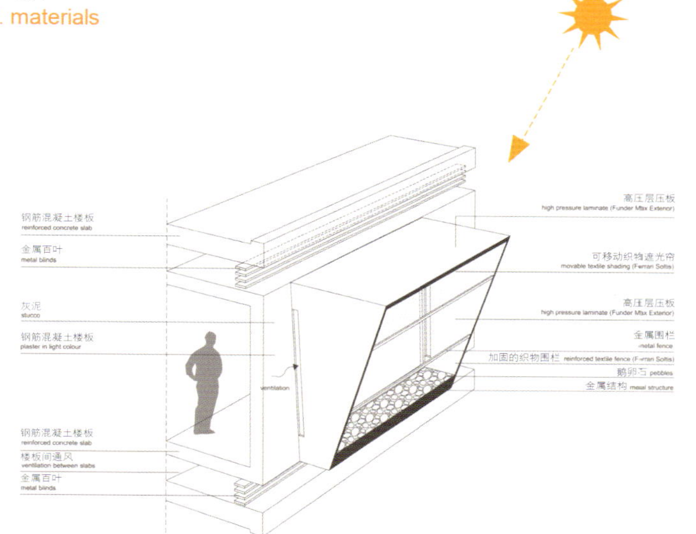

4 凉廊类型

4. loggia typologies

鄂尔多斯的(X)住宅
(X) HOUSE, ORDOS

建筑师：*multiplicities
项目团队：Daniel Holguin, Issei Suma, Perla Pequeño, Joanna Park Sohn, Christopher Chan, Nicole Rodríguez
甲方：Jiang Yuan Water Engineering Ltd.
工程师：OVE ARUP NY / Brian Markham, Roger Chang, Rajesh Shah
顾问：Methus Srisuchart + Tatchapon Lertwirojkul

(X)住宅中的“()”是建筑的外壳，“X”是内部的巨大结构，“()”是“X”从内部膨胀和收缩而形成的，创造了一种鲜明的空隙空间和许多高度不同的露台，位于其上可欣赏周围的河边美景。“()”由黑瓦制成，而“X”由白色的石英和石膏制成。白色的核心结构将光线和空气带到住宅内部，就像人的肺部一样，使黑色的建筑外皮保持旺盛的活力。黑色的外皮吸收热量，给白色“肺部”带来温暖与保护，也可举办多种活动，并在必要的地方设计了不同厚度的墙体，制造缓冲地带。

The "()" is imploded by the "X"that expands and contracts from within, creating a dramatic void and multiple terraces at different levels from which to view the surrounding river landscape. The "()" is made of black brick while the "X" is made of white quartz white plaster. The white lung takes and bounces light and air inside the house to keep the black skin healthy. The black skin absorbs heat and gives protection and warmth to the lung as well as varied activities, providing different programmatic thicknesses creating buffer zones wherever it is necessary.

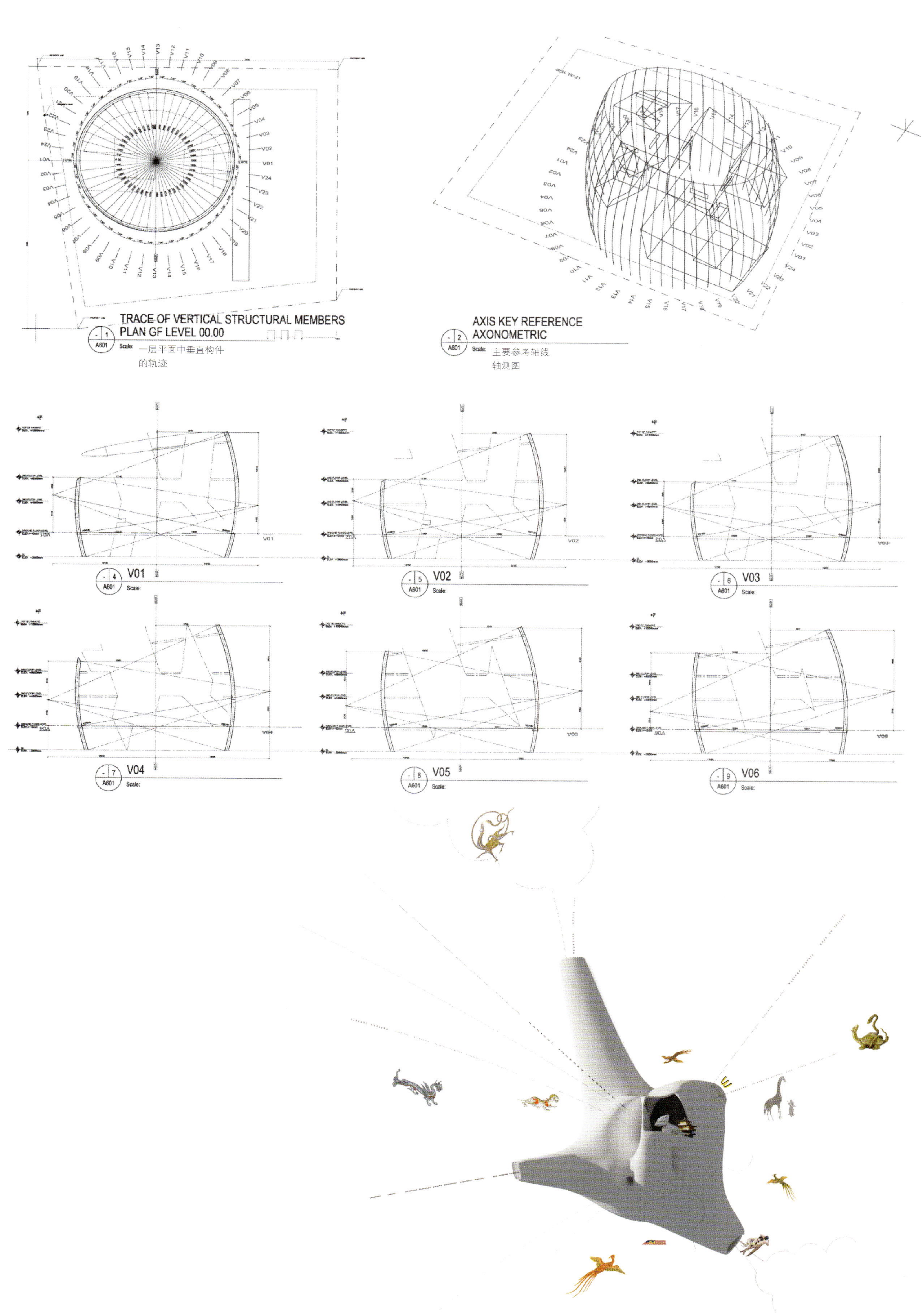
TRACE OF VERTICAL STRUCTURAL MEMBERS
PLAN GF LEVEL 00.00
1
A601
Scale:
一层平面中垂直构件
的轨迹
AXIS KEY REFERENCE
AXONOMETRIC
2
A601
Scale:
主要参考轴线
轴测图
4
A601
V01
Scale:
5
A601
V02
Scale:
6
A601
V03
Scale:
7
A601
V04
Scale:
8
A601
V05
Scale:
9
A601
V06
Scale:

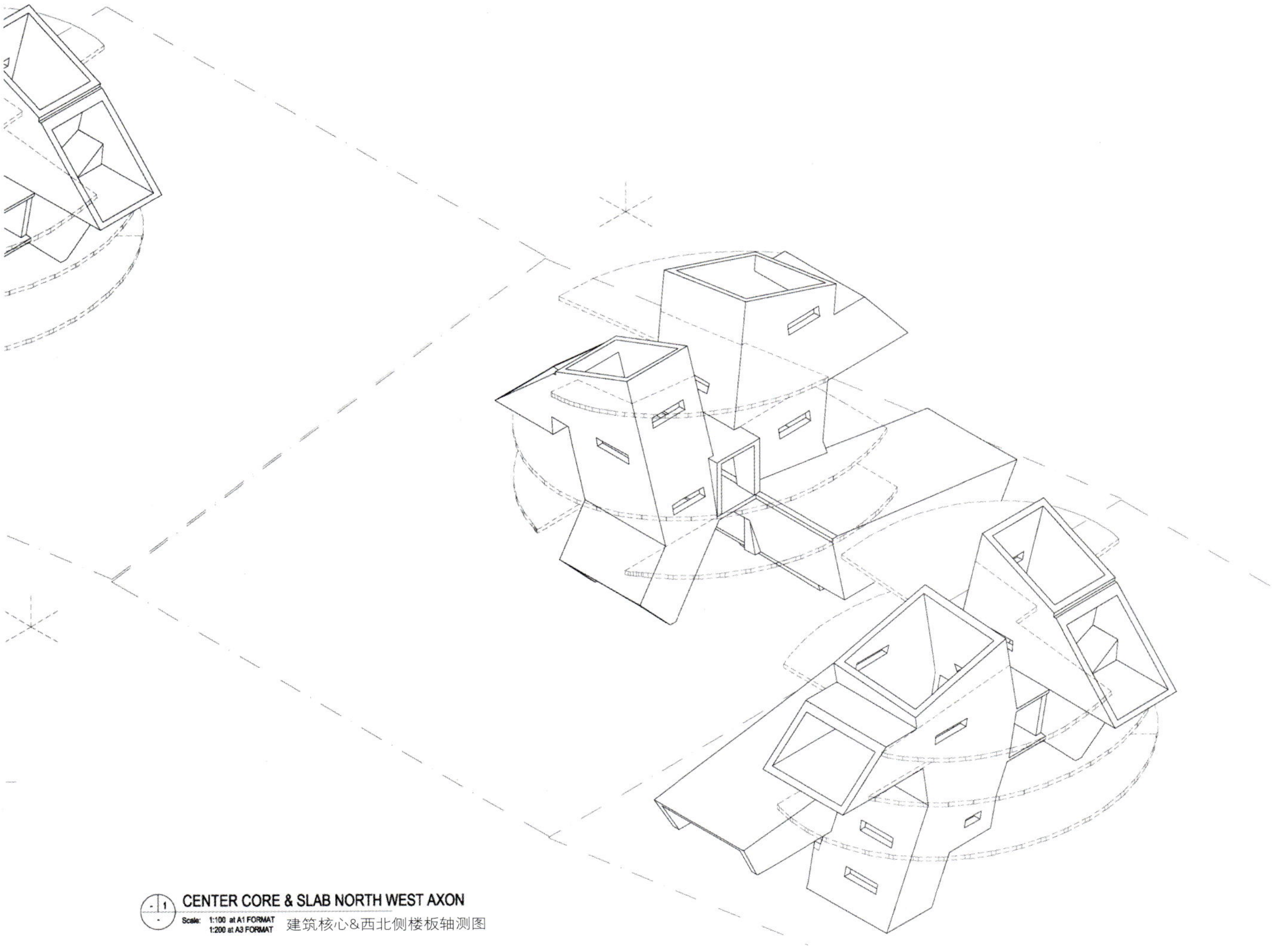

CENTER CORE & SLAB NORTH WEST AXON

Scale: 1:100 at A1 FORMAT
1:200 at A3 FORMAT

建筑核心&西北侧楼板轴测图

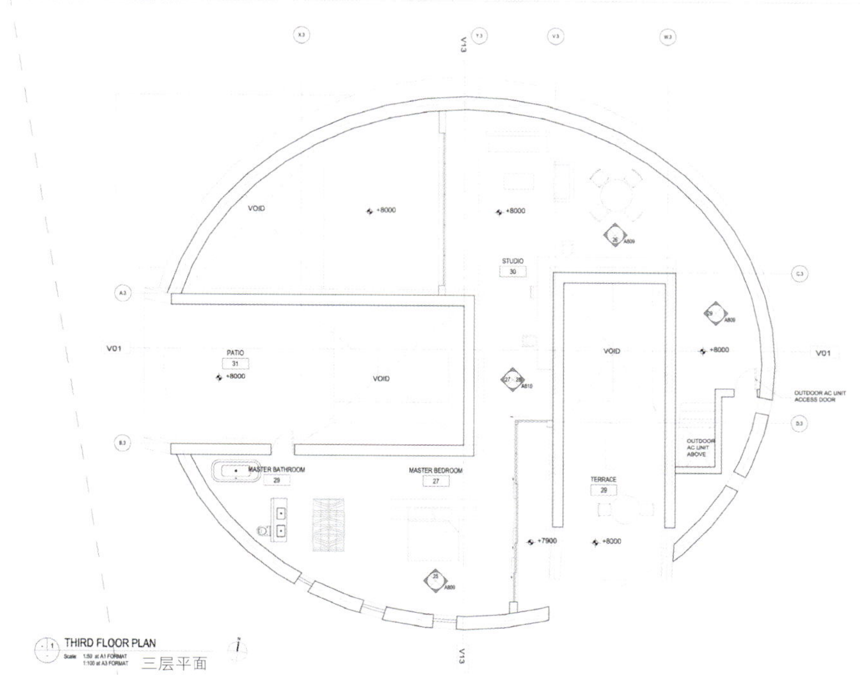
VOID
STUDIO
PATIO
MASTER BATHROOM
MASTER BEDROOM
TERRACE
OUTDOOR AC UNIT ACCESS DOOR
THIRD FLOOR PLAN
三层平面

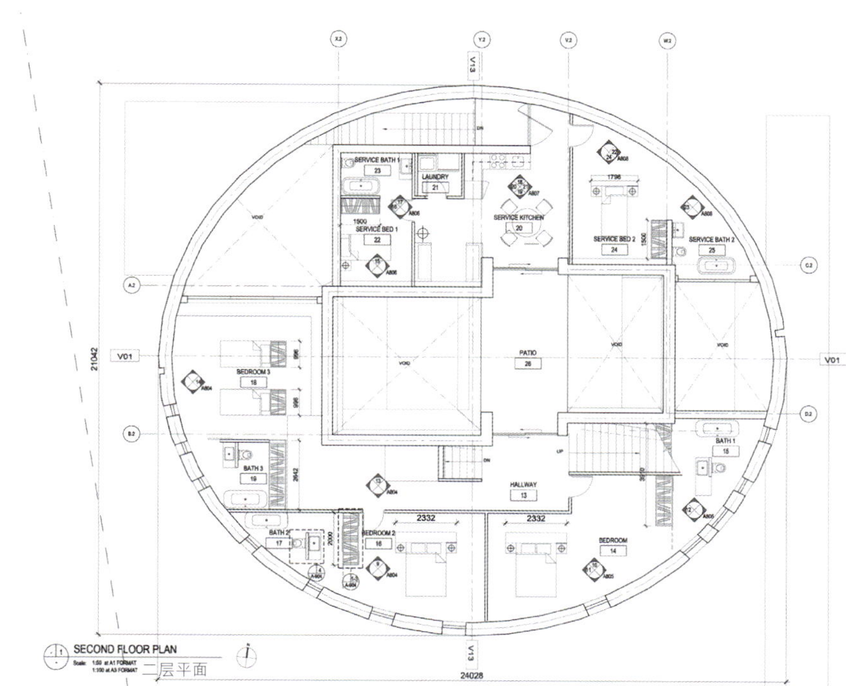
SERVICE BATH 1
LAUNDRY
SERVICE KITCHEN
SERVICE BED 1
SERVICE BED 2
SERVICE BATH 2
BEDROOM 3
PATIO
BATH 3
HALLWAY
BEDROOM 2
BEDROOM
BATH 1
21042
24028
SECOND FLOOR PLAN
二层平面

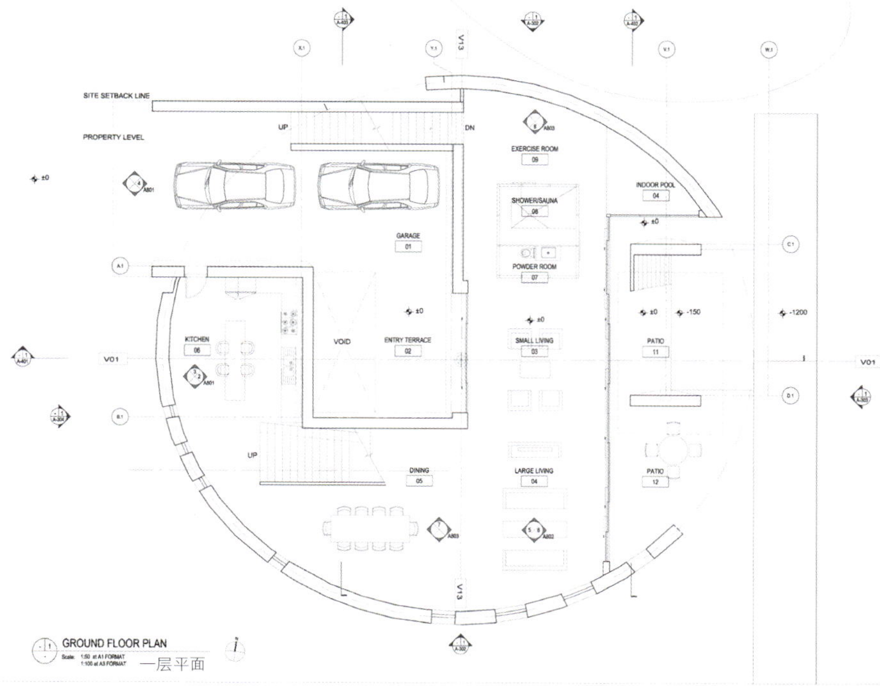
SITE SETBACK LINE
PROPERTY LEVEL
GARAGE
EXERCISE ROOM
SHOWER/SAUNA
INDOOR POOL
POWDER ROOM
KITCHEN
VOID
ENTRY TERRACE
SMALL LIVING
PATIO
DINING
LARGE LIVING
GROUND FLOOR PLAN
一层平面

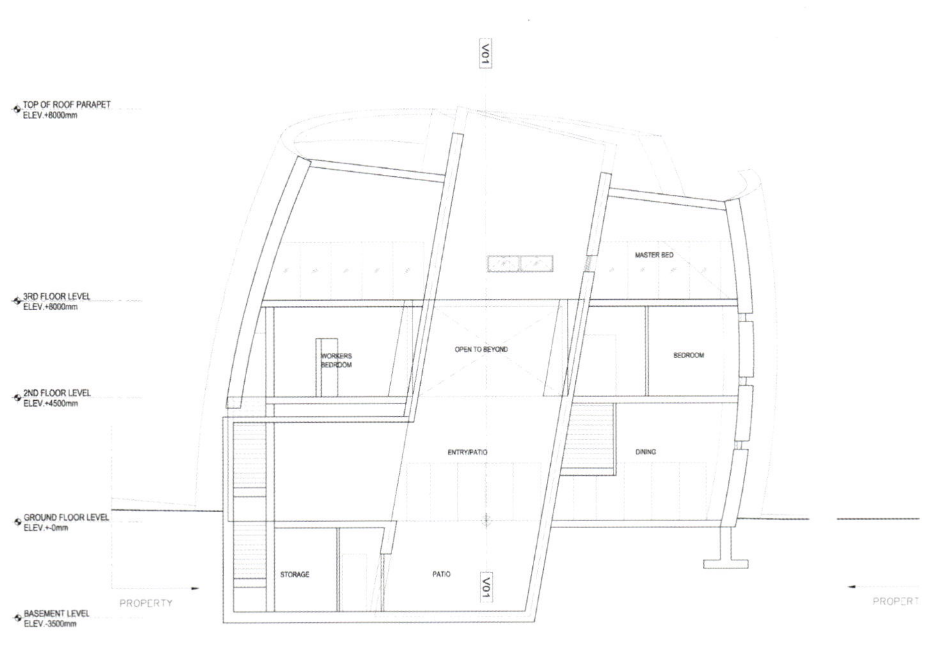
V01
TOP OF ROOF PARAPET
ELEV.+8000mm
3RD FLOOR LEVEL
ELEV.+8000mm
2ND FLOOR LEVEL
ELEV.+4500mm
GROUND FLOOR LEVEL
ELEV.+-0mm
BASEMENT LEVEL
ELEV.-3500mm
MASTER BED
WORKERS
BEDROOM
OPEN TO BEYOND
BEDROOM
ENTRY/PATIO
DINING
STORAGE
PATIO
V01
PROPERTY
SECTION 剖面1
Scale 1:50 at A1 FORMAT
1:100 at A3 FORMAT

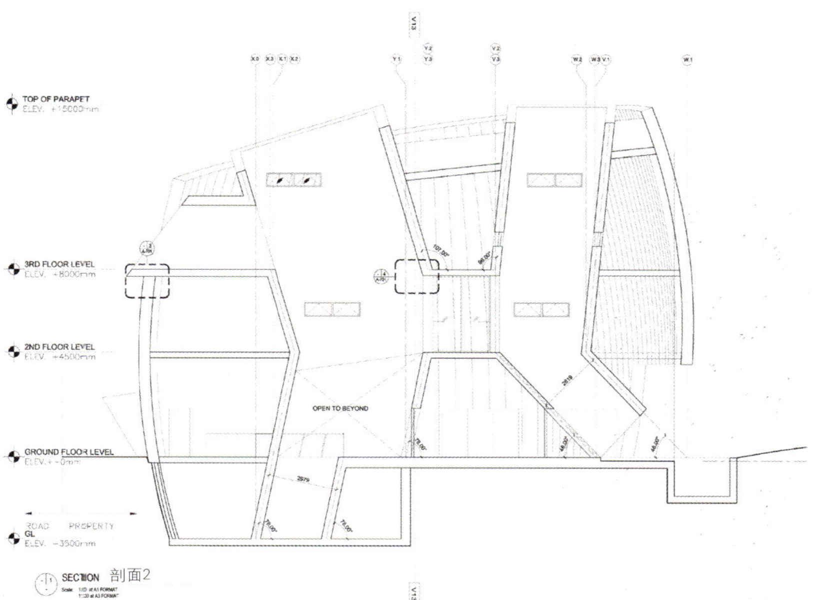
TOP OF PARAPET
ELEV. +15000mm
3RD FLOOR LEVEL
ELEV. +8000mm
2ND FLOOR LEVEL
ELEV. +4500mm
GROUND FLOOR LEVEL
ELEV. +-0mm
ROAD
PROPERTY
GL
ELEV. -3500mm
OPEN TO BEYOND
V13
SECTION 剖面2

荷兰布卢门达尔别墅
VILLA BLOEMENDAAL, THE NETHERLANDS

设计师：Hofman Dujardin Architects, Amsterdam

效果图：A2 Studio, Rotterdam

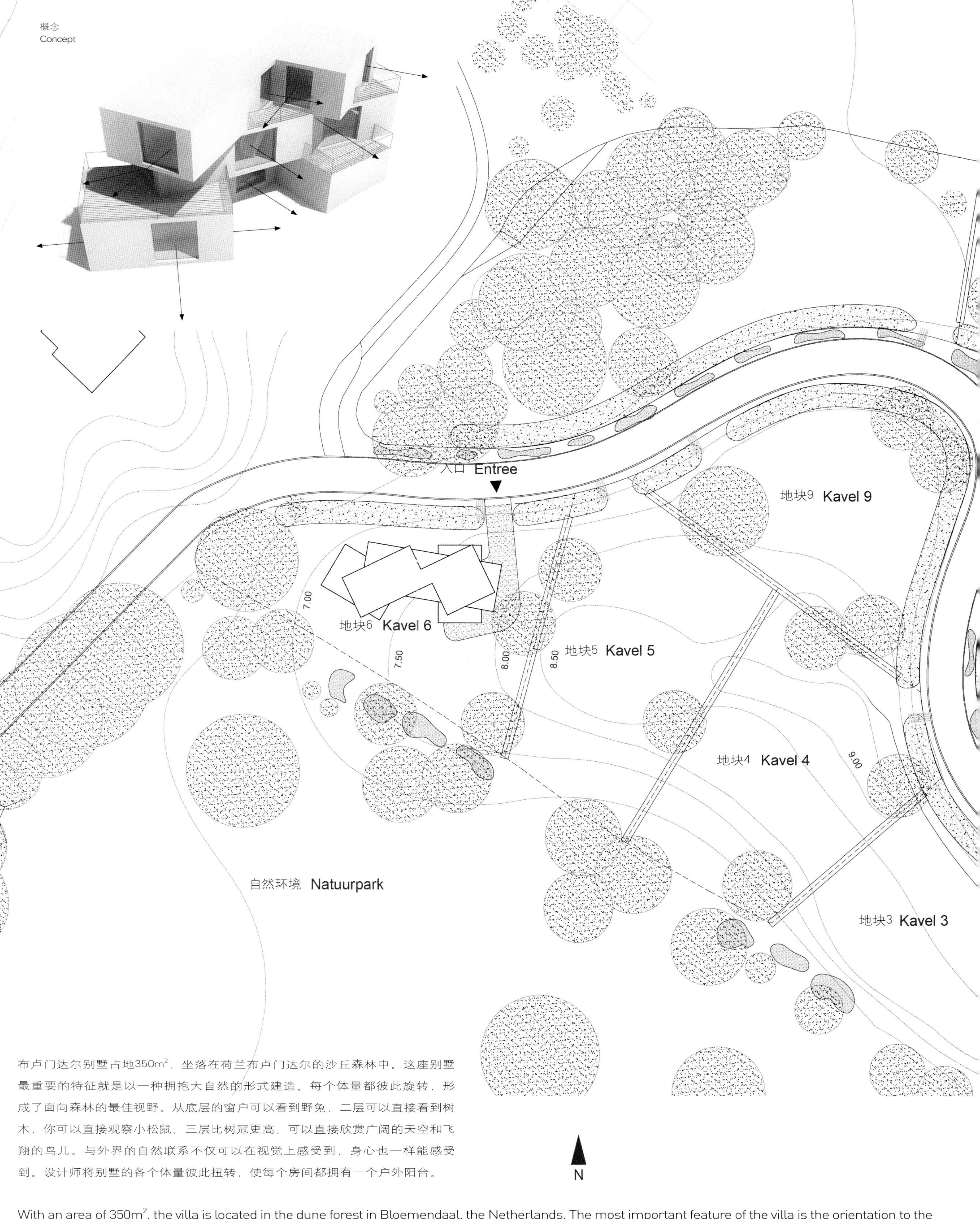

布卢门达尔别墅占地350m²，坐落在荷兰布卢门达尔的沙丘森林中。这座别墅最重要的特征就是以一种拥抱大自然的形式建造。每个体量都彼此旋转，形成了面向森林的最佳视野。从底层的窗户可以看到野兔，二层可以直接看到树木，你可以直接观察小松鼠，三层比树冠更高，可以直接欣赏广阔的天空和飞翔的鸟儿。与外界的自然联系不仅可以在视觉上感受到，身心也一样能感受到。设计师将别墅的各个体量彼此扭转，使每个房间都拥有一个户外阳台。

With an area of 350m², the villa is located in the dune forest in Bloemendaal, the Netherlands. The most important feature of the villa is the orientation to the surrounding nature. Volumes are rotated towards each other creating beautiful views into the forest. The windows on the ground floor create the relation with the rabbit, on the first floor the views are straight into the trees where you can admire the squirrel, the third level rises from the tree level and you can enjoy the open sky with the birds. The relation with the outdoor space is not only visual but also physical. The architect shifting the building volumes from each other, each room has an outdoor terrace.

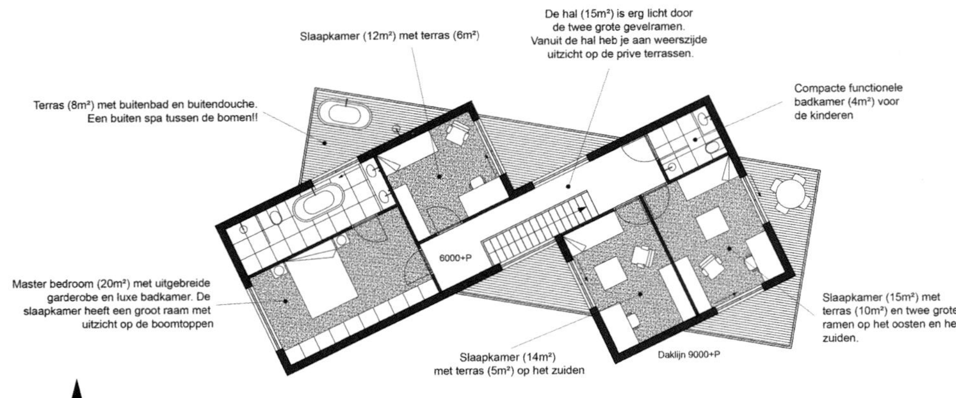

三层
Second floor

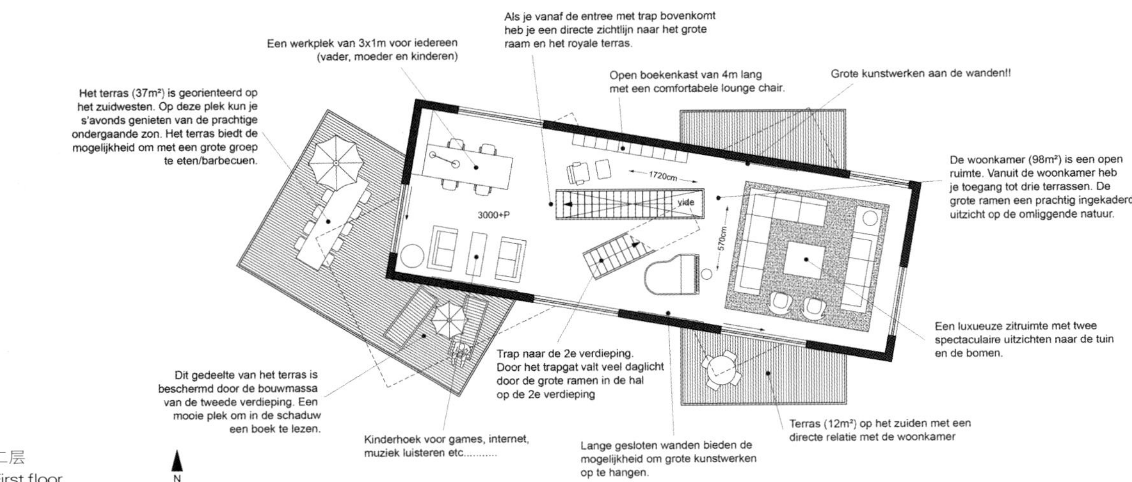

二层
First floor

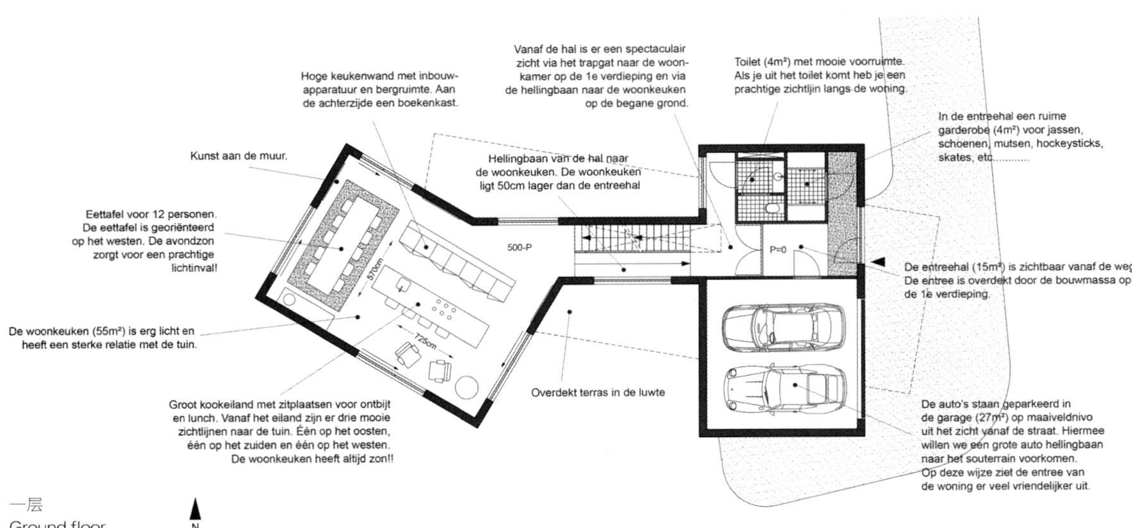

Vanaf de hal is er een spectaculair zicht via het trapgat naar de woonkamer op de 1e verdieping en via de hellingbaan naar de woonkeuken op de begane grond.
Hoge keukenwand met inbouwapparatuur en bergruimte. Aan de achterzijde een boekenkast.
Toilet (4m²) met mooie voorruimte. Als je uit het toilet komt heb je een prachtige zichtlijn langs de woning.
In de entreehal een ruime garderobe (4m²) voor jassen, schoenen, mutsen, hockeysticks, skates, etc..........
Kunst aan de muur.
Hellingbaan van de hal naar de woonkeuken. De woonkeuken ligt 50cm lager dan de entreehal
Eettafel voor 12 personen. De eettafel is georiënteerd op het westen. De avondzon zorgt voor een prachtige lichtinval!
500-P
570cm
P=0
De entreehal (15m²) is zichtbaar vanaf de weg. De entree is overdekt door de bouwmassa op de 1e verdieping.
726cm
De woonkeuken (55m²) is erg licht en heeft een sterke relatie met de tuin.
Overdekt terras in de luwte
Groot kookeiland met zitplaatsen voor ontbijt en lunch. Vanaf het eiland zijn er drie mooie zichtlijnen naar de tuin. Eén op het oosten, één op het zuiden en één op het westen. De woonkeuken heeft altijd zon!!
De auto's staan geparkeerd in de garage (27m²) op maaiveldnivo uit het zicht vanaf de straat. Hiermee willen we een grote auto hellingbaan naar het souterrain voorkomen. Op deze wijze ziet de entree van de woning er veel vriendelijker uit.
N

一层
Ground floor

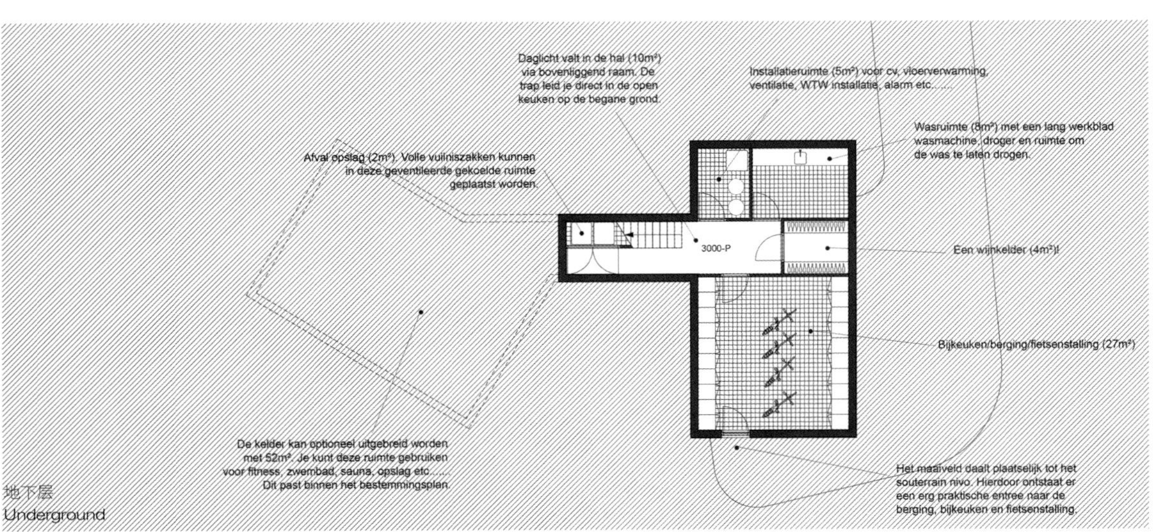

地下层
Underground

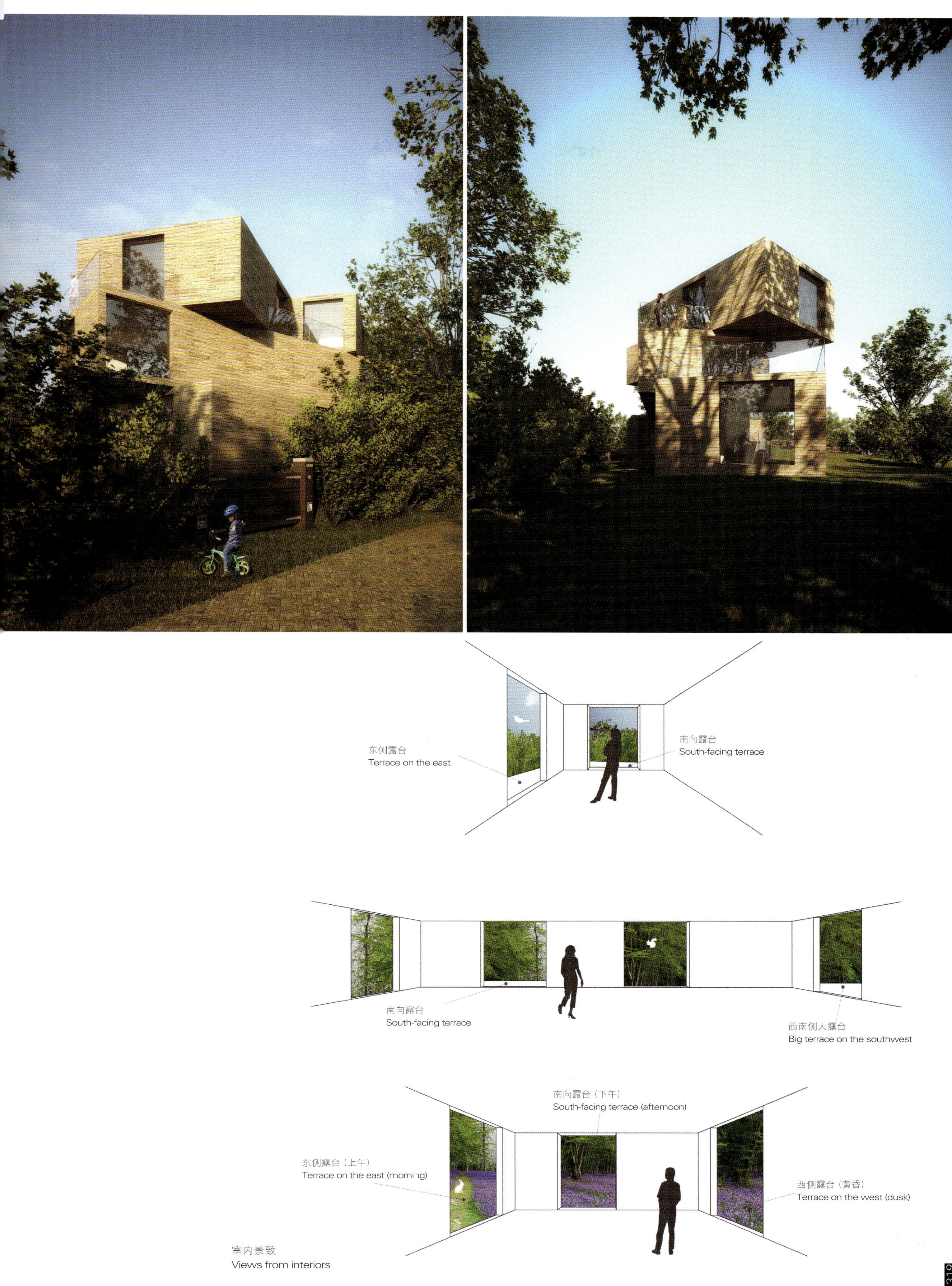

室内景致
Views from interiors

柏林约翰内斯大街3号公寓
JOH 3 – APARTMENT HOUSE JOHANNISSTRASSE 3, BERLIN

建筑师：J. MAYER H. Architects
业主：Euroboden Berlin GmbH, München
项目团队：Jürgen Mayer H., Hans Schneider, Wilko Hoffmann, Filipa Frois Almeida
竞赛团队：Jürgen Mayer H., Thorsten Blatter, Marcus Blum
现场建筑师：Architekturbuero Wiesler, Stuttgart with Thomas Quinten Projektmanagement, Berlin
结构工程师：EiSat GmbH, Berlin
建筑设备：Ingenieurgesellschaft Striewisch mbH
建模：Werk5, Berlin
模型摄影师：Ludger Paffrath, Berlin
效果图：Buenck + Fehse, Berlin

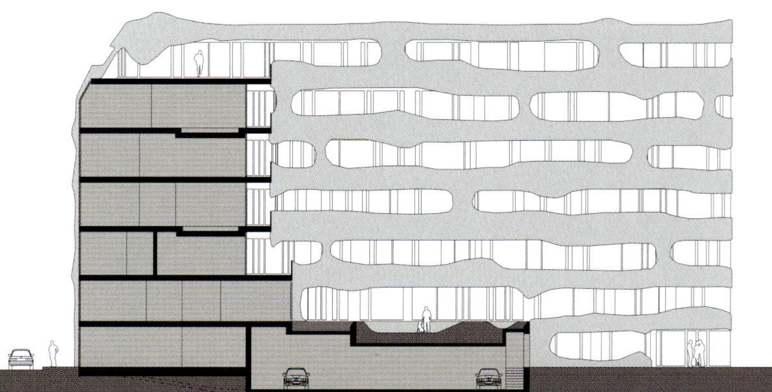

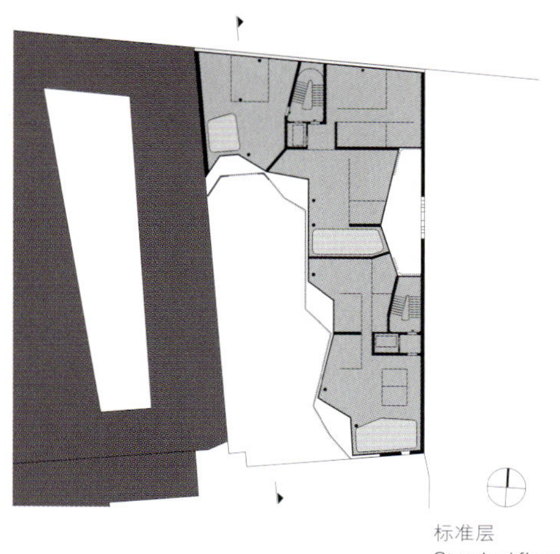

标准层
Standard floor

二层
1st floor

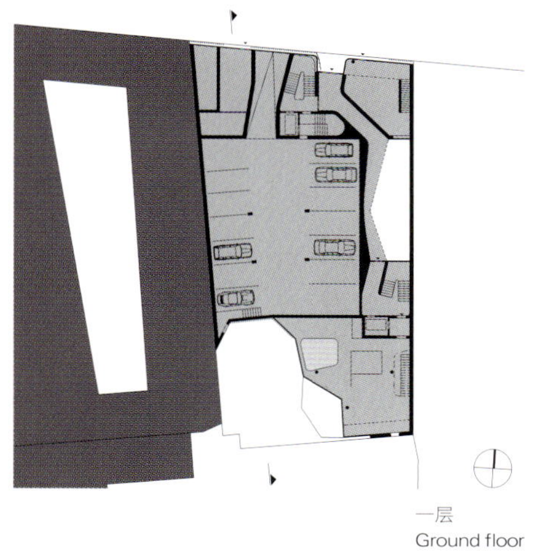

一层
Ground floor

房地产发展集团Euroboden在柏林商业区米特的约翰内斯大街设计了一栋独特的公寓。悬挂板条组成的立面设计显得纹理分明，借鉴了城市中景观的概念，鲜明的特色体现在层层递进的庭院花园和建筑的轮廓和布局中。临街一层还设计有诸多商业空间。这些宽敞的公寓面向西南方，正对着一座精心设计的宁静的花园。公寓由带私人花园的联排住宅、经典公寓以及能够俯瞰弗里德里希古城景色的阁楼组成。总体设计理念着眼于高端设计，用独到的设计眼光为业主提供了一个独特的空间及居住体验。

总平面图
Site plan

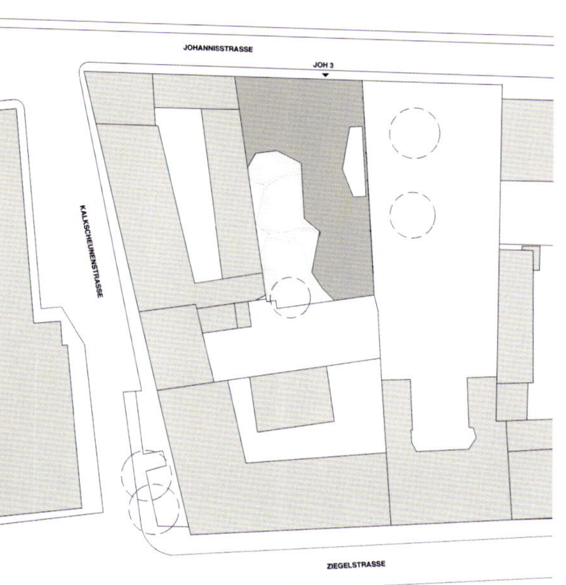

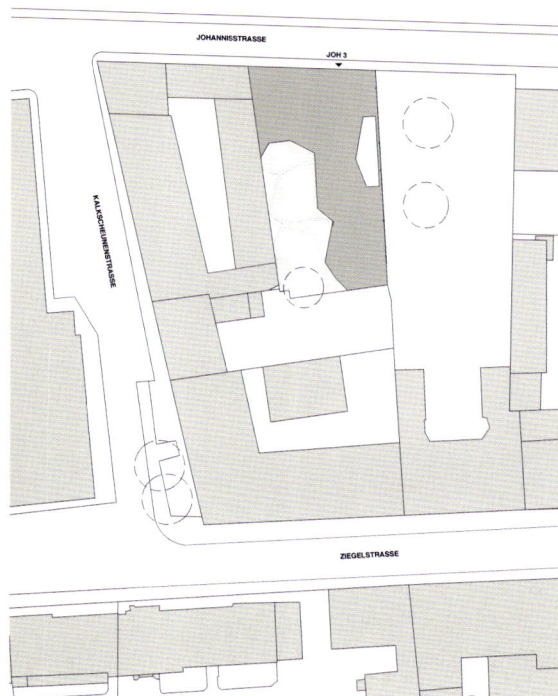

Property development group Euroboden is building a unique apartment house at Johannisstraße in Mitte, Berlin's downtown district. The sculptural design of the suspended slat facade draws on the notion of landscape in the city, a quality visible in the graduated courtyard garden and the building's silhouette and layout. Plans for the ground floor facing the street also include a number of commercial spaces. The generously sized apartments will face south-west, opening themselves to a view of the calm, carefully designed courtyard garden. Condominiums are organized into townhouses with private gardens, classic apartments or penthouses with a spectacular view of the old Friedrichstadt. The integrated design concept promises a unique spatial and living experience with an eye to high design.

索仑霍夫办公与住宅大楼
SONNENHOF BUILDING

建筑师：J. MAYER H. Architects, Berlin
业主：Wohnungsgenossenschaft "Carl Zeiss" eG, Jena
项目团队：Jürgen Mayer H., Jan-Christoph Stockebrand, Christoph Emenlauer, Jens Seiffert, Max Reinhardt, Christian Pälmke
项目与施工管理：Kappes Partner IPG, Berlin
结构工程/结构防火：Ingenieurbüro Dr. Krämer GmbH, Weimar
建筑设备：Scholze Ingenieurgesellschaft mbH, Dresden
建筑物理顾问：Ingenieurbüro Santer Bauphysik, Duisburg
通风顾问：Ingenieurbüro Rau, Heilbronn
交通运输工程顾问：GRI Ingenieure,Berlin
外部设施顾问：Ingenieurbüro Abraham, Berlin
照明工程：Lichttransfer, Berlin
模型制作：Werk 5, Berlin

索仑霍夫是由四栋商住两用大楼组成的新建筑。它位于德国的历史中心耶拿，建在一个由许多小地块组成的地块之上，其独立的结构使人们能够自由地出入，它们安置在地块的外缘，由此形成了一块小规模的室外空间，与中世纪的城市结构风格不谋而合。室外设施延续了建筑的整体设计理念，该项目囊括商业、住宅以及办公设计，实现了小型地块的灵活运作模式，也与周边环境完美融合。

Sonnenhof consists of four new buildings with office and residential spaces. Located on a consolidated number of smaller lots in the historical center of Jena, Germany, the separate structures allow for free access through the grounds. Their placement on the outer edges of the plot defines a small-scale outdoor space congruent with the medieval city structure. Its outdoor facilities continue the building's overall design concept. The planned incorporation of commerce, residence and office enables a small-sectioned and flexible pattern of use that also integrates itself conceptually into the surroundings.

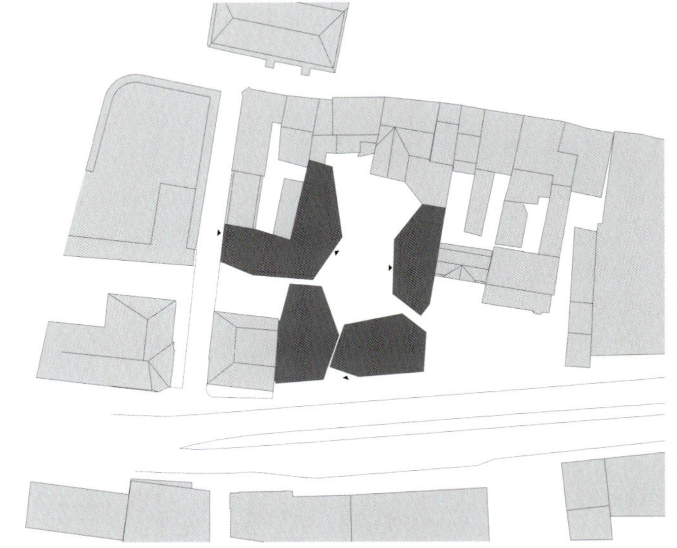

一层
Ground floor

二层
1st floor

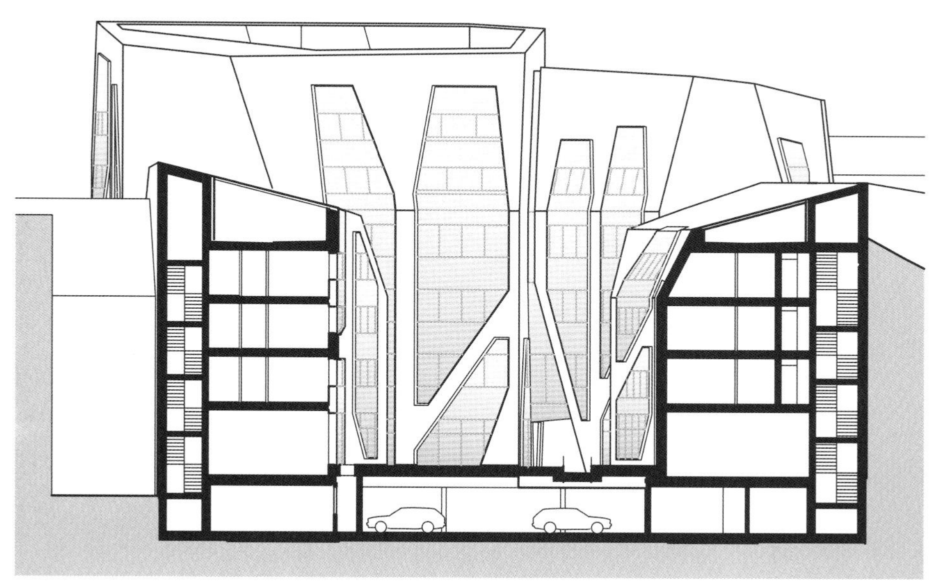
A-A剖面
Section A-A

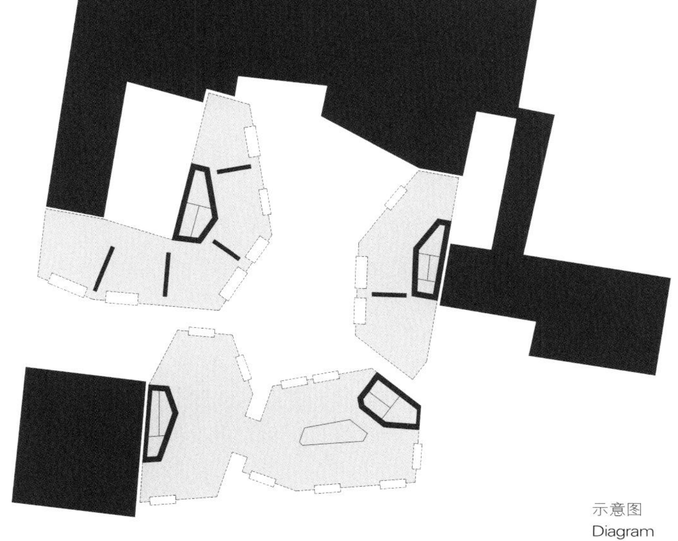
示意图
Diagram

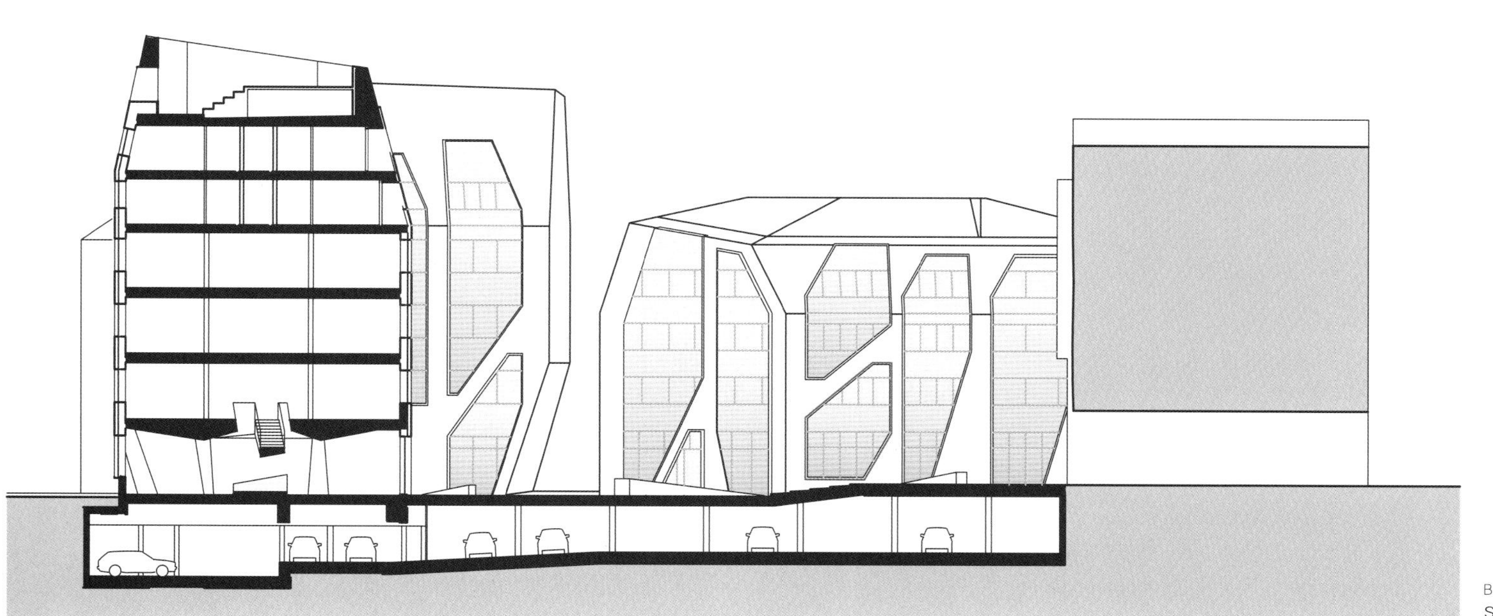
B-B剖面
Section B-B

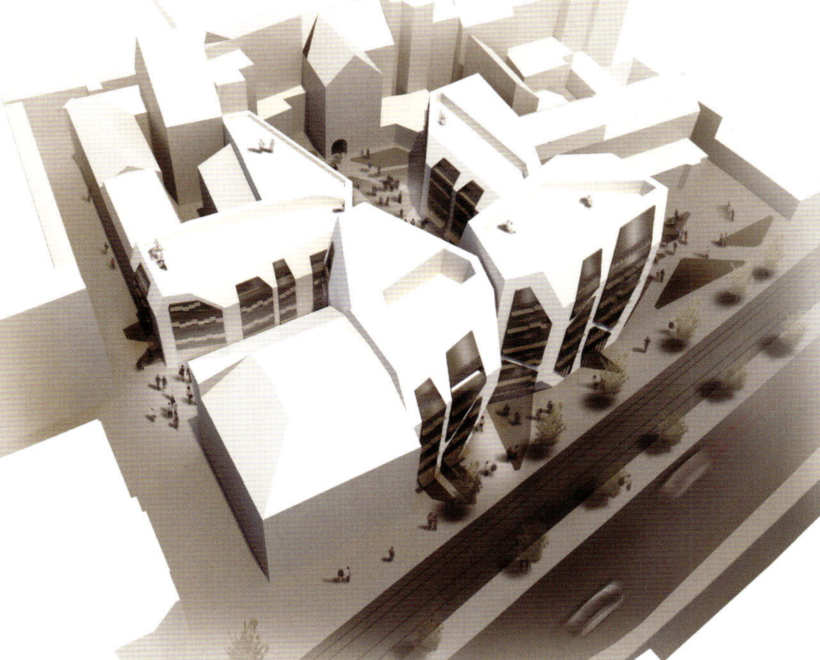

埃科住宅大厦
EKO

建筑师：Piotrowski Sylvain Caroline Diraison
设计机构：10RAISONS（法国）

埃科住宅大厦的设计灵感来自于宏伟的埃菲尔铁塔，它是一栋未来主义风格的住宅大楼，位于迪拜市中心的Zaabeel公园。大楼高24层，安装有8500m²的纳米太阳能板，栽种了350棵棕榈树，其中有酒吧、图书馆及一间巨大的展览厅；这些居所风格并不张扬，其中住着富有的阿拉伯人及家眷，这让所有热爱埃菲尔铁塔的人们都忍不住感到欣羡不已。

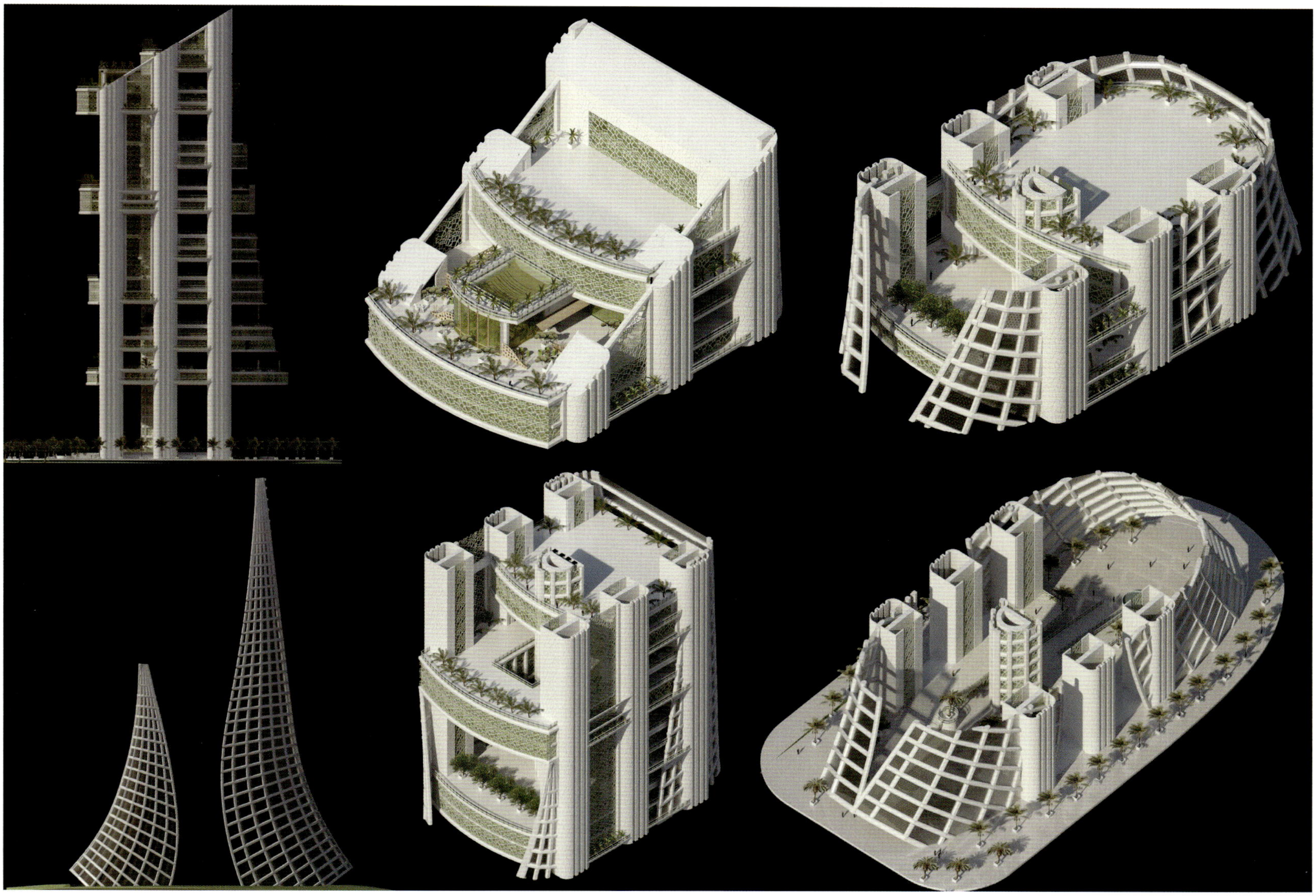

Drawing inspiration from the magnificent Eiffel Tower, the Eko is a futuristic housing high-rise for Zaabeel Park, City Centre in Dubai. With stats like 24 floors, 8500 Square meters of Nano Solar, 350 palms, a bar, a library and a big exhibition hall; this humble living quarters of rich Arabs and their Harem will be the envy of every Eiffel-loving Parisian!

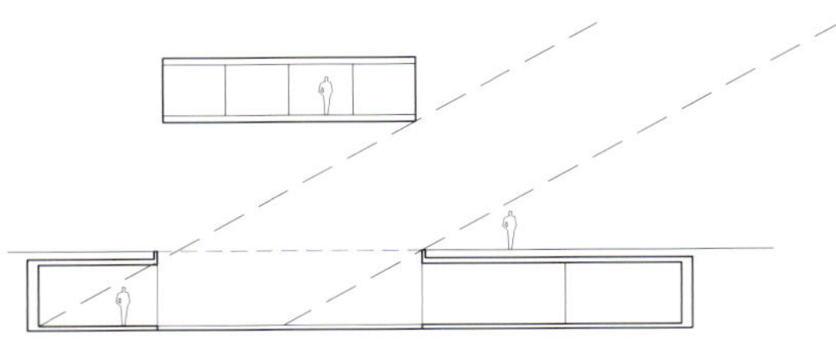

Winter sun
冬天的太阳

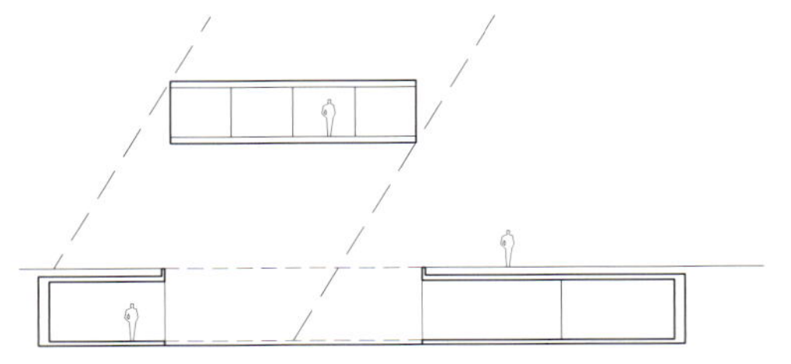

Summer sun
夏天的太阳

鄂尔多斯的树屋
TREE HOUSE, ORDOS

建筑师：FRENTE / Juan Pablo Maza
项目团队：Juan Pablo Maza, Xavier S. Valladares, Juan Carlos Vidals, Efraín Ovando, Andrés Ortega, Adrián De Lucio, Miguel Ocampo, Onnis Luque, Carlos Sanchez, Ernesto Gadea

A volume of land is extracted.
掘土

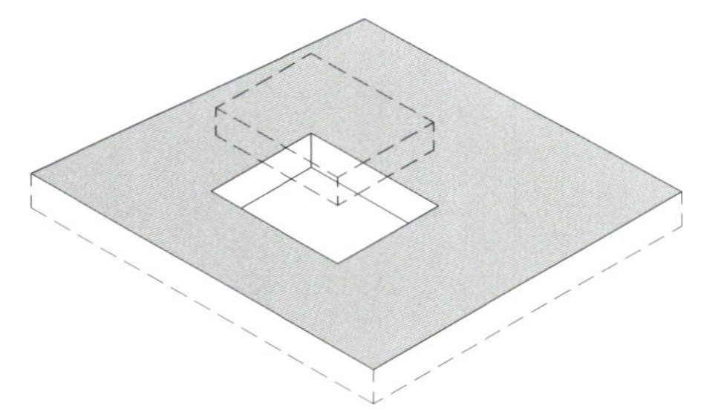

A buried villa which lives to a patio is created.
造成一个地下别墅，住在一个院子里。

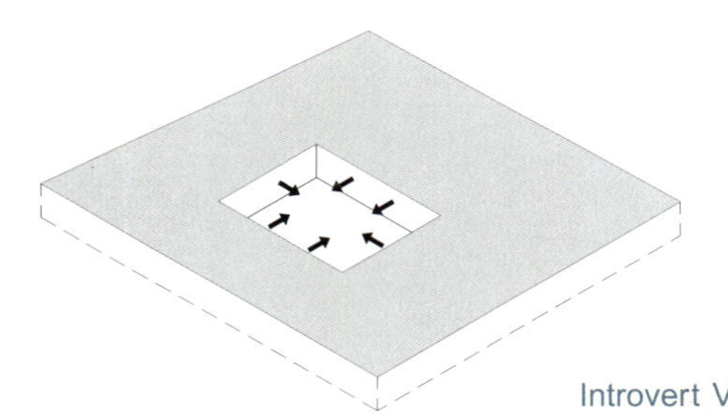

Introvert Villa
内向型别墅

A floating villa which gains the views is created.
形成一个飞起别墅，视野开阔。

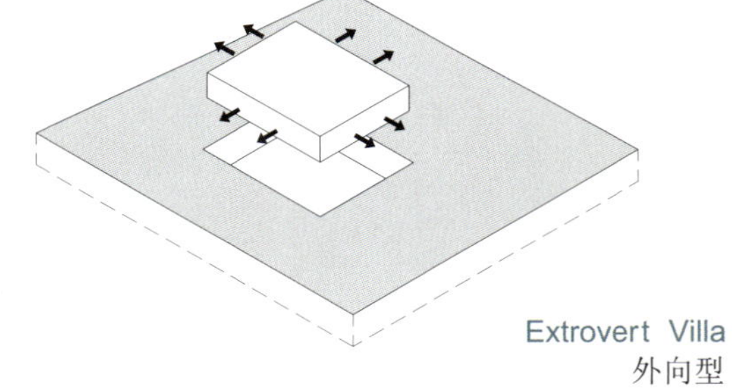

Extrovert Villa
外向型

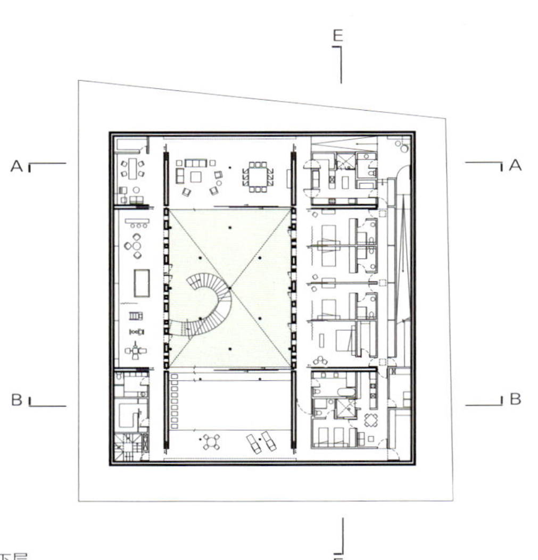

地下层
Lower level

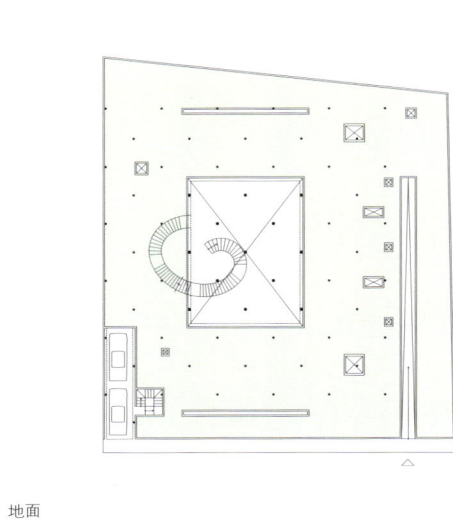

地面
Ground level

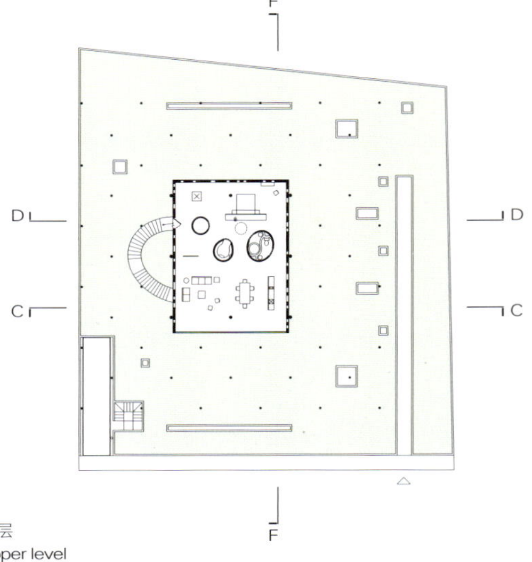

上层
Upper level

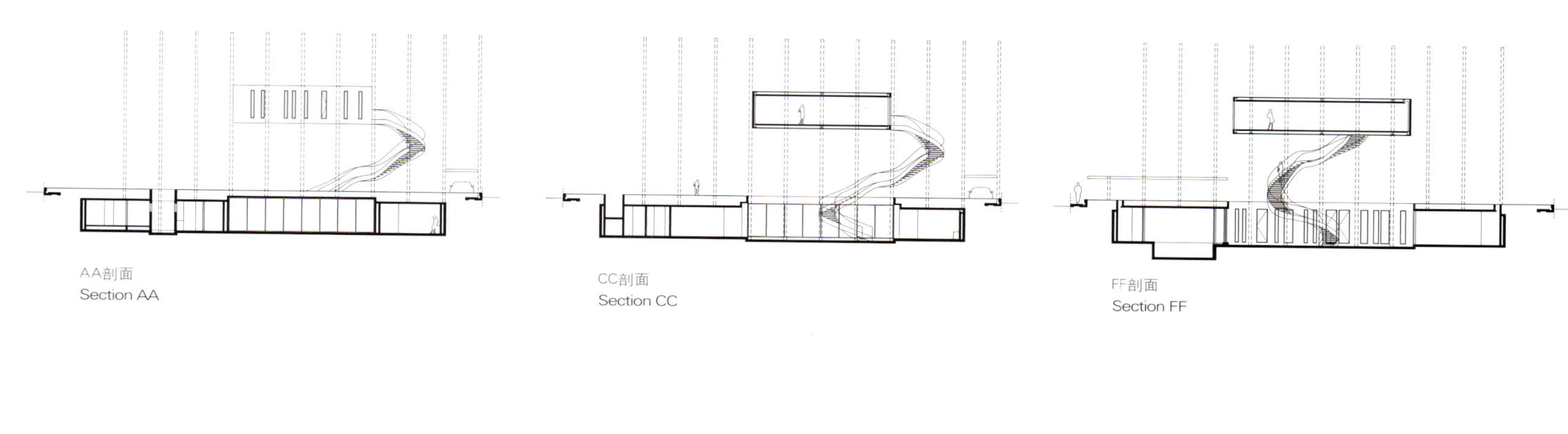

AA剖面
Section AA

CC剖面
Section CC

FF剖面
Section FF

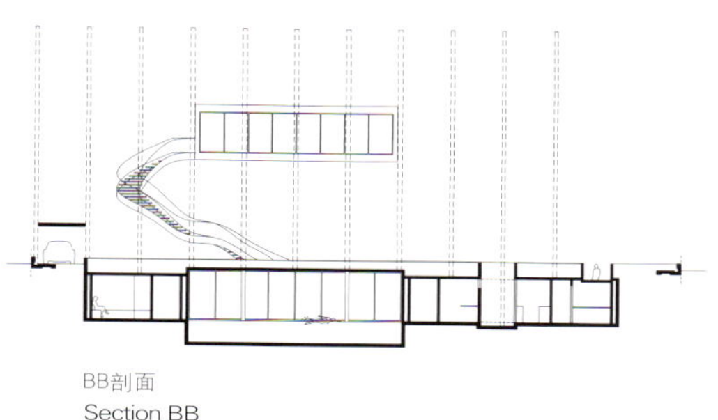

BB剖面
Section BB

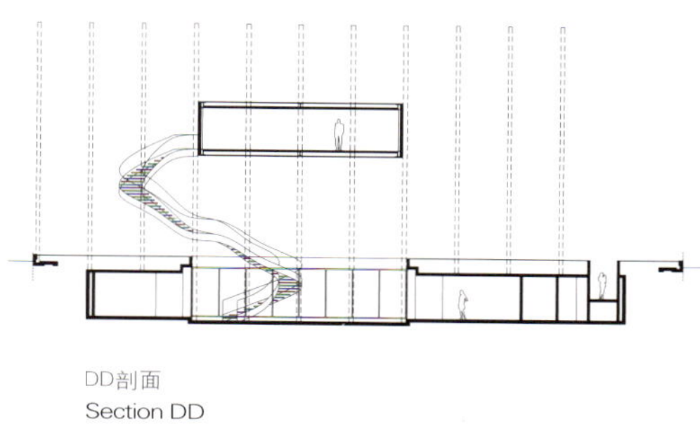

DD剖面
Section DD

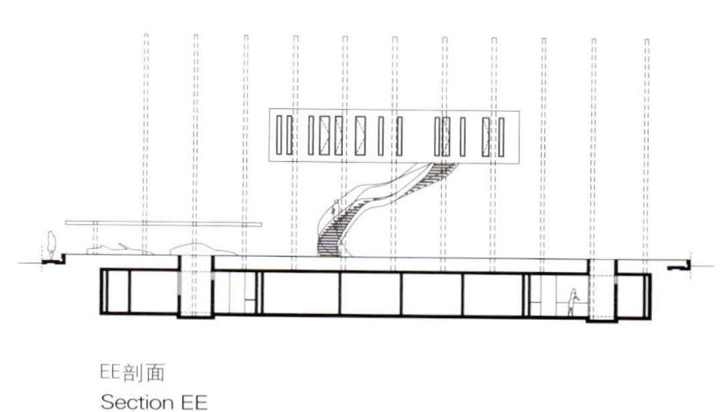

EE剖面
Section EE

该项目位于内蒙古沙漠，当前的背景环境促成了设计向内型别墅的决定。由于内蒙古冬夏的天气极端恶劣，所以选择将房屋建在地下，充分利用丰富的地热资源，有效地缓解了酷暑与严冬。另一方面，内蒙古的春天和秋天气候适宜，考虑到这种双重特点，设计师将别墅的一部分露在外面，形成了一部分完全外向型的结构。所以，这栋别墅是兼具内向型与外向型的双重功能住宅。

The project is situated in the Mongolian desert, therefore the immediate context drove the decision of designing an introvert villa. As a result of the extreme weather, it has been chosen to bury the house taking advantage of the generosity of underground temperatures and therefore neutralizing the harsh winter and summer weather conditions. On the other hand, realizing the good weather during spring and autumn times, the villa responds to this duality by leaving a part of the construction completely exposed, and therefore completely extrovert. This way, the villa celebrates the duality of an introvert-extrovert house.

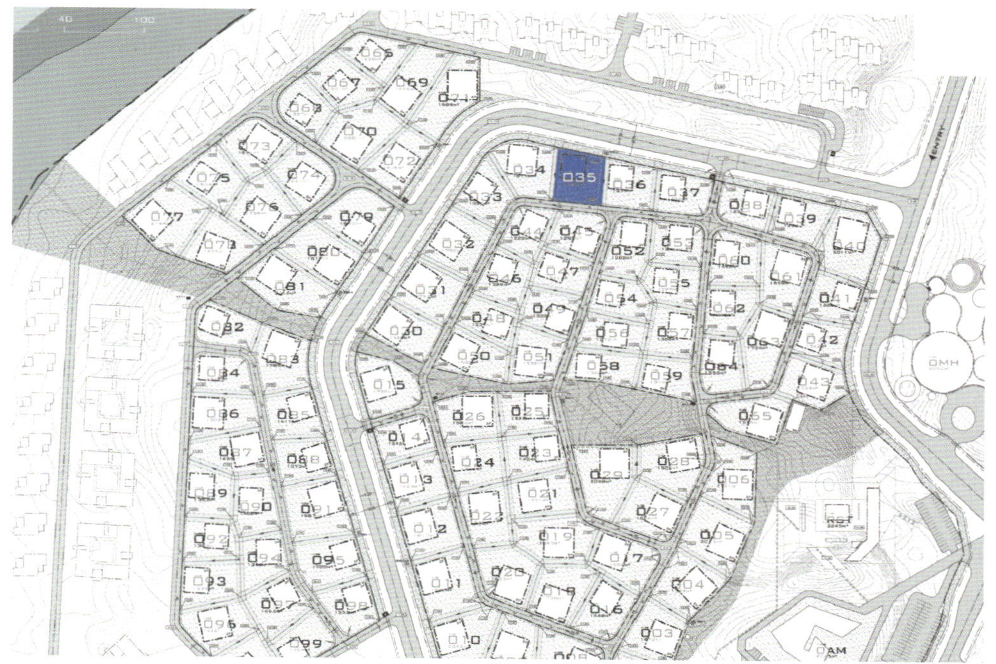

鄂尔多斯的"给我庇护"住所
GIMME SHELTER, ORDOS

建筑师：Rojkind Arquitectos
主建筑师：Michel Rojkind
项目负责人：Agustín Pereyra
项目团队：Juan Carlos Vidals (3D massing), Alejandro Biguria, Moritz Melchert, Mónica Orozco, Phillip Jung, José Moreno, Laura Rodriguez , Roberto Gil Will, Tere Levy, Alan Rahmane
结构工程师：Juan Felipe Heredia
室内设计：Rojkind Arquitectos
景观设计：Rojkind Arquitectos
效果图：Glessner Group
特邀摄影师：Guido Torres

"给我庇护"住所移除了游牧民族四处漂泊的居住习惯，但保留了游牧民族住所的根本原则，即能够阻挡不利的气候条件。该项目的设计不仅回应了特殊的地势，还尝试着为现代游牧民族提供一个独一无二的居住场所。"给我庇护"住所在秀丽的风景中添上了浓墨重彩的一笔，为人们提供了冬暖夏凉的宜人环境。该住所不但能阻挡不利的气候条件，并且在包括私人空间、公共区域以及设备空间在内的动态设计流通路线中带来了一种独特的体验。间隔区作为居住者的交通空间，这里还有许多种植着当地植被的花园。

Gimme Shelter moves away from the temporality of nomad sm but maintains the underlying principal of nomadic dwellings; which is to shelter from detrimental climatic conditions. The Villa responds not only to site specificity but attempts to provide a unique shelter for the modern nomad. Gimme Shelter submerges itself into the landscape, providing warmth through the winter and cool air during the summer. The Villa not only protects its inhabitants from harsh climatic conditions, but provides a unique experience for dynamic-programmatic circulation between private, public, and service spaces. Interstitial space serves as circulation for inhabitants and provides unique opportunities for gardens filled with native flora.

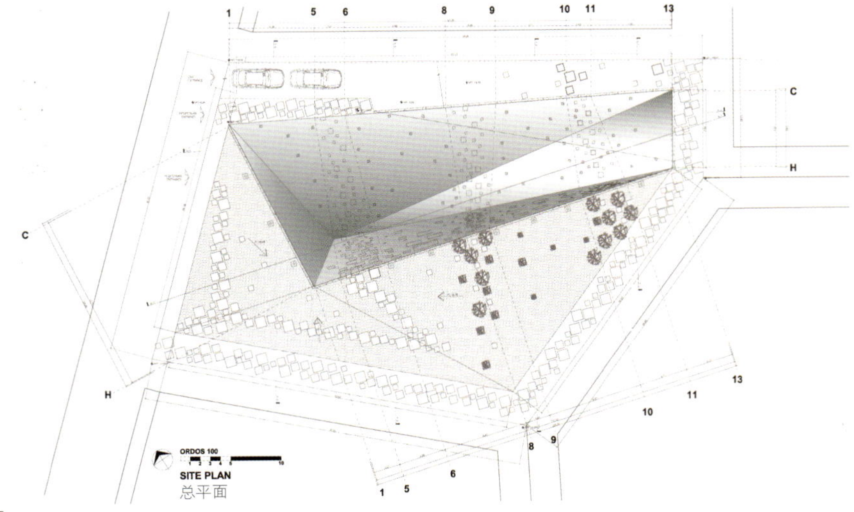

SITE PLAN
总平面

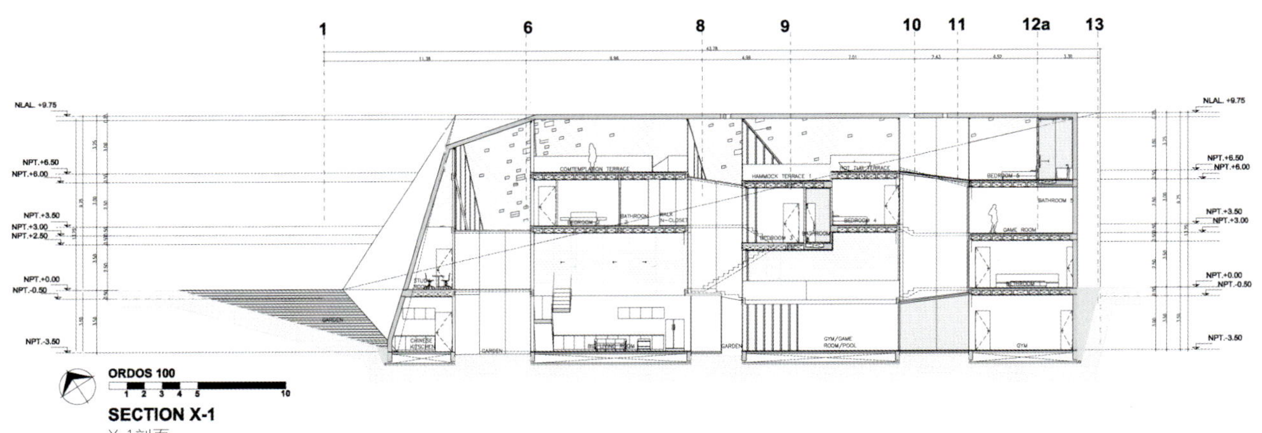

SECTION X-1
X-1剖面

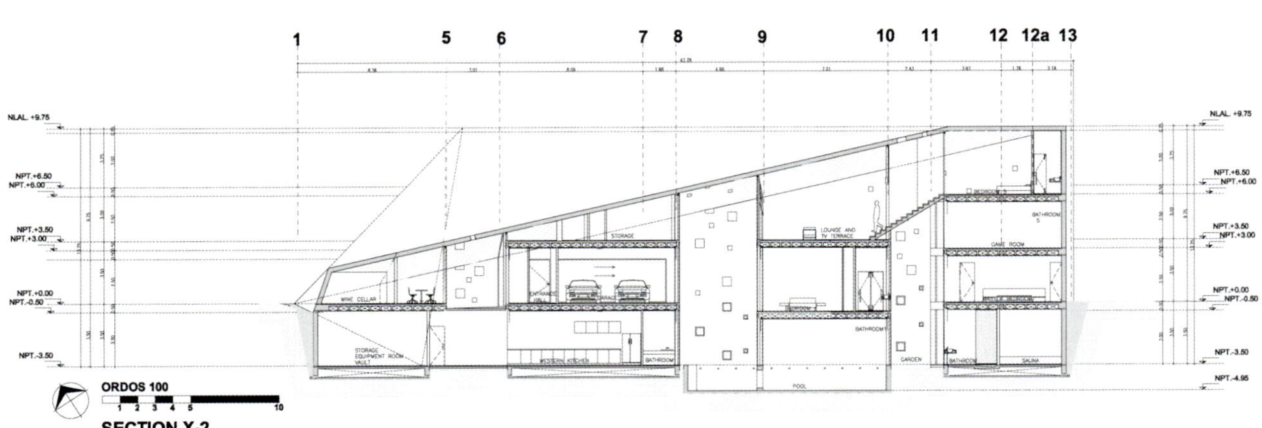

SECTION X-2
X-2剖面

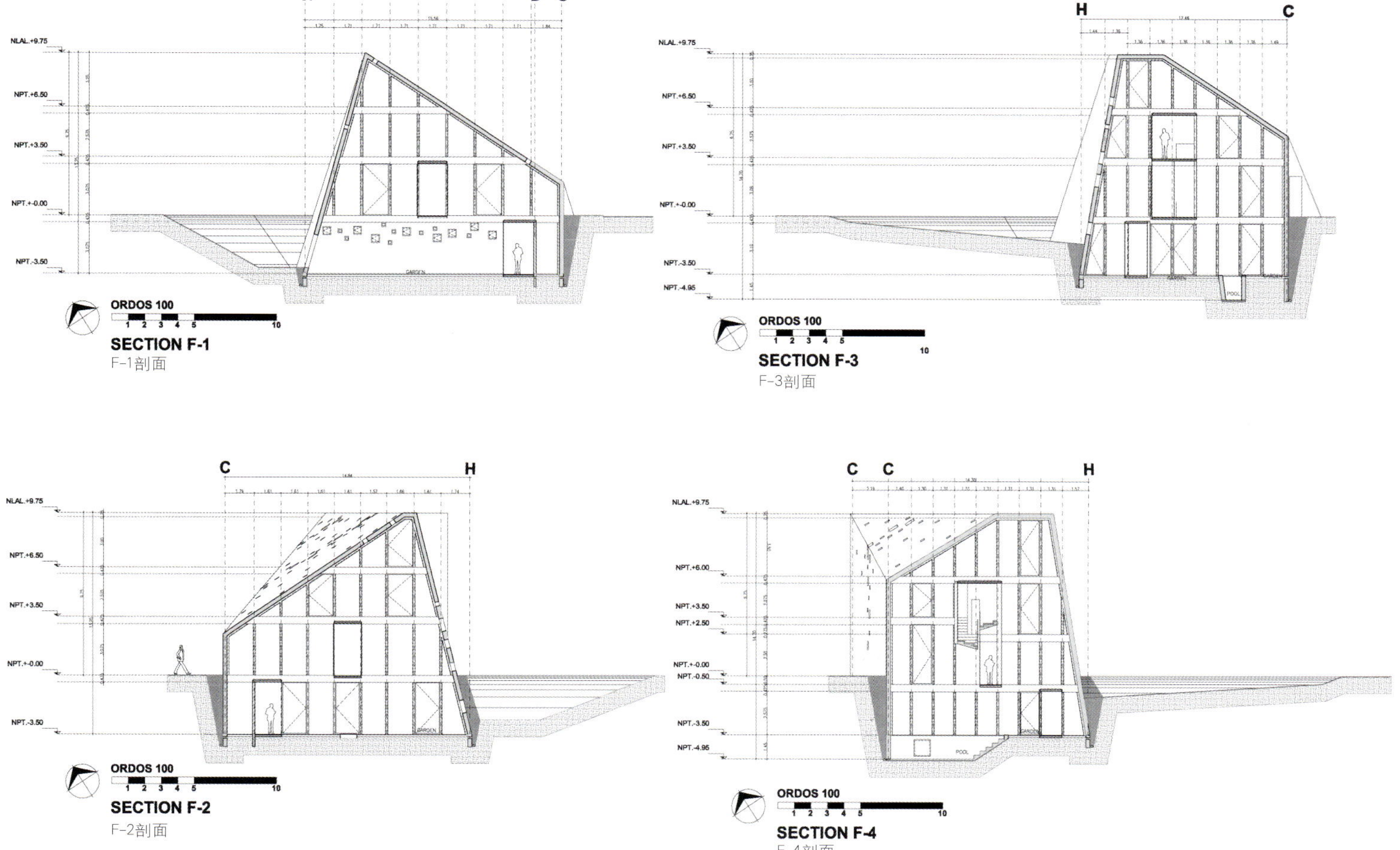
H
D
C
NLAL.+9.75
NPT.+6.50
NPT.+3.50
NPT.+-0.00
NPT.-3.50
ORDOS 100
SECTION F-1
F-1剖面
H
C
NLAL.+9.75
NPT.+6.50
NPT.+3.50
NPT.+-0.00
NPT.-3.50
NPT.-4.95
ORDOS 100
SECTION F-3
F-3剖面
C
H
NLAL.+9.75
NPT.+6.50
NPT.+3.50
NPT.+-0.00
NPT.-3.50
ORDOS 100
SECTION F-2
F-2剖面
C
C
H
NLAL.+9.75
NPT.+6.00
NPT.+3.50
NPT.+2.50
NPT.+-0.00
NPT.-0.50
NPT.-3.50
NPT.-4.95
ORDOS 100
SECTION F-4
F-4剖面

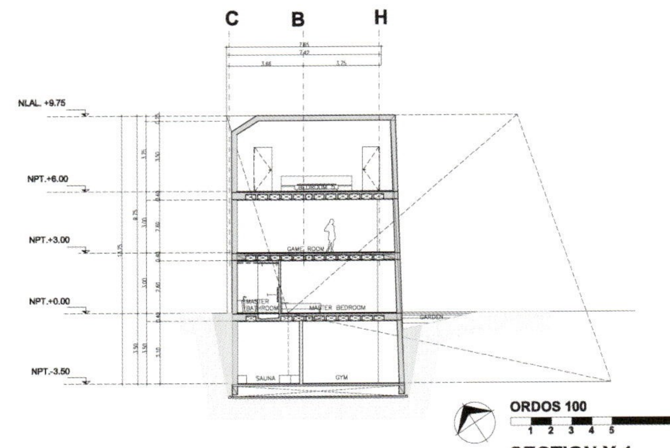

ORDOS 100

SECTION Y-1

Y-1剖面

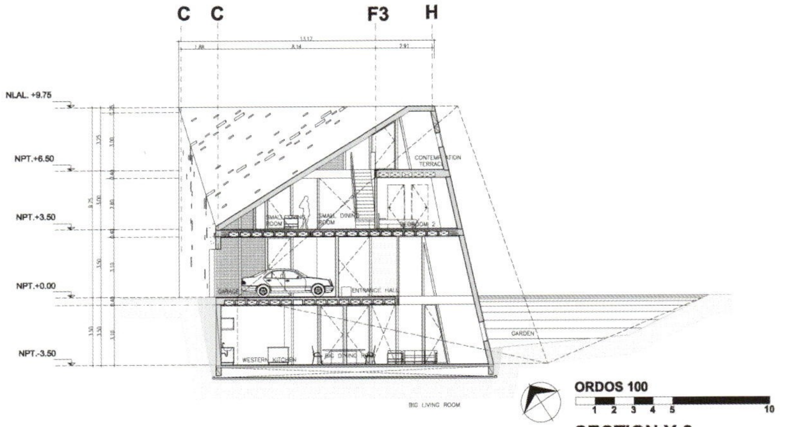

ORDOS 100

SECTION Y-3

Y-3剖面

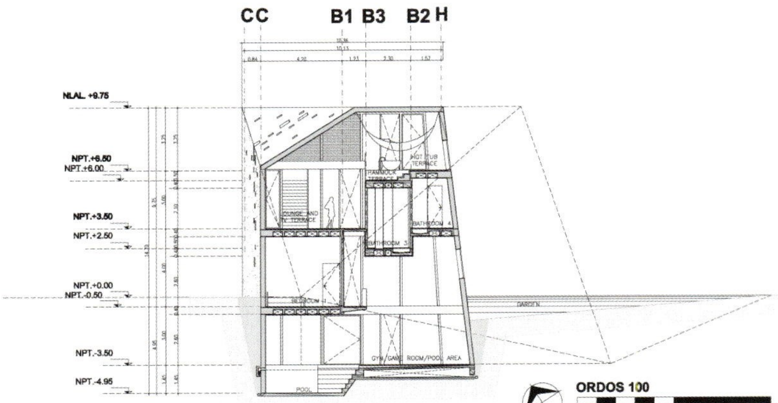

ORDOS 100

SECTION Y-2

Y-2剖面

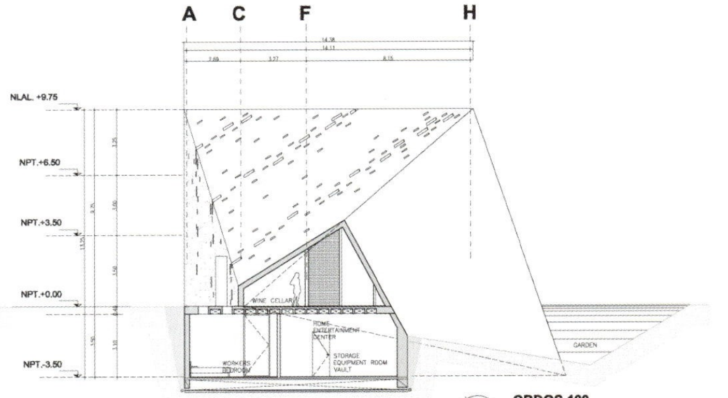

ORDOS 100

SECTION Y-4

Y-4剖面

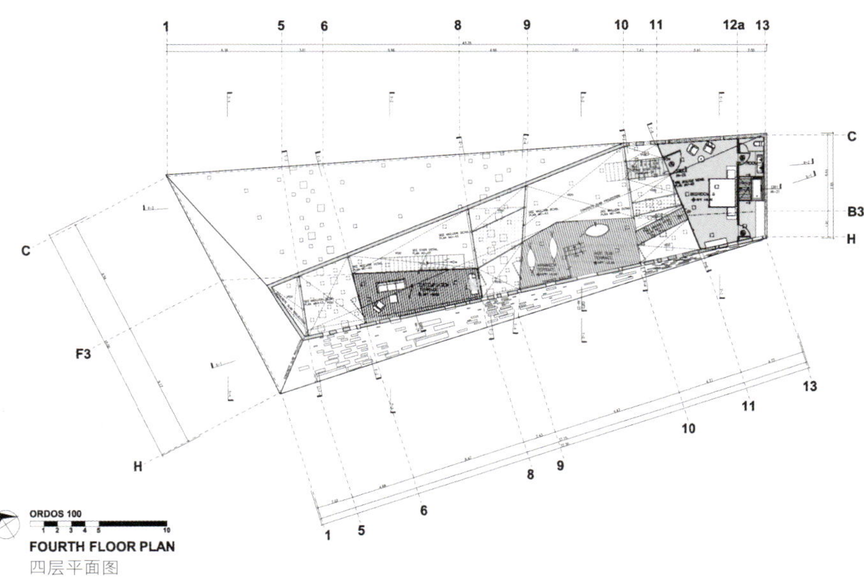

FOURTH FLOOR PLAN
四层平面图

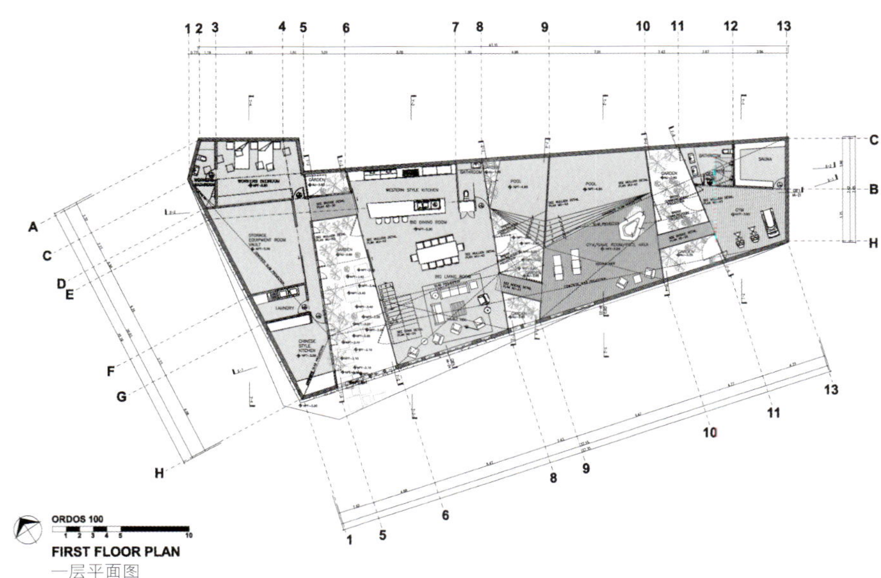

FIRST FLOOR PLAN
一层平面图

Liiva别墅
LIIVA HOUSE

建筑师：Arhitektid Muru&Pere OÜ
项目团队：Peeter Pere, Urmas Muru, Janek Maat, Doris Orasi

该别墅坐落于塔林附近维米斯半岛的一座古老渔村。其造型的设计灵感源自海边堆积的卵石。房屋面向街道的立面坚固而富有传统特色。庭院则更富有表现力，彰显着原生态的开放气息。设计方案的关键为冬景花园蝴蝶结式的主轴。花园四周环绕着爱沙尼亚的传统居家构成：客厅、厨房、卧室、客房、书房、蒸汽浴室。外墙与屋顶采用了沥青落叶松墙面板和屋顶板。南墙以白色混凝土建造而成。院中的木质表面则主要由天然落叶松木制成。

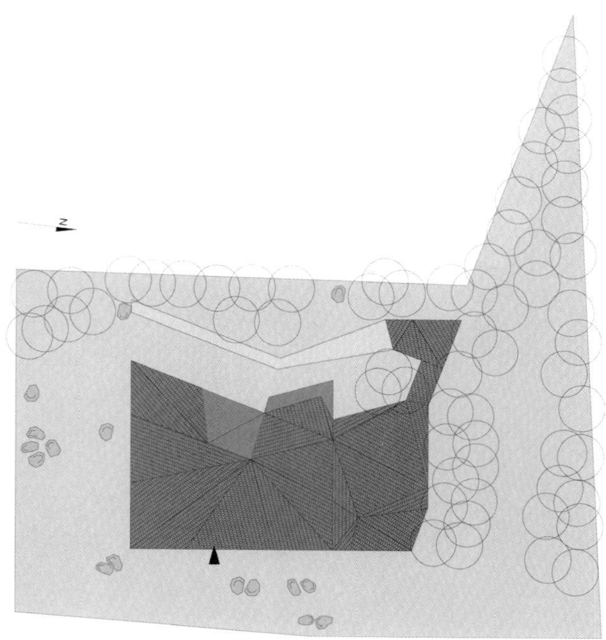
总平面
Site plan

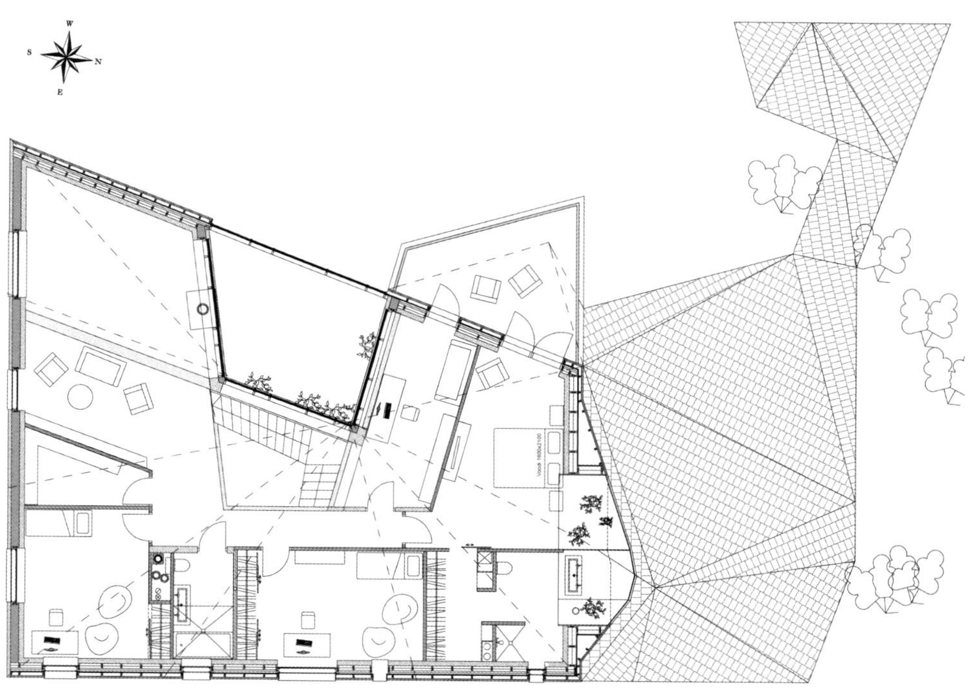
二层
First floor

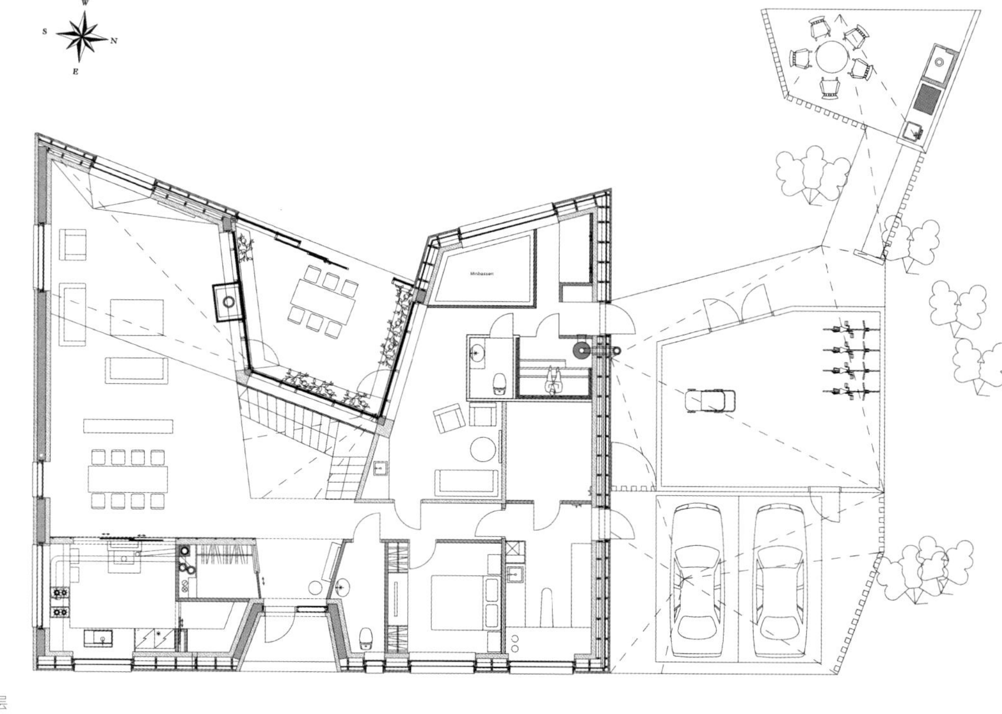
一层
Ground floor

Villa is situated in an old fishermen village near Tallinn in Viimsi peninsula. The shape has been derived from sea- side stack of boulders. More firm and traditional facade is exposed to the street. The courtyard is more expressive, wild, opened, communicating. The master key of the plan solution is the butterfly tie shaped main axes of the winter garden. It is surrounded by the rooms which compose traditional Estonian home- living room, kitchen, bedrooms, guest room, study, Sauna. The external walls and roof use tarred larch shingles. Southern wall is of white concrete. The wooden surfaces in the yard mainly consist of untreated larch.

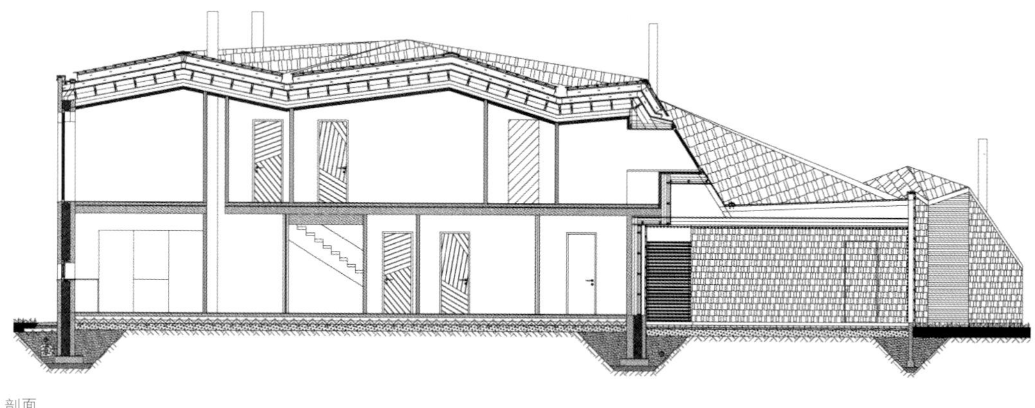
剖面
Section

麦迪逊谷的诺亚方舟之居
ARK HOUSE, MADISON VALLEY, USA

建筑师：Axis Mundi

总建筑师：John Beckmann

项目团队：John Beckmann, Ronald Dapsis, Masaru Ogasawara and Natacha Mankowski

效果图：Ronald Dapsis and Masaru Ogasawara

该住宅的设计可与古代帆船相媲美，帆船停靠在山腰，就好像停泊在古时候的海洋中一样。整体设计为仓库式结构，从中间被一个将近4800m^{2}的巨大瞭望台一分为二，一半是开放式壳体作为入口亭子，其中只有一条楼梯通向瞭望台。通过瞭望台就可以到达另外一半建筑中的主建筑区。在瞭望台下及楼梯后面是一条交错的带顶通道。还有一座60ft（约合18.3m）长、由耐候钢和玻璃焊接而成的桥，其下方为三层的中庭空间。

The design for this residence can be likened to the discovery of an archaic sailing vessel, beached on a mountainside, as if a great ocean receded in the ancient past. The overall design is a long barn-like structure bisected across the centre by an enormous cantilevered observation deck of nearly 4800 sq. ft. Half of the main form is an open shell which serves as an entrance pavilion. It contains only a staircase leading up to the observation deck. From the deck, one can enter the main house in the other half. There is an alternate, sheltered path under the deck, opening behind the staircase. Fabricated from Corten steel and glass, a 60-foot bridge spans a 3-story atrium space below.

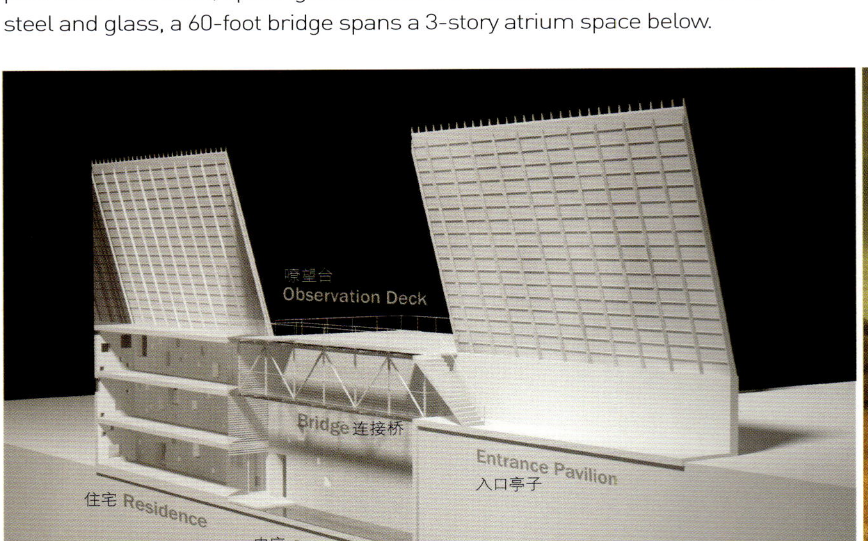

比利时霍尔斯贝克的W别墅
HOUSE W IN HOLSBEEK, BELGIUM

建筑师：dmvA
总建筑师：Mr. and Mrs. Wathion – Van Hellemont
项目团队：David Driesen, Tom Verschueren, Valerie Lonnoy, Sofie Buggenhout, Liesbet de Winter
总承包商：Van Merhaeghe（框架结构）；Alucobel（铝材）
结构工程师：Studie 10
概念：house in disguise
摄影与效果图：dmvA

建筑场所位于比利时一个叫做“Vlaams Brabrant”的群山连绵的村庄，设计目标是在这里的森林中找一处绝妙之所建造房屋，要求建筑风格朴实无华。这是一座与景观相融的住宅，并非典型的弗兰德民居形式，而是一个长方形的体量，横向建造在山丘的轮廓线上，建筑形体与山坡的曲线相吻合。没有使用一砖一瓦，而是掩映在植被和景天属的绿色表皮之下，从而使建筑与自然合而为一。

Building on a marvellous location in the woods, in a hilly country as "Vlaams Brabrant", requires an unpretentious architecture. It is a house going up into the landscape. No typical Flemish house, but an oblong living-object, transversely placed on the contour of the hill, takes over the form of the slope. No bricks, but a camouflaging green skin of plants and sedum. House and nature become one.

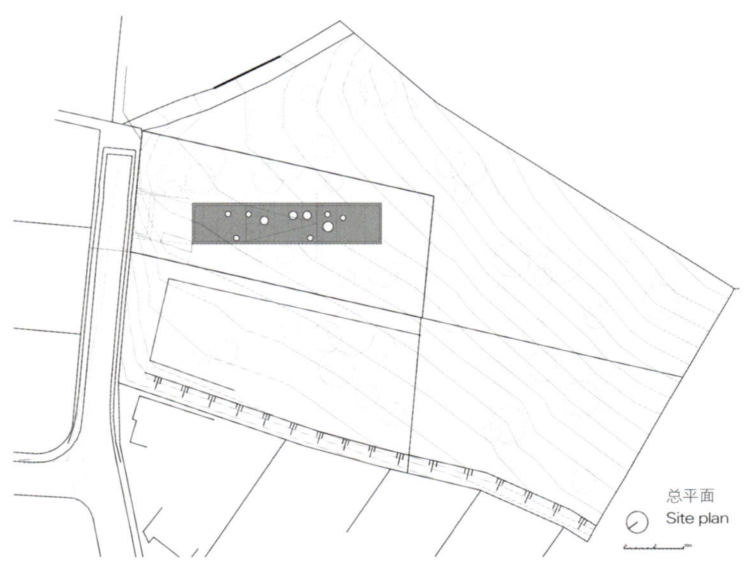

总平面
Site plan

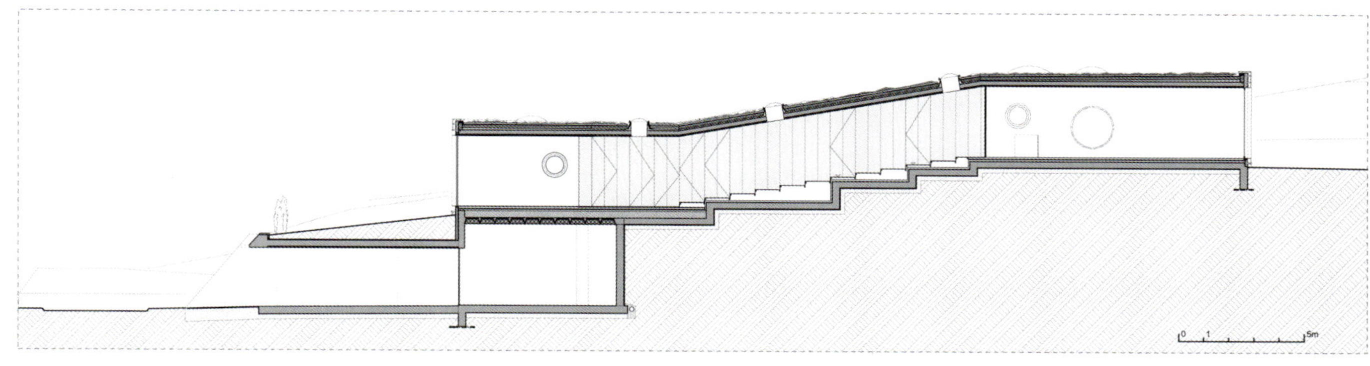

剖面
Section

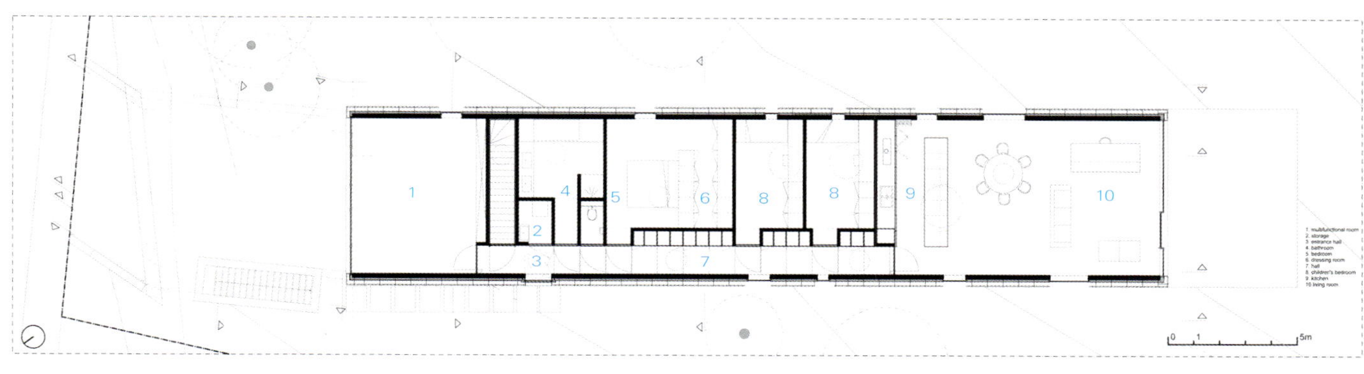

一层
Ground floor

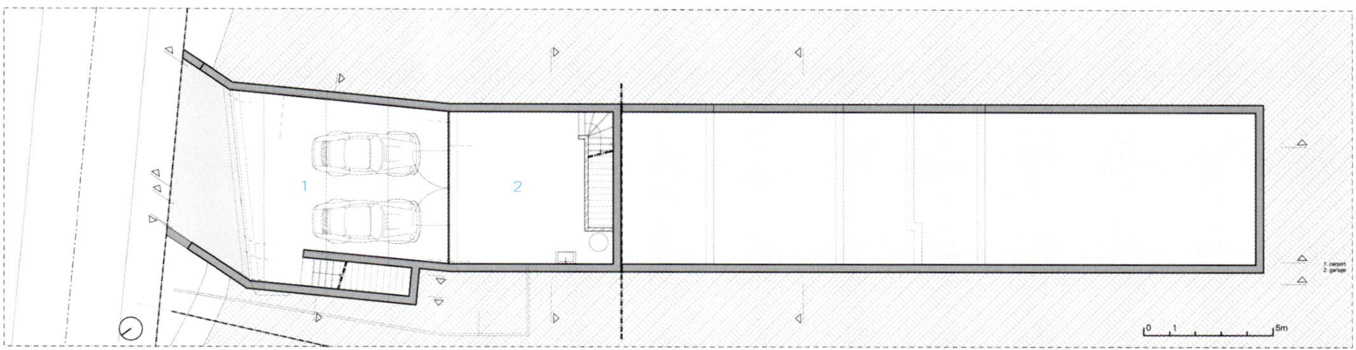

地下层
Basement

1 多功能房间
2 储藏室
3 入口大厅
4 浴室
5 卧室
6 更衣室
7 大厅
8 儿童卧室
9 厨房
10 起居室

1 multifunctional room
2 storage
3 entrance hall
4 bathroom
5 bedroom
6 dressing room
7 hall
8 children's bedroom
9 kitchen
10 living room

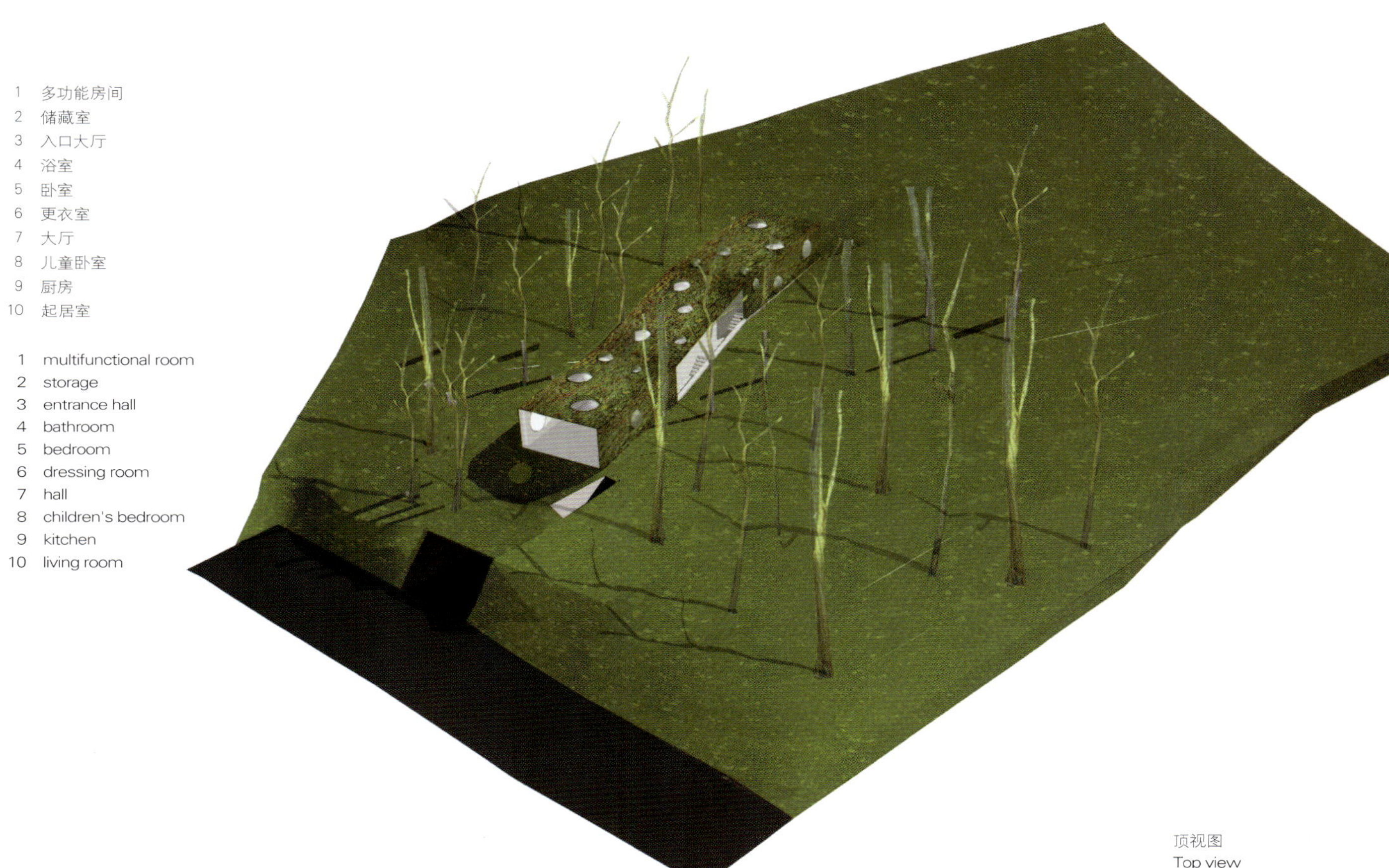

顶视图
Top view

100

100 Global Architectural Schemes

国 际 最 IN 建 筑 设 计

100

其他

OTHERS

4

VTB “发电机”体育场
VTB DYNAMO STADIUM

建筑师：(designed by) Erick van Egeraat

甲方：VTB Bank

项目地点：Petrovsky Park, Moscow, Russia

用途：多功能文化、健康、运动中心，包括一个45 000座的运动场、一个10 000座的运动大厅、零售休闲综合楼、餐厅、停车场和其他设施

基地面积：116 000m²

总楼面面积：335 000m²

艺术表现图：(designed by) Erick van Egeraat

获奖情况：竞赛一等奖

本案项目对莫斯科“发电机”体育场及其周围的公园进行重建。两个新的体育场被设在老体育场内，保留原来体育场外围的立面，在新设计中融入功能性与美观性。项目内还增加了文化及零售设施。运动、文化与零售设施的混合保证体育场能够一年365天、一周七天、一天24小时地创造收益，举办各种充满生机的活动。

VTB Arena Park comprises the redevelopment of the Dynamo Moscow Stadium and its surrounding park. The two stadiums will be situated within the ring of the old stadium, preserving the perimeter facade of the existing arena and integrating it functionally and esthetically into the new proposal. To be financially self sustaining VTB Arena Park will also host cultural and retail facilities. The mix of sports, culture and retail facilities create profitable and vibrant activities 24 hours a day, 7 days a week, 365 days a year.

аллея Театральная
Петровско-Разумовская аллея
VTB Arena Moscow
april 2010

FIFA WORLD CUP
2018
FIFA WORLD CUP
2018
ВТБ

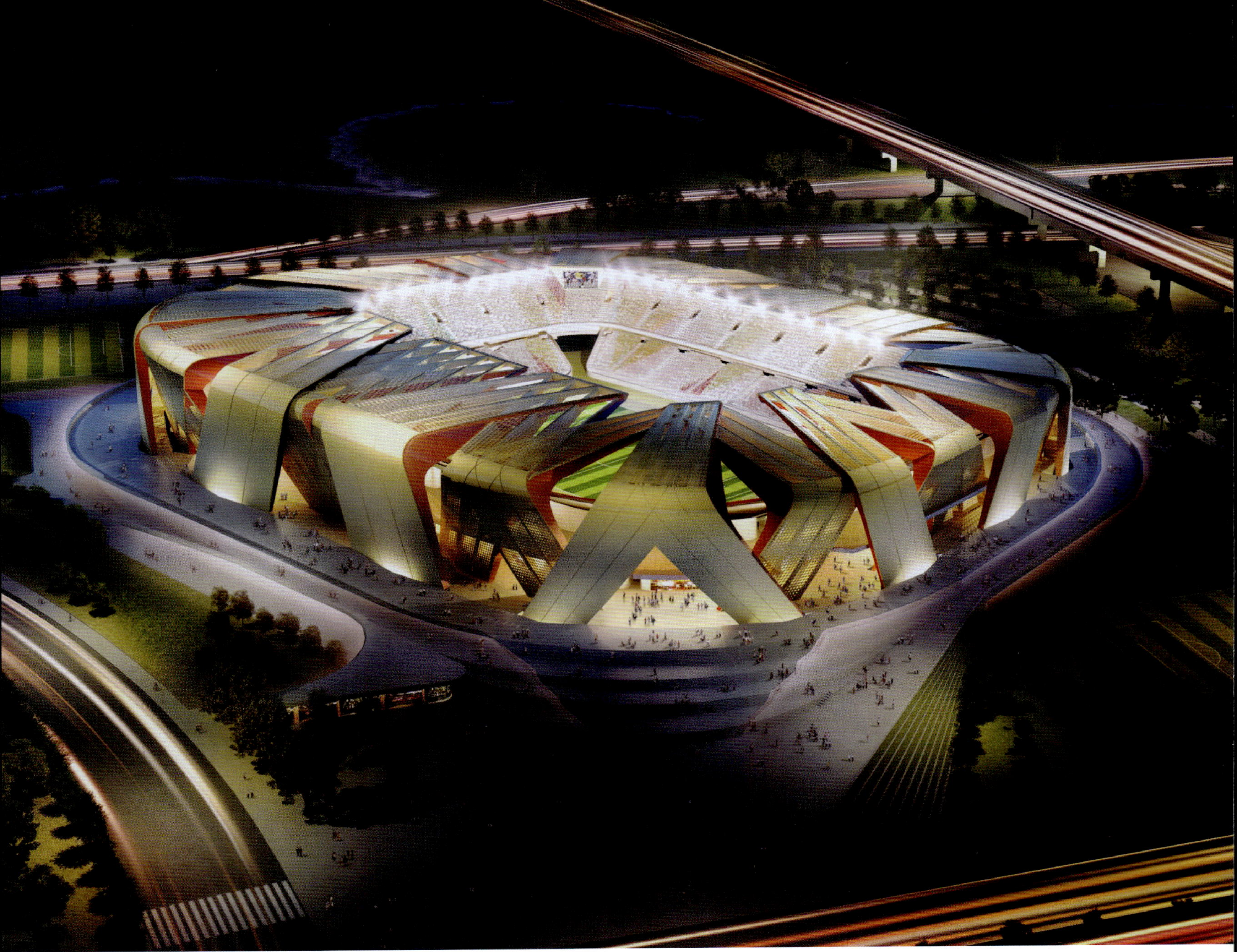

大连足球体育场
DALIAN FOOTBALL STADIUM

建筑师：UNStudio：

Ben van Berkel, Caroline Bos, Astrid Piber with Nuno Almeida, Ger Gijzen and Cynthia Markhoff, Luis Etchegorry, Shu Yan Chan, Ramon van der Heijden, Marcin Koltunski, Fernie Lai, Patrik Noome

结构工程顾问：ARUP Shanghai, China

Arup International Consultants (Shanghai) Co., Ltd

运动场顾问：ASS Planungs GMBH Freie Architekten, Germany

交通顾问：MVA Hong Kong LTD.

效果图：UNStudio / and SZ Silkroad Digi Tech Co. LTD., China

动画效果：IDF Global Pty Ltd.

甲方：Dalian City Bureau of Urban Planning

项目地点：Dalian, China Building surface: 38,500 m^2

基地面积：144 000 m^2

容积：40 000名观众

用途：足球体育场，带有两个训练场地

获奖情况：竞赛一等奖

总平面图

Master plan

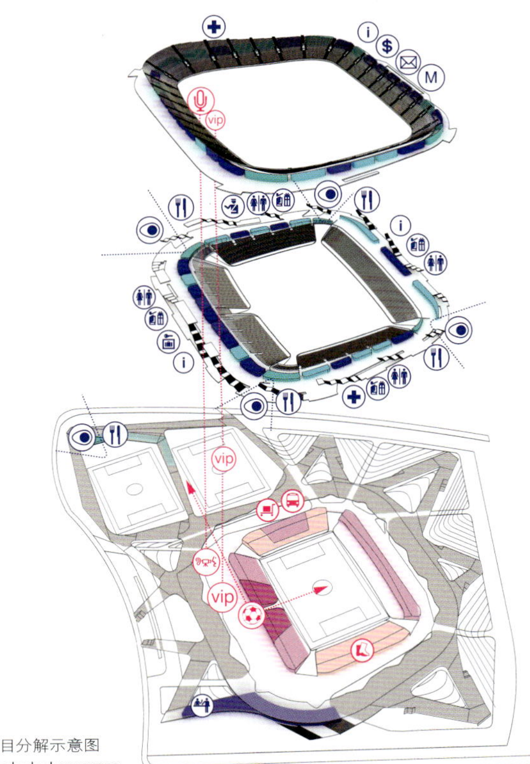

项目分解示意图

Exploded program

场馆设计类型的关键在于观众的亲身体验。除了作为竞技场的体育场要满足观众围绕球场中心运动之外，场馆设计必须考虑到建筑结构、项目、文脉、基础结构以及风格等元素，并要求这些元素相互结合，展现出强大、整体的姿态。场馆基础结构需要考虑通道和疏散路线的灵活、来宾路线和泊车设施的有序。文脉的考虑无论对于场馆与整个城市的关系还是场馆方位和周围交通方式的关系都是重要的因素。

Essential to the stadium typology is the experience of the spectator. Aside from the basic function of a stadium as an arena for spectator sport with one central focal point, stadium design requires the consideration of many essential structural, programmatic, contextual, infrastructural and stylistic elements and the incorporation of these into a strong, integral gesture. Infrastructural considerations include ease of access and evacuation, visitor routing and parking facilities, while contextual considerations form an important element in both the relationship of the stadium to the city, its surroundings and its orientation with regard to nearby transport modes.

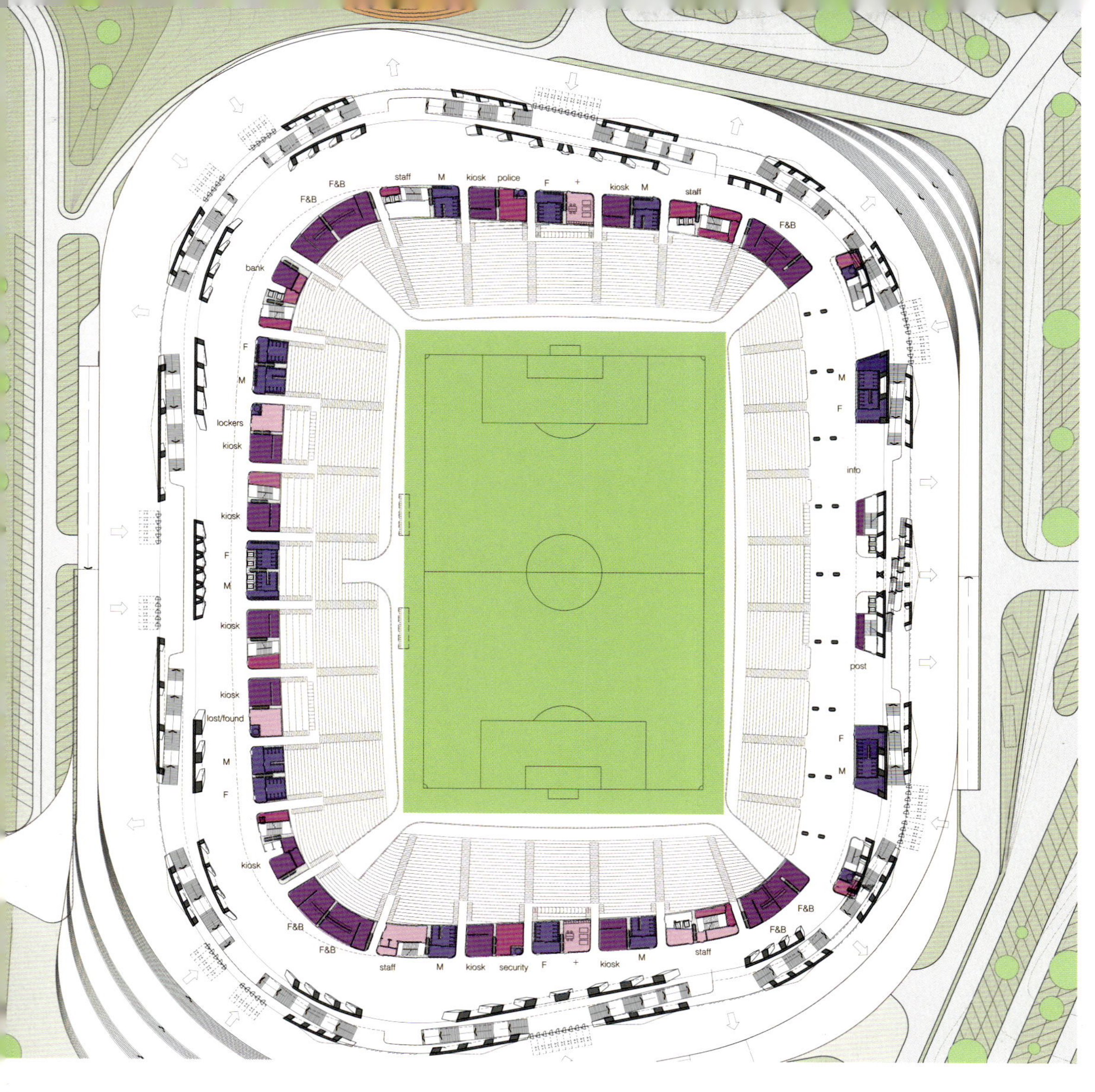

6m高处平面图
Level 6.0 plan

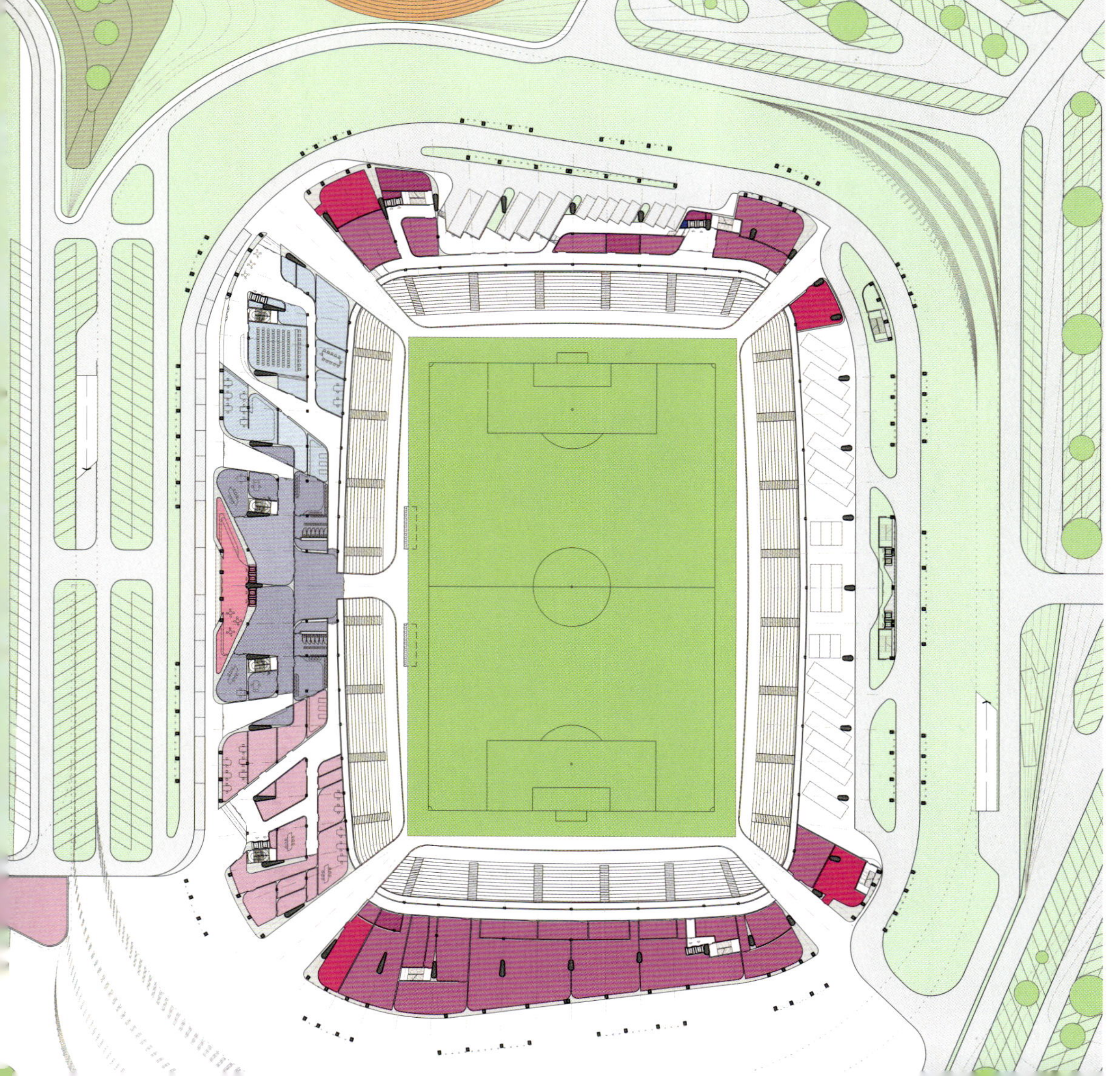

地面层平面图
Level 0.0 plan

运动员用房	Sportsman Facility
新闻媒介用房	Media & Press related
竞赛管理用房	Game Managment/Organizer
贵宾区	VIP Area
安保/警察用房	Police/Security
技术设备用房	BOH/Service
商店/餐饮	Retail/F&B
厕所	Toilet

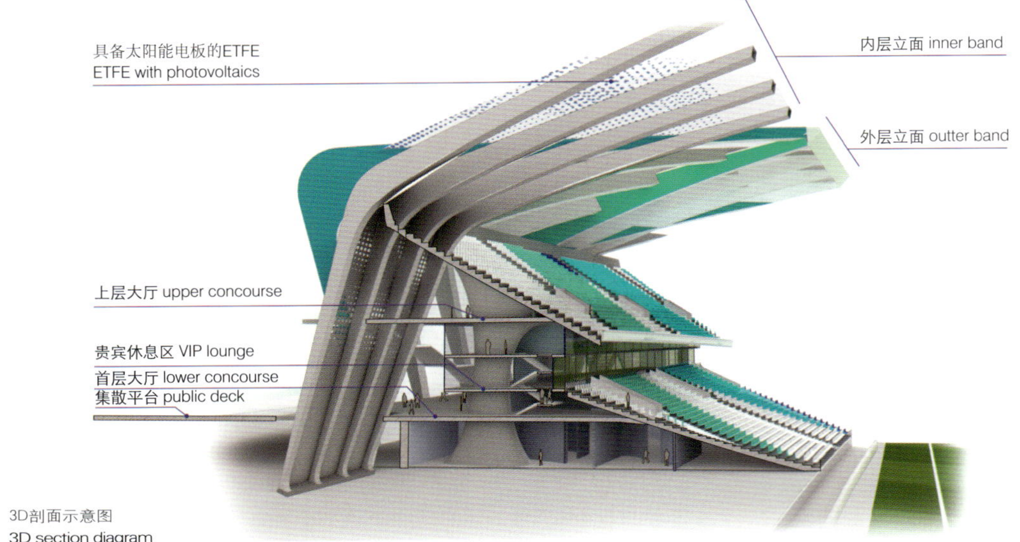

3D剖面示意图
3D section diagram

广泛视野分析
Extensive view line analysis

通过先进的三维工程技术，保证每个座位往球场的最佳视野。

Aided by advanced 3D engineering techniques, each seat has guaranteed optimal view at the pitch.

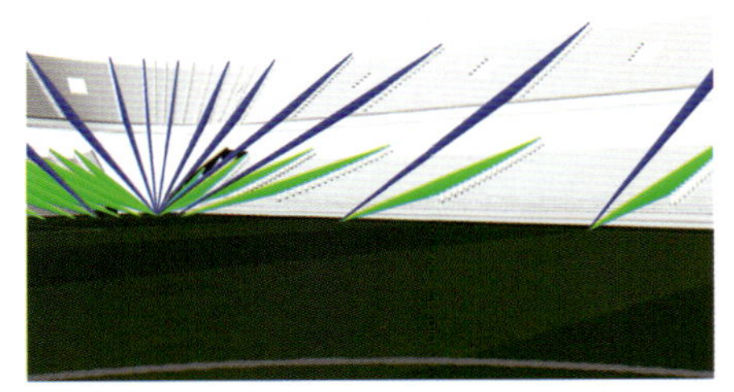

座位的排列与角度根据视野公式的分析所安排。反复研究距离、环境的影响、视角、功能与安全性，得出了最佳的座位布局。

Seat arrangement and geometry are integrally driven by view line formulas. Distances, environmental influences, view angles, program and safety, have been iteratively investigated with an optimal layout as result.

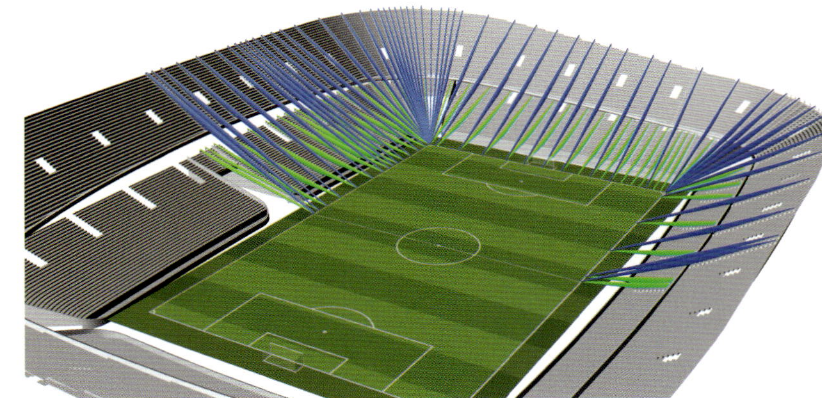

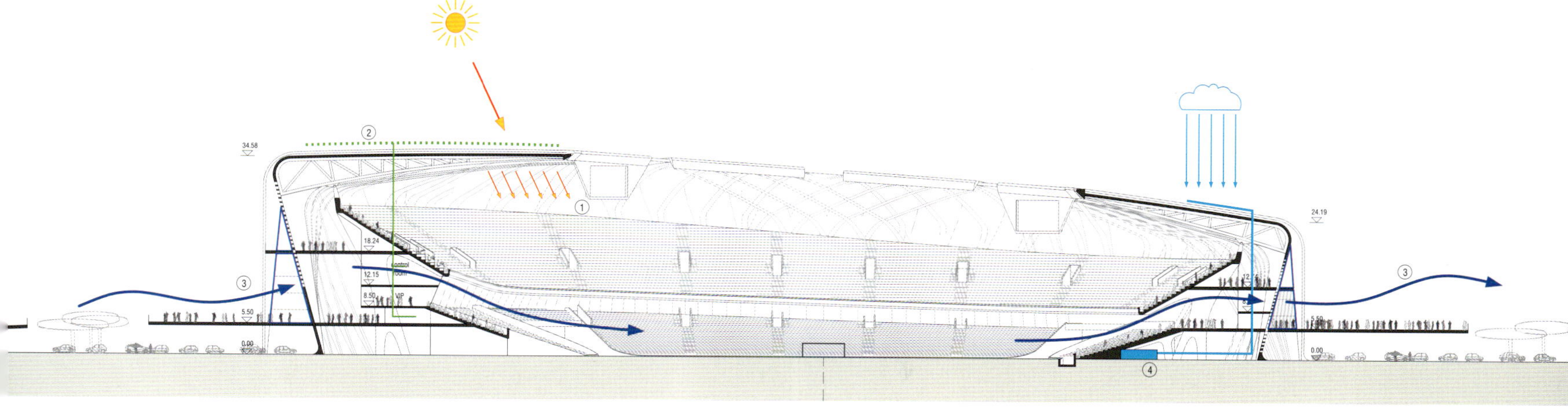

可持续概念
Sustainability concept

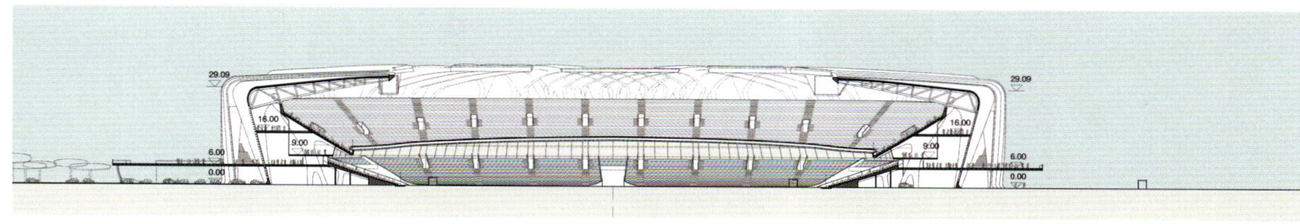

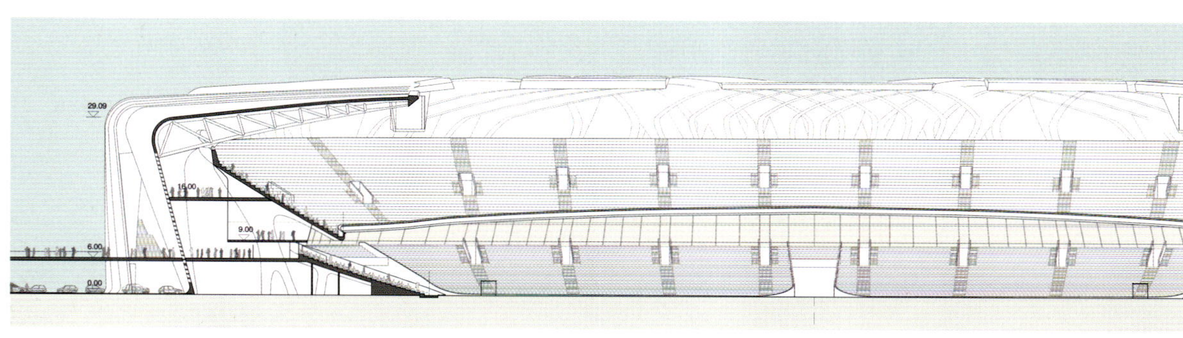

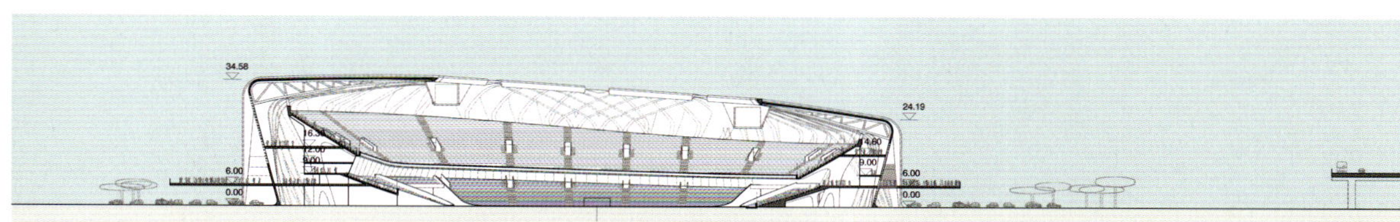

座位色彩概念，雨伞概念
结构线条延伸以确定彩色的座位区域
seats colour concept: umbrella idea
projection of structural lines to frame colour seating areas

彩色座位平面示意图
Colourful seat plan concept diagram

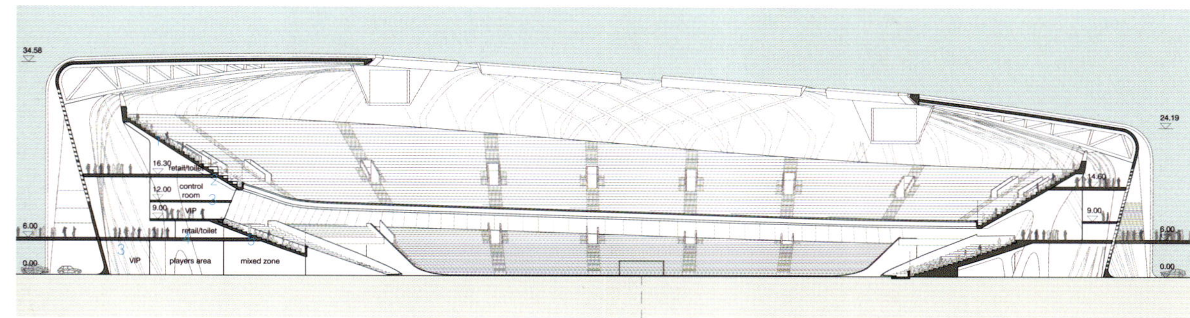

1 商店／厕所
2 控制室
3 贵宾区
4 运动员使用区
5 多功能区

剖面图
Sections

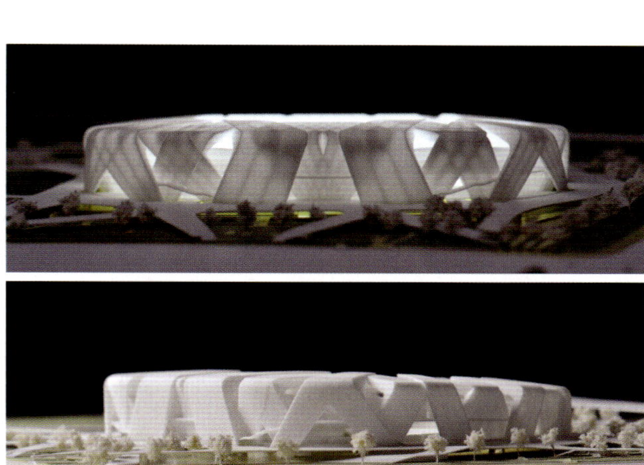

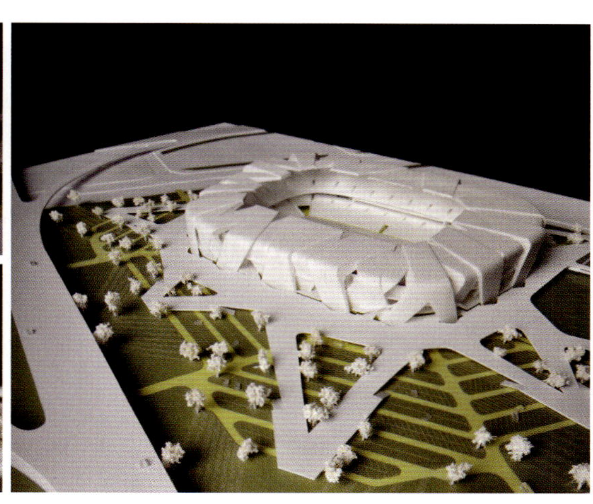

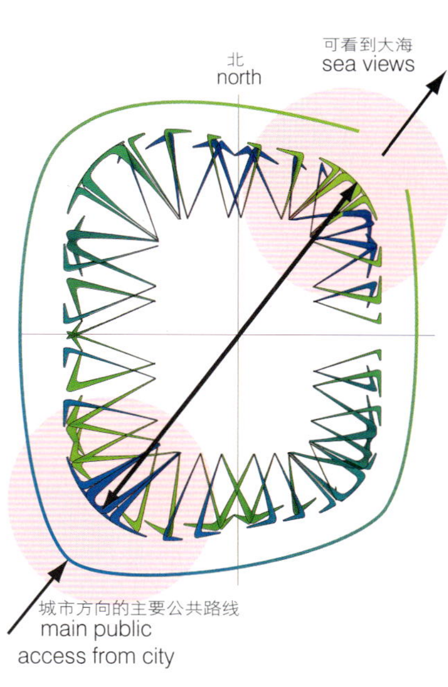

彩色平面概念示意图
Colourful plan concept diagram

巴塞罗那足球俱乐部的诺坎普体育场
CAMP NOU STADIUM FOR FC BARCELONA

建筑师：Foster+Partners

甲方：FC Barcelona

顾问：Ramboll Whitbybird , Gleeds, Jason Bruges, AFL Architects, Pfeiffer, PHA, RWDI, SDG, STRI

巴塞罗那足球俱乐部的诺坎普体育场是世界上最大的足球场地之一，将被大规模重新扩建。它原本已经是欧洲最大的体育场了，将扩建成能容纳106 000名球迷并包括招待区和公共区域的场所。设计师也设计了一种新型看台为球迷遮阳。体育场由明亮的马赛克外皮围绕，它们包围着建筑物并一直延伸到新建的看台顶棚，这种色彩缤纷的外表由具有俱乐部代表色的半透明瓷砖搭接组成。在比赛的夜晚，体育场将为这座城市带来一种崭新的建筑符号。同样，巴塞罗那足球俱乐部将不仅仅是俱乐部，新诺坎普体育场也不仅仅是体育场了。

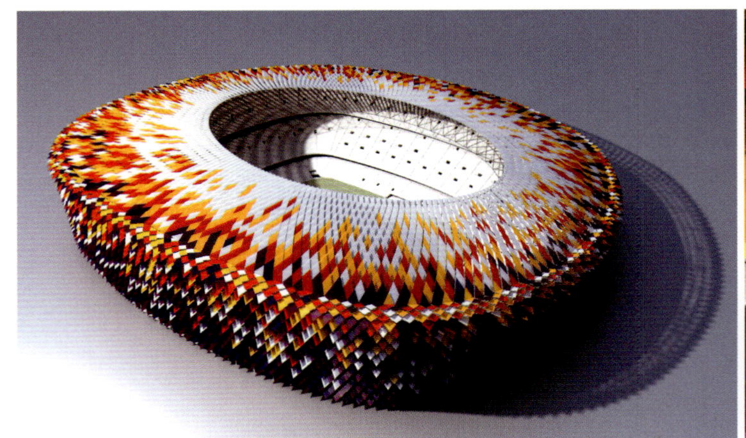

FC Barcelona's Camp Nou Stadium, one of the world's greatest football venues, is to be extensively remodelled. The stadium, already the largest in Europe, will be enlarged to accommodate over 106,000 fans, together with extensive new facilities including hospitality and public areas. A new roof will also be created to shelter the fans. The stadium will be enclosed by a brightly coloured mosaic outer skin that wraps around the building and continues over a new roof. The multi-coloured enclosure comprises overlapping translucent tiles in the club colours. On match nights, the stadium will glow, providing a new architectural icon for the city. In the same way that FC Barcelona is "more than a club", the new Camp Nou will be much more than a stadium.

场地南侧鸟瞰图
Aerial View of the Site from the South

建筑师受马哈拉施特拉板球协会的委托，设计了一个能容纳55 000人的新体育场。低层看台可容纳34 000名观众，上面四层阶梯看台还能容纳21 000名观众坐席、协会会员专用看台和多达80个包厢。印度每年的板球运动时间从11月到第二年的5月底，这个时段太阳高度相对较低，所以使用轻巧的拉伸膜结构作为看台遮篷，可以最大化地起到遮阳作用，并形成一个令人难忘的体育场标识。由于普内处于地震带，所以建筑主体采用了钢筋混凝土结构。施工后期阶段将为5000名会员提供便利设施，这些设施将安置在主要场地北侧的实验用地。

MCA普内国际板球中心
MCA PUNE INTERNATIONAL CRICKET CENTRE

甲方：Maharashtra Cricket Association
项目管理：Buro Happold Management
建筑师：Hopkins Architects
结构工程师：Adams Kara Taylor
MEP顾问："BDSP Engineering London with TCE Consulting Engineers Ltd Mumbai"
成本顾问：Dongre Associates
张拉结构工程师：Tensys Limited
防火工程：SAFE
经营：Jeremy Evans

The architects have been commissioned to design a new 55,000 seat stadium in Pune for the Maharashtra Cricket Association. A lower terrace of 34,000 spectator seats is supplemented by four upper terrace stands providing a further 21,000 seats, a Members' Pavilion and 80 hospitality boxes. Cricket is played in India from November until late May: the sun is thus relatively low during play. Using lightweight tensile membrane structure, roofs to the stands will maximize shading and provide a memorable signature for the stadium. Pune lies within a seismic zone, and this has generated complex braced structural solutions for the predominantly steel and concrete stands. Later construction stages will provide additional amenities for 5,000 members and this will be assembled around a practice ground to the north of the main ground.

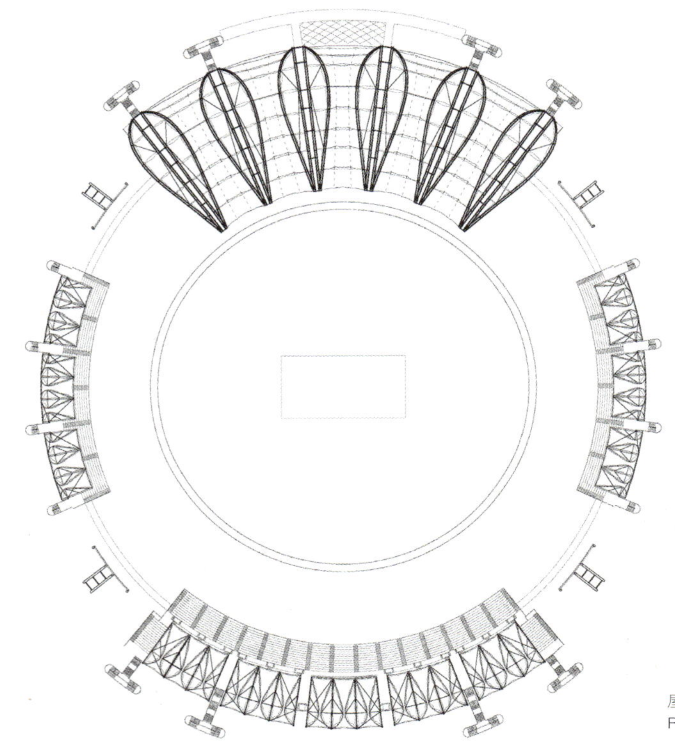

屋顶平面
Roof plan

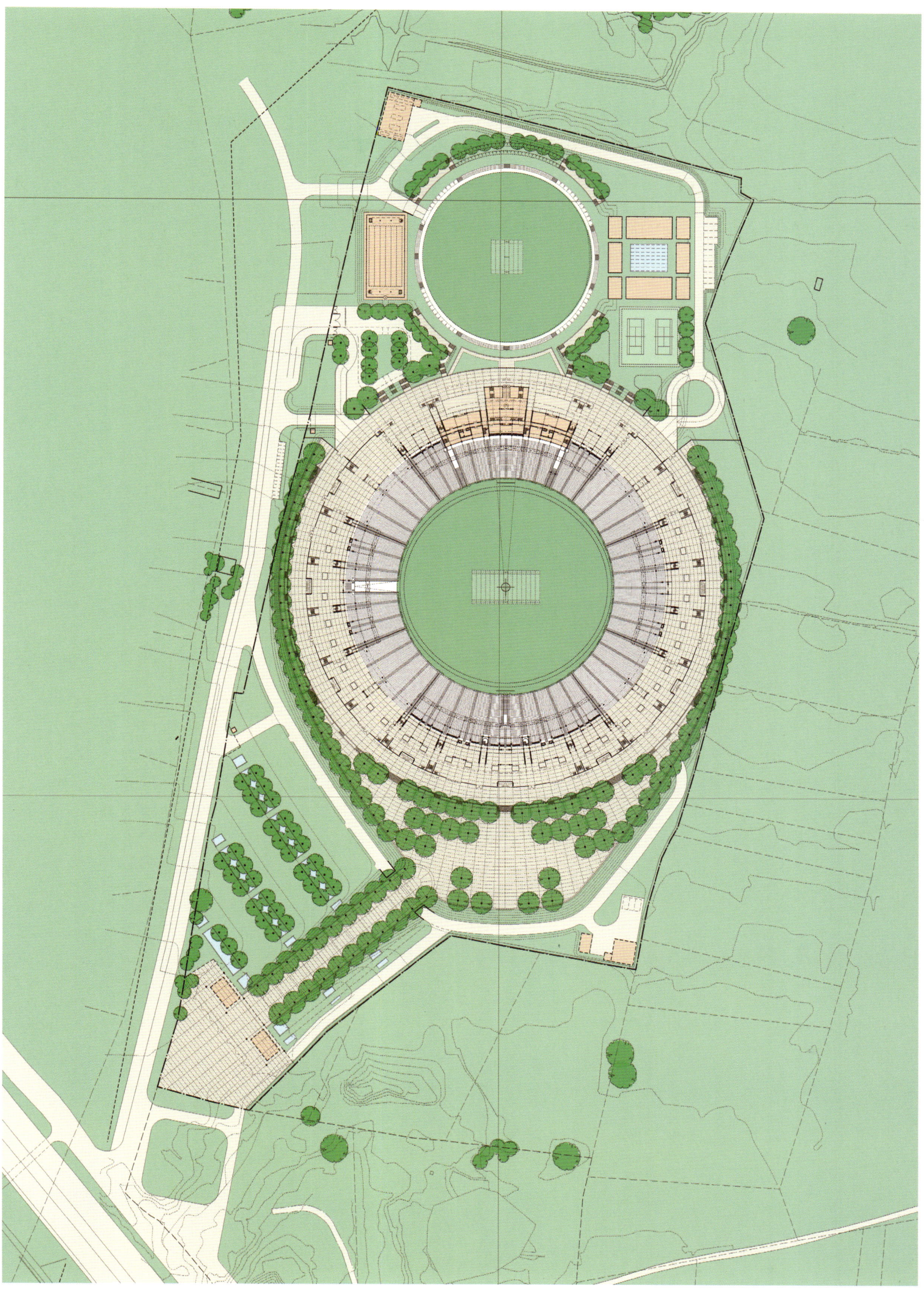

总体规划
Master plan

从大厅看向南看台
View Towards the South Stand From Concourse level

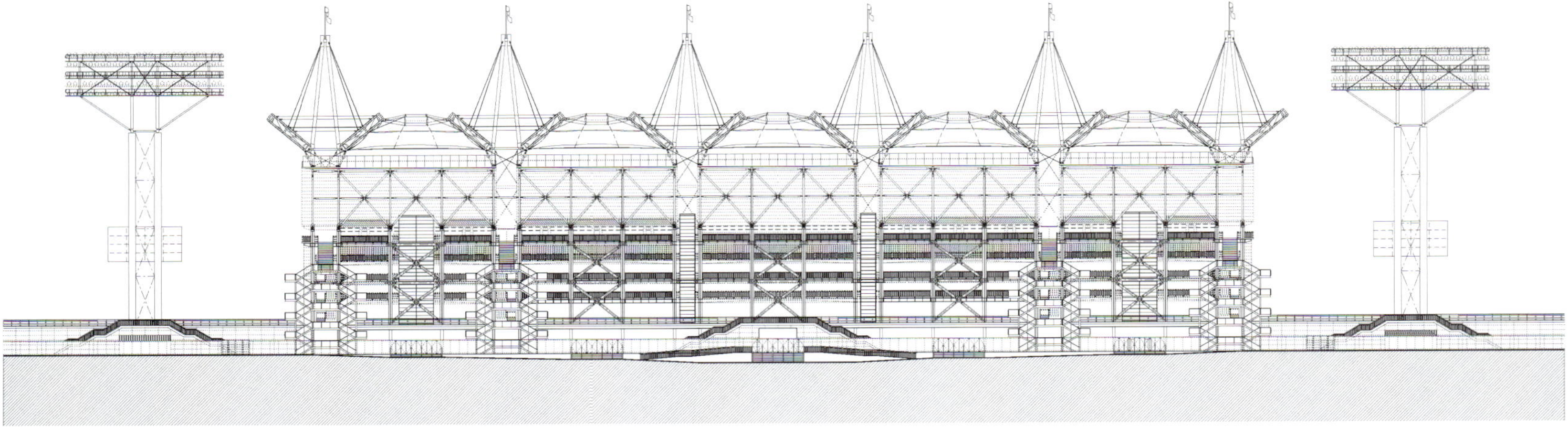

南看台外立面图
External Elevation of South Stand

从主看台入口处看南看台
View of South Stand From the Main Spectator Approach

北看台研究模型，展示协会会员俱乐部设施
Study Model for the North Stand Showing Associated Members Club Facilities

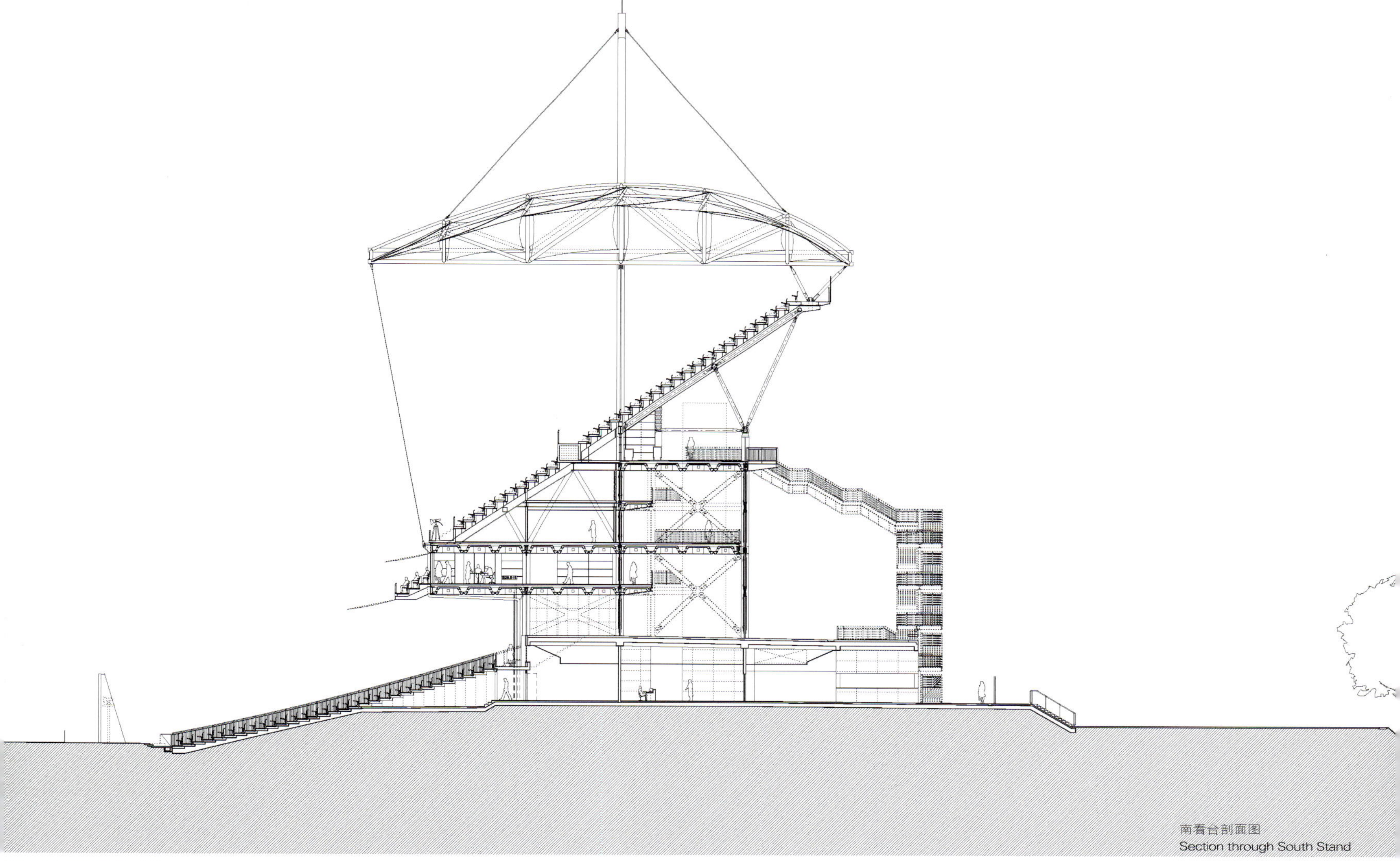

南看台剖面图
Section through South Stand

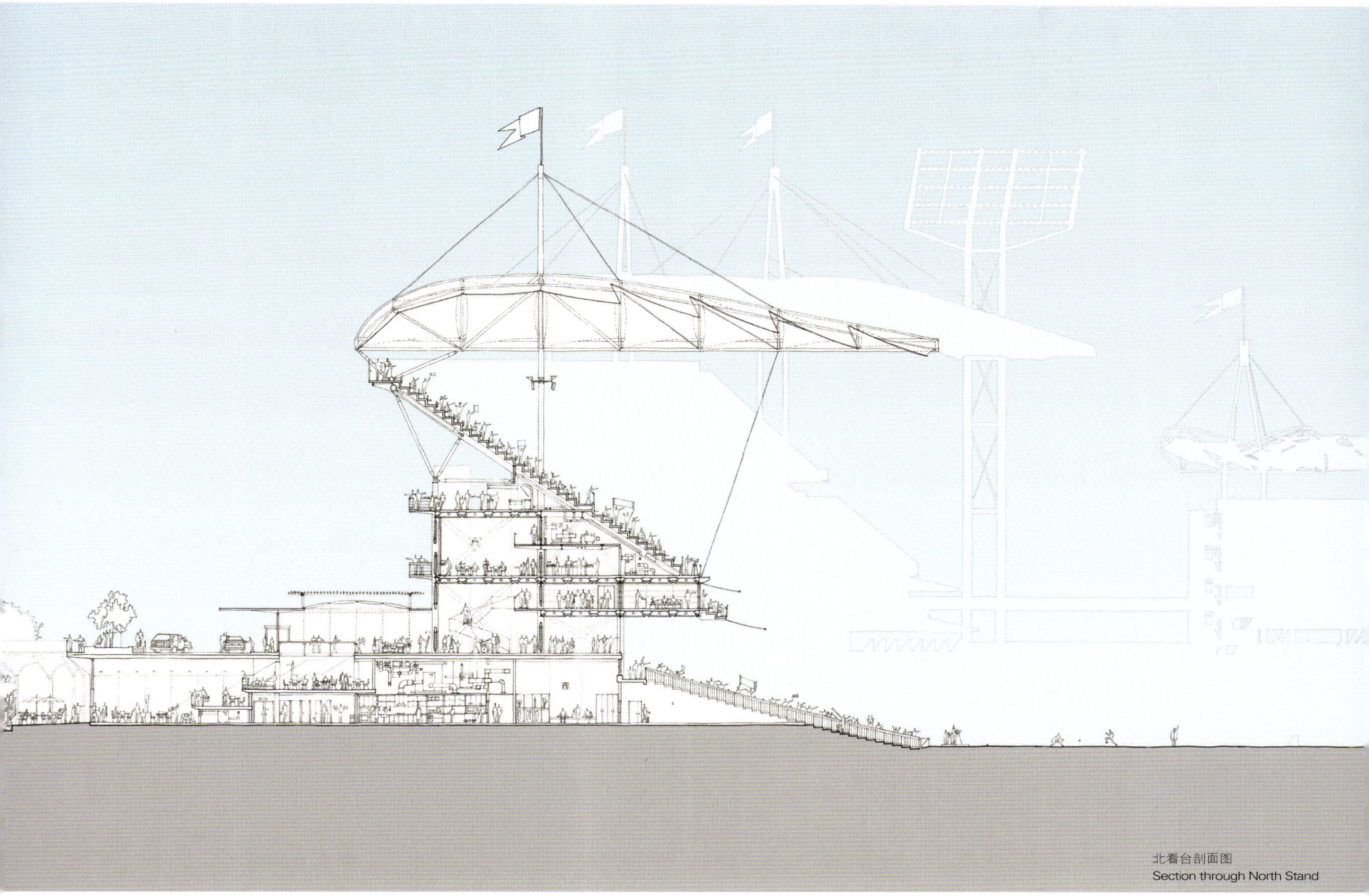

北看台剖面图
Section through North Stand

俄亥俄州托莱多联邦法院，美国
TOLEDO US FEDERAL COURTHOUSE

建筑师：Yazdani Studio of Cannon Design
设计团队：Mehrdad Yazdani（设计主管）；Michael J. Smith（项目主管）；Craig Booth（高级设计师）；
Paul Gonzales（项目经理）；Robert Levine（项目规划与设计）；Hansol Park, Philip Ra, Jessica L. Yi, Joe O'Neill,
Yan Krymsky, Alek Zarifian, Nadine Quirmbach, Sepideh Nabavi, Mimi Lam, Kiduck Kim, Jonathan Moody,
Charles Aweida, Manson Fung, Morgan Newman, Jenny Tse
室内设计师：Felderman & Keatinge Associates
景观设计师：Hargreaves Associates
照明设计：Luminesce
记录建筑师/结构工程师/MEP工程师：URS Corporation
土木工程师：Proudfoot Association
声效顾问：Acoustical Design Group
预算：Project and Construction Services, Inc.
摄影：Tom Bonner, Model Shots

这栋法院建筑面积223 600平方英尺（约合20 773m²），设计理念是提供六间法庭，也为将来地方法院的扩建预留了空间。设计目的旨在建筑的平面布局中掺入民主之感，重新设计了穹顶、柱子及其他典型的杰佛逊式构件。法院的体量设计为L形。拐角处的开放一侧有一个面向东南的双层高中庭；这里的两根支架包裹着圆柱形的法院体量，其中也容纳了法官的办公套间。入口层位于法庭楼层下层，其中有律师服务、地区接待员以及缓刑/审前服务办公室。底层则容纳了最高指挥官停车场和建筑设备房间。

Our design concept for the 223,600 sf courthouse provides six courtrooms and space to accommodate future District Court expansion. Our intent is to imbue a sense of democracy into the plan of the building – by repurposing typical Jeffersonian elements such as domes, columns and the like. The court masses are arranged in a L-shaped configuration. The open side of the angle is a double height atrium facing southeast; the two legs of the angle wrap the cylindrical court volumes and house the judges' chambers. The entry level, one floor below the court floor houses the US attorney, district clerk and probation/pretrial offices. The ground floor houses the US marshal parking and building service.

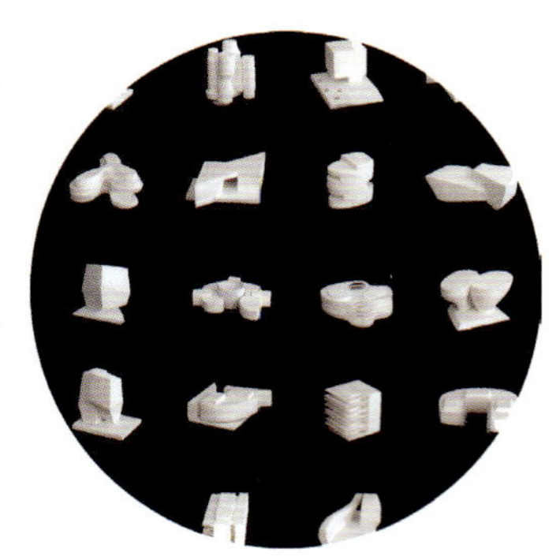

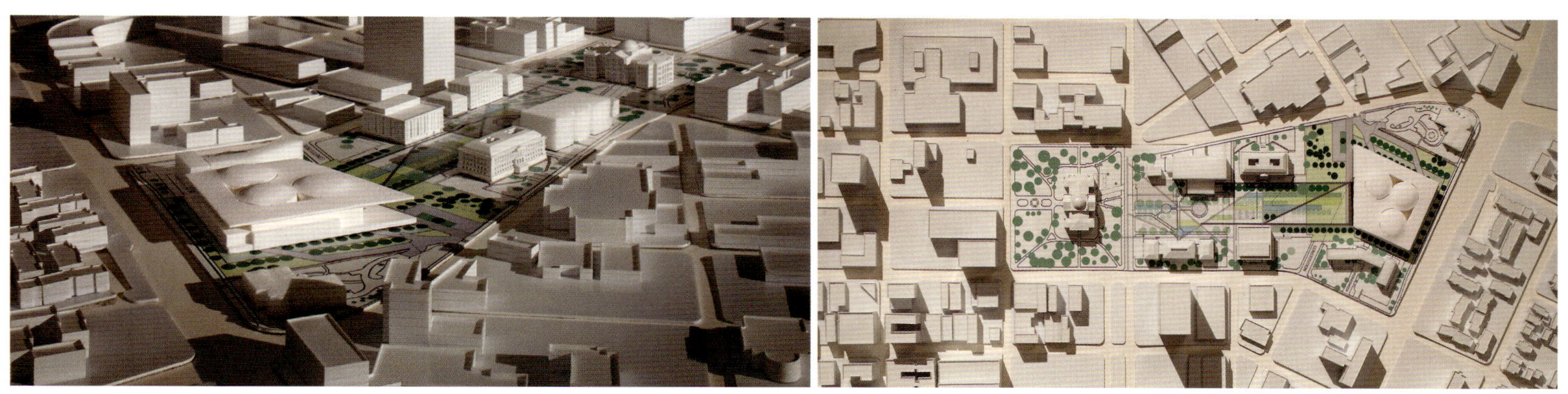

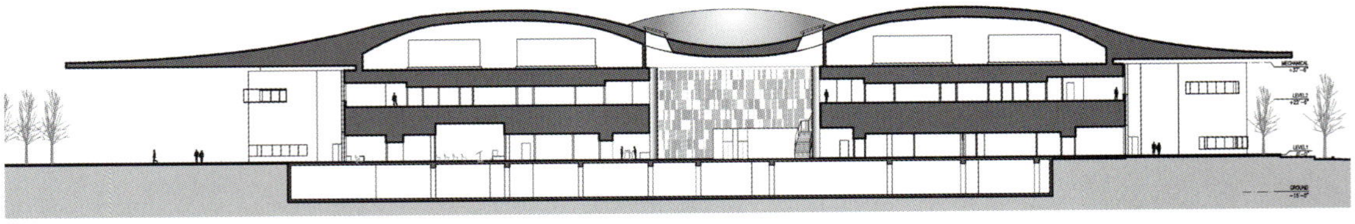

剖面图
Section

一层
First Floor

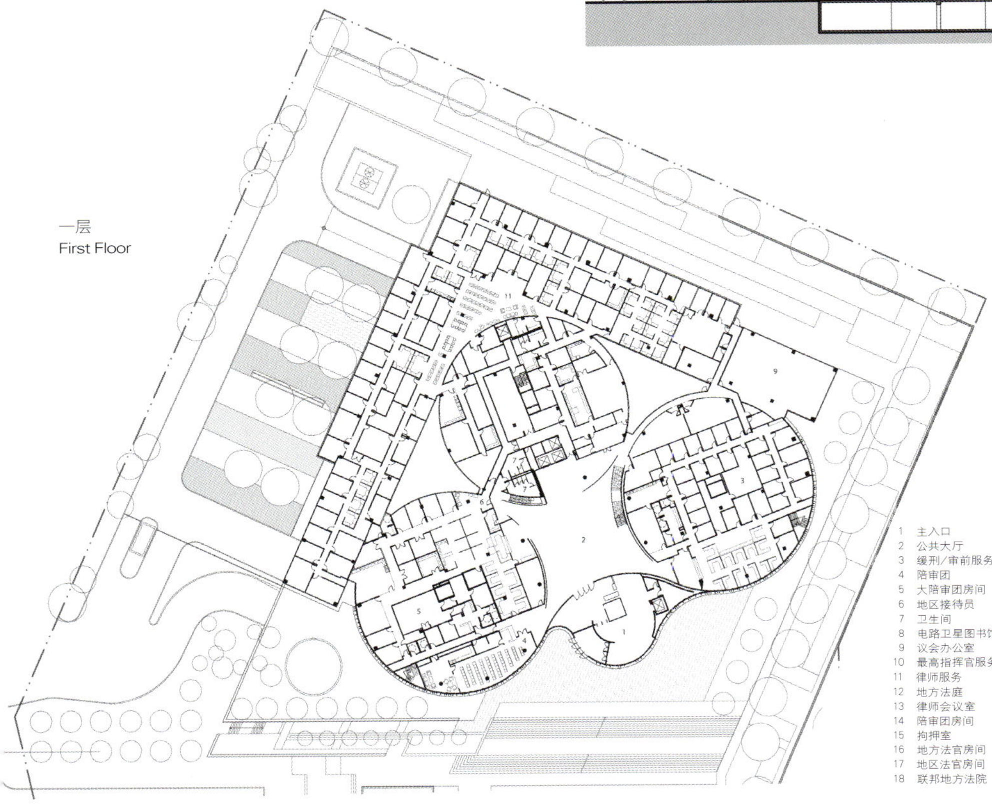

二层
Second Floor

1 主入口
2 公共大厅
3 缓刑/审前服务
4 陪审团
5 大陪审团房间
6 地区接待员
7 卫生间
8 电路卫星图书馆
9 议会办公室
10 最高指挥官服务
11 律师服务
12 地方法庭
13 律师会议室
14 陪审团房间
15 拘押室
16 地方法官房间
17 地区法官房间
18 联邦地方法院

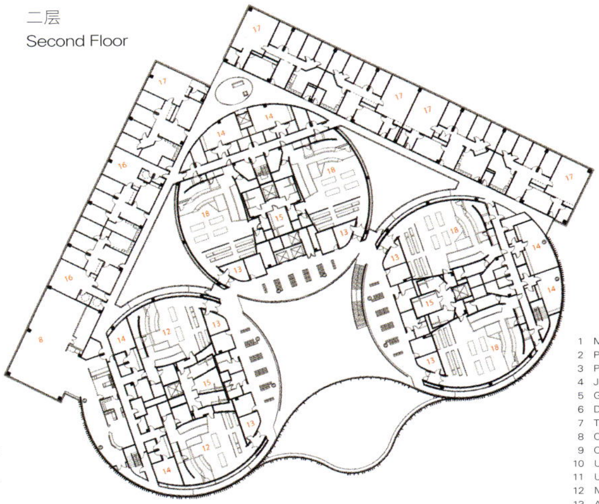

1 Main Entry
2 Public Lobby
3 Probation/ Pretrial Services
4 Juror Assembly
5 Grand Jury Suites
6 District Clerk
7 Toilet Rooms
8 Circuit Satellite Library
9 Congress Offices
10 U.S. Marshal Service
11 U.S. Attorney
12 Magistrate Court
13 Attorney Conference
14 Trial Jury Room
15 Holding
16 Magistrate Judge Chambers Suites
17 District Judge Chambers Suites
18 District Court

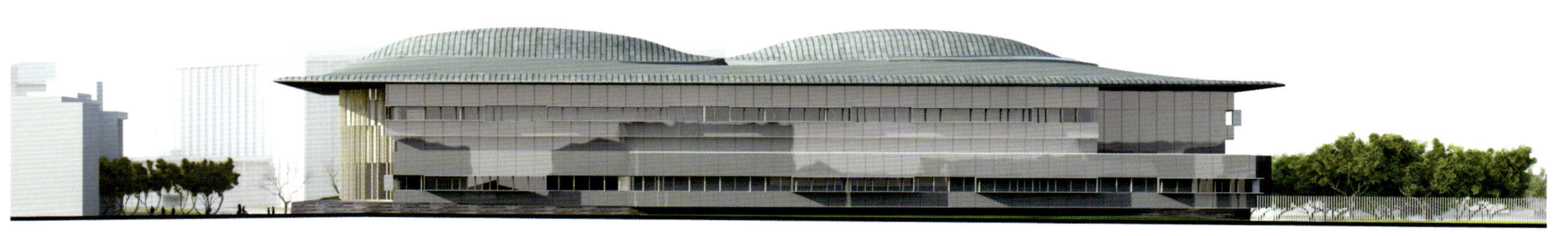

新法庭位于北海之滨，介于自然与城市之间，被设置在海牙周边滚动的沙丘景观中。设计的主要概念是对建筑群体进行如雕塑般的布置，并设计一座地标性建筑以表达国际刑事法庭（ICC）的权威，同时兼顾到人文尺度。schmidt hammer lassen建筑师事务所的夺标设计符合复杂设计纲要，并抓住了ICC的精神主旨。整个建筑形式就像是地平线上波浪状的体量组合，使人想起沙丘景观。

据建筑评审团的说法，该设计为ICC提供了一座具有雕塑构成特点的方形塔楼。评审团就这一设计点评道："这是一个非常吸引人、非常有趣的建筑表达，通过建筑与景观的新颖组合为城市增添了魅力。这一点同样应用在通过宽敞的楼梯"下入"法庭中的设计理念。大而尖的景观切口和较低的一层平面都是非常有趣的设计元素。建筑内部具有友善的使用氛围，尤其是从上部进入室内的美丽天光更是为这一氛围锦上添花。一层平面可以作为室内的私人花园，这促进了ICC员工之间的愉悦和积极的交流。"

通过设置一个深入地下的切口，楼群与周围的沙丘景观形成了对比。建筑的理念是将一层平面的花园（花坛）延展为包裹整个法庭塔楼的"覆层"。"花园一直是所有文化和宗教的一部分。有了来自110个ICC成员国的花卉和植物，花坛将升级为绿色的地标和城市的象征，无论种族和文化有何不同，" Bjarne Hammer说道。他是该事务所的创意总监和创始合伙人之一。

环保性对于建筑的占地面积和建筑材料的选择来说是至关重要的因素。办公楼里面覆有一层复合材料，它入选的理由是其对当地多风和多盐气候的适应性、维护的简易性和功能的安全性。由于其出色的耐用性，该种材料通常用在职业赛车的车身和风车的覆层。设计的目前阶段被 BREEAM评为优秀等级。

海牙国际刑事法庭
INTERNATIONAL CRIMINAL COURT IN THE HAGUE

建筑师：schmidt hammer lassen architects
甲方：The International Criminal Court (ICC)

Located close to the North Sea, the new Court is placed between nature and city, set in the rolling dune landscape at the edge of The Hague. The main concept is the sculptural arrangement of buildings in the landscape and the design of a landmark that conveys the eminence and authority of the ICC while at the same time relating to a human scale. schmidt hammer lassen architects' winning design complies with a complex brief and captures the spirit of the ICC. The overall building form can be seen as an undulating composition of volumes on the horizon, reminiscent of the dune landscape.

According to the Architectural Jury, the design provides the ICC with a sculptural composition of square towers. The Jury quoted this approach as "a very impressive and interesting architectural gesture and a great contribution to the city with an attractive integration into the landscape. This applies also for the idea of 'moving down' to the Court through the spacious staircase. The big and sharp incision in the landscape and the lower ground floor are very interesting elements. The inner atmosphere is confirmed as user-friendly, especially the spacious ground floor with beautiful daylight from above. This ground floor can be seen as an inner private park area which facilitates the interaction between all the ICC employees in a very pleasant and positive way."

By making a sharp incision into the ground the building complex forms a contrast to the surrounding dune landscape. The architectural idea is to continue the gardens in the ground floor (parterre) level of the building as a cladding of the Court Tower. "Gardens have always existed as part of all cultures and all religions. With flowers and plants from each of the 110 ICC member countries, the parterre gardens rise up as a green landmark and a symbol of unity, regardless of nationality and culture," explained Bjarne Hammer. He is by the way one of the founding partners and creative direction at schmidt hammer lassen architects.

Environmental sustainability is a key criterion in terms of the building's footprint and the selection of building materials. The facades of the office buildings are clad in a composite material selected for its suitability to the windy and salty local climate, ease of maintenance and security performance. The material is normally used in the bodywork of professional race cars and in the cladding of windmills due to its durability. The design has at this stage been assessed as BREEAM Excellent.

丹麦腓特烈堡法院大楼
FREDERIKSBERG COURTHOUSE, DENMARK

建筑师：3XN
甲方：Slotsog Ejendomsstyrelsen
工程师：Lemming & Eriksson
景观设计师：Schønherr Landskab

总平面图
Site plan

该法院大楼的扩建将成为一种雕塑和经典的结构，在表达出一种广受欢迎的透明化的同时，也保留了司法系统的庄严和肃穆。新法院大楼将成为现有新古典主义大楼的自然延续，并通过现代的表达方式保留了自己的特性。法院大楼的内部设计焦点在于与法律改革的要求相一致，这种要求即为建筑物的安全性和内部分隔性。因此，建筑物为雇员、被告、目击证人及观众提供了一种开放和友好的环境，使得各种群体都很容易在这种环境里进行工作。一个小中庭将大楼从中间一分为二，将光线引入大楼内部，在各部门之间形成了开放而空气流通的衔接。

东立面
East elevation

西立面
West elevation

南立面
South elevation

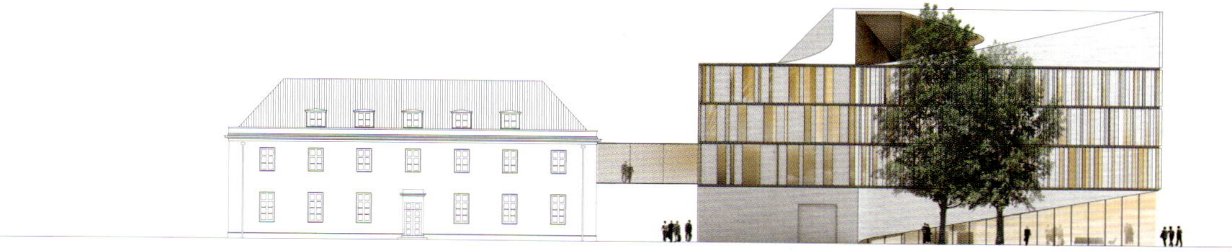

北立面
North elevation

The expansion of the Courthouse will be a sculptural and classic structure, which at once expresses a welcoming transparency while maintaining the justice system's sobriety and seriousness. The new Courthouse will be a natural extension of the existing neo-classical building, yet still maintaining its own identity through a modern and contemporary expression. The interior of the courthouse focuses on compliance with the law reform's requirements on security and internal segregations in the building. Therefore, the building provides its employees, the defendants, witnesses and guests an open and friendly environment in which it is easy for the different user groups to navigate. A small atrium cuts through the middle of the building, drawing light deep into the interior creating an open and airy connection across departments.

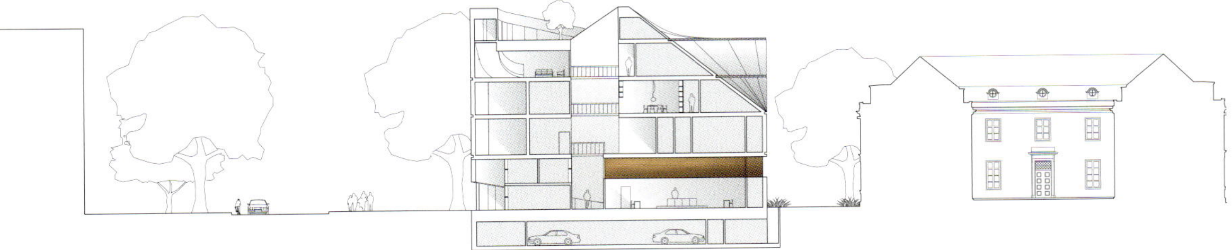

AA剖面
Section AA

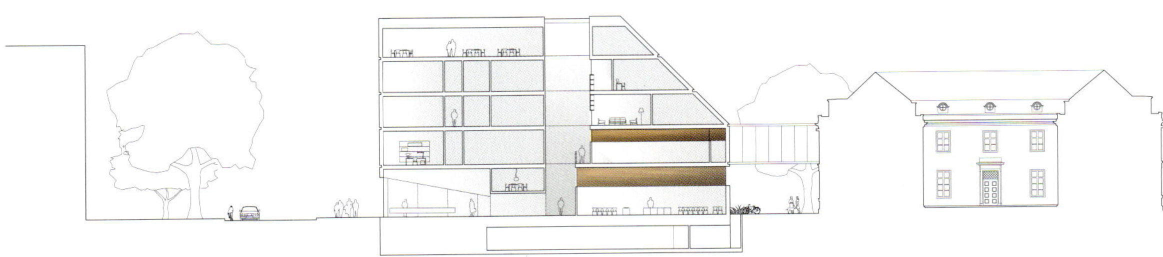

BB剖面
Section BB

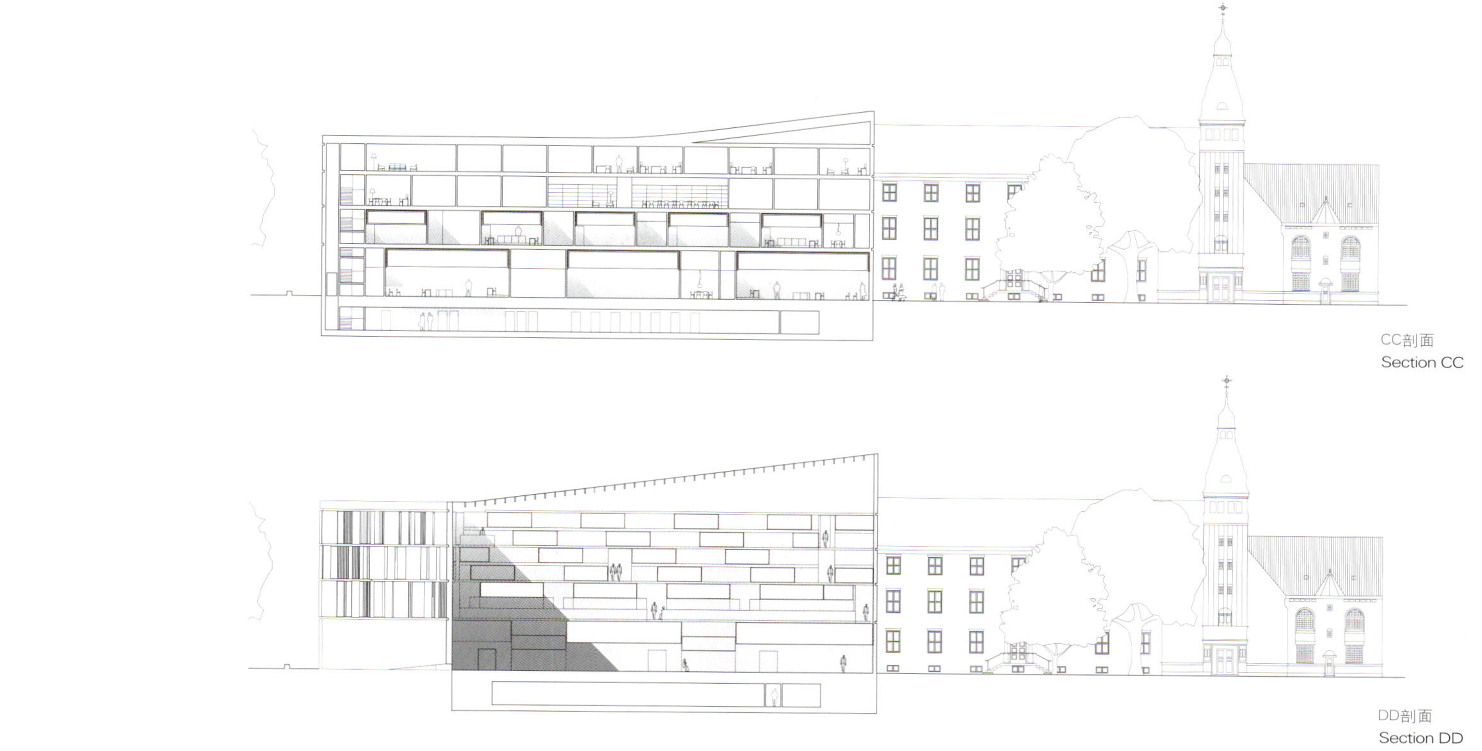

CC剖面
Section CC

DD剖面
Section DD

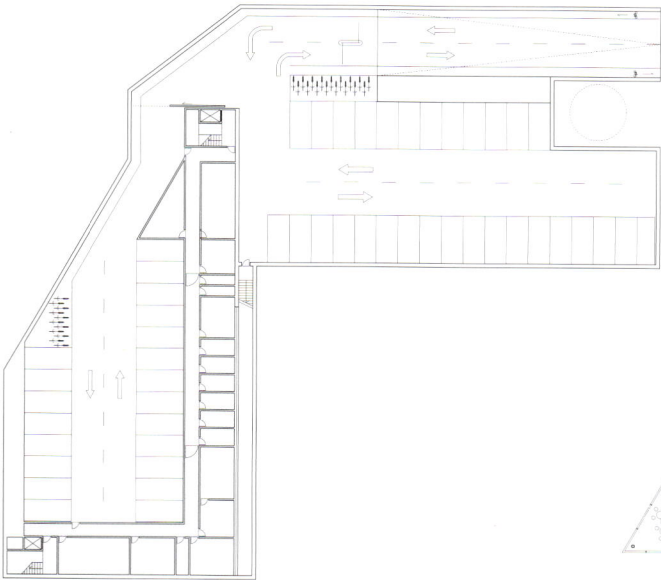

地下一层
Basement

一层
Ground floor

夹层
Mezzanine

二层
First floor

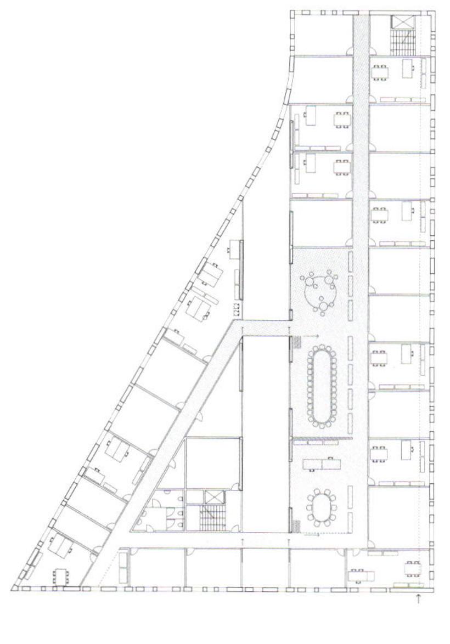

三层
Second floor

四层
Third floor

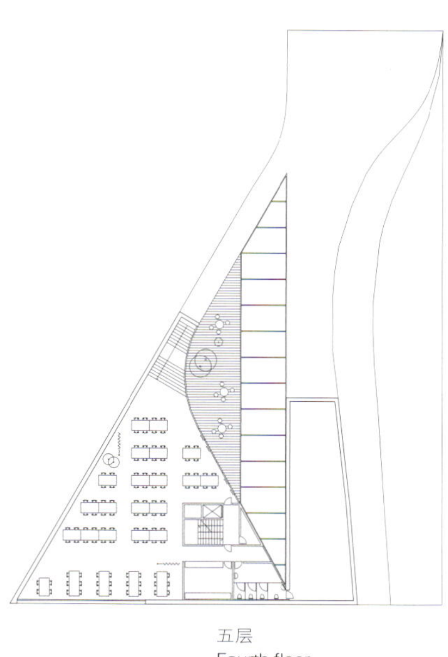

五层
Fourth floor

Bagsværd公园总体设计
BAGSVÆRD PARK MASTERPLAN

建筑师：C. F. Møller Architects
景观设计师：C. F. Møller Architects
工程师：Danish Road Directorate - traffic and environmental planning
图片：C. F. Møller Architects
位置：Grusgraven, Gladsaxe, Denmark
甲方：Gladsaxe Municipality

场地的先决条件、位置、历史痕迹、四分五裂的所有权和现有的状况是整体规划实用布局的基础。为了确保社会、经济和环境在未知的时间规划内实现可持续的发展，场地原有品质提到了提升，潜在的危险被转化为优势。项目的整体目标是要形成都市化的风格，这种风格将随意、绿色郊区生活方式与的密集与动感的都市生活轻松地融为一体。

The site's prerequisites, location, historic traces, fragmented ownership, and current state are the foundation for the pragmatic layout of the master plan. The existing qualities are enhanced, and potential risks turned into advantages, to ensure social, economic and environmental sustainability in a long term implementation within an undefined timeframe. The overall objective is an urbanity, which effortlessly combines the informal, green suburban lifestyle with an urban density and dynamic.

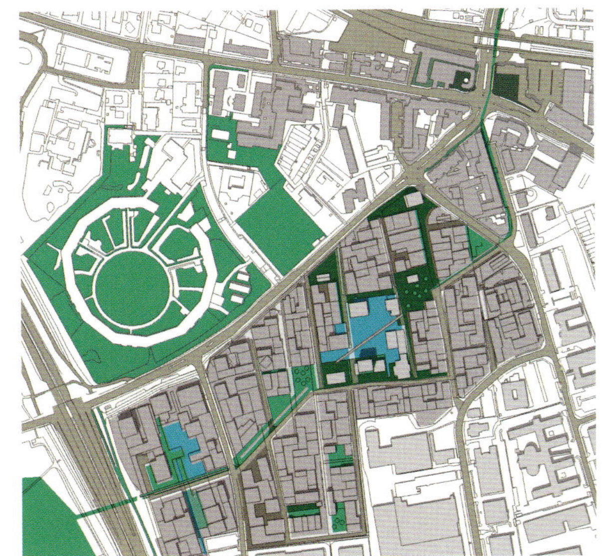

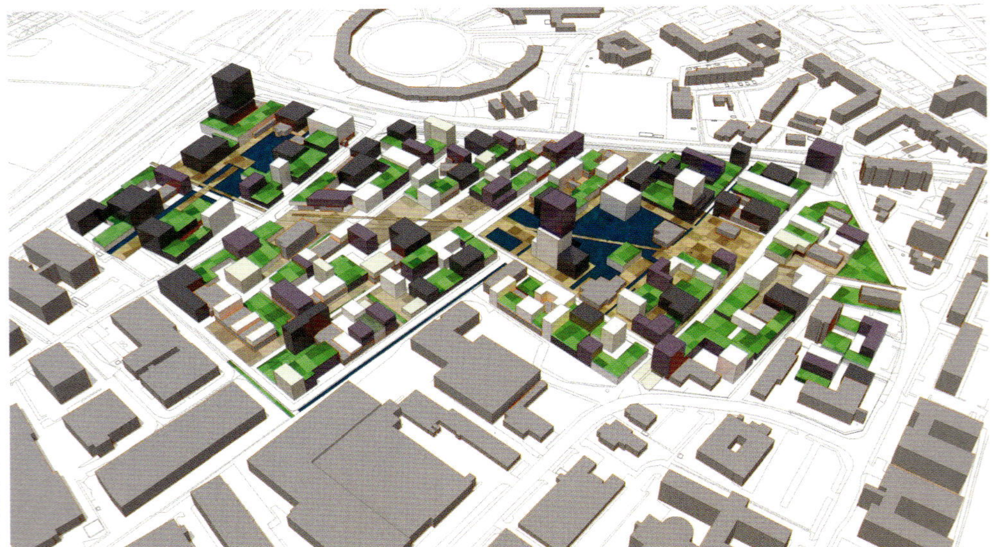

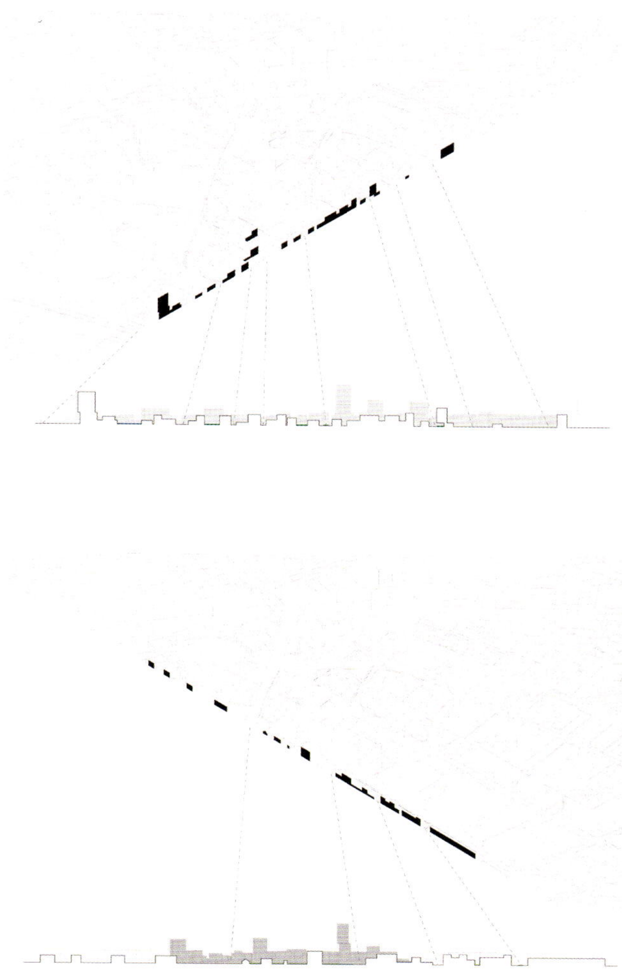

BAGSVÆRD
Bagsværd
Jernhandel a·s

欧登赛的研究与知识公园
RESEARCH AND KNOWLEDGE PARK, ODENSE

建筑师：C. F. Møller Architects
景观建筑师：SLA A/S
工程师：Alectia A/S
参与设计建筑师：NCC Property Development
规模：200000m²
年份：2009-2020
获奖情况：2009建筑竞赛一等奖
图片：C. F. Møller Architects
地点：SDU, Odense, Denmark
甲方：Freja Properties A/S

欧登赛大学的研究与知识公园和入口区域的核心理念是用可持续性的方法建造高密度的群体开发，这样可以保留更多的绿地不被破坏。这个理念意味着将10万平方米研究与科技设备与附加的共10万平方米的公共区域、住宅区、学生住宅、公共机构和旅馆容纳在一片密集的城市环境中。

这一理念提供了明确简单的环保回馈，包括密集的建筑、简短的设备路线、简化的基础设施、最低的土地利用率和良好的城市微气候。该设计同样具有巨大潜力，让不同的用户能够近距离地彼此接触，并将科学研究与周围的环境相融合。

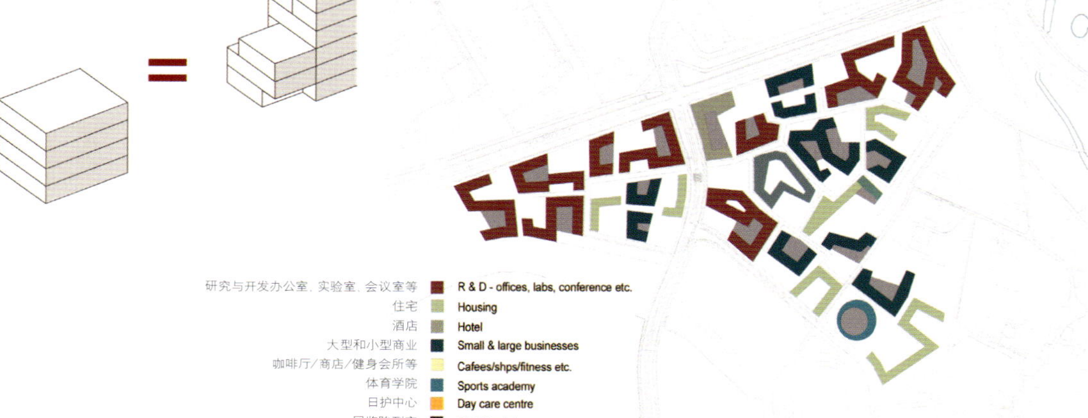

Functional layout - all floors
功能分布图－所有楼层

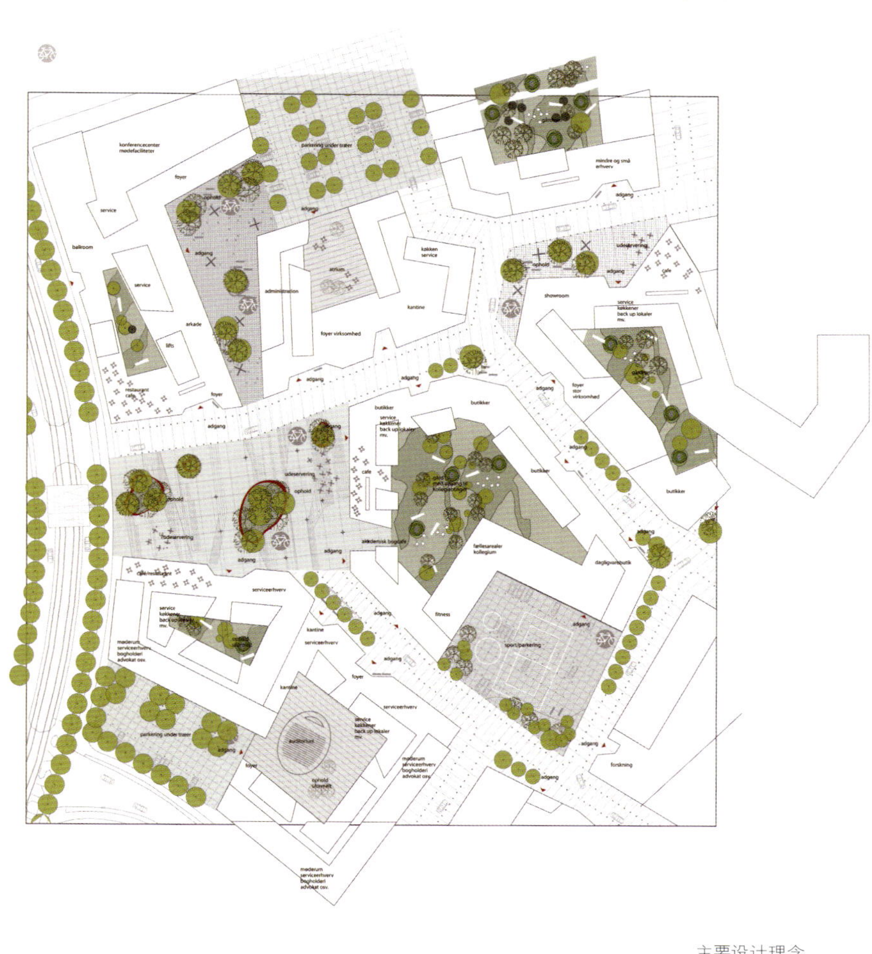

The central idea behind the new Research and Knowledge Park and Portal Zone at the University of Odense is a consequent sustainable approach, creating a dense cluster development, and thus leaving more green nature untouched. The concept means bringing together the 100.000 m^2 research and science facilities with the additional mix-use of 100.000 m^2 public functions, dwellings, student housing, institutions and hotel in a dense urban environment.

This concept provides straightforward, simple sustainability rewards, including compact buildings, short services routings, minimal infrastructure, minimal land-use and a good urban micro-climate. It also has great potential for bringing the various users in close contact to each other, and for integrating scientific research with the surrounding community.

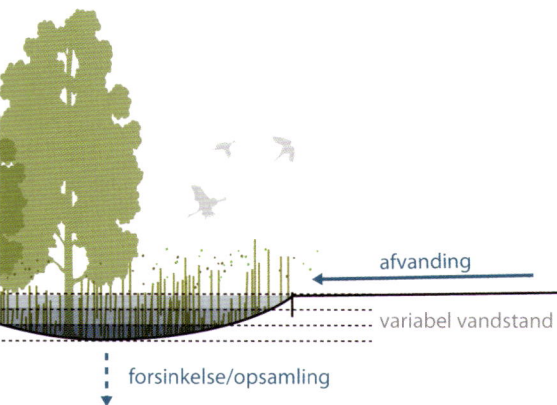

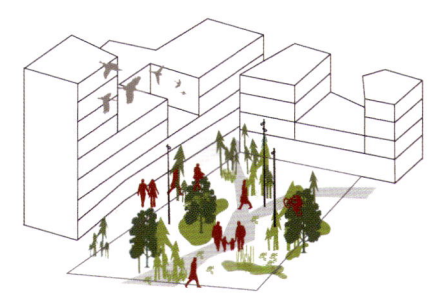

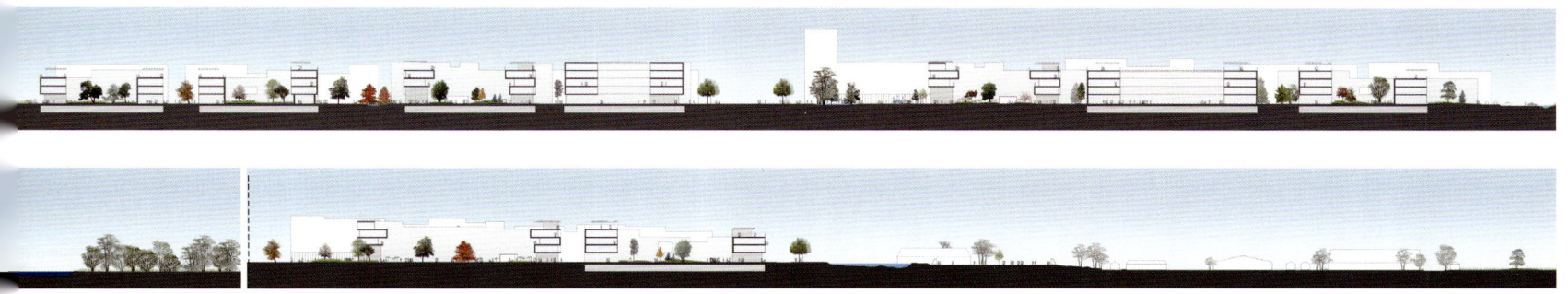

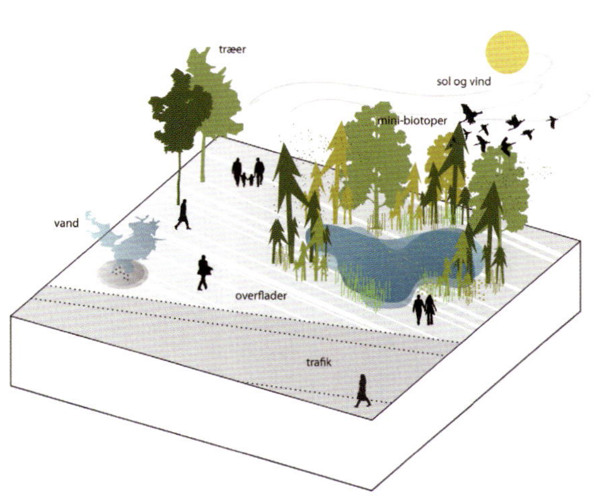

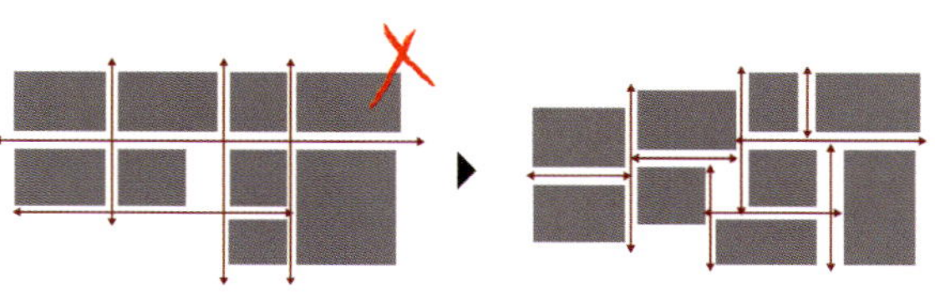

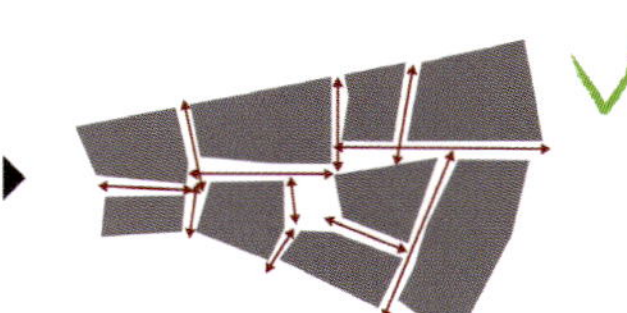

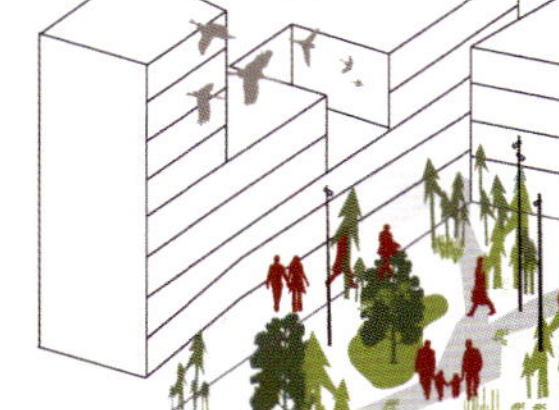

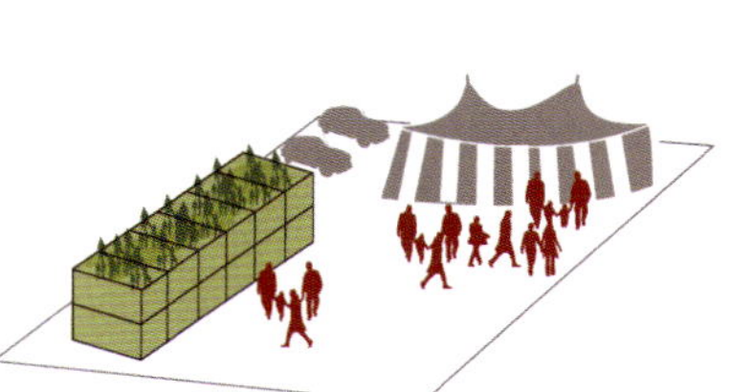

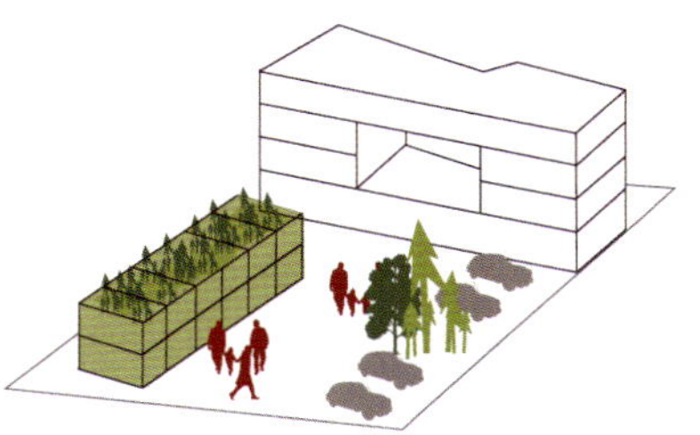

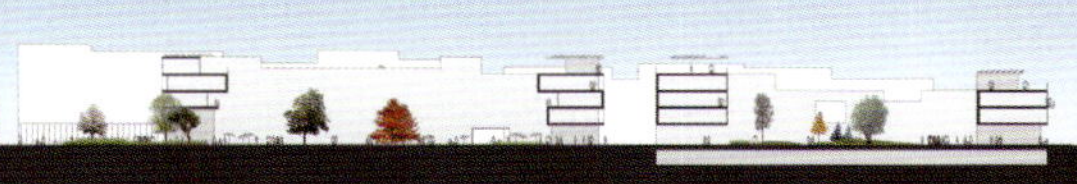

多样空间/共用空间

MULTIPLADS / SHARED SPACE

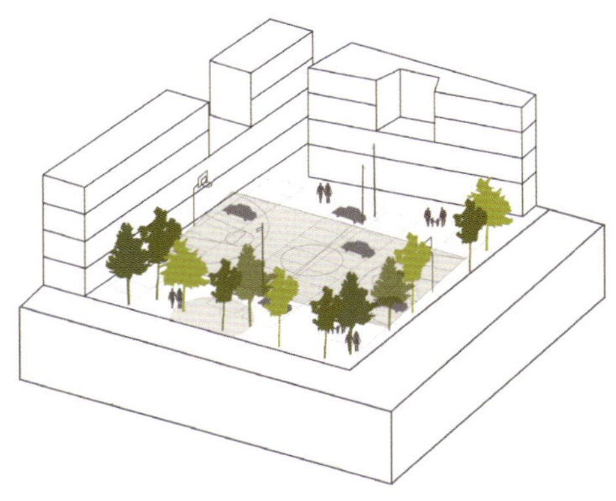

绿色公园

GRØN PARKERINGSLOMME

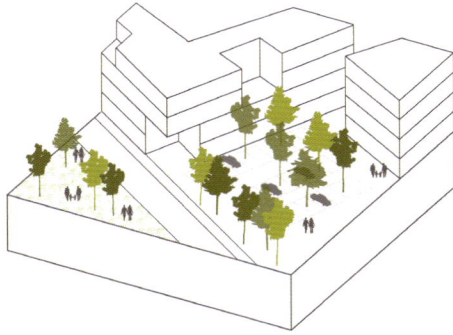

绿色公园

GRØN RANDPARKERING

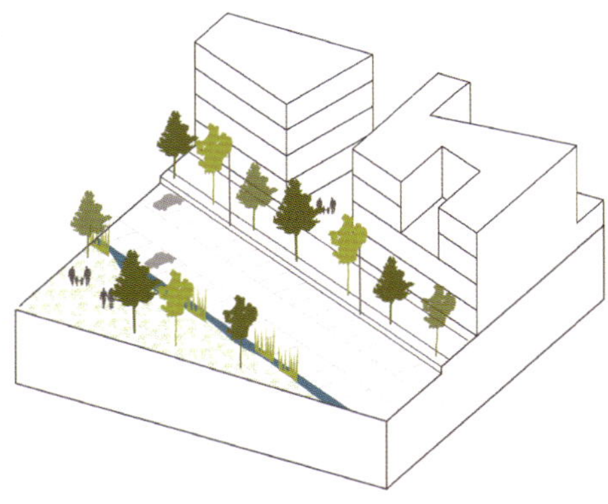

城市空间

BYRUM

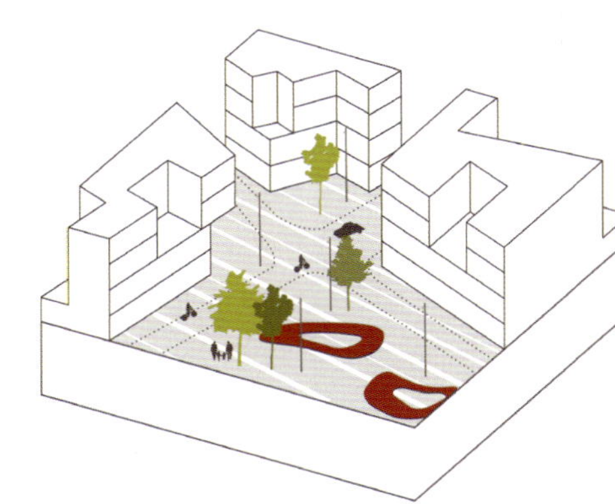

绿色公园

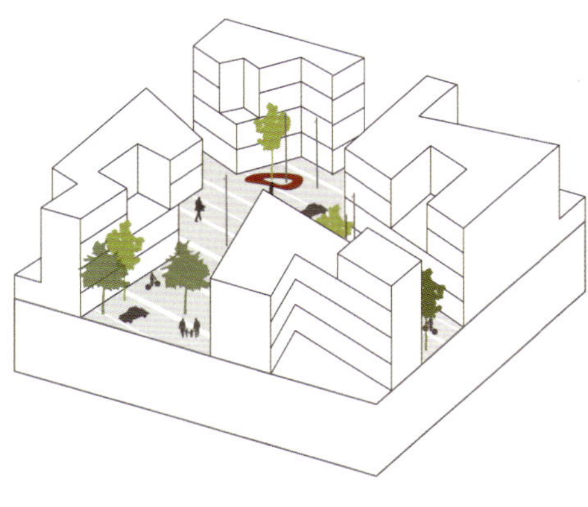

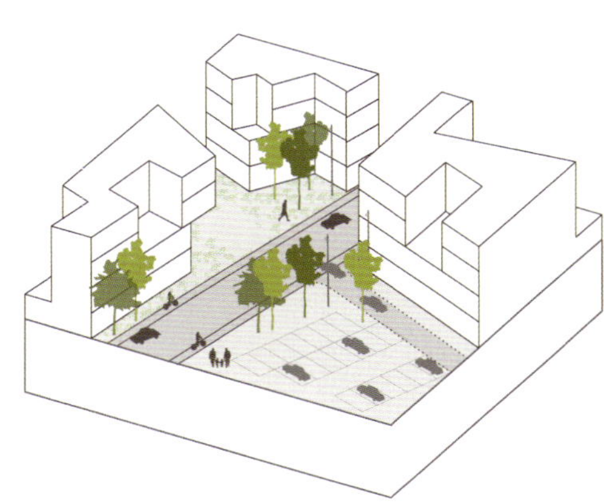

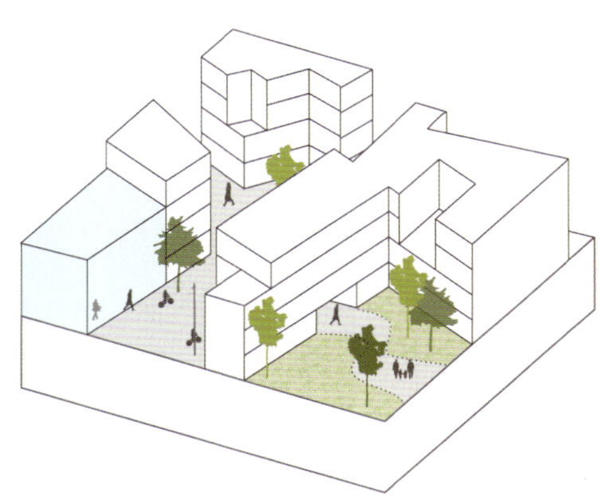

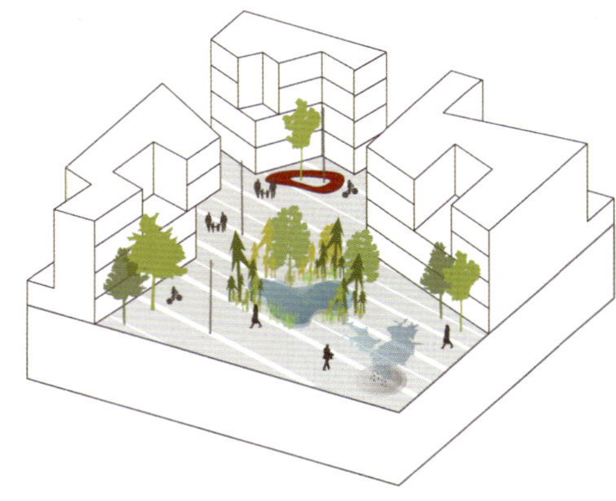

公园生活整体规划
PARK LIFE MASTERPLAN

建筑师：JA Joubert Architecture
总建筑师：Marc Joubert
项目团队：Jeroen de Loor, Marian Dusinsky, Alessandro Guida, Kim Byungchan
甲方：Eurocol
模型：Made by Mistake
面积：51 300 m²＋14 000 m²停车场
项目地点：Tirana, Albania
获奖情况：邀请赛一等奖

建筑师开发了一套灵活的核心筒体系，其中包括一个独立的开放式楼梯以及一部电梯，因此可以灵活地分隔公寓。工作室公寓、跃层公寓、带阁楼公寓以及复式公寓这些不同类型的公寓在同一个结构内都可实现。每个核心筒服务于6～16间公寓，根据需要可以调整电梯的数量，甚至在施工过程中也可以调整。
室外空间包括从斜坡屋顶中切割出来的露台、凉廊和凸窗，虽然类型和尺寸各不相同，但仍然保持了统一的外观。

By developing a flexible system of cores, consisting of a single open-air staircase with an elevator, a flexible division of apartments is possible. Ranging from studio apartments, duplexes, penthouses, split-level apartments, these can be varied within the same structure. With circulation for 6 to 16 apartments per core, the number of apartments and elevators can be varied according to demand and even adjusted during the process.
Outdoor spaces consist of roof terraces, cut into the sloping roofs as well as loggia's and bay windows, allowing for variation in type and size, while still maintaining the appearance of the volumes.

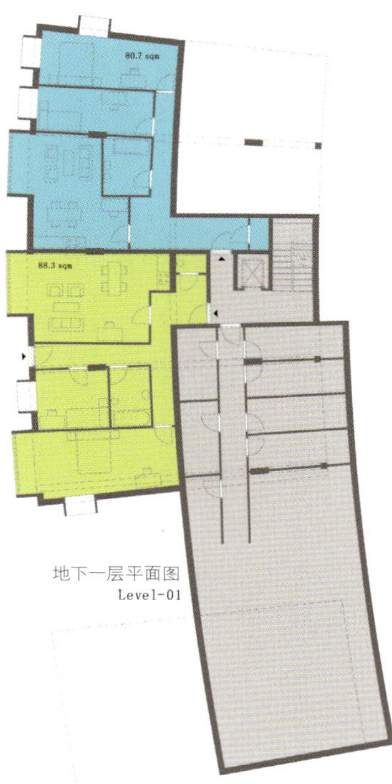

地下一层平面图
Level-01

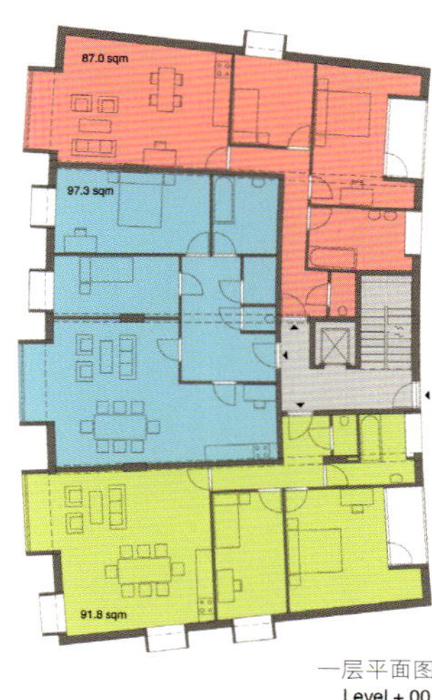

一层平面图
Level + 00

四层平面图
Level + 03 roof

变容天主教堂
CATHOLIC CHURCH OF THE TRANSFIGURATION

建筑师：DOS Architects Ltd
甲方：Catholic Church of the Transfiguration
项目地点：Victoria Garden City，Lekki，Lagos，Nigeria
3D图片鸣谢：Meshroom Ltd
结构工程师：AKT Engineers
设备工程师：DSA Engoneers

虽然设计方案与非专业人士来说看起来非同寻常，但实际上是按照天主教堂设计的传统原则设计的。主圣会大厅的管风琴和圣坛上方有一个天主教十字架。大厅内有一个正殿，两侧各有一个过道，与教堂的主轴一致。天主教十字架设在教堂的最高点上，将成为乐卡市和整个拉各斯的标志。

Even though our design proposal may seem unconventional to the untrained eye, it is actually based on traditional principles of Catholic Church design: The main congregation Hall features a Latin cross above the Organ and altar; The hall has a nave and two aisles at each side which are all coincident with the main axis of the Church; we have placed a Latin Cross on the highest point of the Church's structure, which will become an icon for the city of Lekki and Lagos as a whole.

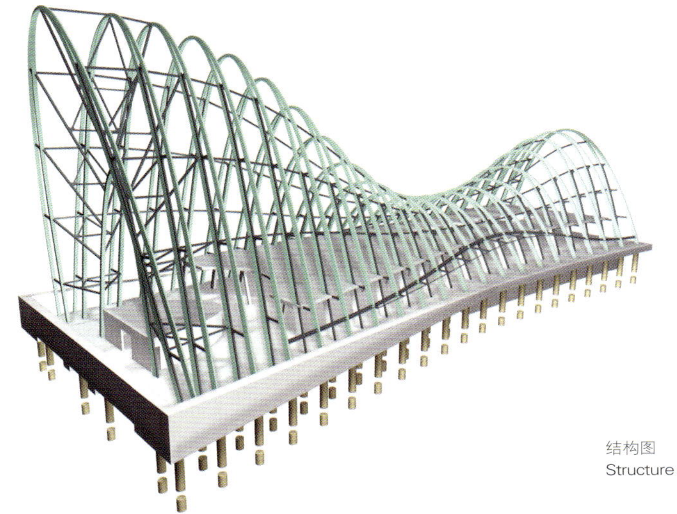

结构图
Structure

剖面图
Sections

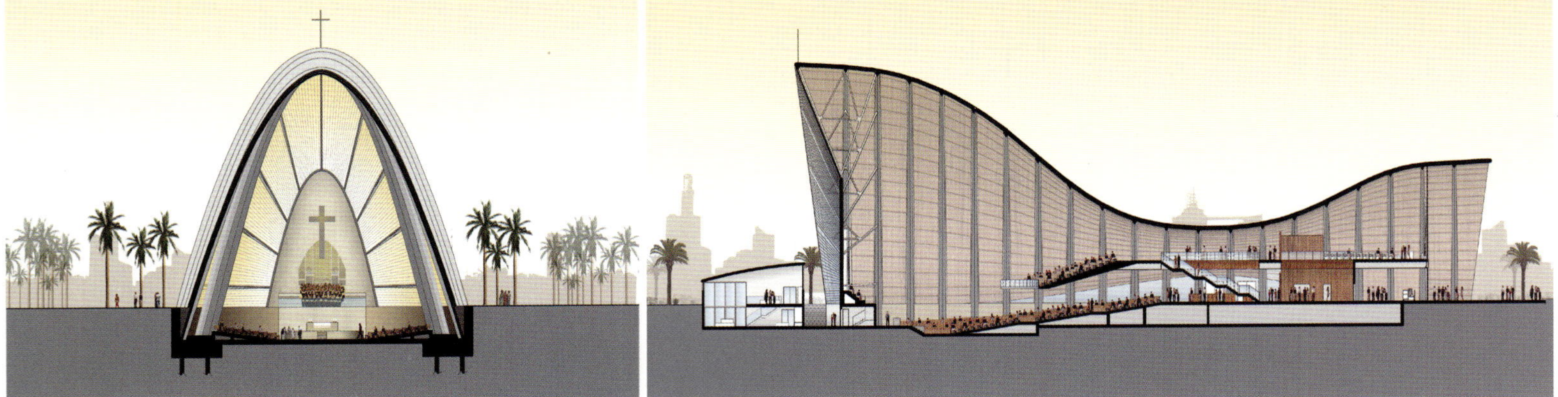

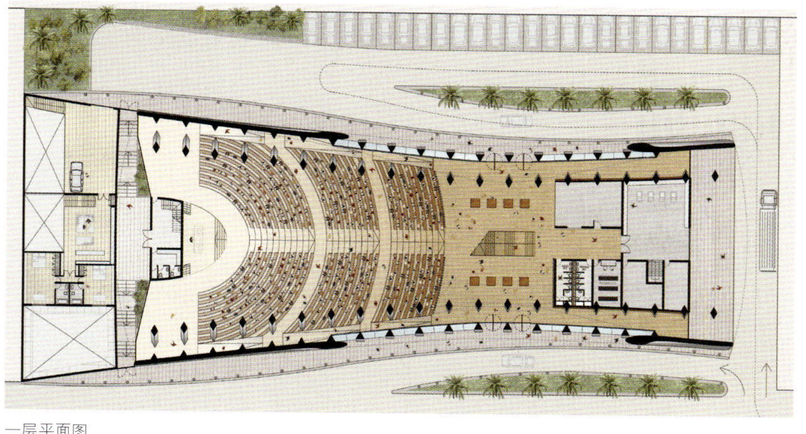

一层平面图
Ground floor plan

二层平面图
First floor plan

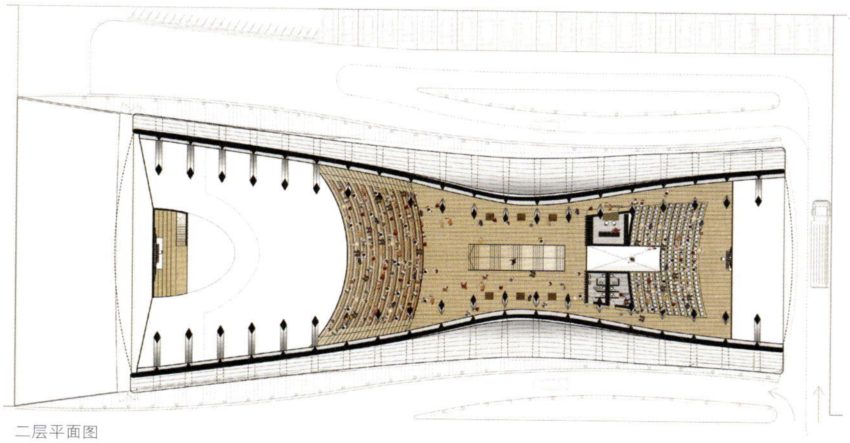

圣经
The Bible
圣三一
The Holy Trinity
圣礼仪：面包与酒
The Eucharist: Bread and Wine
JHS
面包与鱼的奇事
The Miracle of the Bread and Fish
十字架
The Cross

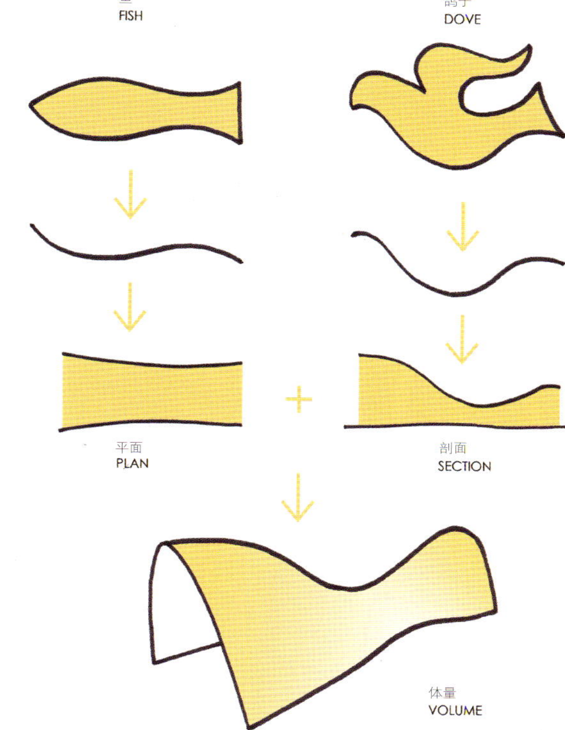
鱼
FISH
鸽子
DOVE
平面
PLAN
剖面
SECTION
体量
VOLUME

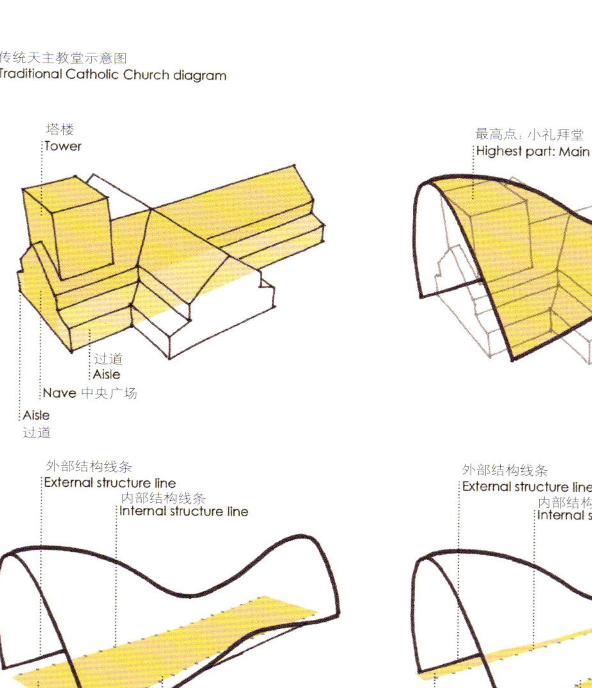
传统天主教堂示意图
Traditional Catholic Church diagram
塔楼
Tower
过道
Aisle
Nave 中央广场
Aisle
过道
最高点：小礼拜堂
Highest part: Main chapel
外部结构线条
External structure line
内部结构线条
Internal structure line
双拱结构在平面上创造了三个不同的空间
Double arch structure creating three different spaces in plan
中央广场
Nave
外部结构线条
External structure line
内部结构线条
Internal structure line
双拱结构在平面上创造了三个不同的空间
Double arch structure creating three different spaces in plan
过道
Aisle

设计灵感
Inspiration

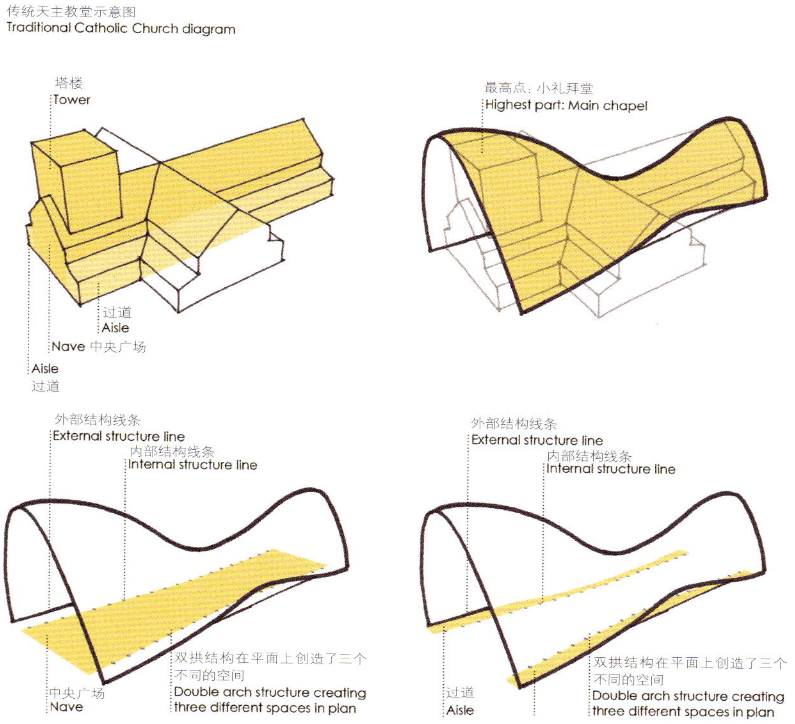

空军教会中心
AIR FORCE VILLAGE SPIRITUAL CENTRE

建筑师：IwamotoScott

主持建筑师：Lisa Iwamoto & Craig Scott

项目团队：Sean Canty, Ryan Golenberg, Shuang Hao, David Swa m, Natsuki Matsumoto

图片提供：IwamotoScott（除了P399页下图由Kory Beig/IwamotoScott提供）

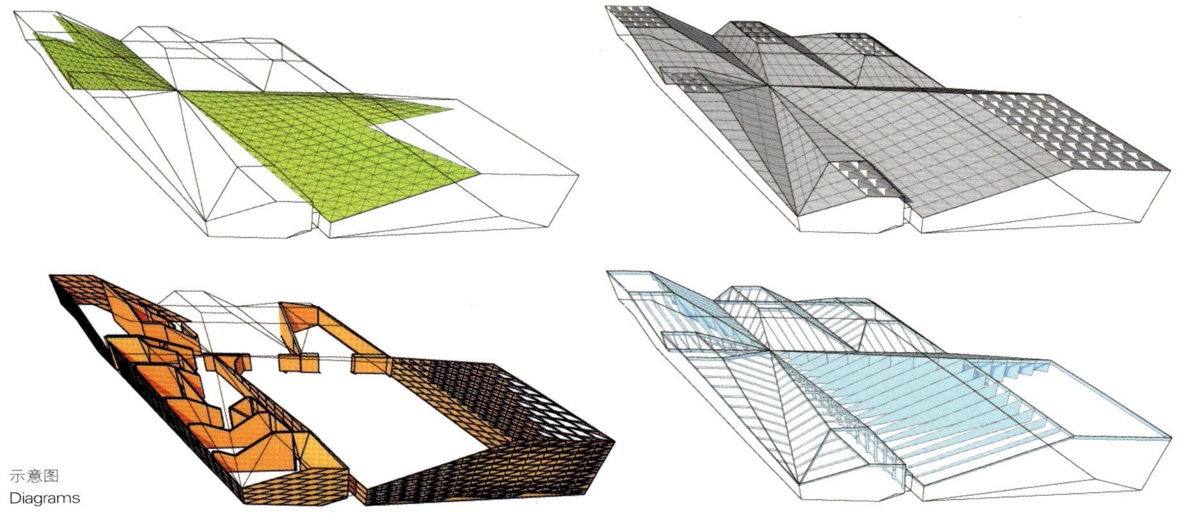

示意图
Diagrams

新建空军教会中心的设计像是由几个巨大的天窗构成的。这座钻石一样的建筑物把天窗在地面上结合为一个整体，但同时也允许它们的屋顶以独立的形式存在，就像一个方队或车队一样，分别代表五种职能不同的军队——空军、陆军、海军、海军陆战队和海岸警卫队。这五个分离的小天窗围绕着一个较大的圣所紧紧聚集在一起，成为一个整体结构，象征着各种信仰的统一。该建筑物也充当着视觉和建筑学上的标志，成为空军教会现有场地布局轴线上的一个独特形象。

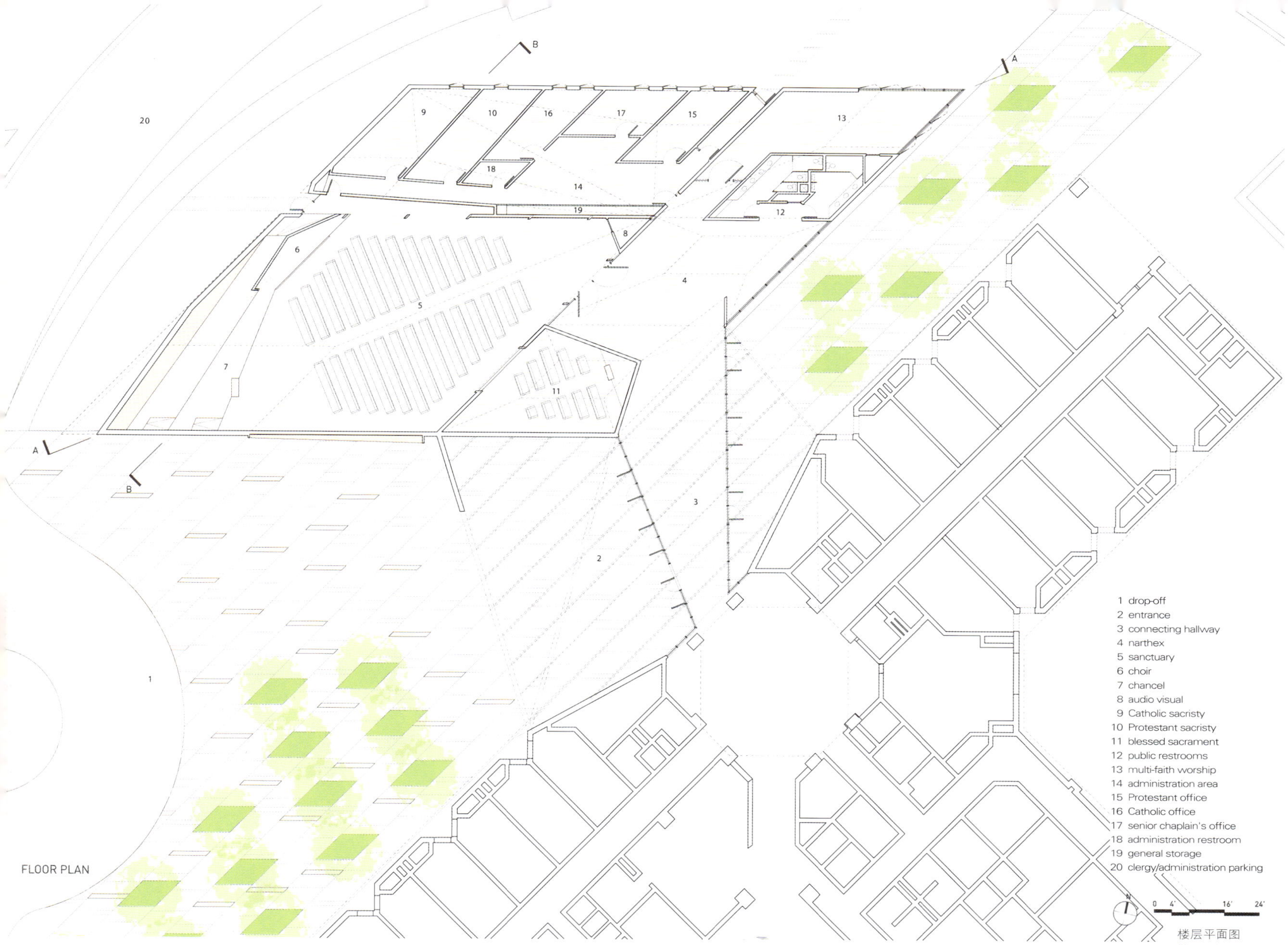

FLOOR PLAN

1 drop-off
2 entrance
3 connecting hallway
4 narthex
5 sanctuary
6 choir
7 chancel
8 audio visual
9 Catholic sacristy
10 Protestant sacristy
11 blessed sacrament
12 public restrooms
13 multi-faith worship
14 administration area
15 Protestant office
16 Catholic office
17 senior chaplain's office
18 administration restroom
19 general storage
20 clergy/administration parking

楼层平面图
Floor plan

1 下车区域
2 入口
3 连接走廊
4 教堂前厅
5 圣所
6 唱诗班
7 高坛
8 视听设备间
9 天主教圣器室
10 新教圣器室
11 圣体室
12 公共卫生间
13 多信仰礼拜室
14 行政管理区
15 新教徒办公室
16 天主教办公室
17 高级牧师办公室
18 管理人员卫生间
19 总储藏室
20 牧师/管理人员停车场

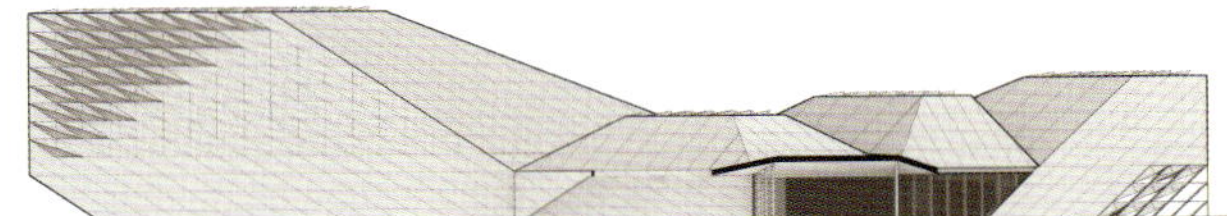

东南立面
Southeast elevtion

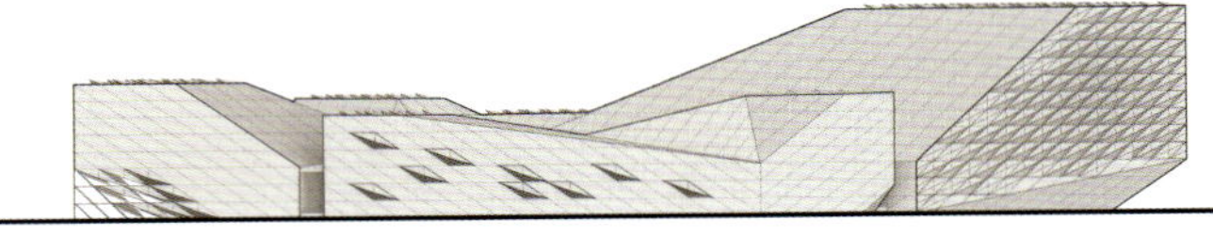

0 8' 16' 32'

西北立面
Northwest elevtion

The new Air Force Village Spiritual Center is designed as a formation of volumetric skylights. The diamond shaped building perimeter binds the skylights together into a singular whole at ground level, but allows them to emerge as independent volumes at the roof, like a squadron or fleet, representing the five branches of the military that the Air Force Villages serve - Air Force, Army, Navy, Marines and Coast Guard. These five smaller skylights are clustered around a larger volume that is the Sanctuary. The multi-denominational nature of the spiritual center is also represented by the separate skylight volumes held together by the pure overall building geometry symbolizing the unity of faith. The building also serves as a visual and architectural landmark, emerging as a unique figure from the existing axes of the Air Force Village site plan.

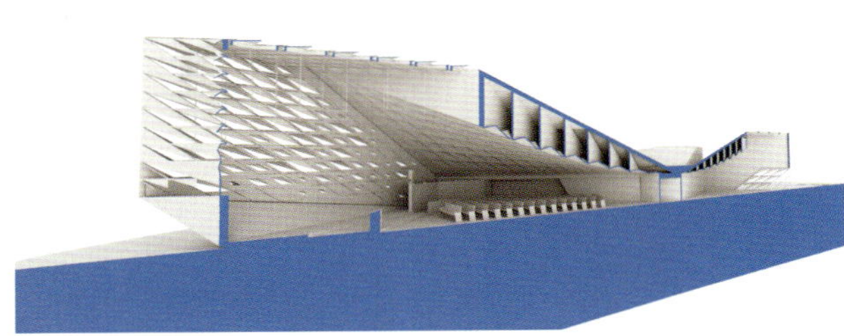

A-A剖面2
Section A-A 2

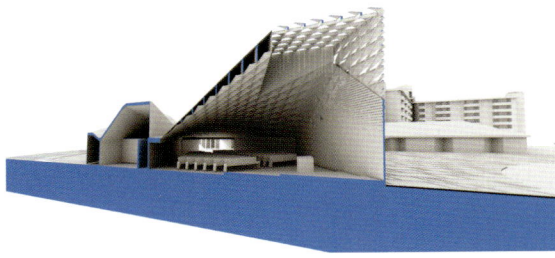

B-B剖面
Section B-B

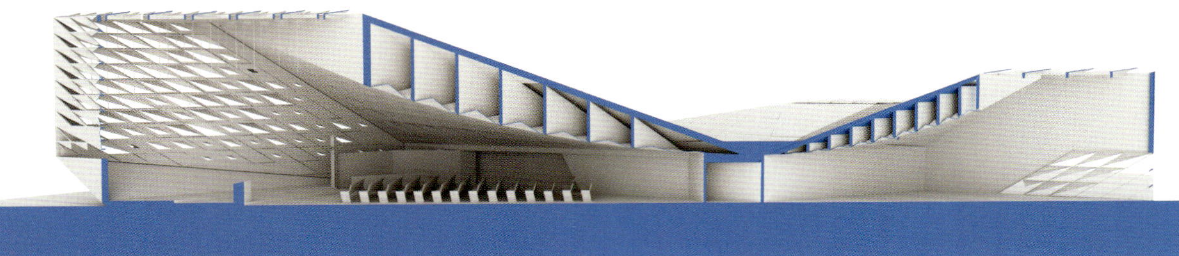

A-A剖面1
Section A-A 1

清真寺项目设计——"一次完美的清真寺建筑竞赛"
THE MOSQUE PROPOSAL – "Ideal Competition on Mosque Architecture"

建筑师：studiOZ
项目团队：Hakki Can Ozkan, Ytu-Yildiz Technical University, Istanbul, Mehmet Yigit Ozturk, Ytu-Yildiz Technical University, Istanbul

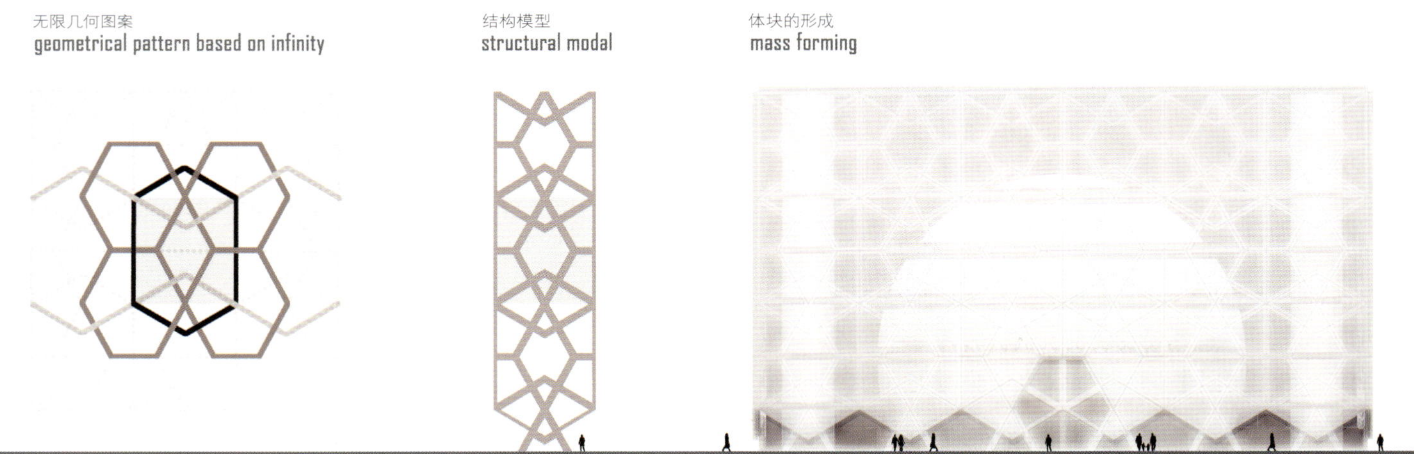

在历史长河中，清真寺是社会生活中必不可少的组成部分。在土耳其的建筑史上，16世纪安纳托利亚的"伟大的锡南"亲手实现了最不可思议、最巧妙的清真寺形式。锡南将一个独立的拱顶置于方形建筑上，从而形成当时那个年代最辉煌的建筑形式。"清真寺"项目运用此种理解作为一种隐喻，并使用最稳定的形式建造圆顶，象征着球体的无穷无尽。该项目批判近来漫无目的地重复建造这种神圣场所，并提倡从有益于城市空间的角度对建筑物的象征性和多功能化进行改革。

Mosques are the indispensable components of our social life upon history. In Turkish architecture, the mosque had reached the most marvellous and smartest form by the hands of "Sinan the Great" at 16th century of Anatolia. Sinan used a single dome placed on the square form to reach the most brilliant form of his era. "The Mosque" approached to this comprehension as a metaphor and built a dome by using the most stable form to symbolize infinity, the sphere. "The Mosque" critisizes this holy and public place which recently turned into the aimless repetitions, and reforming benefits to urban spaces, symbolity and functionality of the building.

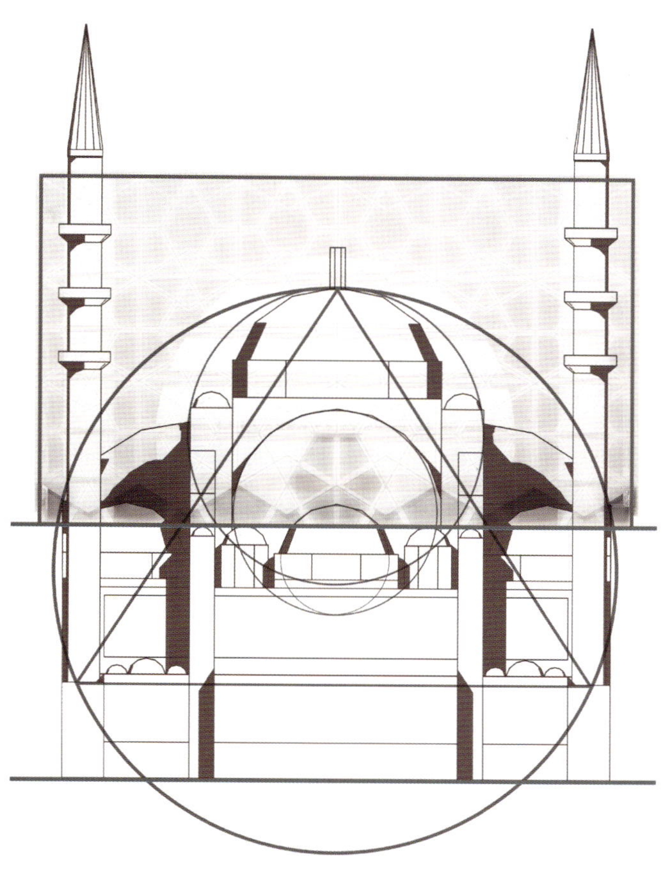

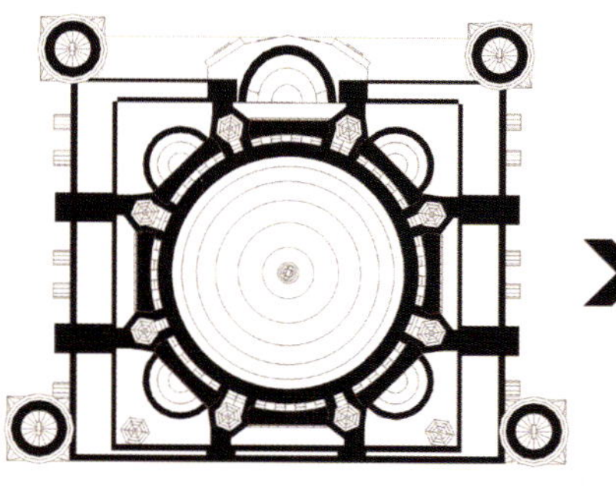

塞利米耶清真寺传统平面设计
Selimiye mosque traditional plan scheme

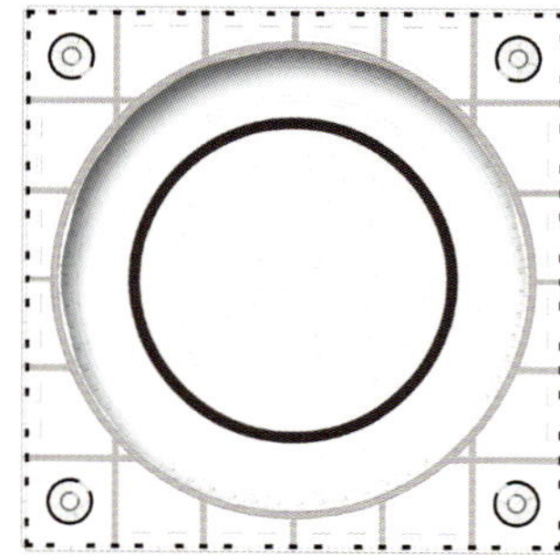

本项目平面设计
"The mosque" plan scheme

1 照明储备
2 混凝土拱顶
3 讲坛
4 米哈拉布（清真寺中朝向麦加方向的壁龛）
5 祈祷大厅（500人）
6 卫生间
7 门厅
8 钢结构图案
9 模块玻璃嵌板立面
10 储藏室
11 观景台
12 媒体中心
13 图书馆
14 祈祷大厅上层
15 讲习班
16 水景花园
17 上空空间
18 主入口
19 遗体处理室
20 会议准备室
21 会议大厅（350人）

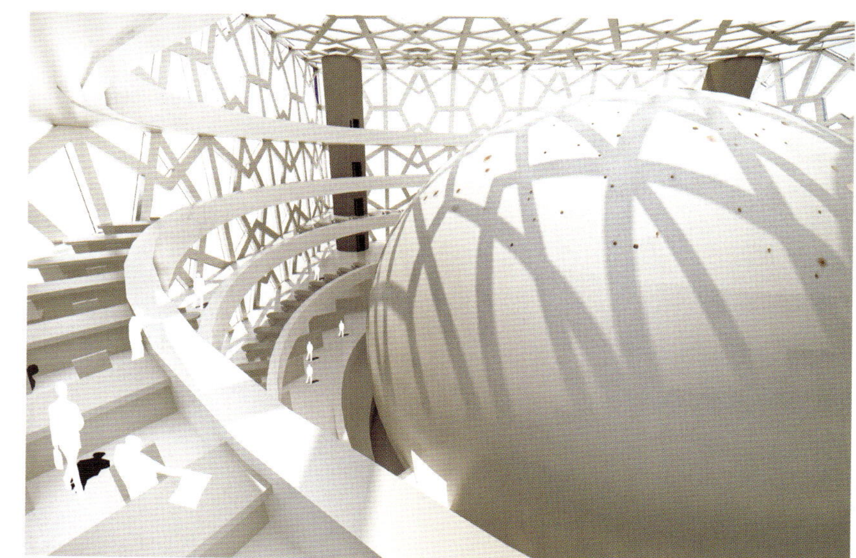

楼层平面
Floor plans

1 上空空间
2 展厅
3 祈祷大厅上层
1 void
2 exhibition hall
3 upper pray hall

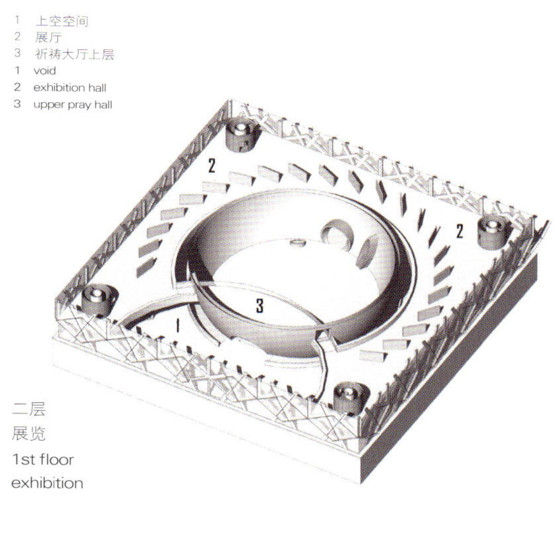

二层
展览
1st floor
exhibition

1 1号阅读大厅
2 2号阅读大厅
3 3号阅读大厅
4 4号阅读大厅
1 reading hall #1
2 reading hall #2
3 reading hall #3
4 reading hall #4

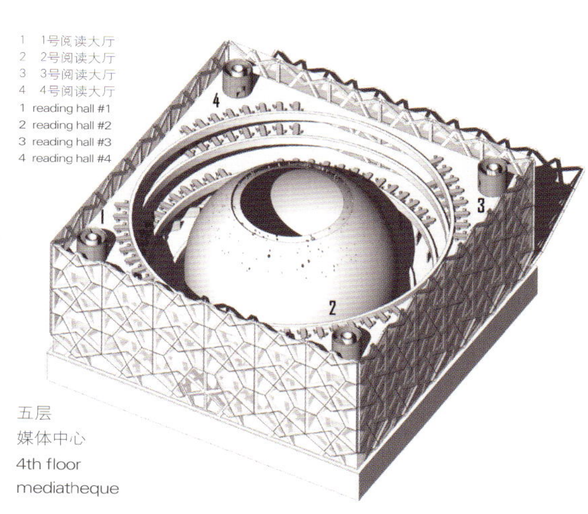

五层
媒体中心
4th floor
mediatheque

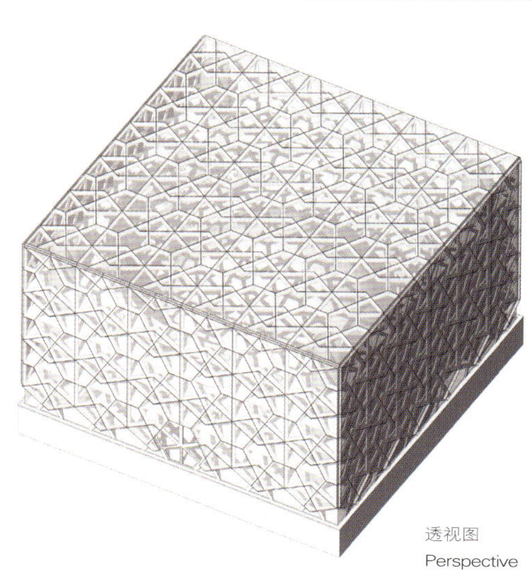

透视图
Perspective

1 入口
2 洗礼墙
3 祈祷大厅（500人）
4 灵柩台
5 布道桌
6 米哈拉布（清真寺中朝向麦加方向的壁龛）
7 讲坛
1 entrance
2 ablution wall
3 prayer's hall□500p□
4 coffin table
5 sermon desk
6 mihrab
7 minber

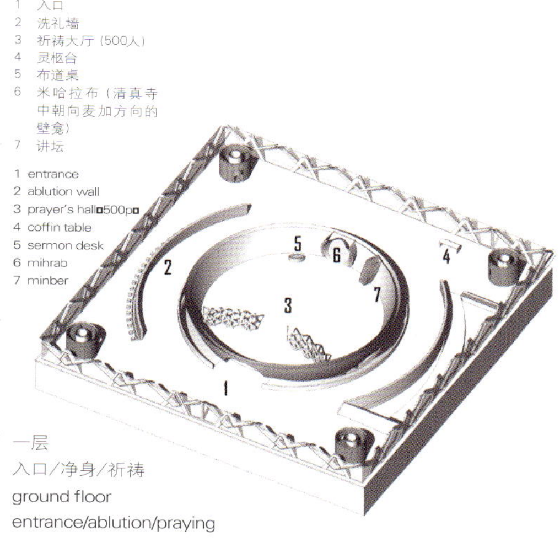

一层
入口/净身/祈祷
ground floor
entrance/ablution/praying

1 1号自修室
2 2号自修室
3 3号自修室
4 4号自修室
1 study hall #1
2 study hall #2
3 study hall #3
4 study hall #4

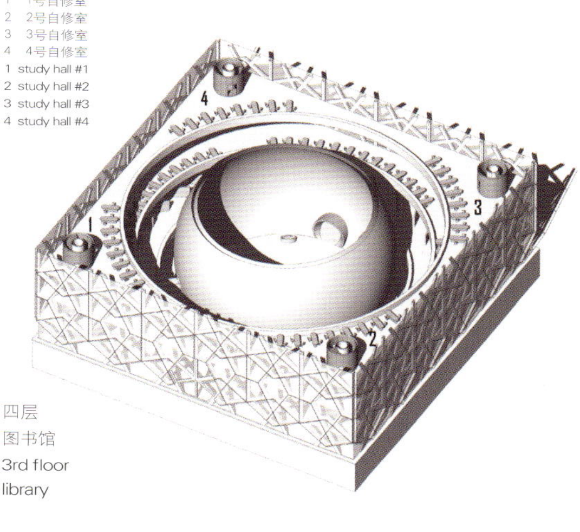

四层
图书馆
3rd floor
library

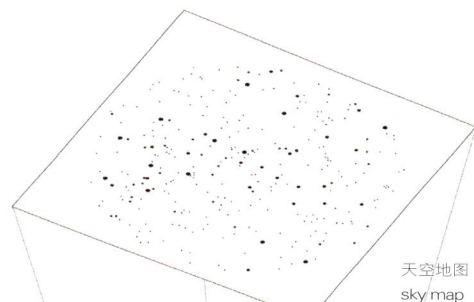

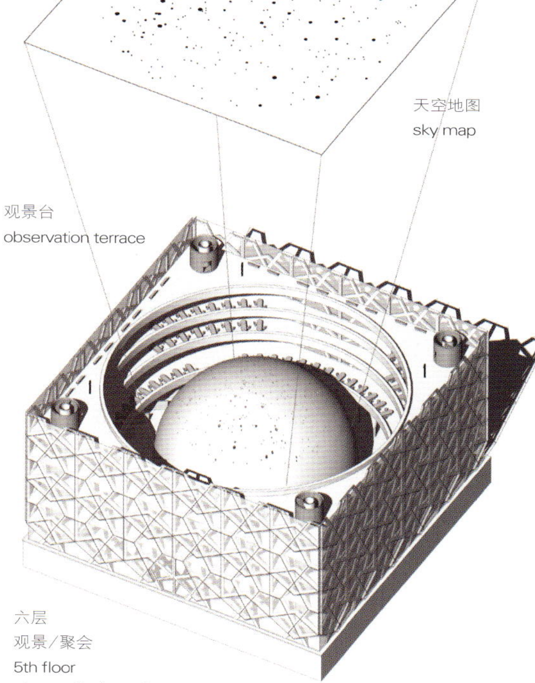

六层
观景/聚会
5th floor
observation/meeting

1 伊玛目和宣礼员室
2 卫生间
3 会议大厅（350人）
4 门厅
5 服务台/寄存室
6 遗体处理室
7 葬礼坡道
1 imam & muezzin rooms
2 wc
3 conventional hall(350p)
4 foyer
5 info/cloak room
6 corpse preparing
7 funeral ramp

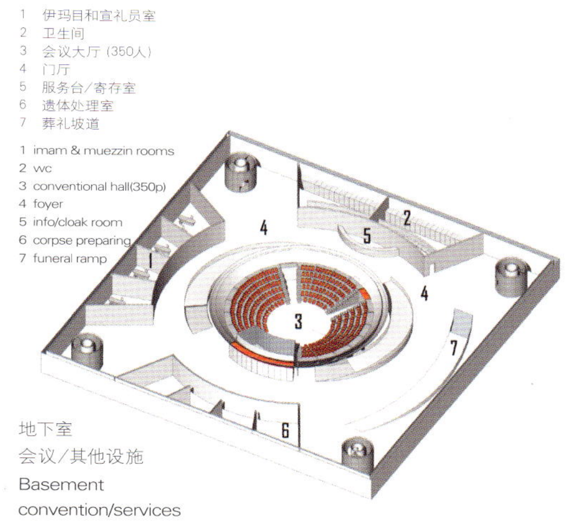

地下室
会议/其他设施
Basement
convention/services

1 1号讲习班
2 2号讲习班
3 3号讲习班
4 4号讲习班
1 workshop #1
2 workshop #2
3 workshop #3
4 workshop #4

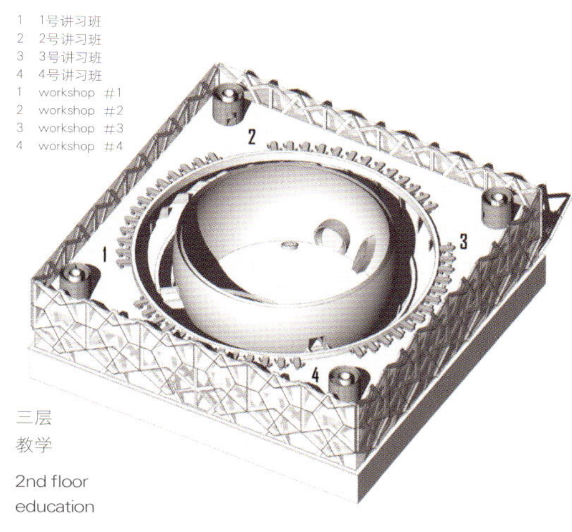

三层
教学
2nd floor
education

费城Breakaway自行车中心
THE BREAKAWAY, PHILADELPHIA

建筑师：Team: We Are You (Linda Heiman and Per Kaatman)

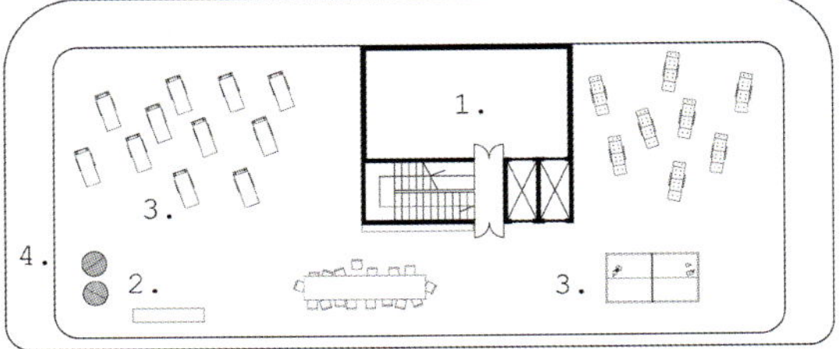

四层/半公共屋顶露台

1 服务区
2 烧烤区
3 休闲区
4 自行车停车立面上的不可攀爬的玻璃屋顶

4th Floor/semi public roof top terrace

1 service area
2 BBQ area
3 leisure area
4 non-climbable glazed roof of bicycle facade parking

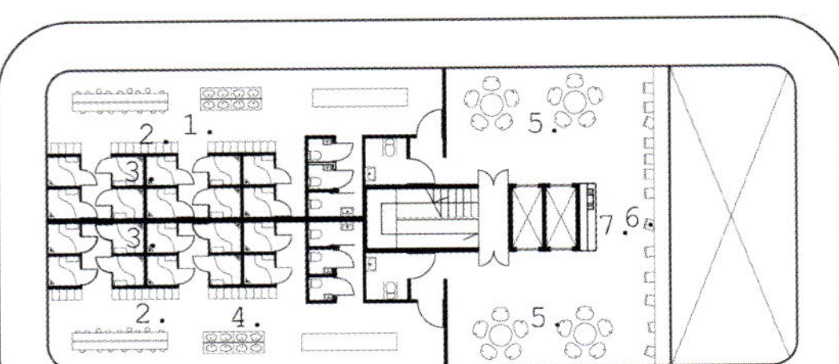

二层

1 女士更衣室
2 寄物柜，带镜子的座位，水槽
3 淋浴
4 男士更衣室
5 休息室座椅
6 非正式工作场所
7 打印/传真设备

2nd Floor

1 changing room – female
2 lockers, seating with mirrors, sink
3 showers
4 changing room – male
5 sit in lounge
6 informal working places
7 printing/fax facilities

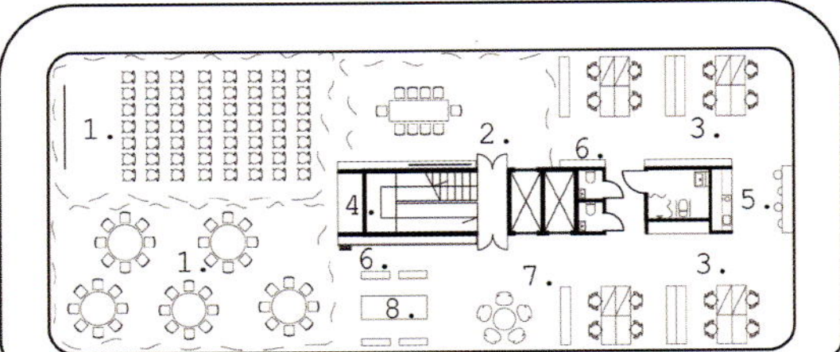

三层

1 会议室/教室，可通过吸音窗帘拆分
2 主会议室
3 开放式座椅
4 贮藏室（比如：椅子）
5 小厨房
6 图书馆（包括办公室）
7 入口大厅
8 走廊

3rd Floor

1 meeting room/classroom – divisible with sound absorbing curtains
2 main conference room
3 open plan seating
4 storage(eg. chairs)
5 kitchenette for social coffee breaks
6 library(spread across the office)
7 entrance lounge
8 gallery space

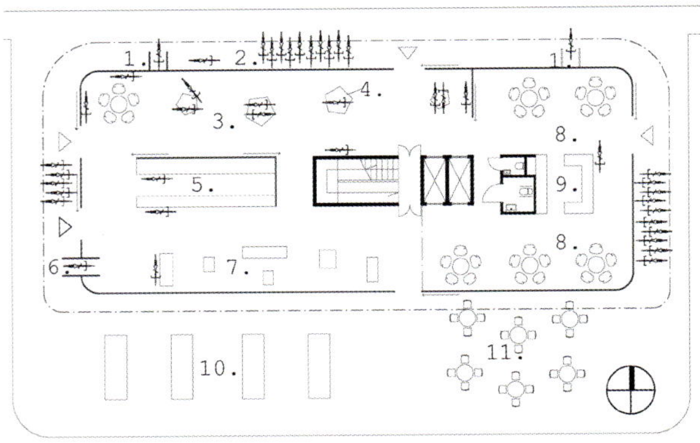

一层

1 立面泊车系统的装载箱
2 露天自行车停车场
3 维修车间——社交和教育空间
4 工具站
5 零售店/自行车租赁
6 店铺自行车停放计时装载箱
7 商铺展示
8 咖啡厅休息区，允许自行车进入
9 设备区
10 室外闲逛和修理区

Entrance Floor

1 loading box for façade parking system
2 exposed bicycle parking
3 repair workshop – social and educational space
4 tool stands
5 retail shop/bicycle rental
6 loading box with shop access and after hours access of rentals
7 shop display
8 café seating area – bikes are allowed
9 service area
10 outdoor hang out and repair area

垂直的自行车停车场
Vertical Bike Parking

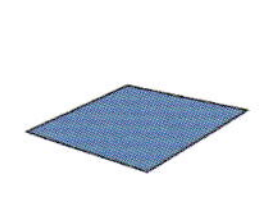

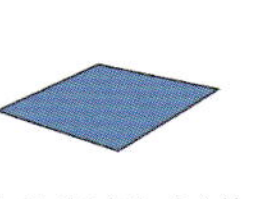

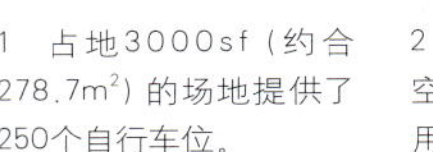

1 占地3000sf（约合278.7m²）的场地提供了250个自行车位。

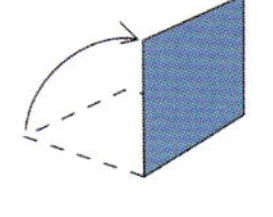

2 垂直停车场不仅释放了空间，还起到了展示的作用。

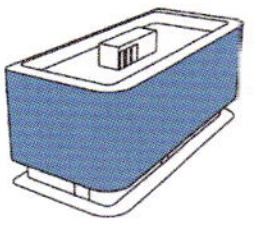

3 Breakaway自行车中心由三部分组成，即自行停放区域，自行车出租以及出售区域。

Breakaway是费城的一个新兴自行车中心。它的使命在于以各种形式推广自行车活动，并成为使用自行车上下班的人群、邮差、自行车旅行者和竞技者的热点区域。该建筑物的立面就是一个竖直的自行车停车场，向人们展现出这座建筑的特色。入口的一层设有公共自行车维修中心，用来停放自行车，与好友谈天说地或上一堂自行车组装课。这里也有出租自行车的商店，可以购买新自行车或仅仅更新旧零件。这里也有可以在上班路上买到三明治、纯净水和牛角面包的地方。其余的楼层为骑自行车上下班的人们提供沐浴和更衣的设施，还为当地自行车拥护群体设立了一间办公室。

The Breakaway is a new bike centre in the city of Philadelphia. Its mission is to promote bicycling in all its forms, and become a social hotspot for bike commuters, messengers, biking tourists and racers. The facade, that is a vertical bike parking, tells the story of the building. The entrance floor holds a public bicycle repair space where you can fix your bike, hang out with your friends and take a bike-building course. There is also a store where you can rent bikes, buy a new commuter friend or just top up the old buddy with some shiny parts. Here you will also find a cafe/urban feed station where you can pick up sandwiches, water bottles and croissants on your way to work. The other floors holds showers/changing facilities for bike commuters and an office for the local bike advocacy group.

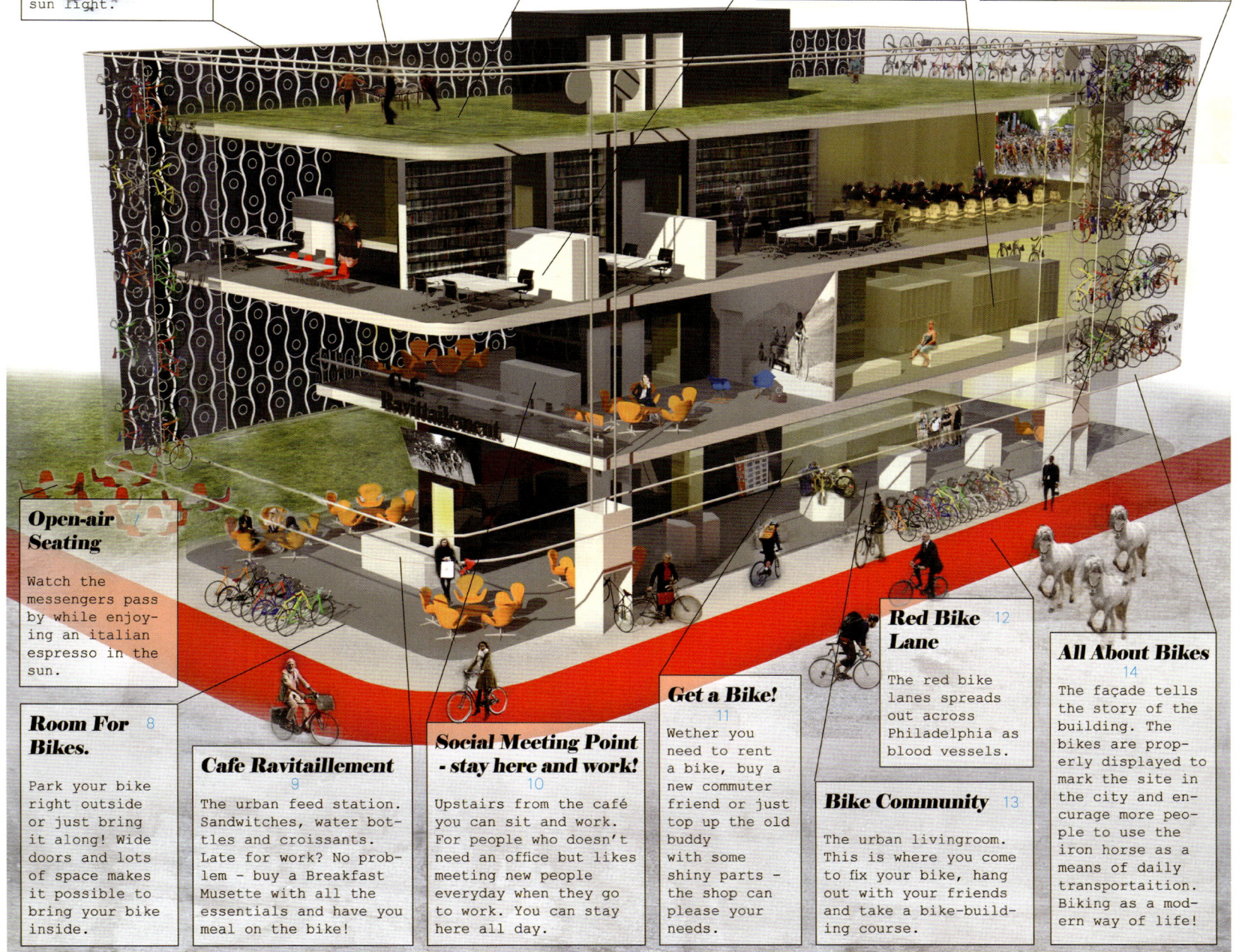

1 光伏板
光伏板像自行车的链条一样，为大楼提供了电力以及引人注目的外观，减少了阳光直射。
2 通风立面
双层立面确保了冬暖夏凉。
3 绿色屋顶
当你打乒乓球时，感受着脚趾间的点点绿意，感觉一定非常惬意。
4 自行车爱好者的办公室
当地的自行车联盟得到了一间支撑他们梦想的美妙场所。
5 快来！快来！
不要害怕上班骑自行车辛苦，享受速度带来的畅快感觉吧，然后在二楼的更衣间冲个热水浴。
6 垂直自行车停车场
把你的自行车停在立面上吧！只要停放到平台上，拿着小票，点击按扭，就能目送你的自行车驶过天空。快捷、安全、简便，能看到你的自行车吗？
7 露天座椅
在阳光的照耀下享受一杯浓咖啡，看着过往的邮差。
8 停放自行车的空间
把自行车停放在室外的右侧或直接随身携带！宽敞的大门和广阔的空间都便于随身携带自行车。
9 咖啡供给点
这就是一处都市食品小站，三明治、瓶装水以及牛角面包。上班要迟到了？没问题——带上所有的早餐，边骑车边享用吧！
10 社交聚会点——待在这儿工作吧！
从咖啡厅上楼，你可以坐在这儿工作。对于那些上班的时候不需要办公室但喜欢结识新朋友的人来说，你可以一整天都呆在这儿。
11 给自己拿一辆自行车！
不管你是需要租一辆自行车，还是买一个能陪你通勤的好伙伴，抑或是用闪亮的配件修理一下老朋友，这个商店都可以满足你的需求。
12 红色自行车路线
红色的自行车道在费城纵横交错，就好像流动的血液。
13 自行车社区
这里相当于都市的客厅。在这儿可以修自行车，和朋友闲逛或上一节有关自行车的课程。
14 所有与自行车有关的东西
建筑立面讲述了大楼的故事。恰当地展示自行车是为了给城市建一个标志性建筑，并鼓励更多人把自行车当作每日的交通工具，把骑自行车当成一种现代的生活方式。

Photovoltaic Panels 1
The photovoltaic panels shaped like bike chains supplies the building with electricity and a striking appearance and reduces direct sun light.
Ventilated Façades 2
The double façades keeps the heat away in the summer and insulates in the winter.
Green Roof 3
Less runoff and it's quite pleasant to feel grassbetween you toes while playing table tennis.
Office For Bike Lovers 4
The local bicycle advocacy group gets fantastic localities to support their future missions in the field of bikes.
Allez! Allez! 5
Don´t be scared to crank those pedals extra hard on your way to work, enjoy the speed - salvation is a warm shower on the second floor changing rooms.
Vertical Bike Parking 6
Park your bike in the façade! Just place it on the platform, hit the button, get your ticket and watch your bike travel to the sky. Quick, safe and easy. Can you see you bike up there?
Open-air Seating 7
Watch the messengers pass by while enjoying an italian espresso in the sun.
Room For Bikes. 8
Park your bike right outside or just bring it along! Wide doors and lots of space makes it possible to bring your bike inside.
Cafe Ravitaillement 9
The urban feed station. Sandwiches, water bottles and croissants. Late for work? No problem - buy a Breakfast Musette with all the essentials and have you meal on the bike!
Social Meeting Point - stay here and work! 10
Upstairs from the café you can sit and work. For people who doesn't need an office but likes meeting new people everyday when they go to work. You can stay here all day.
Get a Bike! 11
Wether you need to rent a bike, buy a new commuter friend or just top up the old buddy with some shiny parts - the shop can please your needs.
Red Bike Lane 12
The red bike lanes spreads out across Philadelphia as blood vessels.
Bike Community 13
The urban livingroom. This is where you come to fix your bike, hang out with your friends and take a bike-building course.
All About Bikes 14
The façade tells the story of the building. The bikes are properly displayed to mark the site in the city and encurage more people to use the iron horse as a means of daily transportation. Biking as a modern way of life!

意大利佩鲁贾的节能屋顶
ENERGY ROOF IN PERUGIA, ITALY

业主：Università degli Sudi di Perugia/ Dipartimento di Ingegneria Civile e Ambientale, Italy
规划设计：COOP HIMMELB(L)AU
Wolf D. Prix / W. Dreibholz & Partner ZT GmbH
主设计师：Wolf D. Prix
项目负责人：Andrea Graser
参与项目建筑师：Giulio Polita
项目团队：Robin Heather, Daniel Reist, Anja Sorger, Jenny Chow, Luis Ferreira
本地合伙人：Heliopolis 21 – Architetti Associati, San Giuliano Terme (Pisa), Italy
土木和结构工程师：B+G Ingenieure, Bollinger und Grohmann GmbH, Frankfurt, Germany
能源设计：Baumgartner GmbH, Germany
建模：Paul Hoszowski
摄影师：COOP HIMMELB(L)AU

走进佩鲁贾市中心的马其尼大街，就能看到一座新的玻璃画廊，其上方覆盖的节能屋顶形成了进入这个考古项目地下通道的入口，这条通道同时还将市中心与Pincetto小型地铁站衔接起来。屋顶的设计目的是想为城市生产能量。西翼的方位最适于吸收太阳能辐射，而东翼则用以捕捉风力。屋顶共分三层：顶部的发电层、中间的结构层和底下的层压玻璃与透明充气垫。顶层安装了透明的光电池，可用于发电和遮阳。每块光电池都通过电脑编程以最优化的角度安装。另外，结构层中还安装了五座风力涡轮机用于发电。这样，屋顶和地下通道就都可以实现能源自给自足了。

The new glass gallery along Via Mazzini in the centre of Perugia covered by the Energy Roof creates the entry point to the archaeological underground passage that connects the city centre with the mini metro station Pincetto. The roof design is driven by the generation of energy for the city. While the orientation of the west wing is optimized in relation to solar radiation, the east wing captures wind. The roof consists of 3 layers: the energy generating top layer, the structural layer in the middle and a layer on the bottom as a combination of laminated glazing and translucent pneumatic cushions. The top layer includes transparent photovoltaic cells to generate electricity and shade the sun. The orientation of the individual cells is generated and optimized by a computer driven scripting program. Furthermore 5 wind turbines that are placed inside the structural layer are generating additional energy. Both the roof and the underground passage are energy self-sufficient.

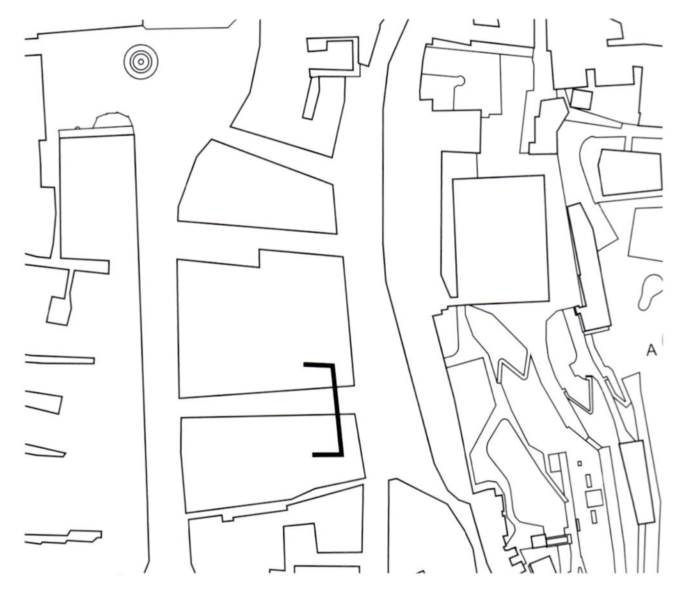

总平面图
Site plan

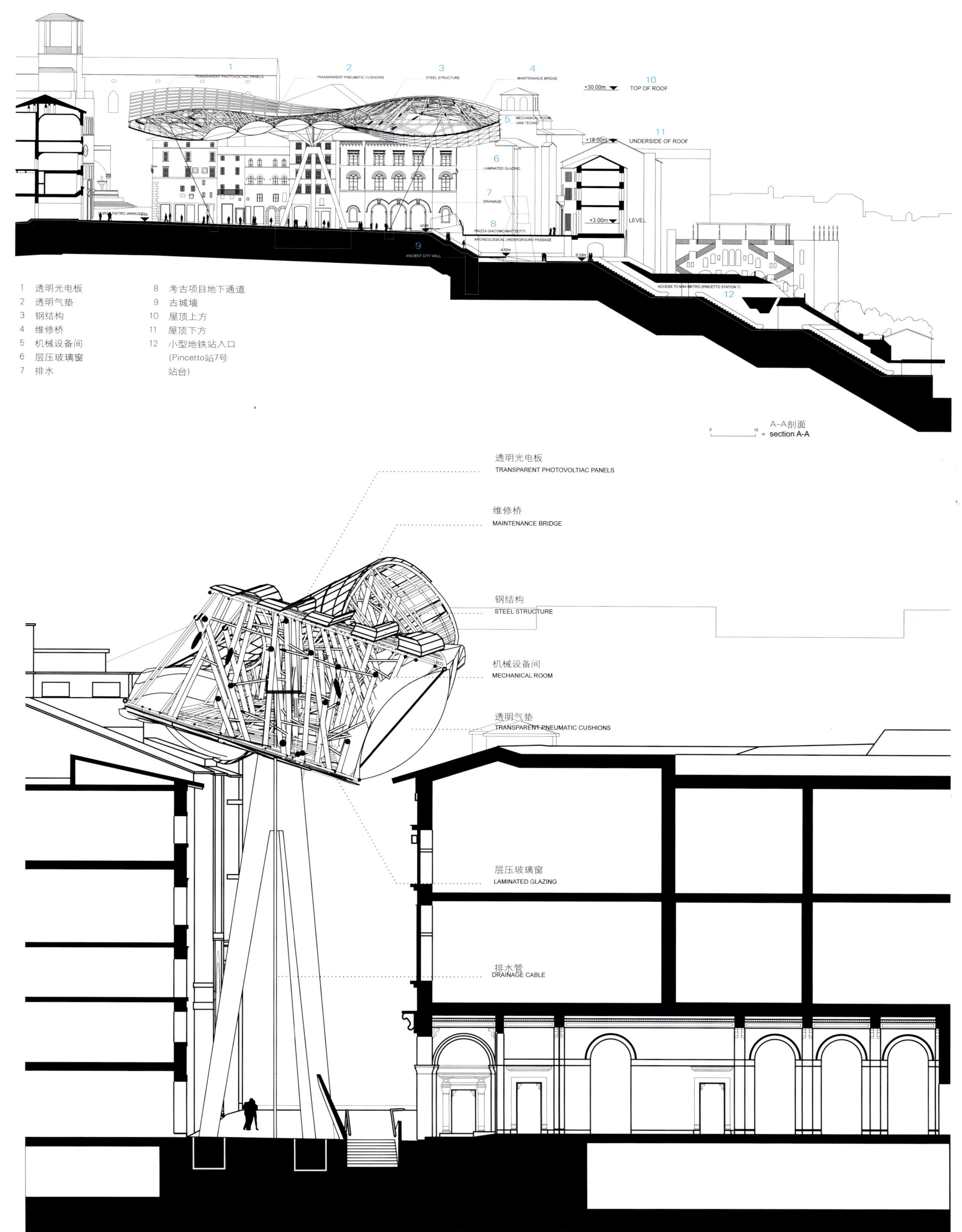

1
TRANSPARENT PHOTOVOLTIAC PANELS
2
TRANSPARENT PNEUMATIC CUSHIONS
3
STEEL STRUCTURE
4
MAINTENANCE BRIDGE
10
+30.00m
TOP OF ROOF
5
MECHANICAL ROOM
VANI TECNICI
11
+18.00m
UNDERSIDE OF ROOF
6
LAMINATED GLAZING
7
DRAINAGE
CORSO PIETRO VANNUCCI
+3.00m
LEVEL
8
PIAZZA GIACOMO MATTEOTTI
ARCHEOLOGICAL UNDERGROUND PASSAGE
±0.00
-4.50m
-6.20m
9
ANCIENT CITY WALL
ACCESS TO MINI METRO (PINCETTO STATION 7)
12
1 透明光电板
2 透明气垫
3 钢结构
4 维修桥
5 机械设备间
6 层压玻璃窗
7 排水
8 考古项目地下通道
9 古城墙
10 屋顶上方
11 屋顶下方
12 小型地铁站入口 (Pincetto站7号站台)
0 10 m
透明光电板
TRANSPARENT PHOTOVOLTIAC PANELS
维修桥
MAINTENANCE BRIDGE
钢结构
STEEL STRUCTURE
机械设备间
MECHANICAL ROOM
透明气垫
TRANSPARENT PNEUMATIC CUSHIONS
层压玻璃窗
LAMINATED GLAZING
排水管
DRAINAGE CABLE

A-A剖面
section A-A

马其尼大街剖面图
Section via Mazzini

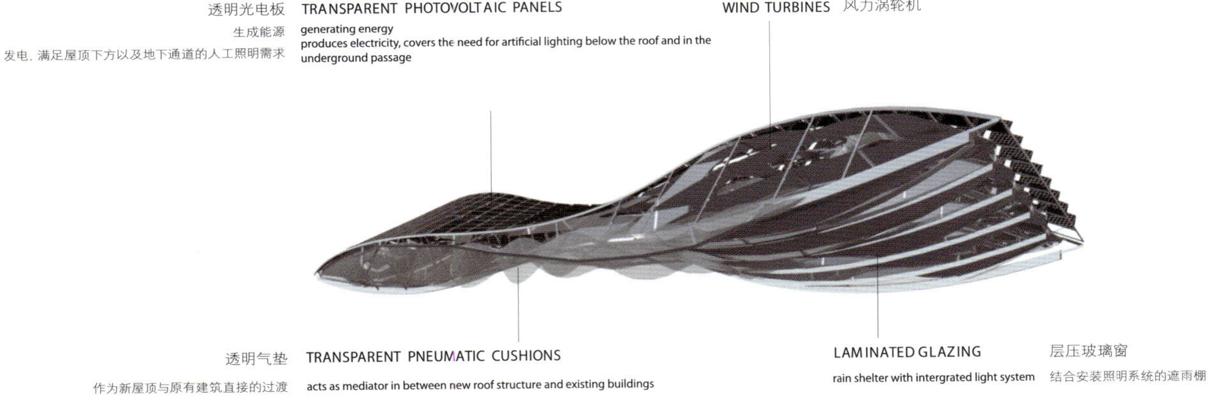
透明光电板 TRANSPARENT PHOTOVOLTAIC PANELS
生成能源
generating energy
发电，满足屋顶下方以及地下通道的人工照明需求
produces electricity, covers the need for artificial lighting below the roof and in the underground passage
WIND TURBINES 风力涡轮机
透明气垫 TRANSPARENT PNEUMATIC CUSHIONS
作为新屋顶与原有建筑直接的过渡
acts as mediator in between new roof structure and existing buildings
LAMINATED GLAZING 层压玻璃窗
rain shelter with intergrated light system
结合安装照明系统的遮雨棚

能源产生
GENERATION OF ENERGY

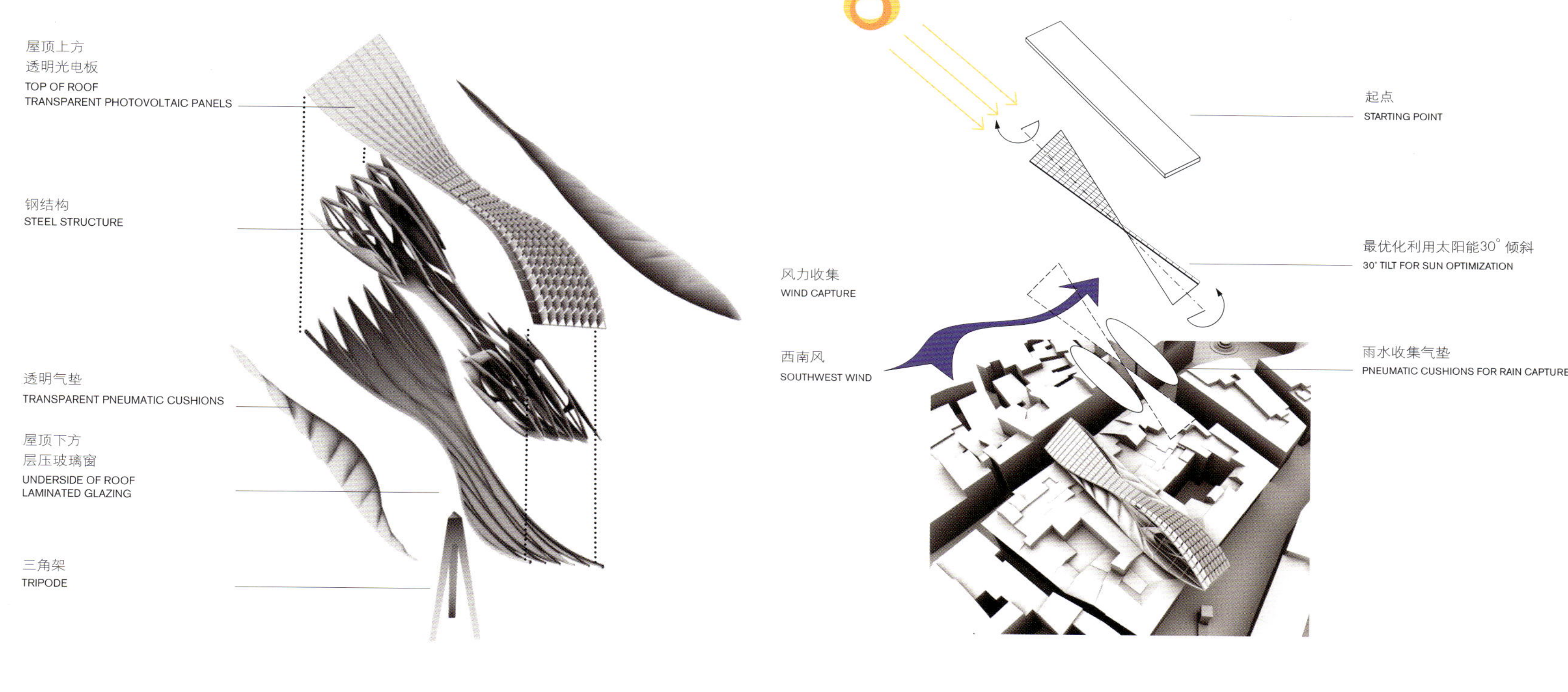

构件轴测图
AXONOMETRIC OF ELEMENTS

形式的生成
FORM DEVELOPMENT

光照条件 SUNLIGHT CONDITIONS

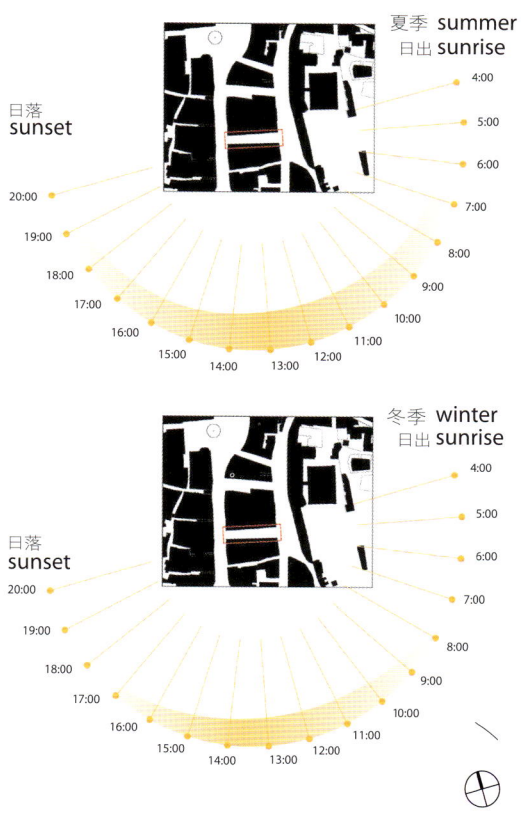

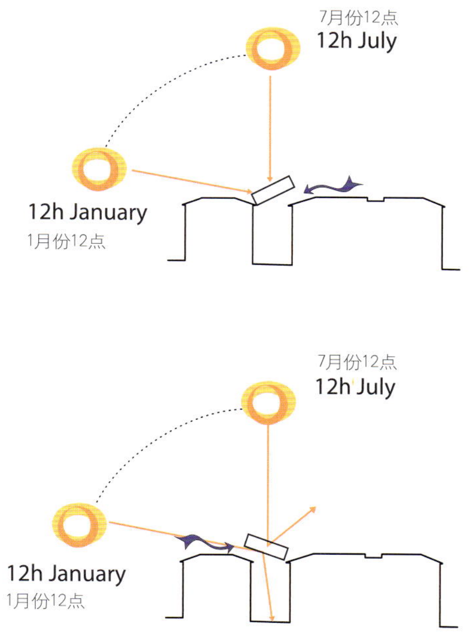

夏季 summer
日出 sunrise
日落
sunset
4:00
5:00
6:00
7:00
8:00
9:00
10:00
11:00
12:00
13:00
14:00
15:00
16:00
17:00
18:00
19:00
20:00

冬季 winter
日出 sunrise
日落
sunset
4:00
5:00
6:00
7:00
8:00
9:00
10:00
11:00
12:00
13:00
14:00
15:00
16:00
17:00
18:00
19:00
20:00

7月份12点
12h July
12h January
1月份12点

7月份12点
12h July
12h January
1月份12点

19h
晚上7点
9h
9点

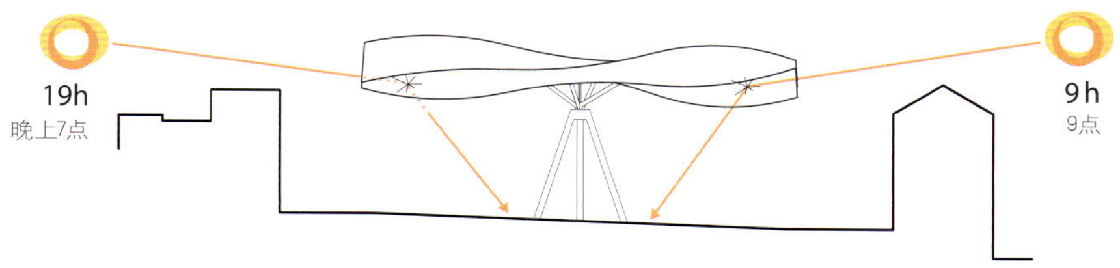

光线通过底面的折射
DEFLECTION THROUGH THE BOTTOM SURFACE

风力条件 WIND CONDITIONS

夏季:
西南风9 km/h
气温12°C～26°C
六月
summer:
southwest 9km/h
temperature 12° - 26°
June

西南风4km/h
气温15°C～30°C
七月
southwest 4 km/h
temperature 15° - 30°
July

东北风4km/h
气温15°C～29°C
八月
northeast 4 km/h
temperature 15° - 29°
August

冬季:
东北风16km/h
气温2°C～9°C
十二月
winter:
northeast 16km/h
temperature 2° - 9°
December

东北风16 km/h
气温0°C～9°C
一月
northeast 16 km/h
temperature 0° - 9°
January

东北风15km/h
气温2°C～11°C
二月
northeast 15 km/h
temperature 2° - 11°
February

伦敦阿塞洛·米塔尔轨道
ARCELORMITTAL ORBIT, LONDON

设计师：Anish Kapoor
合作者：Cecil Balmond/ARUP
照片提供：ARUP

获奖的伦敦艺术家Anish Kapoor获得了千载难逢的设计委托机会，为奥林匹克公园设计一处雄伟壮观的新旅游胜地。这一壮观的作品将由连续循环的钢管格子框架构成，被视作英国最高的雕塑。该项目高耸入云，高达115m，比纽约自由女神还要高出22m，在特定观光台上能饱览整个250英亩（约合1011 700m^2）的奥林匹克公园和伦敦天际线。游客可以通过巨大的升降梯登上雕塑结构参观，之后既可原路返回，也可从螺旋楼梯回到地面。

Award winning London-based artist Anish Kapoor has been given the commission of a lifetime to design the spectacular new public attraction in the Olympic Park. The breathtaking sculpture – thought to be the tallest in the UK – will consist of a continuous looping lattice of tubular steel. Standing at a gigantic 115m, it will be 22m taller than the Statue of Liberty in New York and offer unparalleled views of the entire 250 acres of the Olympic Park and London's skyline from a special viewing platform. Visitors will be able to take a trip up the statuesque structure in a huge lift and will have the option of walking down the spiraling staircase.

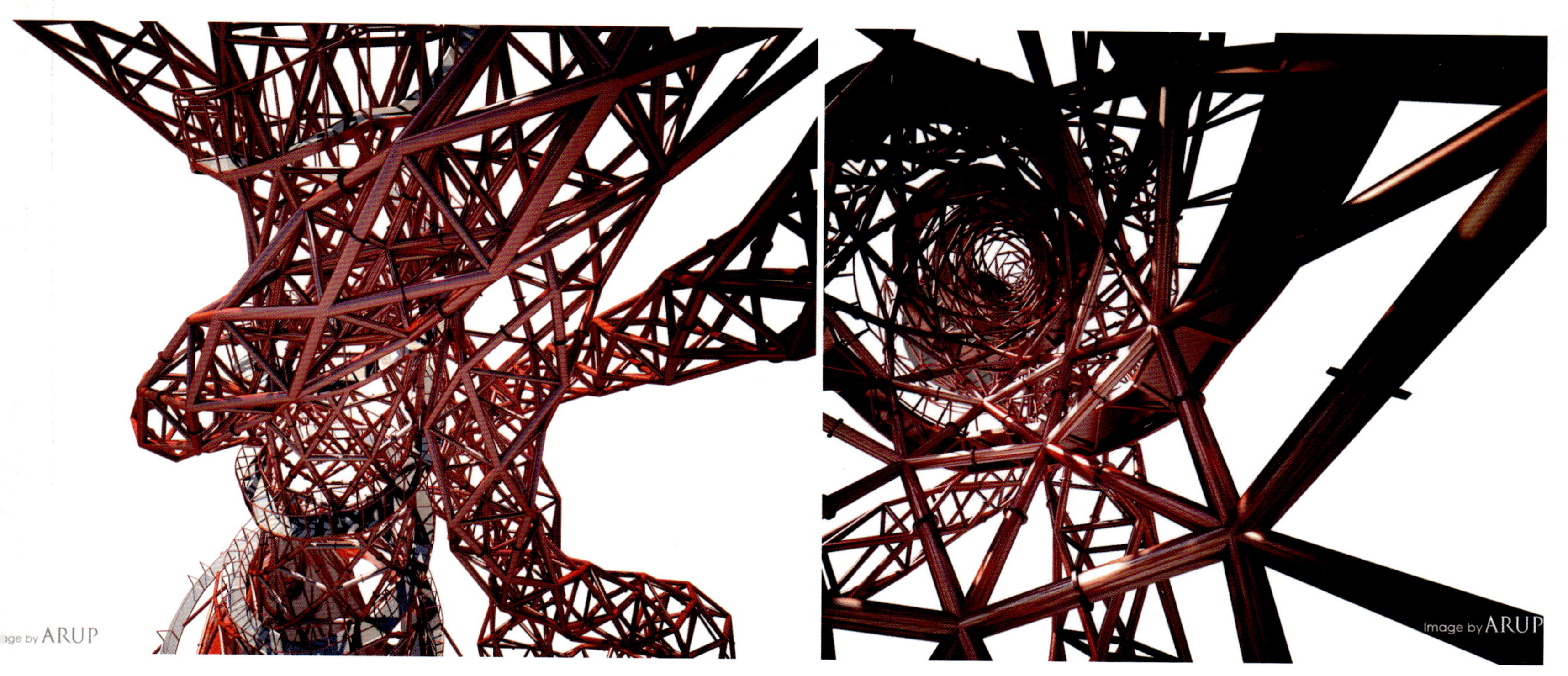

Image by ARUP

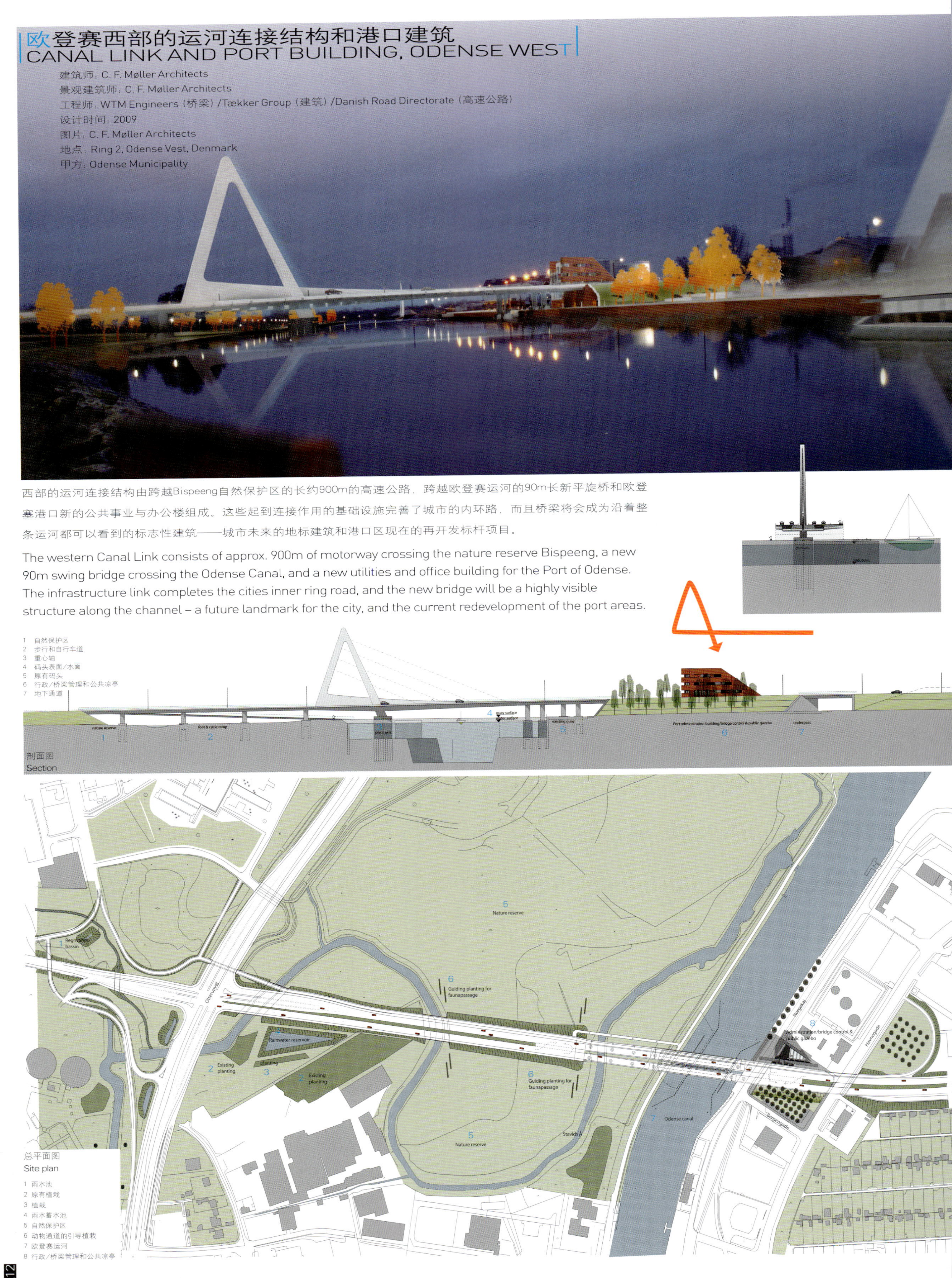

欧登赛西部的运河连接结构和港口建筑
CANAL LINK AND PORT BUILDING, ODENSE WEST

建筑师：C. F. Møller Architects
景观建筑师：C. F. Møller Architects
工程师：WTM Engineers（桥梁）/Tækker Group（建筑）/Danish Road Directorate（高速公路）
设计时间：2009
图片：C. F. Møller Architects
地点：Ring 2, Odense Vest, Denmark
甲方：Odense Municipality

西部的运河连接结构由跨越Bispeeng自然保护区的长约900m的高速公路、跨越欧登赛运河的90m长新平旋桥和欧登塞港口新的公共事业与办公楼组成。这些起到连接作用的基础设施完善了城市的内环路，而且桥梁将会成为沿着整条运河都可以看到的标志性建筑——城市未来的地标建筑和港口区现在的再开发标杆项目。

The western Canal Link consists of approx. 900m of motorway crossing the nature reserve Bispeeng, a new 90m swing bridge crossing the Odense Canal, and a new utilities and office building for the Port of Odense. The infrastructure link completes the cities inner ring road, and the new bridge will be a highly visible structure along the channel – a future landmark for the city, and the current redevelopment of the port areas.

1 自然保护区
2 步行和自行车道
3 重心轴
4 码头表面/水面
5 原有码头
6 行政/桥梁管理和公共凉亭
7 地下通道

剖面图
Section

总平面图
Site plan

1 雨水池
2 原有植栽
3 植栽
4 雨水蓄水池
5 自然保护区
6 动物通道的引导植栽
7 欧登赛运河
8 行政/桥梁管理和公共凉亭

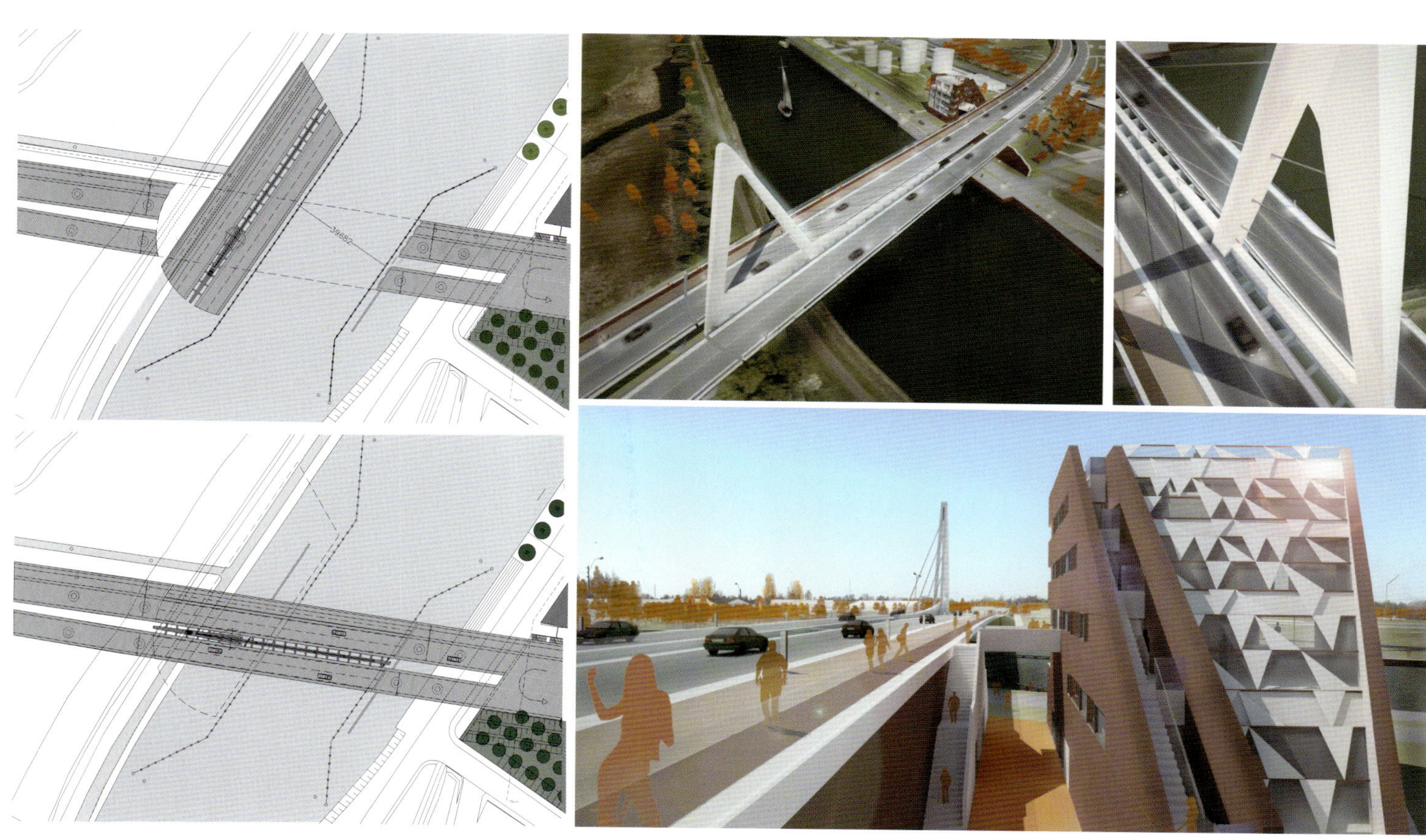

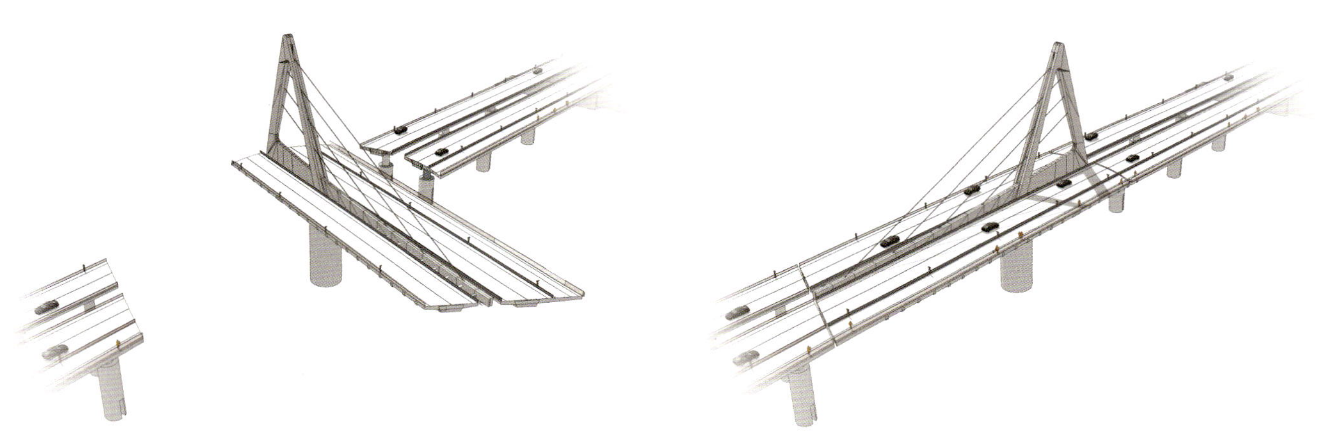

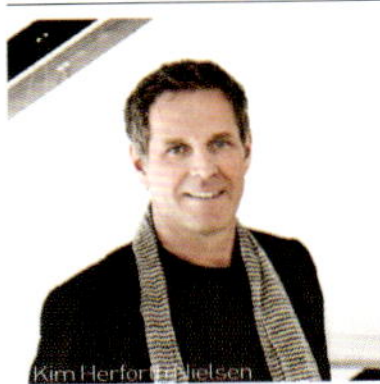
Kim Herforth Nielsen

3XN

3XN was founded as Nielsen, Nielsen and Nielsen in Aarhus in 1986 by the architects Kim Herforth Nielsen, Lars Frank Nielsen (partner until 2002) and Hans Peter Svendler Nielsen (partner until 1992). The three Nielsen architects, often referred to as the Nielsens – and today simply as 3XN – quickly became known for two things: their preference for ground breaking architecture, in defiance of the anti humanistic modernism, and projects demanding a high level of detail and employing workmanship of the highest quality.

*MULTIPLICITIES

Daniel Holguin. Born in Mexico. He earned a Degree in Arch tecture from the Universidad Nacional Autónoma de México and worked as an architect at his design/build studio in Mexico City. He moved to New York City in 2000 where he received a MS in Advanced Architectural Design from Columbia University in 2001 and worked for Bernard Tschumi Architects, TEN Arquitectos by Enrique Norten, and the Rockwellgroup.
Daniel currently lives and works as an architect in New York City where he is actually working at his studio called *MULTIPLICITIES. The range of work going from residential, hospitality and designs for his new line called JEWELLERY 4 ARCHITECTURE done in collaboration with Victoria Simes (SALTALAMACCHIA).

Anish Kapoor

Anish Kapoor is one of the most influential sculptors of his generation.
He was born on 12 March 1954 in Bombay. He moved to London in the early 1970s where he has lived and worked ever since. He studied art at Hornsey College of Art (1973-1977) and at Chelsea School of Art (1977-1978). His first solo exhibition was held at Patrice Alexandre in Paris in 1980.
He quickly gained international attention and acclaim for a series of solo exhibitions at venues including: Tate Gallery, London (1990-01); Tel Aviv Museum of Art (2003); Fondazione Prada, Milano (1995); Hayward Gallery, London (1998); BALTIC Centre for Contemporary Art, Gateshead (1999); Piazza del Plebiscito, Naples (1999); Kunsthaus Bregenz (2003); MAC Grand-Hornu, Belgium (2004); Museo Archeologico Nazionale, Naples (2004); Deutsche Guggenheim Berlin (2008) MAK, Vienna (2009); Royal Academy of Arts, London (2009); Pinchuk Arts Centre, Kiev (2010); and Guggenheim Bilbao (2010).
His recent major solo exhibition at the Royal Academy of Arts in London (26 September – 11 December 2010), showcasing a number of new and previously unseen works, was the most successful ever presented by a contemporary artist in London.
He has participated in many group shows internationally including those at the Whitechapel Art Gallery, Royal Academy of Arts and Serpentine Gallery in London, Documenta IX in Kassel, Moderna Museet in Stockholm and Jeu de Paume and Centre Georges Pompidou in Paris.

Arhitektid Muru&Pere OÜ

Arhitektid Muru & Pere OÜ was established in 1997. Our speciality is architectural designing which is integrated interior design and project management. We have designed private residences, apartment houses and public buildings as recreation centers, libraries, also industrial and commercial facilities.
Owners:

Mr. Peeter Pere (1957-)
Born in Tartu, Estonia.
1980 - Degree in architecture, Estonian Academy of Arts, Tallinn

Mr. Urmas Muru (1961-)
Born in Pärnu, Estonia.
1984 - Degree in architecture, Estonian Academy of Arts, Tallinn

Anna-Maria Erik Urmas Muru Reet Viigipuu Peeter Pere Janek Maat

John Beckmann

Axis Mundi

In a world increasingly dominated by the homogeneity of globalization, we create meaningful designs with a cultural specificity. Axis Mundi is a tight-knit multidisciplinary design firm founded by John Beckmann. Recent projects have been featured in Metropolitan Home, Spaces, C3, Archiworld, and The New York Times, as well as on blogs such as Dezeen, Designboom, Domus, Inhabitat, ArchDaily, The ... read moreArchitect's Newspaper, Curbed, and BLDGBLOG.In 1996, John Beckmann received a grant from the Graham Foundation for Advanced Studies in the Fine Arts for the book The Virtual Dimension: Architecture, Representation and Crash Culture, Princeton Architectural Press, 1998. He is a graduate of Parsons School of Design.

Bjarke Ingels

BIG

BIG is a Copenhagen-based group of architects, designers and thinkers operating within the fields of architecture, urbanism, research and development. BIG has created a reputation for completing buildings that are as programmatically and technically innovative as they are cost and resource conscious. In our architectural production we demonstrate a high sensitivity to the particular demands of site context and program.

Cecile Brisac Edgar Gonzalez

Brisac Gonzalez

Brisac Gonzalez is a multi-lingual and multi-national architecture firm established in London in 1999 by Cécile Brisac and Edgar Gonzalez.
The practice has won numerous awards and competitions across the world, and is currently working on projects in the UK, France, Russia, Scandinavia and the Middle East. By working on a variety of building types in numerous locations we continuously enrich the knowledge in our practice and the buildings we produce. We are concerned with all things architectural in the broadest sense, be it urban or interior design, cultural, commercial or sociological matters.

C. F. Møller Architects

C. F. Møller Architects is one of Scandinavia's oldest and largest architectural practices. Their work involves a wide range of expertise that covers programme analysis, town planning, master planning, all architectural services including landscape architecture, as well as the development and design of building components. Simplicity, clarity and unpretentiousness, the ideals that have guided our work since the practice was established in 1924, are continually re-interpreted to suit individual projects, always site-specific and based on international trends and regional characteristics. Over the years, they have won a large number of national and international competitions. Their work has been exhibited locally as well as internationally at places like RIBA in London, the Venice Biennale, and the Danish Cultural Institute in Beijing.Today C. F. Møller Architects has app. 275 employees. Their head office is in Århus and they have branches in Copenhagen, Aalborg, Oslo, Stockholm and London, as well as a limited company in Iceland.

Marek Chalupa

Chalupa Architekti

Chalupa Architekti is an office for architecture, urbanism and related fields. It is dedicated to creating and implementing designs with distinctive concepts and strong contextual ties to the surroundings. The emphasis is put on an intensive analysis of every task that leads to powerful conceptual solutions.
Successfully completed projects of the team include a.o. the Czech Pavilion for the World EXPO 2000 in Hannover, the vestibule of the Prague Metro station Kolbenova, an extension of a vineyard gazebo in Prague or the Metropol Hotel in Prague. Some of the realizations were published internationally (e.g. in The Phaidon Atlas of Contemporary Architecture, The Phaidon Atlas of 21st Century World Architecture) or nominated for awards (Mies van der Rohe Award – Metro station Kolbenova, Metropol Hotel).

Stepan Chalupa

Wolf D. Prix

COOP HIMMELB(L)AU

COOP HIMMELB(L)AU was founded by Wolf D. Prix, Helmut Swiczinsky and Michael Holzer in Vienna, Austria in 1968, and is active in architecture, urban planning, design, and art. In 1988, a second studio was opened in Los Angeles, USA. Further project offices are located in Frankfurt/Germany, Paris/France and Hongkong/China. Today the studio is directed by Wolf D. Prix, Wolfdieter Dreibholz, Harald Krieger, Karolin Schmidbaur and Project Partners. COOP HIMMELB(L)AU employs currently 150 team members from nineteen nations.
Over the course of the past four decades, COOP HIMMELB(L)AU has received numerous international awards including the RIBA International Award for the Akron Art Museum and the RIBA European Award for the BMW Welt.
The work of COOP HIMMELB(L)AU has been published in numerous books such as Get Off of My Cloud (Hatje Cantz, 2005), Dynamic Forces. BMW Welt (Prestel, 2007), HS#9 (Prestel, 2010), Pavillon 21 MINI Opera Space (Himmelprint, 2010), himmelblau no es ningún color (Editorial Gustavo Gili, 2010) and COOP HIMMELB(L)AU Complete Works 1968-2010 (TASCHEN, 2010).
COOP HIMMELB(L)AU has been featured in many museums and collections worldwide including the solo retrospectives Construire le Ciel at the Centre Georges Pompidou in Paris, France (1992), the exhibition Deconstructivist Architecture at MoMA (1988), New York or COOP HIMMELB(L)AU: Beyond the Blue at the Museum of Applied Arts/ Contemporary Art, Vienna (2007) and Wexner Center for the Arts, Columbus (2009).

dbEVE

Erick van Egeraat graduated from the Technical University of Delft, Department of Architecture, with honourable mention in 1984. During his over 25 years of successful practice, Erick van Egeraat (Amsterdam 1956) built a highly diverse portfolio containing ambitious and high-profile projects in the Netherlands, Europe and the Russian Federation. He has led the realisation of over 100 projects in more than 10 countries ranging from buildings for public and commercial use to luxury and social housing projects, projects for mixed use and master plans for cities and even entire regions. Each of these projects represents his very personal and expressive vision on architecture and urban development.

While Erick van Egeraat formulates unique architectural solutions for each individual project, his basic principles remain flexibility, sensibility and sustainability. His broad vision on sustainability includes the aim to create and maintain architecture with a timeless value. This vision manifests itself in buildings with a strong character that receive broad support and offer a high level of flexibility in use. By this definition, the designs of Erick van Egeraat are built to last a multitude of the usual 30-50 year lifespan, thus resulting in a significantly lower social, economic and environmental burden.

Erick van Egeraat works on projects ranging from entirely new buildings, including interior and masterplanning, to interventions within historic structures. Common denominator in this diverse portfolio is our expertise which lies in striking visionary balances between the intended identity, level of ambition and overall quality of the projects.

dmvA

David Driesen Tom Verschueren

Tom Verschueren

Born 1970 in Hoogstraten, Belgium. Studied architecture at the Henry van de Velde College in Antwerpen. Graduated in 1993.

Post-graduated in Monument- and Landscape in Antwerpen, 1995.

Started the office for architecture dmvA in 1997, together with his friend David Driesen.

David Driesen

Born in 1968 in Duffel, Belgium. Graduated architecture, with town-planning as specialty, in 1992 at Sint Lucas, Brussels.

Post-graduated in Monument- and Landscape in 1995, Antwerpen, where he met Tom Verschueren.

Since a few years examiner at Hogeschool voor Wetenschap en Kunst Sint Lucas, Brussels

Both architects want to express themselves by means of architecture, which is where the Dutch abbreviation dmvA stands for.

Together with their team of architects they have completed several private and public projects, such as offices, apartments, sport accommodation, schools, social housing, private houses, shop-interiors and art projects.

Occasionally design-objects have been created.

Commissions are frequently won in competition and many of them were (inter)nationally nominated or received national awards.

Several publications in special books and magazines.

DOSarchitects

DOSarchitects is an award-winning practice with a growing reputation for consistent innovation at the highest level of cutting-edge design. In recent years, they have firmly established themselves as a leading light among young European architecture firms, with a comprehensive portfolio of projects, awards and publications underpinning their ongoing success. This success rests, first and foremost, on the ability to fuse inspired design work with a close and intuitive understanding of their clients' needs. A strong relationship with worldwide specialists – from structural engineers to costing consultants and lighting designers – permits them to deliver firmly tailored projects, through active collaboration and committed communication at every stage.

DOS Architects' directors, Lorenzo Grifantini and Tavis Wright, boast impressive credentials working with the world's most prestigious architectural firms, including Foster and Partners, Atelier Philippe Starck, Zaha Hadid Architects and Ron Arad Associates.

The formation of DOS Architects in 2004 brought together these two careers in an environment in which their personal creative visions could thrive. Since this time the company has undertaken numerous projects across Europe, Africa and the Middle East, expanding its size and scope with every new challenge undertaken.

DOS Architects pledge in every case to value as paramount the individual needs of its clients. They respect equally the creative potential and the practical constraints of every project, and are committed to delivering seamless, detailed and punctual service to private clients and developers alike. Last but not least, they will devote to every project, however great or small, a passionate creative energy and flair which sets them firmly at the forefront of the industry.

Juan Pablo Maza

FRENTEarquitectura

Juan Pablo Maza, was born in Mexico City in 1974. In 1997 he graduated with a degree in architecture and urbanism at –Universidad Iberoamericana– getting the highest grade point average among his fellow students.

He has been a scholar at the most prestigious architecture schools in Mexico:

Facultad de Arquitectura –Universidad Nacional Autónoma de México– (UNAM)

Departamento de Arquitectura –Universidad Iberoamericana– (UIA)

His work has been exhibited in many venues such as the "Centre of Architecture" in New York, USA (2005), the "Venice Biennial", Italy (2008) or "Art Basel", Switzerland (2009).

In 2006, he won the contest, in collaboration with Fernando Romero, for the expansion of the experimental museum –El Eco– designed by Mathias Goeritz in 1953.

In 2009 he received an Honorable Mention for the design of the Bicentennial Memorial of the Independence of Mexico.

Since 2002 he is the director of –FRENTE arquitectura– advocated to create innovative and responsible architecture.

Among some of his most important designs are: "Teoloyucan" low income housing development (Mexico), "Doctor-G" social apartment building (Mexico), "Mixcoac" House (Mexico).

Foster and Parners

Foster + Partners is an international studio for architecture, planning and design led by its founder and chairman, Lord Foster, together with chief executive, Mouzhan Majidi, two heads of design, Spencer de Grey and David Nelson, five design directors and seven senior partners. The practice's work ranges in scale from the largest single building on the planet, Beijing International airport, to its smallest commission, a range of door furniture. The scope of its work includes master plans for cities, the design of buildings, interior and product design. There is also a strong interest in city planning and the infrastructure of communication. Projects can be found on six continents throughout the world: Europe, North and South America, Africa, Asia and Australasia.

Henning Larsen Architects

Henning Larsen Architects is the most international architectural company in Denmark. The company was founded by Henning Larsen in 1959 and today occupies 140 employees.

In the World Architecture Survey of the recognised British building magazine BD, Henning Larsen Architects features as number 86 on this year's Top 100 where the overall performance of the world's architecture companies is measured on the basis of capacity, turnover and number of employees. In January 2010, the same magazine placed Henning Larsen Architects as an impressive number six on this year's Top 10 in the market sector of culture.

Henning Larsen Architects has many years' experience of cultural projects and currently works on Reykjavik Concert and Conference Centre in Iceland, the Massar Children's Discovery Centre in Syria, Lanzarote Congress Centre and Moesgaard Museum in Aarhus.

Henning Larsen Architects has worked on several projects in Georgia, including a ballet school in Tbilisi, a Russian cultural centre, building consultancy in connection with the planning of a large university project as well as the current competition for a four-star hotel resort in Abastumani.

Herzog & de Meuron

Jacques Herzog and Pierre de Meuron established their office in Basel in 1978. Today the practice is led by the two Founding Partners alongside Senior Partners Christine Binswanger, Ascan Mergenthaler and Stefan Marbach. Herzog & de Meuron have designed a wide range of projects, from private houses to hospitals, factories, office buildings, museums, libraries and stadiums. Current projects include the Elbphilharmonie Hamburg, Germany (projected completion 2012), the new Miami Art Museum, Florida, USA (projected completion 2013) and the Porta Volta Fondazione Feltrinelli, Milan, Italy (projected completion 2013). The practice has been awarded „The Pritzker Architecture Prize" in 2001, and the „Praemium Imperiale" in 2007. Jacques Herzog and Pierre de Meuron are both teaching at ETH Studio Basel – Contemporary City Institute, Switzerland and at Harvard University, USA.

HofmanDujardin Architects

Barbara Dujardin

Michiel Hofman

Hofman Dujardin Architects was founded in 1999. Since then, we have been working on a wide variety of architectural, interior and product design projects with a team consisting of approximately eight architects. This wide scope of projects has been a well considered decision. The diversity in design issues entails an enrichment of any project. Our team switches easily from innovative product development to e.g. pragmatic dwelling floor plans. Our clients mainly are developers, housing corporations, multi-nationals, law firms and private clients. Our main goal is to create inspiring buildings, interiors and products that enhance life at large. A surrounding in which people can live and work in an optimized way and where the investments made are fully effected.

For office buildings we design creative working environments in which the employees can work highly efficient and they are inspired continuously. Condominiums and dwellings we transform into interiors with maximum comfort and fully adjusted to ideas and wishes of the clients. And by executing product designs such as the BloomframeŸ balcony, we express our relentless effort to innovate, surprise and challenge.

Michael Hopkins

Hopkins Architects

Our goal at Hopkins Architects is to design innovative, cost effective and beautiful buildings that enable clients to make the most of their site, programme and budget.
We create logical and clear designs starting from our clients' needs, using the principles of "truth to materials" and the expression of structure, from which stems the aesthetic quality, efficiency and popular appeal of our building.
Since we started in 1976 we have pioneered a series of strategies including; fabric roofs, lightweight structures, energy efficient design, weaving new structures into existing ones, and recycling brown field sites. Our contribution to architecture has been recognised in numerous design awards, the RIBA's Royal Gold Medal in 1995, and a knighthood for Michael Hopkins.
In over 30 years our Practice has expanded to develop the variety of skills we require to serve the diverse needs of our clients, and we have become one of the largest architectural firms in the UK. Since 1984 we have worked from our site in Marylebone in London, which we have developed into an office campus with drawing studios, offices and a model shop. We enjoy working in buildings we have designed ourselves.

Lisa Iwamoto   Craig Scott

IwamotoScott

IwamotoScott Architecture is a San Francisco based practice led by Lisa Iwamoto and Craig Scott, committed to pursuing architecture as a form of applied design research, with projects ranging in scale from installations to urban design. Conceptual themes of the work focus on intensifying the experiential and performance based qualities of architecture by rethinking the very terms of its production – program, form, space, structure, material and fabrication technique – particularly in relation to site and environmental contexts.
IwamotoScott have received numerous awards and honors including: participation in the Cooper-Hewitt 4th Annual National Design Triennial, Grand Prize for "Hydro-Net" in the History Channel's City of the Future competition, inclusion in Ordos 100 development in Inner Mongolia, finalist in the MoMA/PS1 YAP competition, Emerging Voices and Young Architects awards from the Architectural League of New York, California Council AIA Emerging Talent Award, and multiple AIA Design Awards and I.D. Design Awards. IwamotoScott's work has been exhibited at SFMOMA, Vitra Design Museum, MoMA, Guggenheim Museum, The Architecture Center, Artists Space Gallery and Smithsonian Cooper-Hewitt National Design Museum.
Iwamoto and Scott both received Master of Architecture degrees with Distinction from Harvard University's Graduate School of Design, and both have maintained teaching careers alongside their practice. Iwamoto is Associate Professor in the Architecture department at University of California Berkeley, and Scott is Associate Professor in Architecture at California College of the Arts in San Francisco. They have also taught at SCIArc, Sydney University, Harvard University, University of Michigan and Yale University.

JA Joubert Architecture

Marc Joubert founds JA in 2007, working on a wide range of projects and scales from interior design and architectural projects to master and regional planning. Notable projects are the Novi Sad market, Korça masterplan and ZonE in Albania, Holland Casino Amsterdam, the Supertubes villa, a Carbon free neighbourhood in Rotterdam and the project for National Parks in southern Holland.
JA is currently working on an energy atlas for the Netherlands, a research project on water-management and sustainable energy, a 35 ha masterplan in Bordeaux with MVRDV, as well as on a neighbourhood with 600 dwellings in Tirana, Albania.
Marc Joubert is (co)-author of KM3-excursions on capacity (MVRDV), NL28-Olympic Fire, REAP-Rotterdam Energy and Planning and Regional Landscapes for southern Holland and is teaching at the University of Novi Sad, Serbia.

XU Yang   JIANG Yuan

姜元・徐洋/JIANG Yuan · XU Yang

姜元：法国国家建筑师
法国巴黎美丽城建筑学院（2010）
西安建筑科技大学（2005）
2010，美国eVolo摩天楼竞赛 特别奖
2003，实体空间建造站，策展，中国・西安

徐洋：建筑学硕士
法国巴黎拉维莱特建筑学院(2009)
西安建筑科技大学(2005)
2010，美国eVolo摩天楼竞赛 特别奖
2007，UIA，日本建筑师学会大奖

JIANG Yuan :
DIPLOME D'ETAT D'ARCHITECTE.
Ecole Nationale Supérieure d'Architecture de Paris-Belleville. (2010)
Xi'an University of Architecture and Technology. (2005)
"SPECIAL MENTION" eVolo, 2010 Skyscraper Competition, United States. (2010)
Exposition of installation and architecture, curator, Xi'an.China. (2003)

XU Yang :
Diplôme de Spécialisation et d'Approfondissent en Architecture
Ecole Nationale Supérieure d'Architecture de Paris-Lavillette (2009)
Xi'AN University of Architecture and Technology (2005)
"SPECIAL MENTION" eVolo, 2010 Skyscraper Competition, United States. (2010)
"JAPAN INSTITUTE OF ARCHITECTS (JIA) PRIZE". UIA. (2005)

Jurgen Mayer H.

J. MAYER H. Architects

Photographed by Jens Passoth
Founded in 1996 in Berlin, Germany, J. MAYER H Architects' studio, focuses on works at the intersection of architecture, communication and new technology. Recent projects include the Town Hall in Ostfildern, Germany, a student centre at Karlsruhe University and the redevelopment of the Plaza de la Encarnacion in Sevilla, Spain. From urban planning schemes and buildings to installation work and objects with new materials, the relationship between the human body, technology and nature form the background for a new production of space.

Lacoste + Stevenson

Lacoste + Stevenson is a design based architectural and urban design practice established in 1997 by Thierry Lacoste and David Stevenson. As a result of an international partnership, the practice is able to take advantage of both local conditions together with global influences to produce a distinctive architecture.
Rather than produce projects in a particular style, each project generates its own building based on site conditions, client brief, budget, environmental considerations and creative input resulting in a unique project each time. Each new commission presents an opportunity to explore design-ideas that produce architecture of intellectual rigour and delight for the client and users. Landscape and building are often seen as one, with a seamless integration.
Lacoste + Stevenson have developed a reputation for innovative and responsive solutions to a range of projects, both public and private, large and small including an award-winning industrial project in Sydney and an award-winning house on the Noosa River. Urban design projects include Parramatta Road Competition, Sydney Olympic Park Masterplan and Hobart Waterfront.
Their first major project was a large industrial complex of office and warehouse units for a private client ($5M). This award-winning building is acknowledged as providing an alternative model for this speculative building type. It contributes to the streetscape as well as creating an atmosphere that is more akin to a business park than an industrial park.
More recently they have undertaken the award-winning refurbishment of the City of Sydney Library at Customs House, Circular Quay ($7M). The library is spread over 3 levels and is a mix of contemporary design on the ground floor and a more conventional feel on Levels 1 and 2 reflecting different library environments.
Other recent projects include "Barcode" the new $20M storage facility for Recall, a suite of amenities buildings for the National Parks and Wildlife Services in the Royal and Botany Bay National Parks and the refurbishment of historic Bellevue Cottage and Jubile Pavilion for the City of Sydney as part of the Glebe Foreshore Walk.
In all projects, Lacoste + Stevenson aim to make more of a design opportunity than might first be apparent. Both of the principals are directly involved in all their projects, which ensure an innovative and high standard of design together with a strong commitment to client service. The office has a high level of technological skills in design, presentation and documentation.

Francine Houben

Mecanoo

Multi-disciplinary
The Delft based office of Mecanoo, officially founded in 1984, consists of a multi-disciplinary highly professional staff of around 90 people. This includes architects, interior designers, urban planners, landscape architects as well as architectural technologists for quality and technique. The practice is led by founding architect / director Francine Houben and technical director Aart Fransen and is joined by fellow partners, architects Francesco Veenstra and Ellen van der Wal. The extensive experience gained over more than 25 years, together with a structured planning process, means that designs are realized with a high level of technical quality and a lot of attention to detail. Communication and a clear and strict time planning are central to the approach.
Creative and innovative
The work of Mecanoo covers all sectors and scales combining urban planning, landscaping and architecture. The projects of Mecanoo range from houses, complete neighbourhoods and skyscrapers, cities and polders, schools, theatres and libraries, hotels, museums and even a chapel. The broad spectrum of Mecanoo's projects allows the office to be creative and innovative in the search for new solutions, approaching each design problem with new vigour and enthusiasm.
Prizes
The projects Francine Houben and her colleagues at Mecanoo receive a great deal of attention. This finds expression in the various prizes the firm has been awarded, among which are the Maaskant prize for young architects, (1987), the Jhr. Victor de Stuers medal for Herdenkingsplein in Maastricht (1994), the School Building Prize (1996) for Isala College in Silvolde (1998), the National Steel prize (1998) and the Corus Construction Award for the Millennium for the Delft Library of Technology (2000), the Building Quality Award for Nieuw Terbregge, Rotterdam (2000), the TECU Architecture Award (2001) and the Dutch Building Prize (2003) for the National Heritage Museum in Arnhem, the A.M. Schreuders Prize for Office Villa Maliebaan 16 in Utrecht (2001), the International Highrise Award, Dedalo Minosse and Building Quality Award (2007) for Montevideo in Rotterdam, the Brick Award (2007) for Marnix Sports Centre and Swimming Pool in Amsterdam. FiftyTwoDegrees in Nijmegen was awarded with the Dedalo Minosse for Sustainability (2008).
Warm and tangible
Mecanoo's work shows a balance between pragmatic considerations and a strong landmark quality. The three words in the title of Francine Houben's book: composition, contrast and complexity, sum up the basis of Mecanoo's architecture but say little about its nature, which in all respects is the complete opposite of cool, abstract and minimalist. Maximalist might be an appropriate neologism for this architecture that is warm and tangible and always offers a rich sensory experience. For Francine Houben architecture should stir all the senses and is never a purely intellectual, conceptual or visual game. Architecture is about bringing all of the separate elements together in

a single concept. With Mecanoo the sensory aspect is not only determined by form and space, but by the lavish use of materials. Mecanoo excels in subtle combinations of the most diverse materials, including wood, concrete, copper, bamboo, brick, pebbles, zinc, stone, vegetation, glass and planes of saturated colour.

Context

Mecanoo regards every assignment as a new challenge for seeking innovative solutions. Every design assignment consists of looking for a solution that perfectly matches the specific situation and the wishes of the user. The Mecanoo designs respond to their broad environmental context. Each design is considered in terms of its cultural setting, place and time. As such Mecanoo treats each project as a unique design statement.

International

Mecanoo is an international operating office, with personnel from all over the world. Mecanoo regards itself as a laboratory where new ideas about architecture can develop. This is influenced by the architects of different cultural backgrounds, as well as being influenced by the ideas of Mecanoo's international clients. Mecanoo has been working on the master plans of Gdansk in Poland and Tirana in Albania, and is currently working on housing projects in Sheffield and Manchester, the new Library of Birmingham, U.K. In Spain Mecanoo is building major public buildings, such as Theatre and Congress Centre La Llotja in Lleida and a Palace of Justice in Córdoba. In Taiwan the National Performing Arts Centre will arise in Kaohsiung.

Organisation

Architect/director Francine Houben is assisted by a technical director, partners, associated architects and architects. Each team member has specific talents and experience to bring to the design process. The project architect is responsible for all design related issues. The project manager deals with the technical aspects in all design stages up until realisation stage. Throughout the development stages of the project Mecanoo organizes regular internal meetings to review the design with the project team to ensure that the highest standard of design and quality will be reached.

MVRDV

MVRDV was set up in Rotterdam (the Netherlands) in 1993 by Winy Maas, Jacob van Rijs and Nathalie de Vries. MVRDV produces designs and studies in the fields of architecture, urbanism and landscape design. MVRDV develops its work in a conceptual way, the changing condition is visualised and discussed through designs, sometimes literally through the design and construction of a diagram.

The office pursues its fascination and methodical research on density using a method of shaping space through complex amounts of data that accompany contemporary building and design processes. MVRDV first published a cross section of these study results in FARMAX (1998), followed by a.o. MetaCity/Datatown (1999), Costa Iberica (2000), Regionmaker (2002), 5 Minutes City (2003), KM3 (2005), which contains Pig City and more recently Spacefighter (2007) and Skycar City (2007), the latter two were exhibited at the 2008 Biennale of Venice. MVRDV deals with global ecological issues in large scale studies like Pig City as well as in small scale solutions for flooded areas of New Orleans.

The work of MVRDV is exhibited and published worldwide and receives international awards. The 60 architects, designers and staff members conceive projects in a multi-disciplinary collaborative design process and apply highest technological and sustainable standards.

NL Architects

NL Architects is an Amsterdam based office since 1997. The 3 principals, Pieter Bannenberg, Walter van Dijk, and Kamiel Klaasse, were educated at Delft University while living in Amsterdam. NL's "commuting" office started while carpooling between these cities. Often, projects focus on ordinary aspects of everyday life, including the unappreciated or negative, which are enhanced or twisted in order to bring to the fore the unexpected potential of the things that surround us. NL Architects currently employs an international staff of about 25 people. Their monograph Life Logo, published by Hust Press, is now available.

Uldis Luksevics

NRJA

NRJA is a young Riga based architectural practice established in 2005 by Uldis Luksevics (40). RJA is currently involved in many different design proposals starting with a high-rise Z Towers in Riga, several large scale multifunctional housing and mixed-use projects, as well as development of a new town square and wharf in a seaside town Pavilosta, residential housing in Kipsala, on the bank of river Daugava, as well as other commercial and residential buildings for our friends.

The average age of architects working in the office is 25. Our approach is active and we get inspired by creativity and competence of the world around us. Most of our work we get through invited competitions where we always try to propose more than is allowed or required. That is what NRJA stands for – No Rules Just Architecture.

In our opinion there are two very important "P" 's, which are essential in architecture – positive attitude and professionalism. The main thing is to try to feel and love truthfully and to do things which we really want to do, and to be and to work together with people with whom it is good to be together. That is why there is a place for good music and good books, good relations and good meals in the process of architecture. The process itself is like a game between us, our client and building institutions.

OFIS

... architectural office based in ljubljana formed by rok oman and spela videcnik in 1998. ofis work negotiates between architectural projects in different scales (from 30m^2 to 50,000 m^2), performing arts and set design.

rok oman (born 1970) studied architecture at the ljubljana school of architecture (grad.oct.1998) and at the architectural association in London (grad.feb.2000).

špela videčnik(born 1971) studied architecture at the ljubljana school of architecture (grad.oct.1997) and at the architectural association in London (grad. feb. 2000).

philip modest schambelan

About

born in Katowice, Poland 1982

raised in Germany

graduated in Dresden

German, Polish, English, Spanish, French

Experience

research assistant, fac. architecture, professorship gde, TU Dresden I 2010

organization, graphic / layout, Vorlesungsreihe spann_weiten...

assistant, Galerie zanderkasten / zanderarchitekten, Dresden I 2009

organization, graphic / layout, architecture, competitions, creating web presence...

student associate, AWB Architekten, Dresden I 2008

architecture, competitions, models, graphic / layout...

student associate, Rosado Recio Arquitectos, Sevilla, Spain I 2007

architecture, competitions, models, graphic / layout...

School

Technische Universität Dresden I Dipl.-Ing. Arch. I 2010

Hohe Landesschule, Hanau I Abitur I 2003

Weapons of choice

ArchiCAD, AutoCAD, Sketchup, Photoshop, Illustrator, Indesign....

rojkind arquitectos

Michel Rojkind was born in Mexico City, where he studied Architecture and Urban Planning at the Universidad Iberoamericana (1989-1994). After working on his own for several years, he teamed up with Isaac Broid and Miquel Adria to establish Adria+Broid+Rojkind (1998-2002).

With the idea of exploring new challenges that address contemporary society, of designing compelling experiences that go beyond mere functionality, and of connecting at a deeper level with the intricacies of each project, in 2002 he established an independent firm: rojkind arquitectos, recognized by Architectural Record in 2005 as one of the ten best "Design Vanguard" firms.

Philosophy

By addressing users' needs directly and seeing them as potential sources of inspiration and strength, rojkind arquitectos seeks new directions in architectural practice, evoking common identities through the exploration of uncharted geometries that address questions of space, function, technology, materials, structure, and construction methods related directly to geography, climate, and local urban experiences.

By pursuing all projects that represent a particular design challenge, rojkind arquitectos has been able to develop a wide and ever-growing spectrum of design initiatives, from the intimacies of small objects to the intricacies of large buildings and master plans.

schmidt hammer lassen architects

With more than 20 years of experience, schmidt hammer lassen architects is one of Scandinavia's most recognised, award-winning architectural practices committed to innovative and sustainable design. The practice has offices located in Aarhus, Copenhagen, Oslo, London and Shanghai.

schmidt hammer lassen architects has established an international reputation for projects that interact with their urban context. The practice places particular emphasis on the use of natural light as an integral part of the design process. The functionality – meeting the specific needs of the users – is also key, as are all aspects of sustainability. Where possible, the practice will explore the vital relationship between art, design and architecture.

Common to all the practice's work is a democratic approach to architecture which creates modern, open and multi-functional spaces that are consistent with schmidt hammer lassen architects' ethical considerations – a building revolves around people and is not merely an architectonic shape. Architecture should be closely integrated with its surroundings, with close consideration of its functions and social context. schmidt hammer lassen architects designs buildings that are essentially open to the outside world.

The practice has a distinguished track record as designers of high-profile cultural buildings, such as art galleries, educational complexes and libraries. Recent projects include the Amazon Court office building in Prague, the City of Westminster College in London, the Thor Heyerdahl School in Norway, Aberdeen University New Library in Scotland and a number of construction projects and master plans in China and Eastern Europe, with a total of approximately two million square metres currently under development.

In Scandinavia, schmidt hammer lassen architects is best known for the prestigious extension to the Royal Library – The Black Diamond, the ARoS Aarhus Museum of Art and the Cultural Centre of Greenland in Nuuk.

The practice is deeply rooted in the Scandinavian architectural traditions based on democracy, welfare, aesthetics, light, sustainability and social responsibility.

schmidt hammer lassen architects was founded in Aarhus, Denmark, in 1986 by architects Morten Schmidt, Bjarne Hammer and John F. Lassen. Today, the practice has grown substantially and employs 140 staff. The group of partners has also grown and now includes Kim Holst Jensen and Stephen D. Willacy along with three associate partners. Day-to-day management of the practice is the responsibility of CEO Bente Damgaard.

SERERO Architects & Urban Planners

Serero Architects was founded in 2000 in New York by David Serero. Now based in Paris, the firm develops projects combining research and design in the fields of architecture, landscape design and urban planning. With a particular interest in generative design, environmental control, and structural efficiency, Serero Architects attempts to explore new paths for architectural design by weaving connections between these fields and architectural practice.High level of sustainability is one of the main targets of Serero Architects projects. Processes found in natural environment are a strong source of inspiration of the office to design architectural components such as natural ventilation systems, intelligent building skins and optimized natural light shaft.

Space Group

Space Group (est. 1999) is an architecture and design office based in Oslo, Norway.
A land of protected differences and hyper-similarity, Norway provides the backdrop and need for investigation and innovation – a laboratory of curiosity open for failure and radical success.
The profession of architecture has exploded into infinitesimal parts and processes.
Our work provides SPACE for the staging of uncertainty – CONDITIONS for friction and co ncidence – FORMULATIONS on the built environment, through a meshwork of people, materials, information/knowledge, and ideas.
Through design research, we actively engage the performance of architecture, autonomously and in the city – what it does, and where it does it. Politics, economics, aesthetics, and culture form our communication platform.
Space Group is a network-based practice – a diverse international base with a compact core that attacks both small and large projects with similar ambition – strategic and inventive, flexible and specific. The office nurtures intense collaborative processes, guiding clients and teams to find new logics and strategies.
Consistency in approach coupled with the specificity and challenges of each task generates diversity in the work, providing unique results.
We operate in the business of intelligence - gathering, combining and disseminating intelligence.
We buy intelligence, trade it, add value to it, deconstruct it, filter it, transform it, and make it operational through a generative process of design. We are agents of intelligence.
Space Group is directed by Gro Bonesmo (NO), Adam Kurdahl (DK), and Gary Bates (USA).

UNStudio

UNStudio, founded in 1988 by Ben van Berkel and Caroline Bos, is a Dutch architectural design studio specializing in architecture, urban development and infrastructural projects. The name, UNStudio, stands for United Network Studio referring to the collaborative nature of the practice.
Throughout more than 20 years of international project experience, UNStudio has continually expanded its capabilities through prolonged collaboration with an extended network of international consultants, partners, and advisors across the globe. This network, combined with our centrally located offices in Amsterdam and Shanghai, enables us to work efficiently anywhere in the world. With already over seventy projects in Asia, Europe, and North America, the studio continues to expand its global presence with recent commissions in among others China, South-Korea, Taiwan, Italy, Germany and the USA.
As a network practice, a highly flexible methodological approach has been developed which incorporates parametric designing and collaborations with leading specialists in other disciplines. The office has worked internationally since its inception and has produced a wide range of work ranging from public buildings, infrastructure, offices, residential, products, to urban master plans. Pivotal UNStudio projects within these fields include; the New Mercedes-Benz museum in Stuttgart (Germany 2006), the large scale mixed-use project Raffles City in Hangzhou (China 2008 - 2012), the Galleria Department Store in Seoul (2005), the urban and architectural plan for 81 residential towers of I'Park City in Suwon (2007 - 2012 KR), department store Star Place in Kaoshiung (TW 2009), private family house VilLA NM in Upstate New York (USA 2007), the Agora theatre in Lelystad (NL 2007) and the Erasmus Bridge in Rotterdam (NL 1996).

WE Architecture

WE Architecture, founded in 2008, is a young innovating architecture office, based in Copenhagen, Denmark.
Our capability spans from architecture, urban strategies, tangible design and utopian ideas.
WE believes that the best result emerge through teamwork and transdisciplinary networks. That is why WE Architecture work across continents as well as across professional borders to enter complex conditions with the best insight and precision.
We create proposals that merge through creative translation of all the information we al get from contexts, conditions and programs.
WE Architecture strive to push innovative architecture forward to improve the condition of the world. No less.

We Are You

We are a network of young architects, with our origins in Sweden.
We work together in different constellations. Sometimes we collaborate and learn from other friends such as fantastic Norway, Kjellgren Kaminsky and arctic studio.
We are not an office.
We are ideas.
We are architecture.
We are storytelling.
We are the songs of Evert Taube.
We are you.

Wiel Arets Architects

Wiel Arets was born 1955 in Heerlen, the Netherlands and graduated from the Technical University Eindhoven in 1983. In the same year he established Wiel Arets Architect & Associates in Heerlen. In 1997 the office was moved to Maastricht. In 2004, a second office was opened in Amsterdam, with a third opening in Zürich during 2008. Bettina Kraus was born 1970 in Nuremberg and graduated from the TU Stuttgart in 1996, after studying at the ETH Zurich and HDK Berlin. She joined Wiel Arets Architects & Associates in 1997 and became partner in 2000. Since 2004 she has been teaching at the UdK in Berlin.
The studio of Wiel Arets Architects has extensive experie1ce in the fields of urbanism, and public, private and utility buildings on every scale. Additionally, the studio develops products for both limited and mass production in collaboration with leading design manufacturers.
The work of Wiel Arets Architects has been widely published in magazines and several monographs, such as El Croquis, have been produced. The firm's work has been recognized through various awards and nominations; the 1988 Charlotte Köhler Award, the 1989 Rotterdam Maaskant Award, the "1994 Mies van der Rohe Pavilion Award for European Architecture" with special mention "Emerging Architect", the 1998 UIA Nomination as one of world's thousandth best buildings of the 20th century for the Academy of Art and Architecture in Maastricht, the '2005 BNA Kubus Award' for the entire oeuvre, the Rietveld Prize given in 2006, the 2010 'Good Design Award' for the Alessi products Salt.it, Pepper.it, Screw.it and Il Bagno dOt, as well as the 2010 "Amsterdam Architecture Prize".

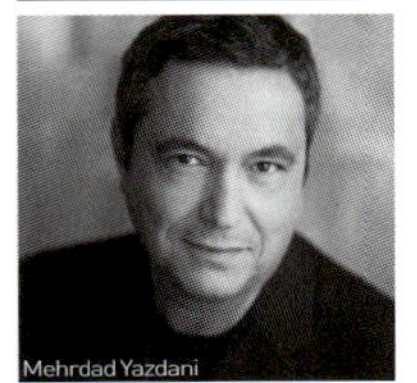
Mehrdad Yazdani

Yazdani Studio of Cannon Design

The Yazdani Studio of Cannon design is a laboratory for exploration and excellence in design. Yazdani Studio is a design studio within Cannon Design, based in Los Angeles, but operating across the platform of the firm. It is redefining the industry's standards of operations, performance, reach, and product. Established upon the reputation and leadership of award-winning designer Mehrdad Yazdani, the Studio integrates the best attributes of a design studio with the resources and reach of a global practice, producing unique environments for progressive institutions and individuals throughout the world.
Yazdani Studio combines the talents of a diverse team of architects, designers, technical specialists, and other thinkers who share a commitment to pushing the boundaries of design – from refining concepts of sustainability to the application of new technologies and urban initiatives. The Studio takes a holistic approach, working to understand the cultural foundations and implications of what we design. From multi-faceted research facilities, to intimate student centers, to campus plans, the Yazdani Studio is engaged in projects across the globe, delivering service in all aspects of architectural design and planning.

Zaha Hadid
©Luke Hayes

Zaha Hadid Architects

Zaha Hadid, founding partner of Zaha Hadid Architects, was awarded the Pritzker Architecture Prize in 2004 and is internationally known for her built, theoretical and academic work. Each of her dynamic and innovative projects builds on over thirty years of revolutionary experimentation and research in the interrelated fields of urbanism, architecture and design.
Working with senior office partner Patrik Schumacher, Hadid's interest is in the rigorous interface between architecture, landscape, and geology as the practice integrates natural topography and human-made systems that lead to experimentation with cutting-edge technologies. Such a process often results in unexpected and dynamic architectural forms.

Zaha Hadid Architects continues to be a global leader in pioneering research and design investigation. Collaborations with artists, designers, engineers and clients that lead their industries have advanced the practice's diversity and knowledge, whilst the implementation of state-of-the-art technologies have aided the realization of fluid, dynamic and therefore complex architectural structures. Zaha Hadid's work was the subject of a critically-acclaimed retrospective exhibition at New York's Solomon R. Guggenheim Museum in 2006 and showcased at London's Design Museum in 2007. Hadid's recently completed projects include the Nordpark Railway stations in Innsbruck, Mobile Art for Chanel in Hong Kong, Tokyo and New York, the Zaragoza Bridge Pavilion in Spain and the Burnham Pavilion in Chicago.

Patrik Schumacher
©Luke Hayes